Biomass for Sustainable Applications
Pollution Remediation and Energy

RSC Green Chemistry

Series Editors:
James H Clark, *Department of Chemistry, University of York, UK*
George A Kraus, *Department of Chemistry, Iowa State University, Ames, Iowa, USA*
Andrzej Stankiewicz, *Delft University of Technology, The Netherlands*
Peter Siedl, *Federal University of Rio de Janeiro, Brazil*
Yuan Kou, *Peking University, People's Republic of China*

Titles in the Series:

How to obtain future titles on publication:
A standing order plan is available for this series. A standing order will bring delivery of each new volume immediately on publication.

For further information please contact:
Book Sales Department, Royal Society of Chemistry, Thomas Graham House, Science Park, Milton Road, Cambridge, CB4 0WF, UK
Telephone: +44 (0)1223 420066, Fax: +44 (0)1223 420247
Email: booksales@rsc.org
Visit our website at www.rsc.org/books

Biomass for Sustainable Applications
Pollution Remediation and Energy

Edited by

Sarra Gaspard
Université des Antilles et de la Guyane, Guadeloupe, French West Indies, France
Email: sarra.gaspard@univ-ag.fr

and

Mohamed Chaker Ncibi
Université des Antilles et de la Guyane, Guadeloupe, French West Indies, France
Email: nmchaker@yahoo.fr

RSC Publishing

RSC Green Chemistry No. 25

ISBN: 978-1-84973-600-8
ISSN: 1757-7039

A catalogue record for this book is available from the British Library

Published by The Royal Society of Chemistry,
Thomas Graham House, Science Park, Milton Road,
Cambridge CB4 0WF, UK

Registered Charity Number 207890

For further information see our web site at www.rsc.org

Foreword

Biomass is now widely recognized as a renewable source of carbon for producing fuels and increasingly for making chemicals. It thus helps address the great challenge we face of replacing non-sustainable fossil resources to make many of the articles we use in modern society. In fact biomass has a much wider range of applications, some of which have been known for many years. Two of the other great challenges the world faces—clean water and renewable energy—can be partly addressed through the use of biomass and these areas are the subject of this book.

The use of biomass to clean water has been used by mostly local communities throughout history. Today the increased levels and increased complexity of pollutants, as well as the increasing demands of a growing population, make the clean water challenge especially great. In particular we need low cost, geographically diverse and effective processes. Biomass can fit these requirements nicely.

Biosorption, along with biocoagulation and bioflocculation, has many advantages over other techniques including renewability and low cost. There are a vast array of bioresources that can be used in this context including seaweeds, fungi, agricultural by-products, woody residues, grasses and bacteria. The book also covers soil remediation— another great societal challenge as our fertile soil levels decline. Phytoremediation—the ability of plants to take up pollutants—is an especially interesting and powerful phenomenon which is already being used to decontaminate soils, even those contaminated with military explosives! The removal of hazardous substances is considered along with the mechanisms of adsorption. Many factors can influence the uptake of pollutants by biomass including pH, temperature and ionic strength. All of these are considered in an attempt to better understand the key biomass–molecule interaction.

Biomass is now a critical part of the renewable energy portfolio. Biomass burning is an important part of the renewable energy strategy in regions such as the European Union and millions of tonnes of wood, straws, grasses and other

RSC Green Chemistry No. 25
Biomass for Sustainable Applications: Pollution Remediation and Energy
Edited by Sarra Gaspard and Mohamed Chaker Ncibi
© The Royal Society of Chemistry 2014
Published by the Royal Society of Chemistry, www.rsc.org

low value biomass will be burnt to produce electricity. Alongside this, biomass as a feedstock for making liquid biofuels can be expected to continue to grow with biodiesel now established in many countries, and bioethanol and biobutanol having especially strong roles in regions such as Brazil and Africa. However, we now need to place strong emphasis on second and third generation biofuels where the feedstocks in no way compromise food production but add to farmers' income though use of low-value by-products or create new businesses through exploitation of other resources such as marine biomass. Some of the most interesting and promising of these technologies are discussed in this book. Bioenergy derived from microorganisms is an exciting opportunity and various metabolic options of bacteria are described here including biohydrogen, biomethane and through the use of microbial fuel cells, bioelectricity.

Hydrogen is one of the most attractive storable energy carriers with a high energy content and non-polluting nature. However, some 96% of the hydrogen generated today comes from fossil fuels, severely reducing its long-term value and sustainability. Biomass is a real option for the production of renewable hydrogen. The efficiency of the biomass conversion becomes critical and the use of heterogeneous photocatalysts to help achieve this is described in this book. The book also extends to more sophisticated energy technologies including high energy density supercapacitors. Here the ability of biomass to be converted into different forms of carbons, including nanotubes, is very important.

Biomass has always been used as a resource but perhaps we now better appreciate its value. Biomass can and surely will play a vital role in the sustainable solutions we now seek to the critical societal challenge of how we can all eat well, have low cost and abundant clean water, use energy to heat our homes, power our devices, and enjoy the benefits of a consumer society. Biomass plus green chemistry is perhaps *the* sustainable solution!

James Clark
York, UK

Preface

Sustainable energy production and good quality water supply are two major challenges facing modern societies nowadays and for decades ahead. In that context, renewable biomass presents both a sustainable energy source and an alternative to expensive pollution remediation technologies. This book gives an overview of the various ways to valorize biomass for energy production and for pollution treatment of contaminated soils and waters. It focuses on the fact that we could produce renewable energy from biomass without using corn, sugarcane or colza oil, but lignocelluloses, bacteria and algae instead. Besides, we could treat any water or soil pollution using renewable and cheap agro-industrial by-product, algae and fungi.

This book was elaborated considering the biological and physiochemical characteristics of these bioresources—the major criteria on which the choice of which application(s) could be reserved for each biomass. Thus, different fields of application, having biomass itself as the link between the different sections of the publication. Because many scientific and engineering disciplines are directly or indirectly involved in the use of biomass for energetic applications and pollution remediation, the book is deeply multidisciplinary in content and treatment of the subject matter. Thus, biologists, chemists, environmentalists, electrochemists and professionals, who are involved in biomass for environmental and energetic applications, will certainly find the whole book or part of it very insightful and useful for discovering possible applications of locally found biomass(es).

The book is divided into two main parts: part 1 made up of chapters 1 to 5 deals with the use of biomass for pollution remediation while part 2 made up of chapters 6 to 9 is dedicated to energy production and storage. **Chapter 1** deals with the use of bioresources as an adsorbant for sustainable water treatment processes. The use of biopolymers produced from different biomasses as natural flocculants for water treatment to remove different kind of organic and inorganic pollutant from water streams without prior treatment (*i.e.* raw

RSC Green Chemistry No. 25
Biomass for Sustainable Applications: Pollution Remediation and Energy
Edited by Sarra Gaspard and Mohamed Chaker Ncibi
© The Royal Society of Chemistry 2014
Published by the Royal Society of Chemistry, www.rsc.org

biomaterials) is also presented. **Chapter 2** describes thermochemical treatments to produce highly efficient activated carbons from various biomasses, their physico-chemical characteristics, and their application for the removal of heavy metals or organic pollutants such as dyes. Thus, the use of locally available biomasses for water and wastewater treatment techniques is a major asset with which to enhance the eco-friendly removal process and to diminish the overall economic charge and energy demand. **Chapter 3** examines the use of plants for soil remediation and pollution attenuation, and even for soil restoration and the prevention of groundwater pollution. The main criteria are their natural abilities of extracting the pollution from the soil and 'processing' it via assimilation, degradation or detoxification. Four main aspects of phytoremediation: phytostabilisation, phytodegradation, phytovolatilisation and phytoextraction are investigated. **Chapter 4** describes the most promising and eco-friendly approaches in soil bioremediation using biomass. The role of fungi, bacteria and biomass-derived chemicals such as biosurfactants in these approaches is discussed. **Chapter 5** presents general approaches of bioreactors used in waste gas stream treatments and details the different biosystems such as biofilters, biological trickling beds and bioscrubbers. The general presentation, operating conditions, yields and industrial applications of these bioprocesses are discussed.

Chapter 6 describes the wide spectrum of bioenergy that can be harnessed through bacterial metabolism. Diverse bioenergy generation processes are depicted: acidogenesis (biohydrogen); methanogenesis (biomethane); electrogenesis through microbial fuel cell (bioelectricity), solventogenesis (bioethanol and biobutanol); and biopolymer synthesis (bioplastics and lipids) through microbial metabolism. **Chapter 7** deals with the valorisation of plantae and marine biomasses, along with agro-industrial wastes, in producing eco-friendly fuels, bioethanol, biodiesel and biomethane. A number of different natural feedstocks are analysed for their aptitude to produce liquid or gaseous fuels including woods, grasses, algae, agricultural residues, industrial by-products and household wastes. **Chapter 8** introduces recent advances in photocatalytic hydrogen production from water and biomass derivatives, such as ethanol, glycerol, sugars and methane, since hydrogen can be produced using solar energy and renewable resources. **Chapter 9** describes nanoporous carbonaceous materials produced from biomass, with a large surface area and pore volume, and their use in electrochemical storage systems such as supercapacitors.

We sincerely hope that our contribution throughout this book will be a valuable asset to researchers, instructors, decision-makers, practising professionals, senior undergraduate and graduate students and others interested in pollution remediation and energy production and storage using renewable and low-cost bioresources. The book could even be used as a textbook or reference book for researchers teaching courses dealing with biomass valorisation.

All the chapters were contributed by professionals from academia and government laboratories from various countries. The editors thank the Royal

Society of Chemistry for believing in our book project. We also gratefully acknowledge all the authors who have contributed to this book for sharing their views and research findings with the scientific and professional communities through our book. The views or opinions expressed in each chapter of this book are those of the authors.

Sarra Gaspard
Mohamed Chaker Ncibi

Contents

RSC Green Chemistry No. 25
Biomass for Sustainable Applications: Pollution Remediation and Energy
Edited by Sarra Gaspard and Mohamed Chaker Ncibi
© The Royal Society of Chemistry 2014
Published by the Royal Society of Chemistry, www.rsc.org

Chapter 5 Biological Waste Gas Treatments   **222**
Pierre Le Cloirec, Abdeltif Amrane, Benoit Anet and Catherine Couriol

CHAPTER 1

Biomass for Water Treatment: Biosorbent, Coagulants and Flocculants

SANDRO ALTENOR AND SARRA GASPARD*

Department of Chemistry, COVACHIM-M2E Laboratory, University of Antilles and Guyane, Pointe à Pitre 97159, Guadeloupe, France
*Email: sarra.gaspard@univ-ag.fr

1.1 Introduction

Pollution by organic and inorganic contaminants is an important environmental problem due to their toxic effects and possible accumulation throughout the food chain and hence in the human body. Many hazardous compounds (metals, dyes, phenolic compounds, *etc.*) have found widespread use in industries such as metal finishing, leather tanning, electroplating, nuclear power, textile, pesticide and pharmaceutical. Thus, water pollution by these contaminants is of considerable concern around the world.

Conventional methods (bioaccumulation, precipitation, reverse osmosis, oxidation/reduction, filtration, evaporation, ion exchange and membrane separation) used for the removal of hazardous compounds from wastewater are expensive and/or inefficient in reducing the effluent concentration to the required levels. The search for new and low-cost techniques is therefore of great importance for the removal of organic and inorganic contaminants from drinking water and wastewater. Biosorption is becoming a potential alternative to traditional treatment processes used for the removal of hazardous metals

RSC Green Chemistry No. 25
Biomass for Sustainable Applications: Pollution Remediation and Energy
Edited by Sarra Gaspard and Mohamed Chaker Ncibi
© The Royal Society of Chemistry 2014
Published by the Royal Society of Chemistry, www.rsc.org

and organic compounds. Biosorption represents a biotechnological innovation as well as a cost-effective and excellent tool for sequestering hazardous compounds from aqueous solutions.

Biosorption is a term that describes the property of some biomolecules or types of biomass to remove and concentrate by passive binding, selected metallic ions or other molecules from aqueous solutions.[1-6] This implies that the removal mechanism is not metabolically controlled. Biomass exhibits this property, acting just like a chemical substance, as for example, an ion exchanger of biological origin. The cell wall structure of certain algae, woody biomass, mosses, fungi and bacteria in particular are found to be responsible for this phenomenon.[1,6-11] In addition, bacteria, fungi, seaweeds, agricultural waste and raw plants can also produce biomolecules having coagulating/flocculating activities. Indeed, the use of biological materials for the treatment of wastewaters containing organic and inorganic contaminants is growing. This relatively new technology has received considerable attention in recent years as it has many advantages over traditional methods. It uses inexpensive and abundant renewable materials with good ability for the recovery of metal pollutants. Hence studies on the use of biomass, such as agricultural wastes, mosses, fungi, bacteria or seaweeds, as a raw material for the production of sorbents is steadily increasing.

Among the many types of biosorbents (*i.e.* algae, fungi, bacteria, yeasts, *etc.*) investigated for their ability to sequester contaminants, algal biomass has proven to be highly effective as well as reliable and predictable in the removal of hazardous compounds from aqueous solutions.[1] Marine algae are a renewable natural biomass and are very abundant in the littoral world. These biomasses have attracted the attention of many investigators as organisms to be tested and used as new supports to concentrate and adsorb hazardous compounds. This chapter is devoted to biosorption, coagulation and flocculation by various biomaterials with an emphasis on algal biomass and the fundamental parameters that come into play during biosorption.

1.2   Biosorption by Biomass

1.2.1   Biosorption by Different Types of Biomass

The removal of hazardous compounds from aqueous solution by biomaterials is an innovative and promising technology. In recent years, research on biosorption mechanisms has been intensified as various biomasses can be employed to sequester organic and inorganic pollutants from industrial effluents. The efficiency of the biomass used depends upon the capacity, affinity and specificity related to its physico-chemical properties. A large variety of biosorbents have been tested for the removal of both organic and inorganic compounds. Among the different biological substrates used for biosorption are fungi, bacteria, yeast, algae and other biowaste materials such as agricultural by-products. The biomasses are taken either in their natural form, or slightly modified by chemical and thermal treatment to increase their sorption capacities.

1.2.1.1 Bacteria

Bacteria (Gram-positive and Gram-negative cells; Figure 1.1(b) and 1.1(c) respectively) are abundant microorganisms and constitute a significant fraction of living terrestrial biomass. In the last decade, some microorganisms were

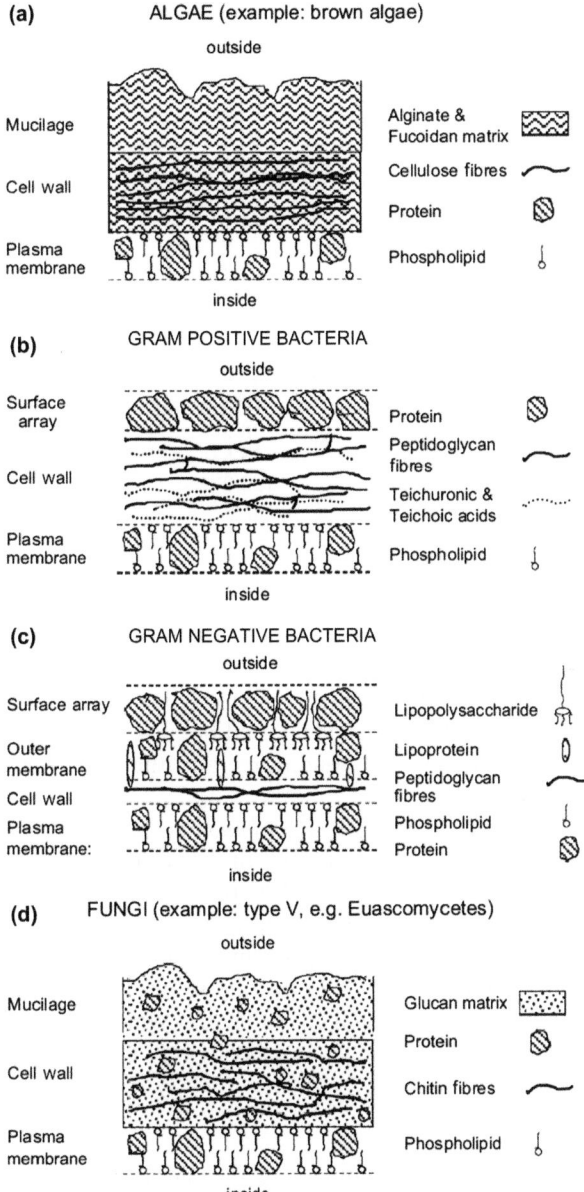

Figure 1.1   Schematic outline of the cell wall structures of: (a) seaweeds; (b) Gram-positive bacteria; (c) Gram-negative bacteria; and (d) fungi.[6]

Table 1.1 Examples of bacterial biomass used for removing hazardous pollutants.

Pollutant	Bacteria species	Biosorption capacity/mg g^{-1}	Reference
Pb	*Bacillus sp.*	92.3	93
	Bacillus firmus	467.0	94
	Pseudomonas aeruginosa	79.5	95
	Streptomyces rimosus	135.0	96
Cr	*Bacillus coagulans*	39.9	97
	Staphylococcus xylosus	143.0	98
	Pseudomonas sp.	95.0	98
	Bacillus thuringiensis	83.0	99
	Escherichia coli	4.6	14
Cu	*Bacillus sp.*	16.3	93
	Pseudomonas aeruginosa	23.1	95
	Pseudomonas putida	96.9	100
	Streptomyces coelicolor	67.7	101
Cd	*Pseudomonas sp.*	278.0	98
	Staphylococcus xylosus	250.0	98
	Streptomyces rimosus	64.9	102
	Pseudomonas aeruginosa	42.4	95
	Escherichia coli	10.3	14
	Streptomyces sp.	38.5	103
Fe	*Escherichia coli*	16.5	14
	Streptomyces rimosus	122.0	104
Ni	*Bacillus thuringiensis*	45.9	105
	Escherichia coli	6.9	14
Reactive Red	*Citrobacter sp.*	–	106
Acid Blue 225	*Paenibacillus macerans*	94.9	17
Reactive red	*Nostoc linckia HA 46*	93.5	76

found to accumulate metallic elements with high capacity.[12] Bacteria were identified as biosorbents; they are widely used because of their small size, their ability to grow under controlled conditions, and their resistance to extreme environmental conditions.[13]

Bacteria species such as *Bacillus*, *Pseudomonas*, *Streptomyces*, *Escherichia* and *Micrococcus* have been the subject of hazardous compound removal studies, for example: removal of Cd(II), Cr(VI), Fe(III) and Ni(II) from aqueous solutions by an *Escherichia coli*;[14] Cr(VI) uptake by *Bacillus thuringiensis*[15] and by Trichoderma species;[16] acid dyes by *Paenibacillus macerans*;[17] and lead and nickel by *Pseudomonas aeruginosa*.[18] Important results of pollutant biosorption using bacterial biomasses are listed in Table 1.1, which draws on references cited by Wang and Chen[13] and Vijayaraghavan and Yun.[12]

1.2.1.2 Fungi

Fungi are another ubiquitous biomass in the natural environment that are important in industrial processes. Fungal cell walls consist mainly of polysaccharide with proteins, lipids, polyphosphates and inorganic ions, with a

Table 1.2 Examples of fungi biomass used for removing hazardous pollutants.

Pollutant	Fungal biomass	Biosorption capacity/mg g^{-1}	Reference
Pb	*Trametes versicolor*	208	107
	Paecilomyces marquandii	505	108
	Mucor indicus	22.1	109
	Lactarius scrobiculatus	56.2	110
Cr	*Aspergillus niger*	–	21
	Aspergillus niger	4.8	23
	Pleurotus ostreatus	10.7	111
Cu	*Pleurotus ostreatus*	8.06	111
	Saccharomyces cerevisiae	144.9	112
Cd	*Trametes versicolor*	166	107
	Lactarius scrobiculatus	53.1	110
Zn	*Paecilomyces marquandii*	308	108
	Pleurotus ostreatus	3.2	111
Ni	*Pleurotus ostreatus*	20.4	111
Reactive Black	*Rhizopus nigricans*	122	20
Crystal Violet	*Ceriporia lacerata*	239	113
Gemacion Red	*Rhizopus arrhizus*	1007	75
Gemazol Blue	*Rhizopus arrhizus*	823	75
Gemactive Black	*Rhizopus arrhizus*	635	75
Reactive Brown 9	*Rhizopus nigricans*	112	20
Reactive Green	*Rhizopus nigricans*	204	20
Reactive Blue 38	*Rhizopus nigricans*	161	20
Phenol	*Schizophyllum commune*	120	22
2-Chlorophenol	*Schizophyllum commune*	178	22
4-Chlorophenol	*Schizophyllum commune*	244	22

chitin network, making up the wall-cementing matrix. Chitin is a common constituent of fungal cell walls (Figure 1.1(d)). The knowledge that metallic ions are very important to fungi metabolism created interest in relating the behaviour of fungi to the presence of metallic ions, particularly heavy metals.

There is much published work on the treatment of water polluted by hazardous compounds such as: the removal of Cr(VI) by *Aspergillus tubingensis*;[19] textile dyes by *Aspergillus niger*, *Aspergillus japonica*, *Rhizopus nigricans* and *Rhizopus arrhizus*;[20] Cr(VI) by *Aspergillus niger*;[21] phenol and chlorophenol by *Schizophyllum commune*;[22] and nickel by *Aspergillus niger*.[23] Indeed, according to some references cited by Wang and Chen,[13] Aksu[24] and Kumar and Min,[22] like bacterial biomass, fungal biomass can also concentrate from aqueous solutions considerable quantities of organic pollutants (such as dyes, phenol and its derivatives, pesticides) by adsorption even in the absence of physiological activity. Both, living and dead fungal cells have a remarkable ability to remove hazardous compounds. Table 1.2 lists some types of fungal biomass used for the uptake of organic and inorganic compounds.

1.2.1.3 Seaweeds

Algae are eukaryotic organisms containing chlorophyll that allows them to carry out the photosynthesis process. The algal cell wall structure is similar to that of the fungal cell wall, being composed of a multi-layered microfibrillar framework generally consisting of cellulose and interspersed with amorphous material in which is embedded different polysaccharides (fucoidan matrix, alginate, protein) (Figure 1.1(a)). Algae abound everywhere in nature but particularly in aquatic habitats, freshwater, marine waters and moist soil.

Among the different biomaterials studied for the biosorption of hazardous compounds, seaweeds have received much attention due to their low cost and low sensitivity to environmental and impurity factors.[25] They are often used for metal ions uptake. Algae achieve generally higher metal uptake than bacteria and fungi.[13] Among the three groups of algae (red, green, brown), brown algae have received the most attention as they have greater pollutant uptake capability than green and red algae. This is due to the presence of alginate, which is present in a gel form in brown algae cell walls. Their macroscopic structure also offers a convenient basis for the production of biosorbent particles suitable for sorption process applications.[26]

In addition, brown algae are cheap and readily available materials. They are commonly used as nutritional supplements, animal feed and fertilizers, and as a source of thickeners such as alginate. Brown algae are the subject of numerous biosorption studies because of their high metal uptake, for example: biosorption of copper by *Ulva fasciata* and *Sargassum* sp.;[2] chromium by *Ulva lactuca* and *Sargassum* sp.;[3,4] cadmium, zinc and lead by *Sargassum filipendula, Laminaria hyperborea, Bifurcaria bifurcata, Sargassum muticum* and *Fucus spiralis*;[5,27] lead and nickel by *Sargassum* sp.[28] acid dyes by *Azolla filiculoides*;[29] phenol and chlorophenol by *Sargassum muticum*;[30] methylene blue by *Turbinaria turbinata* (Figure 1.2),[31] to name a few. The results of most of these studies are very promising. Table 1.3 lists some species of seaweeds biomass used for removal of hazardous contaminants and their performance.

1.2.1.4 Other Types of Biomass

Other biomass such as agricultural by-products can also be used for the removal of hazardous contaminants. Numerous other types of biomass are mentioned in the literature as being used as biosorbent. We list a few, particularly some agricultural by-products in Table 1.4. Examples include: removal of Cd(II) and Pb(II) by rice husk residues;[32] uptake of Pb(II), Cu(II), Cd(II), Zn(II) and Ni(II) by tobacco dust;[33] Pb(II) and Cd(II) by cork waste biomass;[34] removal of Pb(II) by *Gossypium hirsutum* (cotton) waste biomass.[35]

1.2.2 Characterisation of Biosorbents Surface

The ability of these biomaterials to sequester organic and inorganic pollutants is related to the structure and chemical composition of their cell wall which is

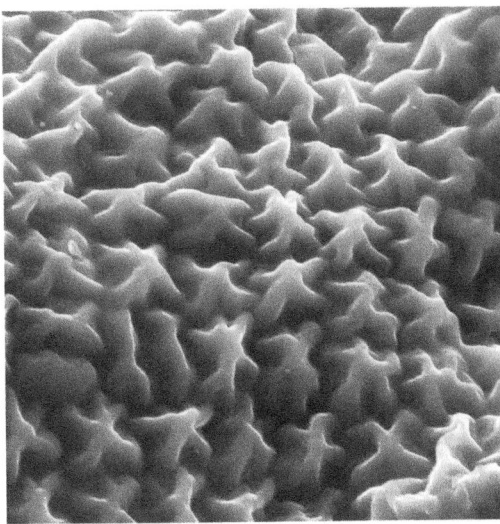

Figure 1.2 Scanning electron microscopy image of raw *Turbinaria Turbinata*.

composed of a fibrous structure and an amorphous matrix, in which is embedded different polysaccharides (Figure 1.1). The main mechanisms of biosorption include electrostatic attraction, ion exchange and complex formation,[1] but these can differ depending on the type of biomass, origin and treatment to which the biomass has been submitted. To understand how contaminants bind to the biomass, it is essential to identify the functional groups responsible for their binding. Most of the functional groups involved in the binding process are found in cell walls.

Fourier transform infrared (FTIR) spectroscopy is the most widely used technique used to characterize biomaterials. The type of binding groups present in the surface of the adsorbents can easily be identified by this method. For all types of biomass (fungi, bacteria, yeast, algae, biowaste materials) used for biosorption, the same functional groups are generally involved in the removal of pollutants. Several chemical groups such as hydroxyl, carboxyl, sulfonate, alcohol, amino and phosphate groups have been proposed as being responsible for the biosorption of hazardous compounds through binding by biomaterials. FTIR spectroscopy allows the determination of different functionalities and offers excellent information on the nature of the bonds between the pollutant and the biomass surface. The level of contaminant uptake depends on factors such as the number of binding sites, their chemical state, accessibility and affinity for the contaminants.

As we can see in the following examples for different biomass such as agricultural waste biomass *Gossypium hirsutum* (Figure 1.3),[35] fungal biomass *Aspergillus niger*,[23] bacterium biomass *Escherichia coli* (Figure 1.4),[14] or seaweed biomass *Lessonia nigrescens* and *Macrocystis integrifolia*,[36] FTIR spectroscopic analysis revealed in all cases that the main functional groups in the

Table 1.3 Examples of seaweed biomass used for removing hazardous pollutants.

Pollutant	Seaweed biomass	Sorption capacity/$mg\,g^{-1}$	Optimal temperature/$°C$	Optimal pH	Ref.
Pb	Fucus spiralis	204.1		3.0	52
	Ulva lactuca	34.7			89
	Cystoseira baccata	186	25	4.5	25
	Ascophyllum nodosum	178.6		3.0	52
	Sargassum sp.	1.16 mmol g^{-1}	22	5.0	37
	Sargassum natans			3.5	28
	Sargassum vulgare			3.5	28
	Ascophyllum nodosum	370		3.5	28
	Laminaria hyperborea	50.3	25	5.0	27
	Bifurcaria bifurcata	52.7	25	5.0	27
	Sargassum muticum	38.2	25	5.0	27
	Fucus spiralis	43.5	25	5.0	27
Ni	Cystoseria indica	47	20	6.0	47
	Nizmuddinia zanardini	50	20	6.0	47
	Sargassum glaucescens	52	20	6.0	47
	Padina australis	23	20	6.0	47
	Ascophyllum nodosum	43		6.0	52
	Fucus spiralis	50		6.0	52
	Sargassum sp	0.61 mmol g^{-1}	22	5.5	37
	Laminaria japonica	1.13 mmol g^{-1}	25	4.3–6.5	48
	Sargassum vulgare			3.5	28
	Sargassum fluitans			3.5	28
	Sargassum natans			3.5	28
Zn	Fucus spiralis	53.2		6.0	52
	Ascophyllum nodosum	42		6.0	52
	Sargassum sp.	0.50 mmol g^{-1}	22	5.5	37
	Laminaria japonica	0.83 mmol g^{-1}	25	4.3–6.5	48
	Fucus vesiculosus			4.5	114
	Laminaria japonica			4.5	114
	Sargassum fluitans			4.5	114
	Laminaria hyperborea	19.2	25	5.0	27
	Bifurcaria bifurcata	30.3	25	5.0	27
	Sargassum muticum	34.1	25	5.0	27
	Fucus spiralis	34.3	25	5.0	27
Cd	Fucus spiralis	114		6.0	52
	Ulva lactuca	29.2			89
	Cystoseira baccata	101	25	4.5	25
	Ascophyllum nodosum	87		6.0	52
	Sargassum sp.	0.75 mmol g^{-1}	22	5.5	37
	Laminaria japonica	0.93 mmol g^{-1}	25	4.3–6.5	48
	Sargassum vulgare	87		4.5	115
	Sargassum fluitans	87		4.5	115
	Sargassum filipendula	74		4.5	115
	Laminaria hyperborea	31.3	25	5.0	27
	Bifurcaria bifurcata	30.3	25	5.0	27
	Sargassum muticum	38.4	25	5.0	27
	Fucus spiralis	42.1	25	5.0	27

Table 1.3 (*Continued*)

Pollutant	Seaweed biomass	Sorption capacity/mg g^{-1}	Optimal temperature/°C	Optimal pH	Ref.
Cu	*Sargassum* sp.	0.99 mmol g^{-1}	22	5.0	37
	Fucus spiralis	70.9		4.0	52
	Ascophyllum nodosum	58.8		4.0	52
	Laminaria japonica	0.97 mmol g^{-1}	25	4.3–6.5	48
	Sargassum vulgare	59		4.5	115
	Sargassum fluitans	51		4.5	115
	Sargassum filipendula	56		4.5	115
	Ulva fasciata	73.5	25	5.5	2
	Sargassum sp.	72.5	25	5.5	2
Cr	*Fucus spiralis*	0.68 mmol g^{-1}	21	2.0	39
	Fucus vesiculosus	0.82 mmol g^{-1}	21	2.0	39
	Cystoseira indica	27.9	25	3.0	51
	Turbinaria ornata	31		3.5	116
	Sargassum sp.	19.0	25	2.0	4
U	*Sargassum fluitans*	150		4.0	117
Se	*Cladophora hutchinsiae*	74.9	20	5.0	57
As	*Maugeotia genuflexa*	57.5	20	6.0	56
Methylene blue	*Turbinaria turbinata*	63.0	25	5.0	31
Acid red	*Azolla filiculoides*	109.0	30	7.0	29
Acid green	*Azolla filiculoides*	133.5	30	3.0	29
Basic Orange	*Azolla filiculoides*	833.3	30	7.0	74
Reactive Black 5	*Laminaria* sp.	101.5	40	1.0	54
Acid orange	*Azolla filiculoides*	109.6	30	3.0	29
Phenol	*Sargassum muticum*	4.6	25	1.0	30
Chlorophenol	*Sargassum muticum*	251.0	25	1.0	30

biomass surface are carboxyl, carbonyl, hydroxyl, sulfonic, amino, phosphate and alcoholic groups.

X-ray photoelectron spectroscopy (XPS) is another widely used method for the biomaterials surface analysis.[37,38] XPS analysis has the ability to determine the elemental composition on the surface of materials. XPS data give an idea of the local oxidation states and chemical bonding environment of the bio-material. XPS can be applied to determine the interactions between the organic functional groups on the biomass surface and the contaminant adsorbed. Figure 1.5 shows the XPS spectrum of *Spirogyra* sp. before and after sorption of fluoride in aqueous solution.[38] In virgin biosorbent, the observed C$_{1s}$ peak can be convoluted into three peaks at 284.6, 286.6 and 287.9 eV, which can be attributed to the presence of C–C/C–H and a carboxylic (–O–C–O) group re-spectively. After fluoride sorption the C1s XPS high resolution narrow scan showed peaks at 284.6, 286.5, 287.4 and 288.9 eV. The peak observed at 288.9 eV was attributed to –CH$_2$–CF$_2$– bond formation. Thus, like the FTIR analysis method, the compounds adsorbed on biomass surface could be also analysed

Table 1.4 Examples of agricultural wastes used as biomass for removing hazardous pollutants.

Pollutant	Biomass	Sorption capacity/ $mg\,g^{-1}$	Optimal temperature/ $T^{\circ}C$	Optimal pH	Ref.
Hg	Carica papaya (wood)	165.5		6.5	92
Pb	tobacco dust	39.6	25	4.0–5.0	33
	orange peel	476.1	30	5.5	118
	Carica papaya (seeds)	1666.6	25		119
	Moringa oleifera (leaves)	209.5	40	5.0	120
	Zea mays (stalk)	106.45		6.0	121
Cu	peanut shell	25.39	20	5.0	122
	tobacco dust	36.0	25	4.0–5.0	33
Cd	tobacco dust	29.6	25	4.0–5.0	33
	orange peel	293.3	30	5.5	118
	Carica papaya (seeds)	1000.0	25		119
Zn	tobacco dust	21.5	25	4.0–5.0	33
Ni	tobacco dust	24.5	25	4.0–5.0	33
	orange peel	162.6	30	5.5	118
Cr	peanut shell	27.86	20	5.0	122
	Posidonia oceanica	–	30	2.0	53
	pistachio hull	116.3			123
	Parthenium hysterophorus (weed)	24.5	20	1.0	91
Reactive Blue 49	Capsicum annuum (seeds)	96.35	25	2.0	124
Basic Green 4	Ananas comosus (leaf)	54.64	25	9.0	125
Basic Red 46	Pine leaves	71.94	25	6.0	126
Methylene blue	Posidonia oceanica	5.6	30	6.0	85
2,4,6-Trichlorophenol	Acacia leucocephala (bark)	256.4	30	5.0	120
Phenolic compounds	banana peel	689.0			87

by XPS analysis. Therefore, FTIR and XPS are two analytical techniques that can be used to get information on contaminant binding mechanisms on biomass.

The physical properties of biomaterials such as surface area, pore volume and pore size distribution are also studied because physical biosorption can be influenced by these characteristics of the biomaterial. In this case, the Brunauer–Emmett–Teller (BET) method based on nitrogen adsorption at 77 K is often used. However, in the most cases, the results obtained indicate non-porous or macroporous materials. For example, the BET surface area and pore volume obtained by N_2 adsorption/desorption isotherms at 77 K are 2.86 $m^2\,g^{-1}$ and 0.003 $cm^3\,g^{-1}$ respectively for *Sargassum muticum* seaweed,[30] similarly to what has been published (Table 1.3) for *Schizophyllum commune* fungus biomass for which the BET surface area and pore volume are 3.95 $m^2\,g^{-1}$ and 0.0041 $cm^3\,g^{-1}$ respectively.[22]

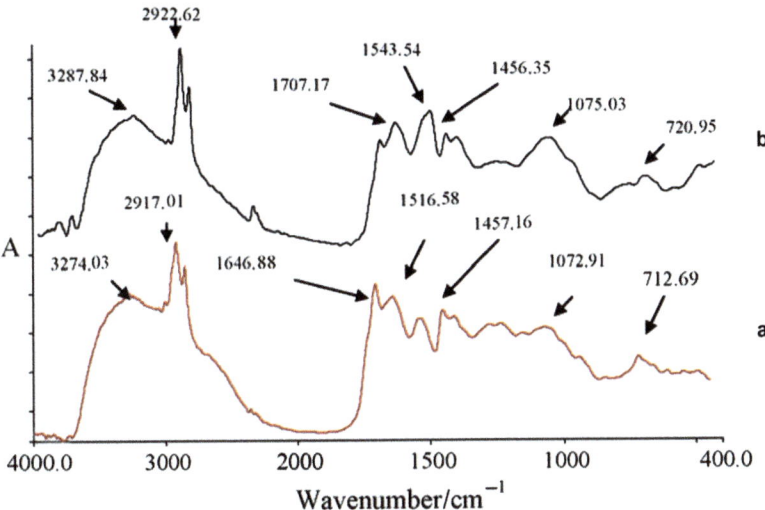

Figure 1.3 FTIR spectra of *Gossypium hirsutum* seed cake biomass: (a) before; and (b) after Pb(II) sorption.[35]

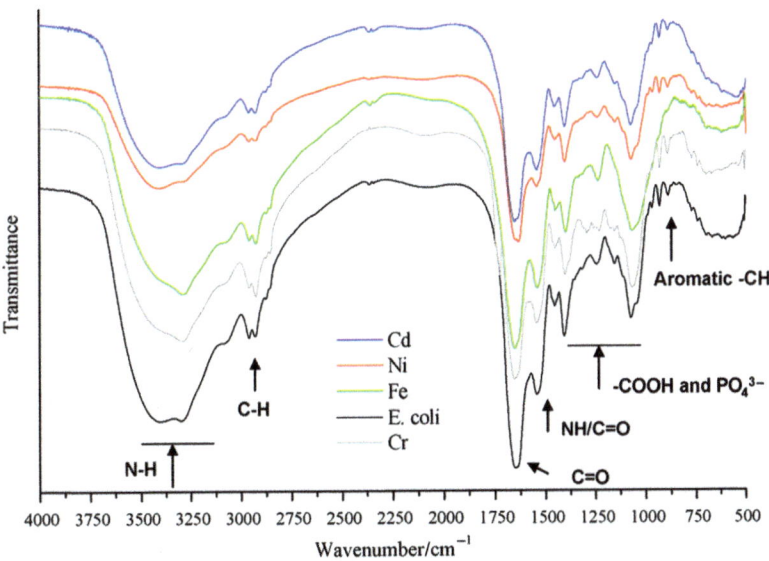

Figure 1.4 FTIR spectra of an *Escherichia coli* bacterium before and after metal loading.[14]

Scanning electron microscopy (SEM) is an analysis method that can clearly reveal the surface texture and morphology of the biosorbent before and/or after biosorption (Figure 1.6). As can be seen when comparing SEM images of cork waste before the biosorption process (Figure 1.6(a)) and after its use

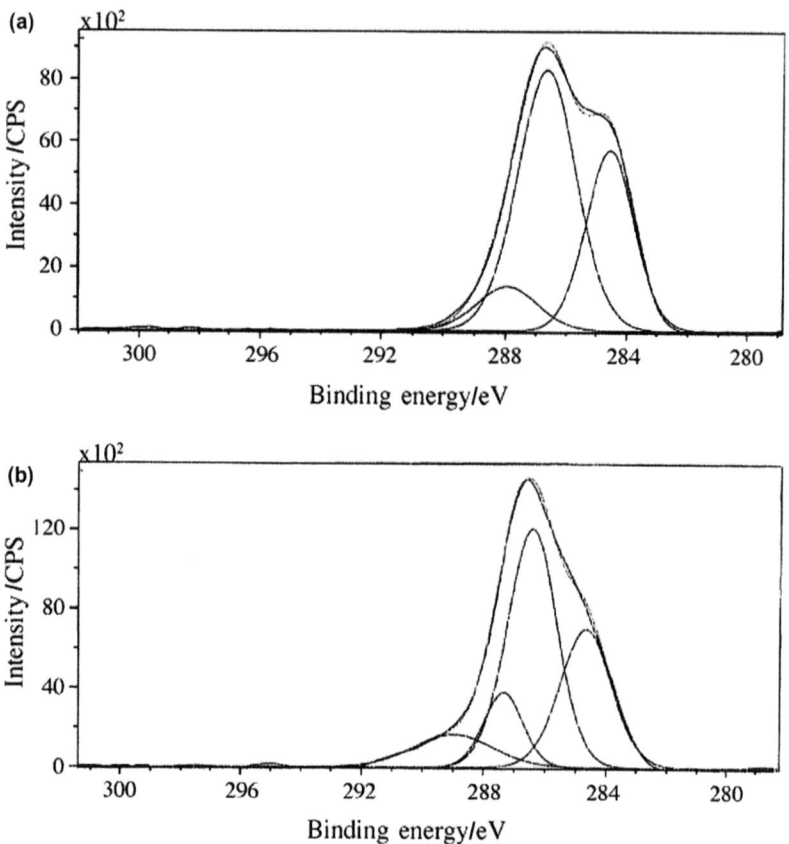

Figure 1.5 XPS spectra of algal biomass *Spirogyra* sp. (a) before and (b) after sorption.[38]

(Figure 1.6(b)), there are no changes in the cork waste morphology,[34] confirming that the morphological structure of the cork has not been affected by the metal biosorption experiments.

To select the optimum pH range for contaminant sorption, it is useful to determine the pH at zero point of charge (pH_{zpc}) of the selected biosorbent, *i.e.* the pH value at which the sorbent surface is globally neutral.[39–41] Indeed, when $pH < pH_{zpc}$, the biosorbent surface becomes positively charged and metal sorption is inhibited due to electrostatic repulsion between metal ions and functional groups. Conversely when $pH > pH_{zpc}$, the number of negatively charged sites on the biosorbent increases, and metal sorption is mostly favoured.

1.2.3 Mechanism of Biosorption

The mechanism of the biosorption of hazardous compounds is a complicated process. In order to understand how pollutants bind to the biomass, it is

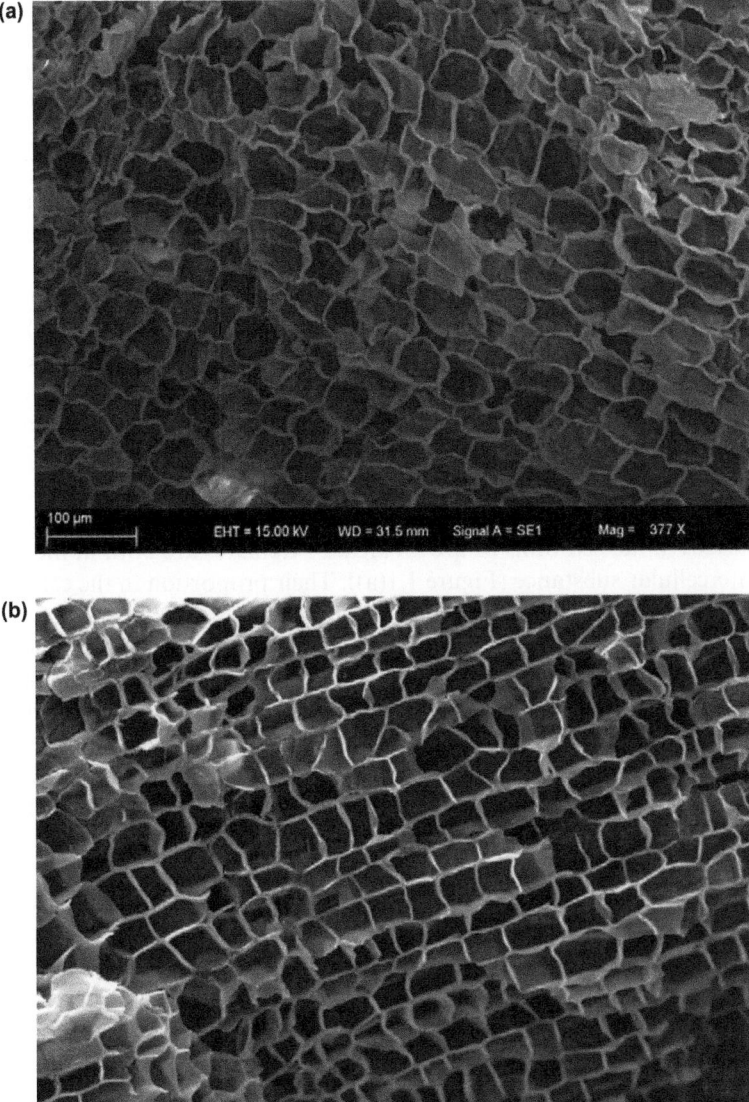

Figure 1.6 SEM images of cork before (a) and after (b) the biosorption process.[34]

essential to identify the functional groups responsible for pollutant binding. Most of the functional groups involved in the binding process are found in cell walls. Biosorption is generally based on physico-chemical interactions between pollutant and functional groups present on the cell surface, such as electrostatic interactions, ions exchange, and metal ion chelation and complexation.

Biomaterial surfaces can be regarded as a mosaic of different functional groups which are responsible for binding of organic or inorganic ions, including amide (–NH2), carboxylate (–COO–), thiols (–SH), phosphate (PO_4^{3-}) and hydroxide (–OH). Therefore, the identification of functional groups is very important for understanding the mechanisms responsible for the binding of certain compounds. In the case of metal uptake, the status of biomass (living or non-living), types of biomaterials, properties of metal solution chemistry, ambient/environmental conditions, *etc.* also influence the mechanism of metal biosorption.[1,42]

For example, alginates and sulfated polysaccharides are important components of the algae cell wall, particularly brown algae (Figure 1.1(a)). These cell wall polysaccharides contain hydroxyl (–OH), sulfate ($-SO_4^{2-}$) and carboxyl (–COO–) groups that serve as binding sites for metallic ions.[1,11,28] These biomass functional groups are strong ion exchangers, and therefore important complexation sites for hard and or transition metal cations.[43] It was reported that alginates have a strong affinity for divalent cations, whereas the binding of trivalent cations has been attributed mainly to the presence of sulfated polysaccharides in the cell walls of brown algae.[1] Alginates are both in the cell wall and intercellular substance (Figure 1.1(a)). Their proportion in the cell wall can be assessed at 40% dry matter.[1,37]

Alginate, a major brown algae polysaccharide, is an important source of carboxylic groups,[1] which are generally the most abundant acidic functional group in brown algae. The carboxylic groups constitute the highest percentage of titratable sites (greater than 70%) in dried brown algal biomass. They have a very important role in metal biosorption being responsible for around 90% of the process of the metal immobilization.[44] The role of carboxylic groups in biosorption process has been clearly demonstrated by the reduction of cadmium and lead uptake by dried Sargassum biomass following partial or complete esterification of the carboxylic sites.[45] FTIR analysis of biomasses after the removal of cadmium and lead by *Cystoseira baccata* shows carboxyl groups participation in metal uptake.[25] This confirmation is found in the most of cases mentioned by the literature.

1.2.4 Factors Affecting Biosorption

The study of parameters influencing the biosorption process is crucial in the evaluation of biosorption potential of any biomaterial. The mechanism of binding by a biomass may depend not only on the chemical nature of contaminant, type of biomass, its preparation and its specific surface properties, but also on environmental conditions such as pH, temperature and ionic strength. Other factors influencing reactions include biomass dosage, initial solute concentration, size of biosorbent, existence of competing organic or inorganic ligands in solution, and agitation rate and period. However, of all these parameters, the aqueous phase pH is the most important in the biosorption of hazardous compounds from aqueous solutions.[3,28,40,42,46,47]

1.2.4.1 pH

Sorption equilibrium is pH dependent as pH affects the dissociation of the different chemical groups involved in the biosorption process.[44] The biosorption of a compound depends on protonation state of functional groups (such as amino, carboxyl, thiol, sulfydryl and phosphate) at the biomaterial cell wall surface.[38]

pH not only influences the binding site dissociation state, but also the solution chemistry of the target metal in terms of hydrolysis, complexation by organic and/or inorganic ligands and redox potentials.[42] The biosorption capacity of an alga depends strongly on equilibrium solution pH and thus a study of its effect on adsorption is necessary for an accurate evaluation of equilibrium parameters. Therefore, if the metal binding groups are weakly acidic or basic, the availability of free sites is dependent on the solution pH. Thereby, the biosorption process is dependent on aqueous phase pH and on the nature of the functional groups on algal cell walls. Their ionic states determine the extent of biosorption.[38] There is an optimum pH for each metal uptake, usually within the range 3.0–7.0, in which the competition of hydrogen ions minimized, thereby enhancing metal sorption. Similarly, each biomass has its own optimum pH for metal ion uptake.[46]

Liu *et al.*[48] have studied the biosorption of Cd^{2+}, Ni^{2+}, Cu^{2+} and Zn^{2+} by the marine brown algae *Laminaria japonica*. The metal adsorption was strictly pH dependent. The optimal removal efficiencies of four metals were observed for the pH range 4.3–6.5 for chemically modified and raw brown algae. *Laminaria japonica* was also used by Luo *et al.*[49] to study the effect of pH on lead biosorption. The lead equilibrium uptake increased as the pH increased from pH 1.4 to 3.0, but then no change was observed as the pH increased 3.0 to 5.3. The maximum Pb^{2+} uptake (84%) was obtained in the pH range. Similar results were obtained by other works, such as Lodeiro *et al.*,[25] who found that the best fixation rates of cadmium and lead on *Cystoseira baccata* were obtained at pH 4. The metal sorption was found to increase with the solution pH reaching a plateau above pH 4 (Figure 1.7). This curve has been interpreted as a result of the change in the ionic state of the acid functional groups involved in the metal binding. This has also been observed for the equilibrium uptake of cadmium by *Sargassum baccularia*[50] when the equilibrium uptakes were similar in the pH range 3–5 over the entire range of the solution phase equilibrium concentration of cadmium. However, the equilibrium uptakes of cadmium at pH 2 were much smaller than the uptakes at pH 3–5. At low pH, protons can compete effectively with cadmium for the binding sites on the biomass surface. The protonated binding sites are thus no longer available to bind cadmium ions from solution.

The pH dependence of metal adsorption can largely be related to type and ionic state of these functional groups on the biosorbents and also on the metal chemistry in solution.[51] At low pH values, cell wall ligands are protonated and are closely associated with hydronium ions, and hence access of the metal ions to the ligands is restricted. With increasing pH, carboxyl groups would be exposed, leading to attraction between these negative charges and the metals

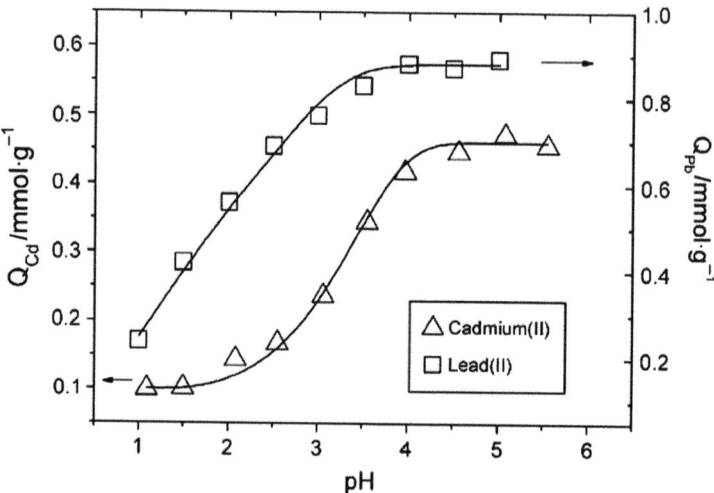

Figure 1.7   Effect of pH on the biosorption of cadmium(II) (triangles) and lead(II) (squares) by *Cystoseira baccata*.[25]

and hence increases in biosorption onto the cell surface. The pH effect may be further explained in relation to the competition between H_3O^+ and metal ions. At low pH values, the concentration of H_3O^+ far exceeds that of metal ions and hence, they occupy the binding sites on the cell walls leaving metal ions un-bound. When the pH increases, the competing effect of H_3O^+ decreases and the positively charged metal ions take up the free binding sites. The metal uptake capacity is hence increased.

Pahlavanzadeh *et al.*[47] studied the pH effect on biosorption of nickel(II) by brown algae such as *Cystoseria indica* (Cys), *Nizmuddinia zanardini* (Niz), *Sargassum glaucescens* (Sarg) and *Padina australis* (Pad) (Figure 1.8). The adsorption capacity of nickel increases as the pH increases and reaches a maximum at pH 6, before decreasing as the pH increases further. Because H^+ compete with metal ions at lower pH, the sorbent surface takes up more H^+, consequently reducing the metal ions binding on the sorbent surface. At higher pH levels, the sorbent surface takes more negative charges, thus attracting more metal ions. With a further increase in pH, the formation of anionic hydroxide complexes reduces the concentration of free metal ion and thus metal ion adsorption capacity decreases.

Similarly, other biosorption experiments were performed by Romera *et al.*[52] using a brown algae (*Fucus spiralis*). The optimum pH for recovery of Cd(II), Ni(II) and Zn(II) was 6. While, in the case of Cu(II) and Pb(II) recovery, the optimum pH was less than 5. This behaviour has also been shown by Basha *et al.*[51] for removal of chromium(VI), where the maximum uptake of Cr(VI) on *Cystoseira indica* were observed at pH 3 (Figure 1.9).

Similar results were obtained by Ncibi *et al.*[53] They used some *Posidonia oceanica* fibres, an aquatic flowering plants for the uptake of hexavalent chromium, where the maximal adsorption capacity of Cr(VI) was found at

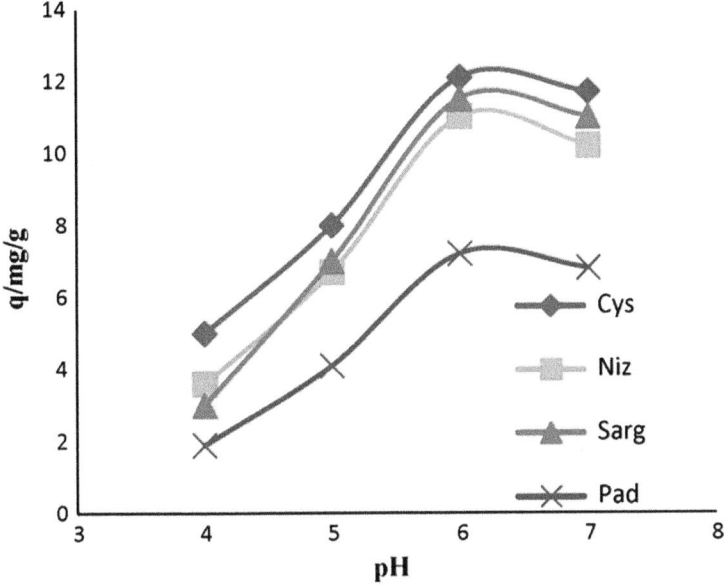

Figure 1.8 Effect of pH on the biosorption of nickel(II) at 30 °C.[47]

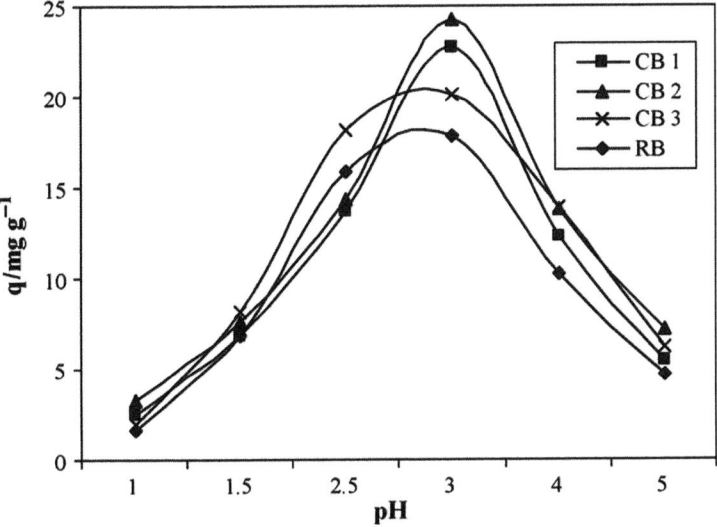

Figure 1.9 Effect of pH on Cr(VI) sorption by chemically modified and raw *Cystoseira indica*.[51]

pH 2. This increase in chromium uptake could be due to the ionization of carboxyl groups present in the seaweeds at this pH, which can result in higher interaction with chromium. Table 1.3 presents some marine algae with the optimum pH for removal of different compounds.

1.2.4.2   Temperature

Biosorption experiments are usually performed at temperatures ranging between 20 and 35 °C. Under these conditions, the temperature seems not to have
a significant effect on the biosorption process. However, according with some
reviews such as that by Mack *et al.*[46] and Vijayaraghavan and Yun,[54] there are
some exceptions. Temperature can affect the interactions between the biosorbent and the contaminant ions, usually by influencing the stability of the
contaminant ions in solution, the stability of the contaminant–sorbent complex
and the ionization of the cell wall chemical moieties.

According to the system biomass–contaminant, one or other of the effects of
temperature can be observed. In some cases increasing the temperature lead to
an increase of biomass biosorption capacity, and in other cases, the opposite
effect can be observed.

The effect of aqueous phase temperature on fluoride ion removal by algal
biosorbent *(Spirogyra IO2)* was investigated by Mohan *et al.*[38] at five different
temperatures (10, 20, 30, 40 and 50 °C), keeping all other experimental conditions
constant. They found an increase in sorption capacity with increasing temperature from 10 to 50°C. In this case, the biosorption process was endothermic. The
sorption increase with temperature was attributed to either an increase in the
number of active surface sites available for sorption on the adsorbent or a decrease in the boundary layer thickness surrounding the sorbent, so that the mass
transfer resistance of absorbate in the boundary layer decreased. Whereas, for
Vijayaraghavan and Yun,[55] the enhanced sorption at high temperatures was due
to the increasing surface activity and kinetic energy of the solute, although
physical damage to the biomass can be expected at higher temperatures.

However, the temperature effect on biosorption of As(III) onto algal biomass
(Maugeotia genuflexa) was investigated by Sari *et al.*[56] who showed that increasing the temperature led to a lowering of the biomass biosorption capacity.
The biosorption percentage decreased from 96% to 60% when temperature was
increased from 20 to 50 °C. These results indicate the exothermic nature of
As(III) biosorption onto this biomass. Similar results were obtained for Se(IV)
uptake by *Cladophora hutchinsiae* biomass.[57] This decrease in biosorption efficiency was attributed to parameters such as the relative increase in the es-
caping tendency of the metal ions from the solid phase to the bulk phase and
the destruction of some active sites on the biomass surface due to bond rup-
tures. Thus, it is always desirable to conduct biosorption process at room
temperature, because this condition is easy to replicate.

1.2.4.3   Ionic Strength

According to a reviews by Das[42] and Vijayaraghavan and Yun,[54] ionic strength
is another important parameter affecting the adsorption of solute to the biomass surface.[58–60] The effect of ionic strength may be ascribed to the competition between ions, changes in the metal activity, or in the properties of the
electrical double layer. When two phases, *e.g.* biomass surface and solute in

aqueous solution are in contact, they are surrounded by a double layer owing to electrostatic interaction. Thus, adsorption decreases with increase in ionic strength.[61] Some inorganic ions, such as chloride, may form complexes with some metal ions and therefore, affect the sorption process.[58,62]

1.2.4.4 Biomass Dosage

The dosage of a biosorbent strongly influences the extent of biosorption. In many instances, the removal efficiency (%) of pollutants increases with increase in adsorbent dosage, whereas the adsorption capacity (*i.e.* amount of pollutants per unit weight of biosorbent) decreases with increasing biosorbent dosage. This was observed by Vijayaraghavan *et al.*[63] using crab shell particles for the removal of copper and cobalt.

An increase in the biomass concentration generally increases the amount of solute biosorbed due to the increased surface area of the biosorbent, which in turn increases the number of binding sites.[64] However, according to Kumar *et al.*[65] and Romera *et al.*[52] the increase in the biosorption with increasing biomass concentration can be attributed to the availability of adsorption sites for the solute, resulting in much electrostatic interactions and interferences between binding sites.

Indeed, the decreasing quantity of biosorbed solute per unit weight of biosorbent with increasing biosorbent dosage may be due to a complex interaction of several factors. An important factor at high sorbent dosages is that the available solute is insufficient to completely cover the available exchangeable sites on the biosorbent, usually resulting in low solute uptake.[66]

1.2.4.5 Biosorbent Particle Size

The size of the biosorbent also plays a vital role in biosorption. Smaller particles have a higher surface area, which in turn favours biosorption and results in a shorter equilibration time.[12]

Pavasant *et al.*[67] studied the sorption of Cu^{2+}, Cd^{2+}, Pb^{2+} and Zn^{2+} by a dried green macroalga, *Caulerpa lentillifera*. The effect of biomass particle size on the biosorption process was investigated. They found that uptake of the ground biomass of each metal ion was faster, and the equilibrium reached faster, than those achieved with the whole thallus. This was because smaller particles allow more rapid contact between the metal ion and the binding sites. However, the particle size has no significant effect on the total sorption capacity. This indicates that grinding does not result in a deterioration in the sorption integrity of the alga.

The review of Aksu[68] gives an example of the effect of biosorbent size on biosorption process. Chu and Chen[69] studied the effect of biomass particle size on the biosorption of Basic Yellow 24 using dried activated sludge biomass. They observed that the biosorption capacity of biomass increased with decreasing particle size. This situation is explained by the greater total surface area of smaller particles for the same amount of biomass.[69]

1.2.4.6   Initial Solute Concentration

It is necessary to identify the maximum saturation potential of a biosorbent, for which experiments should be conducted at the highest possible initial solute concentration.[12,42] The initial solute concentration has also an impact on biosorption, as a higher concentration results in a high solute uptake.[70] This is because, at lower initial solute concentrations, the ratio of the initial moles of solute to the available surface area is low; subsequently, the fractional sorption becomes independent of the initial concentration. However, at higher concentrations, the sites available for sorption become fewer compared with the moles of solute present and, hence, the solute removal is strongly dependent upon the initial solute concentration.

1.2.4.7   Agitation Rate and Period

The biosorption of hazardous compounds can also be influenced by the agitation rate and period.

The rate of the biosorption process can be influenced by external film diffusion. With appropriate agitation, this mass transfer resistance can be minimized. When the agitation rate is increased, the diffusion rate of a solute from the bulk liquid to the liquid boundary layer surrounding particles becomes high due to enhanced turbulence and the decrease in thickness of the liquid boundary layer.[71] Under these conditions, the value of the external diffusion coefficient becomes larger.[72] Finally, at higher agitation rates, the boundary layer becomes thin, which usually enhances the rate at which a solute diffuse through boundary layer.

Chu and Chen[69] investigated the effect of shaking rate on the biosorption of Basic Yellow 24 using dried activated sludge biomass. They observed that the uptake capacity of biomass increased from 18 to 53 mg g^{-1} with increasing shaking rate from 40 to 160 rpm. The results showed that there is a boundary layer surrounding the biomass particles and a decrease in its effect with increasing shaking rate.

The effect of agitation period on the adsorption of Pt(IV), Pd(II) and Au(III) by L-lysine modified crosslinked chitosan resin was studied by Fujiwara *et al.*[73] The results demonstrated that adsorption increases with increasing agitation time and attained an equilibrium at around 120 min for Pt(IV), Pd(II) and Au(III). Over 75% adsorption occurred within 30 min and equilibrium was attained within 120 min.

1.2.5   Biosorption of Pollutants

For many years, numerous studies have focused on some biomaterials including bacteria, fungi, yeast, algae, waste wood and agricultural by-products that are able to bind organic and inorganic compounds (metals, dyes, phenolic compounds, pesticide, *etc.*). The following paragraphs describe some work on the biosorption of hazardous compounds. To investigate contaminant biosorption on a given biomass, most researchers study the influence of the mentioned above, *viz.*, pH, temperature, initial pollutant, biomass concentrations, *etc.*

1.2.5.1 Dyes

Synthetic dyestuffs, one group of organic pollutants, are used extensively in the textile, paper and printing industries, and in dyehouses. The dyes vary greatly in their chemistries, and therefore the interactions of a particular dye with the biomass surface depend on its chemical structure and characteristics, and the specific biomass chemistry. A wide variety of biomass has been used to test the biosorption of a broad range of dyes. Depending on the dye and the biomass used different binding capacities have been observed (Tables 1.1–1.4).

Biosorption of Acid Red 88 (AR88), Acid Green 3 (AG3) and Acid Orange 7 (AO7) on the macro alga, *Azolla filiculoides*, in batch mode was investigated by Padmesh *et al.*[29] Langmuir and Freundlich adsorption models were used to describe mathematically the batch biosorption equilibrium data and the model constants were evaluated. The adsorption capacity was pH dependent with a maximum values of 109.0 mg g^{-1} at pH 7 for AR88, 133.5 mg g^{-1} at pH 3 for AG3 and 109.6 mg g^{-1} at pH 3 for AO7. The ability of *Azolla filiculoides* to biosorb AG3 in packed column was also investigated. The column experiments were conducted to study the effect of important design parameters such as initial dye concentration (50–100 mg L^{-1}), bed height (15–25 cm) and flow rate (5–15 mL min^{-1}) to the well-adsorbed dye. At an optimum bed height (25 cm), a flow rate (5 mL min^{-1}) and an initial dye concentration (100 mg L^{-1}), *Azolla filiculoides* exhibited 28.1 mg g^{-1} for AG3.

Azolla filiculoides biomass was also used by Tan *et al.*[74] to remove Basic Orange (BO) dye from aqueous solution. The maximum biosorption capacity for BO was 833.33 mg g^{-1} based on the Langmuir equation at 30 °C, pH 7.0, biosorbent dosage of 5 g L^{-1} and a contact time of 4 h. The thermodynamic values, including Gibbs free energy (ΔG), enthalpy (ΔH) and entropy (ΔS), obtained from the calculation suggest that the biosorption of BO on the dried *Azolla filiculoides* biomass is feasible, spontaneous and endothermic. FTIR spectroscopy showed that the amino, carboxyl and hydroxyl groups may be responsible for the biosorption of BO on the biomass.

Aksu *et al.*[75] used *Rhizopus arrhizus* as a biosorbent for the removal of three reactive dyes: Gemacion Red H-E7B (GR), a monochlorotriazine mono-azo type reactive dye; Gemazol Turquoise Blue-G (GTB), a vinyl sulfone mono-azo type reactive dye; and Gemactive (Reactive) Black HFGR (GB), a vinyl sulfone di-azo type reactive dye from aqueous solution. The effect of operating parameters such as flow rate and inlet dye concentration on the sorption characteristics of *Rhizopus arrhizus* was investigated at pH 2.0 and at 25 °C for each dye. They found that the total amount of sorbed dye decreased with increasing flow rate and increased with increasing inlet dye concentration for each dye. The biosorption capacity of dried *R. arrhizus* was 1007.8 mg g^{-1} for GR dye, 823.8 mg g^{-1} for GTB dye and 635.7 mg g^{-1} for GB dye at the highest inlet dye concentration of approximately 750 mg L^{-1} and at the minimum flow rate of 0.8 mL min^{-1}.

Mona *et al.*[76] used a cyanobacterium, *Nostoc linckia* HA 46, for biosorption of a textile dye, Reactive Red 198 from aqueous solution. The interactive effects of initial dye concentration (100–500 mg L^{-1}), pH (2–6) and temperature (25–45 °C) on dye removal was examined using the Box–Behnken design. The maximum adsorption capacity of the immobilized biomass was 93.5 mg g^{-1} at 35°C temperature and pH 2, with an initial concentration of 100 mg L^{-1}. In these conditions, 94% of the dye was removed. FTIR studies revealed that biosorption was mainly mediated by functional groups such as hydroxyl, amide, carboxylate, methyl and methylene groups present on the cell surface.

Akar *et al.*[77] studied the biosorption of Acid Blue 40 (AB40) on the cone biomass of *Thuja orientalis* when varying some parameters pH, contact time, biosorbent and dye concentration and temperature to estimate the equilibrium, thermodynamic and kinetic parameters. The AB40 biosorption was fast and equilibrium was attained within 50 min. The equilibrium data fitted well to the Langmuir isotherm model in the AB40 studied concentration range and at various temperatures. The maximum biosorption capacity for AB40 was 97.06 mg g^{-1} at 20 °C. The changes in Gibbs free energy, enthalpy and entropy of biosorption were also evaluated, with the results indicating that the biosorption was spontaneous and exothermic.

Biosorption studies of Acid Blue 225 (AB 225) and Acid Blue 062 (AB 062) from aqueous solution onto *Paenibacillus macerans* were conducted by Colak *et al.*[17] Results showed that a pH value of 1 was favourable for the biosorption of dyes and the maximum adsorption efficiency of AB 225 and AB 062 was 94.98 mg g^{-1} and 95.08 mg g^{-1} respectively. Thermodynamic parameters such as enthalpy, entropy, and Gibb's free energy changes were also calculated and it was found that the biosorption of dyes by *Paenibacillus macerans* was a spontaneous process. FTIR spectroscopy confirmed the presence of carboxylic, hydroxyl and $-NH_2$ groups of the biomass structure involved in the dyes biosorption.[78,79]

The biosorption of Acid Blue 290 and Acid Blue 324 on *Spirogyra rhizopus*, a green algae growing on fresh water, was studied by Ozer *et al.*[80] The optimum initial pH and temperature values for AB 290 and AB 324 biosorption were found to be 2.0 and 30 °C, and 3.0 and 25 °C, respectively. The adsorbed AB 290 and AB 324 amounts increased with increasing initial dye concentration up to 1500 and 750 mg L^{-1}, respectively. The sorption capacities of *S. rhizopus* for AB 290 and AB 324 dyes were found to be 1356.6 mg g^{-1} and 367.0 mg g^{-1}, respectively. Thermodynamic studies showed that the biosorption of AB 290 and AB 324 on *S. rhizopus* was exothermic in nature.

Removal of Acid black 1 (AB1) by a brown alga biomass, *Cystoseira indica*, and a red alga biomass, *Gracilaria persica*, was investigated by Kousha *et al.*[81] The variables examined were the biomass dosage (0.5–1.5 g L^{-1}), initial AB1 concentration (10–50 mg L^{-1}), initial pH (2–6) and time (20–80 min). Due to the saturation of the available binding sites on the biosorbent, removal efficiency decreased with increasing initial dye concentration. The dye removal increased with increasing contact time, and the sorption process reached equilibrium within the first 60 min. At the optimum conditions (pH = 2), the maximum removal efficiencies of AB1 achieved for *Cystoseira indica* and *Gracilaria persica* were 90.76 and 98.18%, respectively. These results also imply that the used brown and red algal biomasses are adequate biosorbents for the removal of Acid black 1 from aqueous solutions.

The sorptions of three basic dyes, Astrazon Blue FGRL (AB), Astrazon Red GTLN (AR) and methylene blue (MB) onto green macroalga, *Caulerpa lentillifera*, were investigated by Marungrueng and Pavasant.[82] The results were compared to the sorption performance of a commercial activated carbon. The results revealed that the alga exhibited greater sorption capacities than activated carbon for the three basic dyes investigated. The sorption capacities were 38.9 mg g^{-1}, 47.6 mg g^{-1} and 417 mg g^{-1} for alga/AB, alga/AR and alga/MB respectively, against 21.6 mg g^{-1}, 6.2 mg g^{-1} and 238 mg g^{-1} for carbon/AB, carbon/AR and carbon/MB respectively. In the experimental conditions (pH 7 at 25 °C), *Caulerpa lentillifera* could sequester Astrazon Red (GTLN) more rapidly than activated carbon, but the process was much slower for Astrazon Blue (FGRL) sorption.

Waranusantigul *et al.*[83] studied the performance of *Spirodela polyrrhiza* biomass (an aquatic plant) as a biosorbent for the removal of the basic dye, methylene blue, from aqueous solution. A series of experiments were undertaken in an agitated batch adsorber (at 25 °C) to assess the effect of system variables, *i.e.* sorbent dosage, pH and contact time. The results showed that, as the amount of dried *S. polyrrhiza* increased, the percentage of dye sorption increased accordingly. At pH 2.0 the sorption of dye was not favourable, while the sorption at other pH (3.0–11.0) was remarkable:119 mg g^{-1} (at pH 7) and 145 mg g^{-1} (at pH 9). There was no significant difference in the dye concentration remaining when the pH was increased from 3.0 to 11.0. The dye removal time was influenced by the initial dye concentration.

Belala *et al.*[84] investigated the biosorption potential of date stones (DS) and palm tree waste (PTW) for the removal of methylene blue (MB) from aqueous solution. The biosorption capacity of DS and PTW biomass for MB was found to be 43.47 and 39.47 mg g^{-1}, respectively. The calculated thermodynamic parameters (ΔG, ΔH and ΔS) showed that the biosorption of MB on both agriculture waste biomasses was spontaneous and endothermic under examined conditions.

The adsorption of methylene blue from aqueous solution was investigated by Ncibi *et al.*[85] using raw fibres of *Posidonia oceanica*, a marine plant. They found the maximal biosorption capacity to be 5.6 mg g^{-1}. The capacity of *Posidonia oceanica* fibres to remove methylene blue depended on the initial pH of the solution, contact time, initial dye concentration and fibre dosage.

Methylene blue adsorption has also been investigated with some brown seaweeds. The biosorption performance of methylene blue on *Turbinaria turbinata* was investigated by Altenor *et al.*[31] Their experiments were performed at pH 5 and 25 °C, and the results showed that the maximum sorption capacity of

Turbinaria biomass was 63 mg g^{-1} for methylene blue. Comparing this result with other biomass, this study indicated that *Turbinaria turbinata* fibres have a good sorption capacity for basic dyes.

Rubin *et al.*[86] used *Sargassum muticum* to evaluate the adsorption behaviour of methylene blue at 25 °C. Uptake was unaffected in the pH range 4–10 and for pH values below 2, sorption of methylene blue was less favourable. However, the adsorption capacity for protonated alga, which is the most pH dependent, at the lower pH was still about 45 mg g^{-1}. The equilibrium binding has been described in terms of Langmuir and Freundlich isotherms. The results obtained showed that *Sargassum muticum* has a high adsorption capacity for methylene blue, with up to 90% being removed.

The biosorption of Reactive Black 5 (RB5) from aqueous solution using the brown seaweed, *Laminaria* sp., was investigated by Vijayaraghavan and Yun.[54] FTIR spectra confirmed the participation of amine groups in the biosorption of RB5 and the mechanism was proposed to be electrostatic interaction between the positively charged amine groups and negatively charged RB5. Biosorption isotherm experiments, under different pH and temperature conditions, revealed that decreasing the pH and increasing the temperature favoured biosorption. According to the Langmuir model, the maximum RB5 uptake of 101.5 mg g^{-1} was obtained at pH 1 and a temperature of 40°C. Thermodynamic parameters indicated that the biosorption process was spontaneous and endothermic.

1.2.5.2 Phenolic Compounds

In recent years, a number of studies have focused on biomaterials able to biosorb phenolic compounds and derivatives. Depending on the phenolic compound and the biomass used, different adsorption capacities are obtained (Table 1.1–1.4). The researchers have determined that the pH, initial pollutant and biomass concentrations and pretreatment method are important parameters affecting the removal efficiency of phenolics.

Biosorption of phenol, 2-chlorophenol (2-CP) and 4-chlorophenol (4-CP) on *Sargassum muticum* was investigated by Rubin *et al.*[30] The efficiency of this alga was determined by measuring the equilibrium uptake using a batch technique. A chemical pretreatment with CaCl$_2$ was employed to improve the stability as well as the sorption capacity of the algal biomass. The pH influence on the equilibrium binding and the effect of the algal dose were evaluated. The experimental data at pH $= 1$ were analysed using Langmuir and Freundlich isotherms. It was found that the maximum sorption capacity of chlorophenols ($q_{max} = 251$ mg g^{-1} for 4-CP and $q_{max} = 79$ mg g^{-1} for 2-CP) was much higher than that of phenol ($q_{max} = 4.6$ mg g^{-1}). Finally, biosorption of the phenolic compounds on *Sargassum muticum* biomass was observed to be correlated with the octanol–water partitioning coefficients of phenols. This result allowed them to postulate that hydrophobic interactions were the main interactions for the binding.

Kumar and Min[22] used the fungus, *Schizophyllum commune*, to remove phenolic compounds (phenol, 2-CP and 4-CP). The effect of experimental parameters such as pH, contact time, initial concentration of adsorbate and amount of biosorbent dosage was evaluated. The maximum monolayer adsorption capacity of *S. commune* fungus for phenol, 2-CP and 4-CP was found to be 120, 178 and 244 mg g^{-1}, respectively, at $25 \pm 2°C$ according to Langmuir model.

Achak *et al.*[87] investigated the potential application of banana peel as a biosorbent for removing phenolic compounds from olive mill wastewaters. The results showed that increasing the banana peel dosage from 10 to 30 g L^{-1} significantly increased the phenolic compounds adsorption rates from 60 to 88%. Increasing the pH to above neutrality resulted in an increase in the adsorption capacity for the phenolic compounds. The banana peel showed a high adsorption capacity of phenolic compounds (689 mg g^{-1}), revealing that it could be employed as a promising adsorbent for phenolic compounds adsorption.

Biosorption of 2,4,6-trichlorophenol (2,4,6-TCP) from aqueous solution was investigated by Kumar *et al.*[88] on *Acacia leucocephala* bark, an agricultural solid waste. The effect of experimental parameters such as contact time, effect of pH (2–10), initial concentration of adsorbate (50–200 mg L^{-1}) and amount of biosorbent dosage was evaluated. The removal is found to be pH dependent, with maximum removal at pH 5.0. The equilibrium time was found to be 3 h. Increasing the biosorbent dose increased and the percentage removal of 2,4,6-TCP, while the adsorption capacity at equilibrium (mg g^{-1}) decreased. The maximum monolayer biosorption capacity of *Acacia leucocephala* bark for 2,4,6-TCP was found to be 256.4 mg g^{-1} at $30 \pm 1°C$ according to the Langmuir model.

1.2.5.3 Metals

In almost all biosorption experiments involving biomass of different origins (algae, fungi, bacteria, yeasts, wood, agricultural by-products, *etc.*), it is found that they have high capacity for heavy metal removal. Toxic heavy metals such as Pb(II), Cd(II), Hg(II), Cu(II), Ni(II), Cr(III), Cr(VI), *etc.* have been successfully removed from contaminated industrial and municipal waste waters using different biomaterials. Studies dealing with these results are described in the following paragraphs. Tables 1.1–1.4 show the removal of numerous metals by various biomaterials species.

The marine macro algae, *Laminaria hyperborea*, *Bifurcaria bifurcata*, *Sargassum muticum* and *Fucus spiralis*, were shown to be effective for removing the toxic metals Cd(II), Zn(II) and Pb(II) from aqueous solutions.[27] Kinetic studies revealed that the metal uptake rate was rather fast, with removal of 75% of the total amount occurring in the first 10 min for all algal species. The experimentally determined biosorption capacities of cadmium, zinc and lead ions were in the ranges of 23.9–39.5, 18.6–32.0 and 32.3–50.4 mg g^{-1}, respectively. These results indicate that all the macro algae species studied provide an efficient and cost-effective technology for eliminating heavy metals from industrial effluents.

Sari and Tuzen[89] studied the biosorption of Pb(II) and Cd(II) ions from aqueous solution using a green alga, *Ulva lactuca*. The biosorption capacity of this biomass for Pb(II) and Cd(II) ions was found to be 34.7 mg g^{-1} and 29.2 mg g^{-1}, respectively. The average free energy calculated from the Dubinin–Radushkevich (D–R) isotherm model, was 10.4 kJ mol^{-1} for Pb(II) and 9.6 kJ mol^{-1} for Cd(II), indicating that biosorption of both metal ions had taken place by chemisorption.

Biosorption of arsenic(III) on a green algae, *Maugeotia genuflexa*, from aqueous solution was investigated by Sari *et al.*[56] Optimum biosorption conditions were determined under the optimum pH, biomass concentration, contact time and temperature. From the Langmuir model, the maximum monolayer biosorption capacity of the biosorbent was found to be 57.48 mg g^{-1} at pH 6, biomass concentration 4 g L^{-1}, contact time 60 min and temperature 20°C. The calculated average free energy (10.2 kJ mol^{-1}) using the D–R model indicated that the biosorption process was carried out *via* an ion exchange mechanism. Thermodynamic parameters showed that the biosorption of As(III) onto algal biomass was feasible, spontaneous and exothermic under studied conditions.

Pahlavanzadeh *et al.*[47] used four brown seaweeds (*Cystoseria indica, Nizmuddinia zanardini, Sargassum glaucescens and Padina australis*) to investigate the removal of nickel(II) ions from aqueous solution. Experimental parameters affecting the biosorption process such as pH level, contact time, initial metal concentration and temperature were studied. They found that the pH level, time, temperature and initial metal ion concentration had a major effect on the biosorption capacity of the sorbent. The optimum pH level for biosorption of nickel(II) is found to be 6 for all four algae under study. Biosorption capacity increased as metal ion concentration and temperature increased. The calculated thermodynamic parameters (ΔG, ΔH and ΔS) show that the biosorption of nickel(II) ions was feasible, spontaneous and endothermic at the temperature range of 20–40 °C.

Four species of red seaweeds, *Corallina mediterranea, Galaxaura oblongata, Jania rubens and Pterocladia capillacea*, were studied by Ibrahim[90] to remove Co(II), Cd(II), Cr(III) and Pb(II) ions from aqueous solution. The experimental parameters that affect the biosorption process such as pH, contact time and biomass dosage were examined. The maximum biosorption capacity of the metal ions was 105.2 mg g^{-1} at a biomass dosage of 10 g L^{-1}, pH 5 and contact time 60 min. The highest metal ion removal efficiency was obtained for *Galaxaura oblongata* biomass followed by *Corallina mediterranea, Pterocladia capillacea and Jania rubens* biomasses with mean metal ion biosorption efficiencies of 84%, 80%, 76% and 72%, respectively.

Karthikeyan *et al.*[2] investigated the biosorption of copper with two different marine algae: *Ulva fasciata* (green algae) and *Sargassum* sp. (brown algae). Equilibrium isotherms and kinetics were performed to evaluate the relative ability of the two algae for Cu(II) sorption from aqueous solutions. The maximum biosorption capacity was 73.5 mg g^{-1} for *U. fasciata* and 72.5 mg g^{-1} for *Sargassum* sp. at a solution pH of 5.5 ± 0.5. A significant fraction of the total copper(II) uptake was achieved within 30 min.

The sorption of Cu^{2+}, Cd^{2+}, Pb^{2+} and Zn^{2+} by a dried green macroalga, *Caulerpa lentillifera*, was investigated by Pavasant *et al.*[67] They found that the removal efficiency increased with pH. FTIR analysis bands corresponding to O–H bending, N–H bending, N–H stretching, C–N stretching, C–O, S=O stretching, and S–O stretching indicating possible functional groups involved in metal sorption by this alga. The sorption process of all metal ions rapidly reached equilibrium within 20 min. The sorption isotherm followed the Langmuir isotherm with the maximum sorption capacities in the following order: Pb^{2+} > Cu^{2+} > Cd^{2+} > Zn^{2+}.

The biosorption study of cadmium(II) and lead(II) by the brown seaweed *Cystoseira baccata* was studied experimentally by Lodeiro *et al.*[26] Kinetic experiments demonstrated a rapid metal uptake; the maximum metal uptake values were around 101 and 186 mg g^{-1} for cadmium(II) and lead(II), respectively) at pH 4.5. The presence and participation of carboxylic groups in metal uptake was confirmed by FTIR analysis. Moreover, some other groups such as: amines, alcohol, hydroxyl and sulfonate, which may be involved in metal binding, were also identified as components of algae structure. Metal sorption was found to increase with the solution pH reaching a plateau above pH 4. Calcium and sodium nitrate salts in solution were found to affect considerably metal biosorption.

An agricultural waste, tobacco dust, was investigated by Qi and Aldrich[33] for its heavy metal binding efficiency. The tobacco dust exhibited a strong retention capacity for heavy metals such as Pb(II), Cu(II), Cd(II), Zn(II) and Ni(II), with a respective biosorption capacity of 39.6, 36.0, 29.6, 25.1 and 24.5 mg g^{-1}. Zeta potential and surface acidity measurements showed that the tobacco dust was negatively charged over a wide pH range (pH > 2), with a strong surface acidity and a high OH^{-} adsorption capacity. However, FTIR analysis showed no substantial change in the chemical structure of the tobacco dust subjected to biosorption. They suggested that heavy metal uptake mechanism by tobacco dust may involve metal–H ion exchange or metal ion surface complexation or both.

Venugopal and Mohanty[91] used a weed, *Parthenium hysterophorus*, to remove Cr(VI) ions from aqueous solution. They investigated the effect of various parameters such as contact time, temperature, pH, agitation speed, biosorbent dose and initial Cr(VI) concentration on the biosorption process. FTIR analysis established that carboxyl, amine and alkane groups were the leading Cr binding groups. The biosorption capacity of *Parthenium* for Cr(VI) was found to be 24.5 mg g^{-1} at pH 1.0, 0.1 g biomass dose, 200 rpm, 160 min equilibrium time and 20 °C. Thermodynamic study confirmed that Cr(VI) biosorption onto *Parthenium* weed was feasible, spontaneous and exothermic within 20–60 °C. The magnitude biosorption heat was calculated to be 7.83 kJ mol^{-1}, confirming that biosorption is a physical process.

Carica papaya wood was evaluated by Basha *et al.*[92] for the sorption of Hg(II) from aqueous solution under varying conditions regarding contact time, metal ion concentration, sorbent dose and pH. The results indicated that sorption equilibrium was established in about 120 min. Hg(II) sorption is strictly pH dependent, and maximum removal (155.6 mg g^{-1}) was observed at pH 6.5. The FTIR spectra of raw and mercury loaded papaya wood indicated that amide groups were major binding sites of mercury.

1.3 Biomass-based Flocculants and Coagulants

Coagulation and flocculation are essential steps in the treatment of surface water, municipal wastewater and industrial wastewater to remove suspended particles by destabilization. To achieve coagulation, especially for drinking water treatment, conventional chemical-based coagulants, namely alum (AlCl$_3$), ferric chloride (FeCl$_3$) and polyaluminium chloride (PAC), are generally used for turbidity removal during the coagulation–flocculation step. Although the effectiveness of these chemicals as coagulants is well-recognized,[127,128] some problems are associated with their use. They have a relatively high cost, are ineffective for low-temperature water treatment,[129] produce large sludge volumes, modify the raw water pH and are detrimental for human health. Moreover, there is also strong evidence linking aluminium-based coagulants to the development of Alzheimer's disease in human beings.[130]

Due to their biodegradability, plant-based flocculants could be a valuable alternative to those classically used chemical coagulants. In addition, such non-conventional coagulants should not be dangerous to health, may not induce strong water pH modification and could be produced by local communities.[131] Plant-based coagulants for turbid water treatment have been used for several millennia ago and, thus far, environmental scientists have identified several plant types for this purpose. Their use nowadays could bring new resources to local communities and allow the use of a sustainable process for water treatment.

So far, four plant-based coagulant have been well described (nirmali seeds, *Moringa oleifera*, tannin and cactus), with only a few works reporting data on other biological coagulants:[131] common bean (*Phaseolus vulgaris*) seed extract;[132] guar gum;[133] Opuntia mucilage;[134] Enteromorpha is a kind of green algae;[135] maize (scientific name: Zeemays);[136] Psyllium husk, the seed of the *Plantago ovata*;[137] other agro residues such as rice husk;[138,139] horse chestnut (*Aesculus hyppocastanum*) from the family Sapindaceae;[140] common oak (*Quercus robur*), Turkey oak (*Quercus cerris*), Northern red oak (*Quercus rubra*) and European chestnut (*Castanea sativa*) from the family Fagaceae;[140] dragon fruit;[141] Okragumfrom seedpods of *Hibiscus esculentus*;[142] Fenugreek mucilage (*Trigonella foenum-graecum*);[143] *Tamarindus indica* seed mucilage;[144] and *Malva sylvestris* (mallow) mucilage.[145]

Four coagulation mechanisms of particulates in a solution can occur:[131] (a) double layer compression due to the presence of salts or suitable coagulants; (b) sweep flocculation due to suspended particulates encapsulation in soft colloidal floc; (c) adsorption and charge neutralization of ions with opposite charges; and (d) adsorption and interparticle bridging when a coagulant provides a polymeric chain which sorbs particulates. Like polymeric coagulants, plant-based coagulants are generally associated with mechanisms (c) and (d) as their long-chained structures (especially polymers with a high molecular weight) greatly increase the number of unoccupied adsorption sites.

The following section presents some of the main classes of biomass-based flocculants; when reported in the literature, data on their coagulation mechanism are also described.

1.3.1   Terrestrial Plant Based Bioflocculants

1.3.1.1   Moringa Oleifera

Moringa oleifera (horseradish or drumstick tree) is a fast growing tree that ranges in height from 5 to 12 m, and is found in India, Asia, sub-Saharan Africa and Latin America.[146,147] The different plant parts such as leaves, flowers, seeds, roots and bark are traditionally used as food and medicinal sources.[148] Crude seed extracts are utilized by rural communities in African countries to clear turbid river water. The seeds contain an edible oil and water-soluble substance.[149] In 1995, Muller[150] and Jahn[147,151] studied the use of *M. oleifera* as a natural coagulant. Ndabigengesere *et al.*[152] reported dimeric cationic proteins with molecular mass of 12–14 kDa and isoelectric point (pI) between 10 and 11, as the active coagulating agents. The coagulation mechanism is adsorption and charge neutralization.

Nevertheless, the exact nature of the Moringa seed active coagulant remains controversial. The coagulating agent was reported to be a protein with a molecular mass of 6.5 kDa and pI > 10 by Gassenschmidt *et al.*[153] Ghebremichael *et al.*[154] described a cationic protein with pI > 9.6 and molecular mass < 6.5 kDa. However, Okuda *et al.*[155] found an organic polyelectrolyte, being either a polysaccharide or lipid with a molecular weight of approximately 3.0 kDa. It was then proposed that Moringa extracts may contain several unidentified coagulating agents in which coagulating capability may be less than the cationic protein.[155] Much work has been carried out to continue the treatment of turbid waters or even industrial wastewaters by Moringa. Okuda *et al.*[155] showed that coagulating capability was enhanced by the addition of bivalent cations, *e.g.* Ca^{2+} and Mg^{2+}. This was explained by the adsorption of cations on the active components extracts to form an insoluble net-like structure able to capture suspended kaolin particles (Figure 1.10).

M. oleifera active coagulating compounds have been reported to remove 99% Chicago Sky Blue 6B[156] and 80% Carmine Indigo dyes.[157] Moringa seed were shown to be an effective coagulant–flocculant in the treatment of high turbid effluent such as coal washing effluent. The process was found to be rapid with a high rate constant and low coagulation period, with a removal of 95% within the first three minutes of treatment. The optimum parameters were a pH of 8, 4 mg L^{-1} dosage and 25 min settling time.

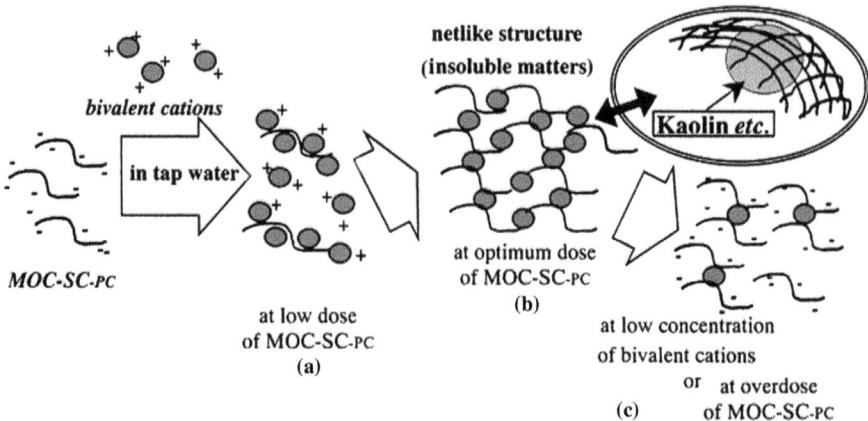

Figure 1.10   Model for the structures of coagulation active component of a NaCl purified *Moringa oleifera* coalulant solution (MOC-SC-PC).[155]

Moringa oleifera seeds have been used as a coagulant for the treatment of an effluent containing surfactants such as long-chain anionic detergents such as polyoxyethylene (3.5) and sodium lauryl ether sulfate (SLES).[158,159] *M. oleifera* was shown to be an effective coagulant since it was capable of reaching a co-agulation capacity of up to 0.245 mg mg^{-1} with an initial coagulant dosage and surfactant concentration of 234 mg L^{-1} and 76 mg L^{-1}, respectively.

1.3.1.2   Tannins

Tannins are large polyphenol compounds obtained from natural materials such as the organic extract from bark and wood,[131,158] of trees such as Acacia, Castanea or Schinopsis.[160] These polymers, traditionally used as tanning agents in the leather industry, have molecular weights ranging from hundreds to tens of thousands.

Ozacar and Sengil[131,159–161] used tannin extracted from valonia obtained from the corn cup of an oak growing in Asia Minor. It was shown to be an excellent substitute for chemical coagulants; its effectiveness for water treat-ment was influenced by the chemical structure of the tannins extracted or by the degree of tannin modification.[159] Tannin was shown to have phenolic groups, making them anionic and a good hydrogen donor. A higher coagulation cap-ability was related to a high number of phenolic groups in the tannin structure. Aboulhassan *et al.*[162] evaluated the activity of Polysep3000, P3000, an indus-trial grade organic polymer prepared with acidified vegetable tannic substances for wastewaters from a textile plant treatment. They follow chemical oxygen demand (COD) reduction, colour removal and the amount of sludge produced for the determination of optimal conditions. P3000 was shown to reduce the pollutants in the textile effluent, with a colour removal higher than 90% but was less effective for COD reduction as values between 40 and 50% were obtained. The volume of decanted sludge produced was lower than those produced when $Al_2(SO_4)_3$ and $FeCL_3$ were used as coagulants.

Sánchez-Martín *et al.*[163] used Tanfloc, a new tannin-based coagulant obtained from *Acacia mearnsii* de Wild at a pilot plant level, to treat four types of waters: surface water (collected from a river); municipal; textile industry (simulated by a 100 mg L^{-1} aqueous solution of an acid dye); and laundry (simulated by a 50 mg L^{-1} aqueous solution of an anionic surfactant). Coagulation, sedimentation and filtration steps were realised during the pilot plant process. For experiments with the surface water, the coagulant dosage was 2 mg L^{-1}; an average coagulant dosage of 92.2 mg L^{-1} was used for the other water samples. The removals obtained were 50% for colour, 75% for surfactant, 40% for COD removal and 60% for biochemical oxygen demand over a 5-day period (BOD5) removal. Later the same group used Tanfloc for the treatment of several dye-polluted aqueous solutions containing triphenylmethane (Eriochrome Cyanine R), indigoid (Carmine Indigo), azoic (Chicago Sky Blue 6B, Palatine Fast Black WAN, or Acid Red 88) or anthraquinonic dyes (Alizarin Violet 3R).[164] Tanfloc presented a high coagulating activity at a low dose of 200 mg·L^{-1}. It was able to remove up to 80% Alizarin Violet 3R at an initial concentration of 0.16 mmol L^{-1}, and almost 100% in the case of Palatine Fast Black WAN at an initial concentration of 0.06 mmol L^{-1}. The coagulating activity was lower when the pH increased due to destruction of Tanfloc's chemical structure, while temperature had no effect on final dye removal. In another study,[165] two tannin-based coagulants *Silvaf loc* and *Acquapol* S5T derived from *Acacia mearnsii* de Wild, and *Schinopsis balansae*, respectively, were used for AlizarinViolet 3R dye removal. It was shown that an acidic pH and highly polluted wastewater samples led to higher efficiencies in both cases. In particular, the *Silvaf loc*–Alizarin Violet 3R system presented an optimum adsorption capacity (q) value at 0.49 mg mL^{-1}.

1.3.1.3 Cactus

Cactaceous opuntia is colloquially known as 'nopal' in Mexico or 'prickly pear' in North America. It grows generally in torrid zone and subtropics, and is used for its medicinal properties[131,166] and dietary food sources.[167] This cactus species have been successfully used as a natural coagulant. Opuntia contains mucilage, which is a viscous and complex carbohydrate stored in cactus inner and outer pads containing such as L-arabinose, D-galactose, L-rhamnose, D-xylose and galacturonic acid.[168,169] It was hypothesized that galacturonic acid, in its polymeric form (polygalacturonic acid),[170] a major constituent of pectin in plants, provides a 'bridge' for particle adsorption. Carboxylic functional group deprotonation of polygalacturonic acid leads to an anionic form allowing chemisorption between charged particles and the COO^- moiety.

Carpinteyro-Urban *et al.*[171] used Opuntia mucilage as coagulant–flocculant aid in the treatment of a high-load cosmetic industry wastewater (WW). COD removals as high as 38.6 % were observed when mucilage at 500 mg L^{-1} was employed with the lowest COD load. Regarding the amount of sludge produced, the maximum value corresponded to mucilage at 500 mg L^{-1} for the highest COD load WW, with a value of 500 mL L^{-1}. Zhang *et al.*[172] used Opuntia coagulant to treat a synthetic water, and showed that the optimum dosage of cactus coagulant was about 50 mg L^{-1}, and that turbidity removal could reach 94, 93 and 98%, respectively, for a synthetic water prepared with kaolin at pH 7, a raw source water at pH 8.1 and a high turbidity seawater at pH 8.

Idriss *et al.*[173] studied the coagulating ability of the foliage of dragon fruit, a round and often red coloured fruit, from the Cactaceae family of the genus *Hylocereus*, known as pitaya, and native of South America. The treatment of a concentrated latex effluent by dragon fruit foliage led to high COD, suspended solids and turbidity removal of 94.7, 88.9 and 99.7%, respectively, at pH 10. It was concluded that this foliage could be efficiently used in the pretreatment stage of latex effluent.

1.3.1.4 Strychnos potatorum (nirmali)

S. potatorum (nirmali), is a tree of moderate size, growing in India and Sri Lanka. In Sanscrit writing, seed extracts of *S. potatorum* (nirmali) were described to clarify turbid surface water over 4000 years ago.[174] The extracts contain anionic polyelectrolytes that destabilize particles in water by means of interparticle bridging.[174,175] It seed extract coagulation capability is due to the presence of lipids, carbohydrates and alkaloids containing –COOH and free –OH surface groups.[174,175] Adinolfi *et al.*[176] reported that a mixture of polysaccharide fraction extracted from *S. potatorum* seeds contained galactomannan and galactan capable of reducing up to 80% turbidity of a kaolin solution.

1.3.1.5 Other Plant-based Coagulants

Many recently discovered biological coagulants are quite uncommon and represent new varieties of plant-based active coagulant extracts alongside the established plant coagulants described above. In 2010, Antov *et al.*[177] showed that the proteins of the common bean *Phaseolus vulgaris* had an interesting coagulation activity. Common bean seed is known for its high protein content ($\sim$20–30%). It contains globulin phaseolin, a trimer with subunit of molecular weight (MW) $\sim$50 kDa representing more than half the total protein content, while albumin, prolamin and glutelin proteins represent the other seed protein fractions. The highest coagulation activity of crude common bean extract was above 50%, for a dosage of 1 mL L^{-1} at pH 9.5. It was about 2.4 and 1.5 times lower at pH 9 and 10, respectively. Turbidity removal from the crude extract and fraction prepared by ultrafiltration was lower than that obtained for natural coagulants from *M. oleifera* and other plant materials.[178–181] Šciban *et al.*[182] extracted the active components of ground seeds of horse chestnut and acorns of some species of the Fagaceae family: common oak, Turkey oak, Northern red oak and European chestnut. At the lowest coagulant dose of 0.5 mL L^{-1}, the highest coagulating activities for a kaolin-containing water were about 80% and 70% for seed extracts from European chestnut and common oak acorn, respectively, at both low and medium turbidities.

Psyllium husk, derived from the seed of the *Plantago ovata* plant, is an annual herb native of Asia that has a long history of being used in traditional medicine. Al-Hamadini *et al.*[183] showed that psyllium husk was an effective coagulant aid in conjunction with polyaluminium chloride for removal of COD, colour and total suspended solids (TSS) of landfill leachate. Patel and Vashi[184] studied the treatment of textile dyeing industry wastewaters by natural coagulants such as Surjana seed powder (SSP), maize seed powder (MSP) and chitosan. The maximum percentage of Congo Red (CR) removal was found to be 98.0, 94.5 and 89.4 for SSP, chitosan and MSP, respectively, at pH 4.0 and 340 K, with a coagulant dose of 25 mg L^{-1}, and a flocculation time of 60 min. Sanghi *et al.*[185] reported that seed guar gum was capable of removing up to 65% of Direct Orange dye. Blackburn[186] reported that galactomannan-containing plant gums (locust bean gum, guar gum, cassia gum) could remove more than 70% of dyes (Reactive Red 238, Direct Black 22 and Acid Blue 193).

1.3.1.6 Compounds Derived from an Agro Residue, Rice Husk

Rice husk is a rice mill by-product used as an energy source in many industries including biomass power plants. Rice husk contains cellulose, lignin and polysaccharides,[187] and can be considered an economically valuable raw material. Fu *et al.*[188] prepared a new polymeric biomass-based flocculant using rice husk. The raw material was soaked with a sodium hydroxide solution and then etherificated using octadecyl trimethyl ammonium chloride which introduced quaternary ammonium groups into the rice husk molecules in the presence of potassium permanganate as initiator. The conditions affecting the flocculation of a 1g L^{-1} diatomaceous earth suspension used as model wastewater was studied. The optimal flocculation temperature was shown to be 50 °C, with turbidity removal increasing with temperature until the temperature reached 50 °C. At temperatures lower than 50 °C, the cellulose was further expanded when the temperature heated up, while at temperatures above 50 °C, the macromolecule chain structure deteriorated. Turbidity removal of the synthetic water could reach above 90% under optimal conditions.

Activated silica and sodium silicate are basic types of coagulants aids that increase the stability of coagulant by fixation to positively charged aluminium or to iron flocs. This leads to larger and denser lamellar flocs that can settle faster and enhance enmeshment. Silica can be obtained by burning rice husk to produce rice husk ash (RHA) which contains high amounts of natural silica.[189]

Abo-El-Eneina *et al.*[187] prepared new materials, RHA1 and RHA2 by burning rice husk for 2 h at 850 and 650 °C, respectively, followed by washing in 10% hydrochloric acid for 1 h in a reflux at 100 °C. RHA3 and RHA4 were prepared by burning rice husk for 2 h at 850 and 650 °C, respectively, and then washing with distilled water and RHA5 was prepared by burning of rice husk for 2 h at 650 °C. They selected then sodium silicate (SSi5) obtained from RHA5 for the preparation of polysilicic acid solution (PSi) by the reaction of sodium hydroxide with amorphous activated silica derived from RHA5. Four new polyinorganic coagulants were also prepared: polyaluminium chloride silicate (PACSi), polyhydroxy aluminium sulfate silicate (PAHSSi), polyferric chloride silicate (PFeClSi) and polyferric aluminium chloride. PAlFeClSi showed maximum percentages removal of 99% and 97% for Fe^{2+} and Mn^{2+} ions contained in groundwater, respectively. When PFeClSi was used, 97% of Pb^{2+} ion was removed from an industrial wastewater. In addition, using PAlFeClSi, the maximum percentages removal of COD, BOD and TSS in sewage wastewater reached 90%, 92% and 93%, respectively. Overall, the efficiency of these rice-husk based coagulants was found to be more than three-fold that of conventional coagulants.[187]

1.3.2 Bioflocculants from Marine Biomass

Enteromorpha[190] is a green, fast growing opportunistic macroalgae that can tolerate salinities varying from freshwater to seawater. Zhao *et al.*[191] used enteromorpha as a coagulant aid in conjunction with aluminium sulfate. They demonstrated the high efficiency of this coagulant and the perfect cooperation between the two compounds. The coagulation effect of aluminium sulfate was improved by around 30% when 0.3 mL L^{-1} enteromorpha was added. They also showed that the flocs produced by the addition of both enteromorpha and aluminium sulfate to the water grew faster and were bigger than those with aluminium sulfate alone. They proposed a main coagulation mechanism involving precipitate charge neutralization of aluminium sulfate with adsorption bridging when enteromorpha was added, leading to better coagulation performance when the compounds were combined.

Alginic acid or alginate is a natural polysaccharide obtained from marine brown algae, formed of two monomeric units: α-L guluronic acid (G) and β-D mannuronic acid (M). This polysaccharide (Figure 1.11) contains naturally carboxyl groups in each constituent residue,[192] allowing it polymers.[193,194] It forms the so-called 'egg-box' structure with calcium ions. Due to its ability to form this gel with calcium and other with metal cations, alginate is an industrially important biopolymer with many applications. Indeed, Devrimci[193] studied the use of calcium alginate as a coagulant for treating 500 mL of

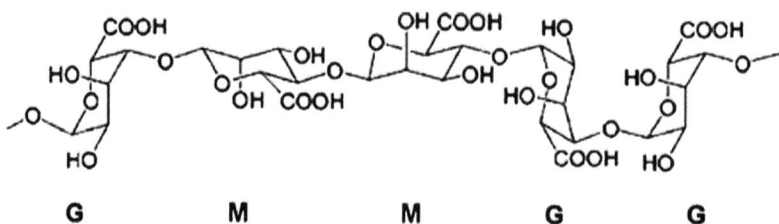

Figure 1.11   Chemical structure of alginic acid. M: β-D-mannuronic acid; G: α-L-guluronic acid.[207]

synthetic turbid water samples with different initial turbidities of 150, 80 and 10 Nephelometric Turbidity Units (NTU). Their study indicated that, at high turbidity values (150 NTU and 80 NTU), calcium alginate was a very effective coagulant, with turbidity removal generally >98%.

Chitin is widely produced from the exoskeleton of crustaceans such as shrimp and crabs. Chitosan is a copolymer of glucosamine and *N*-acetyl glucosamine of high molecular weight obtained from chitin deacetylation. Chitosan contains two reactive hydroxyl groups and a very high reactive amino group that gives the coagulation capacity of chitosan. Altaher[194] used chitosan as a coagulant for the removal of sea water turbidity at doses between 0 and 370 mg L^{-1}, and initial pH between 2 and 11. Its turbidity removal efficiency was found to be greater than ferrous sulfate and comparable with that of alum. Chitosan worked well at alkaline pH; the highest turbidity of 97.5% was obtained at an initial pH of 8.1 with an optimum dose of 18 mg L^{-1}. The removal efficiencies values were comparable for alum and chitosan. Nevertheless, the use of alum required much higher doses. Chang *et al.*[195] prepared a macromolecule flocculant, mercaptoacetyl chitosan, by reacting chitosan with mercaptoacetic acid. They reported that this new flocculant could not only remove turbidity but could also remove heavy metals in wastewater.

1.3.3 Bioflocculants from Microorganisms

Many microorganisms, such as bacteria and fungi, have been found to produce bioflocculants.[196] Bioflocculants produced by microorganisms during growth are various kinds of extracellular biopolymers such as glycoprotein, protein, polysaccharide and nucleic acid.[197] The chemical structure, flocculation mechanism and culture conditions of the bioflocculants have been studied. Researchers have done much work on screening highly efficient strains, on optimization of their culture conditions and on studying their flocculating activity on various polluted waters.[198,199]

Ruditapes philippinarum conglutination mud (RPM) was shown[198] to methylene blue, crystal violet, Malachite Green and ink blue aqueous solutions with efficiencies above 90%. The amount of RPM had a significant effect on the decolourisation of all dye solutions.[200] The temperature had a remarkable effect on ink blue solution decolourisation, but little effect on decolourisation of the three other dye solutions. Decolourisation rate was affected by the initial pH of the malachite green solution, but this had less effect on the other three solutions. The optimum initial pH value was 7, and at an initial pH > 9, the malachite green solution became colourless when the RPM was added. Chang *et al.*[201] found that the dye removal rate increased nearly 2.5-fold as the pH was raised from 5.0 to 7.0. Between pH 7.0 and 9.5, however, they observed that the initial pH value of the malachite green solution had no influence on the dye removal. Contrarily, Deng *et al.*[202] observed that the decolourisation efficiency of a flocculant produced by a strain of *Aspergillus parasiticus* decreased with initial pH increase from 2 to 12 for the four dyes, Acid Blue 45, Direct Blue 1, Acid Orange 8 and Reactive Orange 16, and the decolourisation efficiency decreased rapidly beyond pH 9 or 12 for the different dyes. The bioflocculant consisted mainly of sugar (76.3%) and protein (21.6%).

Gao et al.[203] showed that a bioflocculant produced by a strain growing in *Ruditapes philippinarum* conglutination mud identified as *Rothia* sp. had a flocculating activity of 86.22% for a 5 g L^{-1} kaolin clay suspension at an initial pH of 9 and at 20 °C. It had a flocculating effect in a wide range of pH values between 1 and 13 and at temperatures between 4 and 100 °C. The main component of this strain was a polysaccharide. It was also able to decolourise dye solutions, with a removal efficiency of 86.11%, 97.84% and 99.49% for methylene blue, crystal violet and Malachite Green, respectively. Lower removal efficiency of heavy metal ions of 19.2% and 69.3% respectively, were obtained for Ni^{2+} and Cr$_2$O$_7^{2-}$ containing solutions.

Li et al.[204] purified a bioflocculant produced by *Bacillus licheniformis* for low temperature drinking water treatment. It was identified as a glycoprotein consisting of polysaccharide (91.5%, w/w) and protein (8.4%, w/w) containing amino, amide, carboxyl, methoxyl and hydroxyl groups with an approximate molecular weight of 6.89.10^4 Da. This bioflocculant was used successfully for treatment of low temperature drinking water, with higher maximum COD and turbidity removal than conventional chemical flocculants of 61.2% and 95.6%, respectively. Charge neutralization and a bridging flocculation mechanism were proposed.

Deng et al.[205] studied the flocculating capability of a bioflocculant, MBFA9, produced by a strain identified as *Bacillus mucilaginosus*. The major component of MBFA9 was found to be polysaccharide composed mainly of uronic acid (19.1%), neutral sugar (47.4%) and amino sugar (2.7%) containing carboxyl and hydroxyl groups. A flocculating activity of 99.6% was obtained for a kaolin suspension at a dosage of only 0.1 mL L^{-1}. The removal of suspended solids and COD from a starch wastewater treatment in the presence of Ca^{2+} salt was up to 85.5% and 68.5%, respectively, which is better than conventional chemical flocculants.

Serratia ficaria was shown to produce a bioflocculant that was effective for flocculation of a kaolin suspension over weakly acidic pH (5–7), with a removal of 95.4%.[206] Divalent cations Ca^{2+} and Mg^{2+} increased the flocculating activity, but it decreased when Al^{3+} and Fe^{3+} were co-added. The bioflocculant could treat different real wastewaters, including river water, brewery wastewater, meat processing wastewater and soy sauce brewing wastewater. The treatment of a pulp effluent led to colour and COD removal up to 99.9% and 72.1%, respectively, which was better than conventional chemical flocculants. Zeta potential measurements showed that charge neutralization played an important role in the flocculation mechanism.

1.4 Conclusions

We all recognize that water pollution by hazardous compounds is one of the most important environmental problems throughout the world. Increasing stringent environmental regulations are necessary to supply high quality drinking water and to protect the natural water resources of the various cities and countries in the world. A wide range of treatment methods such as co-agulation, flocculation, bioaccumulation, precipitation, reverse osmosis, oxidation/reduction, evaporation, ion exchange and membrane filtration have been developed and are used for the removal of various contaminants from aqueous solutions. In recent years, however, research has sought to find new alternatives to these traditional methods which are recognized as being generally complex and expensive. This chapter has reviewed the use of inexpensive and efficient biomaterials—bacteria, fungi, seaweeds, agricultural waste and raw plants—for the removal of hazardous contaminants.

Biosorption of pollutants is generally carried out under normal and simple conditions: ambient temperature, relatively shorter contact time and an acidic pH range with low energetic needs. In most cases, a high affinity of the biomass for organic or inorganic ions has been found. All the natural coagulants exhibit highly effective turbidity removal comparable with the capabilities of established chemical coagulants (*e.g.* alum), with some of them removing up to 99% of initial turbidity. Some natural polymers, such as polysaccharides, have been suggested to be moderately efficient due to their low molecular weights and high shear stability; they were noted as being cheap and easily available from reproducible farm and forest resources. Additional advantages of these natural polyelectrolytes include safety for human health, biodegradability, and a wider effective dose range of flocculation for various colloidal suspensions.

So far, identified usage of natural coagulants for industrial wastewater has been limited to academic research. Many findings from these academic studies, however, indicate their good potential for industrial wastewater treatment. In many cases, the natural coagulants can perform at their best when used for treatment of wastewaters with less variety of contaminants. Consequently, the use of biomass for the removal of hazardous compounds may be a new and cost-effective alternative for the treatment of contaminated waters. In addition, the use of biomaterials has some advantages. Almost all biomaterials are renewable natural resources that are abundant and cheap, and most of the biomasses can be used as biosorbents with little or no pretreatment.

References

1. T. A. Davis, B. Volesky and A. Mucci, *Water Res.*, 2003, **37**, 4311.
2. S. Karthikeyan, R. Balasubramanian and C. S. P. Iyer, *Bioresour. Technol.*, 2007, **98**, 452.
3. E.-S. Amany, A. El Nemr, A. Khaled and O. Abdelwehab, *J. Hazard. Mater.*, 2007, **148**, 216.
4. M. G. A. Vieira, R. M. Oisiovici, M. L. Gimenes and M. G. C. Silva, *Bioresour. Technol.*, 2008, **99**, 3094.
5. M. R. Fagundes-Klen, P. Ferri, T. D. Martins, C. R. G. Tavares and E. A. Silva, *Biochem. Eng. J.*, 2007, **34**, 136.
6. B. Volesky, *Water Res.*, 2007, **41**, 4017.

7. M. C. Basso, E. G. Cerrella and A. L Cukierman, *Sep. Sci. Technol.*, 2004, **39**, 1163.

8. E. L. Cochrane, S. Lu, S. W. Gibb and I. Villaescusa, *J. Hazard. Mater.*, 2006, **137**, 198.

9. R. Herrero, B. Cordero, P. Lodeiro, C. Rey-Castro and M. E. Sastre de Vicente, *Mar. Chem.*, 2006, **99**, 106.

10. P. Lodeiro, B. Cordero, Z. Grille, R. Herrero and M. E. Sastre de Vicente, *BioTechnol. Bioeng.*, 2004, **88**, 237.

11. S. Schiewer and M. H. Wong, *Chemosphere*, 2000, **41**, 271.

12. K. Vijayaraghavan and Y. S. Yun, *Biotechnol. Adv.*, 2008, **26**, 266.

13. J. Wang and C. Chen, *Biotechnol. Adv.*, 2009, **27**, 195.

14. C. Quintelas, Z. Rocha, B. Silva, B. Fonseca, H. Figueiredo and T. Tavares, *Chem. Eng. J.*, 2009, **149**, 319.

15. S. Yasemin and A. Ozturk, *Process Biochem.*, 2005, **40**, 1895.

16. V. Padma and S. D. Bajpai, *Desalination*, 2008, **222**, 255.

17. F. Colak, N. Atar and A. Olgun, *Chem. Eng. J.*, 2009, **150**, 122.

18. R. M. Gabr, S. H. A. Hassan and A. A. M. Shoreit, *Int. Biodeterior. Biodegrad.*, 2008, **62**, 195.

19. A. Coreño-Alonso, F. J. Acevedo-Aguilar, G. E. Reyna-López, A. Tomasini, Francisco J. Fernández, K. Wrobel, K. Wrobel and J. Félix Gutiérrez-Corona, *Chemosphere*, 2009, **76**, 43.

20. K. Kumari and T. E. Abraham, *Bioresour. Technol.*, 2007, **98**, 1704.

21. D. Park, Y.-S. Yun, J. H. Jo and J. Moon Park, *Water Res.*, 2005, **39**, 533.

22. N. S. Kumar and K. Min, *Chem. Eng. J.*, 2011, **168**, 562.

23. M. Amini, H. Younesi and N. Bahramifar, *Chemosphere*, 2009, **75**, 1483.

24. Z. Aksu, *Process Biochem.*, 2005, **40**, 997.

25. P. Lodeiro, J. L. Barriada, R. Herrero and M. E. Sastre de Vicente, *Environ. Pollut.*, 2006, **142**, 264.

26. R. H. S. F. Vieira and B. Volesky, *Int. Microbiol.*, 2000, **3**, 17.

27. O. Freitas, M. M. Ramiro, J. E. Martins, M. C. Delerue-Matosand and R. A. R. Boaventura, *J. Hazard. Mater.*, 2008, **153**, 493.

28. Z. R. Holan and B. Volesky, *Biotechnol. Bioeng.*, 1994, **11**, 1001.

29. T. V. N Padmesh, K. Vijayaraghavan, G. Sekaran and M. Velan, *J. Hazard. Mater.*, 2005, **B125**, 121.

30. E. Rubın, P. Rodrıguez, R. Herrero and M. E. Sastre de Vicente, *J. Chem. Technol. Biotechnol.*, 2006, **81**, 1093.

31. S. Altenor, M. C. Ncibi, E. Emmanuel and S. Gaspard, *Biochem. Eng. J.*, 2012, **67**, 35.

32. C. R. Teixeira Tarley and M. A. Zezzi Arruda, *Chemosphere*, 2004, **54**, 987.

33. B. C. Qi and C. Aldrich, *Bioresour. Technol.*, 2008, **99**, 5595.

34. M. Lopez-Mesas, E. Ruperto Navarrete, F. Carrillo and C. Palet, *Chem. Eng. J.*, 2011, **174**, 9.

35. M. Riaz, R. Nadeem, M. A. Hanif, T. M. Ansari and K. Rehman, *J. Hazard. Mater.*, 2009, **161**, 88.

36. A. E. Navarro, R. F. Portales, M. R. Sun-Kou and B. P. Llanos, *J. Hazard. Mater.*, 2008, **156**, 405.
37. P. X. Sheng, Y.-P. Ting, J. P. Chen and L. Hong, *J. Colloid Interface Sci.*, 2004, **275**, 131.
38. S. V. Mohan, S. V. Ramanaiah, B. Rajkumar and P. N. Sarma, *J. Hazard. Mater.*, 2007, **141**, 465.
39. V. Murphy, H. Hughes and P. McLoughlin, *Chemosphere*, 2008, **70**, 1128.
40. L. Seno Ferreira, M. Santos Rodrigues, J. C. Monteiro de Carvalho, A. Lodi, E. Finocchio, P. Perego and A. Converti, *Chem. Eng. J*, 2011, **173**, 326.
41. U. Farooq, J. A. Kozinski, M. Ain Khan and M. Athar, *Bioresour. Technol.*, 2010, **101**, 5043.
42. N. Das, *Hydrometallurgy*, 2010, **103**, 180.
43. M. Teresa, S. D. Vasconcelos, M. Fernanda and C. Leal, *Mar. Chem.*, 2001, **74**, 65.
44. M. C. Palmieri, B. Volesky and O. Garcia Jr, *Hydrometallurgy*, 2002, **67**, 31.
45. E. Fourest and B. Volesky, *Environ. Sci. Technol.*, 1996, **30**(1), 277.
46. C. Mack, B. Wilhelmi, J. R. Duncan and J. E. Burgess, *Biotechnol. Adv.*, 2007, **25**, 264.
47. H. Pahlavanzadeh, A. R. Keshtkar, J. Safdari and Z. Abadi, *J. Hazard. Mater.*, 2010, **175**, 304.
48. Y. Liu, Q. Cao, F. Luo and J. Chen, *J. Hazard. Mater.*, 2009, **163**, 931.
49. F. Luo, Y. Liu, X. Li, Z. Xuan and J. Ma, *Chemosphere*, 2006, **64**, 1122.
50. M. A. Hashim and K. H. Chu, *Chem. Eng. J.*, 2004, **97**, 249.
51. S. Basha, Z. V. P. Murthy and B. Jha, *Chem. Eng. J.*, 2008, **137**, 480.
52. E. Romera, F. Gonzalez, A. Ballester, M. L. Blazquez and J. A. Munoz, *Bioresour. Technol.*, 2007, **98**, 3344.
53. M. C. Ncibi, B. Mahjoub, M. Seffen, F. Brouers and S. Gaspard, *Biochem. Eng. J.*, 2009, **46**, 141.
54. K. Vijayaraghavan and Yeoung-Sang Yun, and, *Dyes Pigm.*, 2008, **76**, 726.
55. K. Vijayaraghavan and Y. S. Yun, *J. Hazard. Mater.*, 2007b, **141**, 45.
56. A. Sari, O. D. Uluozlu and M. Tuzen, *Chem. Eng. J.*, 2011, **167**, 155.
57. M. Tuzen and A. Sari, *Chem. Eng. J.*, 2010, **158**, 200.
58. C. Hui Niu, B. Volesky and D. Cleiman, *Water Res.*, 2007, **41**, 2473.
59. C. J. Daughney and J. B. Fein., *J. Colloid Interface Sci.*, 1998, **198**, 53.
60. D. Borrok and J. B. Fein, *J. Colloid Interface Sci.*, 2005, **286**, 110.
61. G. Donmez and Z. Aksu, *Process Biochem.*, 2002, **38**, 751.
62. M. A. Borowitzka, in *Microalgal Technol.*, ed. M. A. Borowitzka and L. J. Borowitzka, Cambridge University Press, 1988, pp. 456–465.
63. K. Vijayaraghavan, K. Palanivelu and M. Velan, *Bioresour. Technol.*, 2006, **97**, 1411.
64. A. Esposito, F. Pagnanelli, A. Lodi, C. Solisio and F. Veglio, *Hydrometallurgy*, 2001, **60**, 129.

65. Y. Prasanna Kumar, P. King and V. S. R. K. Prasad, *Chem. Eng. J.*, 2007, **129**, 161.
66. J. Tangaromsuk, P. Pokethitiyook, M. Kruatrachue and E. S. Upatham, *Bioresour. Technol.*, 2002, **85**, 103.
67. P. Pavasant, R. Apiratikul, V. Sungkhum, P. Suthiparinyanont, S. Wattanachira and T. F. Marhaba, *Bioresour. Technol.*, 2006, **97**, 2321.
68. Z. Aksu, *Process Biochem.*, 2005, **40**, 997.
69. H. C. Chu and K. M. Chen, *Process Biochem.*, 2002, **37**, 1129.
70. Y. S. Ho and G. McKay, *Water Res.*, 2000, **34**, 735.
71. J. R. Evans, W. G. Davids, J. D. MacRae and A. Amirbahman, *Water Res.*, 2002, **36**, 3219.
72. J. Shen and Z. Duvnjak, *Process Biochem.*, 2005, **40**, 3446.
73. F. K. ujiwara, A. Ramesh, T. Maki, H. Hasegawa and K. Ueda, *J. Hazard. Mater.*, 2007, **146**, 39.
74. C. Tan, Min Li, Y.-M. Lin, X.-Q. Lu and Z.-L. Chen, *Desalination*, 2011, **266**, 56.
75. Z. Aksu, S. Sen Cagatay and F. Gonen, *J. Hazard. Mater.*, 2007, **143**, 362.
76. S. Mona, A. Kaushik and C. P. Kaushik, *Ecol. Eng.*, 2011, **37**, 1589.
77. T. Akar, A. S. Ozcan, S. Tunali and A. Ozcan, *Bioresour. Technol.*, 2008, **99**, 3057.
78. L.-N. Du, B. Wang, G. Li, S. Wang, D. E. Crowley and Y.-H. Zhao, *J. Hazard. Mater.*, 2012, **205–206**, 47.
79. P. Das Saha, S. Chakraborty and S. Chowdhury, *Colloids Surf. B: Biointerfaces*, 2012, **92**, 262.
80. A. Ozer, G. L Akkaya and M. Turabik, *J. Hazard. Mater.*, 2006, **B135**, 355.
81. M. Kousha, E. Daneshvar, M. S. Sohrabi, N. Koutahzadeh and A. R. Khataee, *Int. Biodeterior. Biodegrad.*, 2012, **67**, 56.
82. K. Marungrueng and P. Pavasant, *Bioresour. Technol.*, 2007, **98**, 1567.
83. P. Waranusantigul, P. Pokethitiyook, M. Kruatrachue and E. S. Upatham, *Environ. Pollut.*, 2003, **125**, 385.
84. Z. Belala, M. Jeguirim, M. Belhachemi, F. Addoun and G. Trouvé, *Desalination*, 2011, **271**, 80.
85. M. C. Ncibi, B. Mahjoub and M. Seffen, *J. Hazard. Mater.*, 2007, **B139**, 280.
86. E. Rubin, P. Rodriguez, R. Herrero, J. Cremades, I. Barbara and M. E Sastre de Vicente, *J. Chem. Technol. BioTechnol.*, 2005, **80**, 291.
87. M. Achak, A. Hafidi, N. Ouazzani, S. Sayadi and L. Mandi, *J. Hazard. Mater.*, 2009, **166**, 117.
88. N. S. Kumar, H.-S. Woo and K. Min, *Colloids Surf. B: Biointerfaces*, 2012, **94**, 125.
89. A. Sari and M. Tuzen, *J. Hazard. Mater.*, 2008, **152**, 302.
90. W. M. Ibrahim, *J. Hazard. Mater.*, 2011, **192**, 1827.
91. V. Venugopal and K. Mohanty, *Chem. Eng. J.*, 2011, **174**, 151.
92. S. Basha, Z. V. P. Murthy and B. Jha, *Chem. Eng. J.*, 2009, **147**, 226.
93. S. Tunali, A. Cabuk and T. Akar, *Chem. Eng. J.*, 2006, **115**, 203.

94. H. Salehizadeh and S. A. Shojaosadati, *Water Res.*, 2003, **37**, 4231.
95. J. S. Chang, R. Law and C. C. Chang, *Water Res.*, 1997, **31**, 1651.
96. A. Selatnia, A. Boukazoula, N. Kechid, M. Z. Bakhti, A. Chergui and Y. Kerchich., *Biochem. Eng. J.*, 2004, **19**, 127.
97. T. Srinath, T. Verma, P. W. Ramteke and S. K. Garg, *Chemosphere*, 2002, **48**, 427.
98. M. Ziagova, G. Dimitriadis, D. Aslanidou, X. Papaioannou, E. L. Tzannetaki and M. Liakopoulou-Kyriakides, *Bioresour. Technol.*, 2007, **98**, 2859.
99. Y. Sahin and A. Ozturk, *Process Biochem.*, 2005, **40**, 1895.
100. G. Uslu and M. Tanyol, *J. Hazard. Mater.*, 2006, **135**, 87.
101. A. Ozturk, T. Artan and A. Ayar, *Colloids Surf. B: Biointerfaces*, 2004, **34**, 105.
102. A. Selatnia, M. Z. Bakhti, A. Madani, L. Kertous and Y. Mansouri, *Hydrometallurgy*, 2004, **75**, 11.
103. H.-P. Yuan, J.-H. Zhang, Z.-M. Lu, H. Min and C. Wu, *J. Hazard. Mater.*, 2009, **164**, 423.
104. A. Selatnia, A. Boukazoula, N. Kechid, M. Z. Bakhti and A. Chergui, *Process Biochem.*, 2004, **39**, 1643.
105. A. Ozturk, *J. Hazard. Mater.*, 2007, **147**, 518.
106. H. Wang, J. Qiang, Su, X. Wei Zheng, Y. Tian, X. Jing Xiong and T. Ling Zheng, *Int. Biodeterior. Biodegrad.*, 2009, **63**, 395.
107. M. Venkata Subbaiah, G. Yuvaraja, Y. Vijaya and A. Krishnaiah, *J. Taiwan Inst. Chem. Eng.*, 2011, **42**, 965.
108. M. Slaba and J. Dlugonski, *Int. Biodeterior. Biodegrad.*, 2011, **65**, 954.
109. V. Javanbakht, H. Zilouei and K. Karimi, *Int. Biodeterior. Biodegrad.*, 2011, **65**, 294.
110. R. Altun Anayurt, A. Sari and M. Tuzen, *Chem. Eng. J.*, 2009, **151**, 255.
111. A. Javaid, R. Bajwa, U. Shafique and J. Anwar, *Biomass Bioenergy*, 2011, **35**, 1675.
112. Q. Peng, Y. Liu, G. Zeng, W. Xu, C. Yang and J. Zhang, *J. Hazard. Mater.*, 2010, **177**, 676.
113. Y. Lin, Xi. He, G. Han, Q. Tian and Wenyong Hu, *J. Environ. Sci.*, 2011, **23**, 2055.
114. E. Fourest and B. Volesky, *Appl. Biochem. Biotechnol.*, 1997, **67**, 33.
115. B. Volesky, J. Weber and R. Vieira, *Process Metall.*, 1999, **9**, 473.
116. R. Aravindhan, B. Madhan, J. Raghava Rao and B. Unni Nair, *J. Chem. Technol. Biotechnol.*, 2004, **79**, 1251.
117. J. Yang and B. Volesky, *Water Res.*, 1999, **33**, 3357.
118. N. Feng, X. Guo, S. Liang, Y. Zhu and J. Liu, *J. Hazard. Mater.*, 2011, **185**, 49.
119. U. Adie Gilbert, I. Unuabonah Emmanuel, A. Adeyemo Adebanjo and G. Adeyemi Olalere, *Biomass Bioenergy*, 2011, **35**, 2517.
120. D. Harikishore Kumar Reddy, Y. Harinath, K. Seshaiah and A. V. R. Reddy, *Chem. Eng. J.*, 2010, **162**, 626.

121. G. García-Rosales and A. Colín-Cruz, *J. Environ. Manage.*, 2010, **91**, 2079.
122. A. Witek-Krowiak, R. G. Szafran and S. Modelski, *Desalination*, 2011, **265**, 126.
123. G. Moussavi and B. Barikbin, *Chem. Eng. J.*, 2010, **162**, 893.
124. S. Tunali Akar, A. Gorgulu, T. Akar and S. Celik, *Chem. Eng. J.*, 2011, **168**, 125.
125. S. Chowdhury, S. Chakraborty and P. Saha, *Colloids Surf. B: Biointerfaces*, 2011, **84**, 520.
126. F. Deniz and S. Karaman, *Chem. Eng. J.*, 2011, **170**, 67.
127. J. K. Edzwald, *Water Sci. Technol.*, 1993, **27**, 21.
128. M. Kang, T. Kamei and Y. Magara, *Water Res.*, 2003, **37**, 4171.
129. J. Haaroff and J. L. Cleasby, *J. Am. Water Works Assoc.*, 1988, **80**, 168.
130. T. P. Flaten, *Brain Res. Bull.*, 2001, **55**, 187.
131. C.-Y. Yin, *Process Biochem.*, 2010, **45**, 1437.
132. M. G. Antov, M. B. Šćiban and N. J. Petrović, *Bioresour. Technol.*, 2010, **101**, 2167.
133. R. S. Blackburn, *Environ. Sci. Technol.*, 2004, **38**, 4905.
134. S. Carpinteyro-Urban, M. Vaca and L. G. Torres, *Water Air Soil Pollut.*, 2012, **223**, 4925.
135. S. Zhao, B. Gao, X. Li and M. Dong, *Chem. Eng. J.*, 2012, **200–202**, 569.
136. H. Patel and R. T. Vashi, *J. Saudi Chem. Soc.*, 2012, **16**, 131.
137. Y. A. J. Al-Hamadani, M. S. Yusoff, M. Umar, M. J. K. Bashir and M. Nordin Adlan, *J. Hazard. Mater.*, 2010, **190**, 582.
138. Y. Fu, C. Lin, L. Chen, Y. Song and H. Ma, *Adv. Mater. Res.*, 2012, **415–417**, 1667.
139. S. A. Abo-El-Eneina, M. A. Eissa, A. A. Diafullah, M. A. Rizkc and F. M. Mohamed, *J. Hazard. Mater.*, 2011, **186**, 1200.
140. M. Šćiban, M. Klašnja, M. Antov and B. Škrbić, *Bioresour. Technol.*, 2009, **100**, 6639.
141. J. Idris, A. Md Som, M. Musa, K. Halim Ku, H. Rafidah Husen and M. Najwa Muhd Rodhi, *J. Chem.*, 2013.
142. A. M. Rajani and S. M. A. Rai, *JSP Int. J. Polym. Mater.*, 2003, **52**, 1049.
143. A. Mishra, A. Yadav, M. Agarwal and M. Bajpai, *React. Funct. Polym.*, 2004, **59**, 99.
144. A. Mishra and M. Bajpai, *Colloid Polym. Sci.*, 2006, **284**, 443.
145. K. Anastasakis, D. Kalderis and E. Diamadopoulos, *Desalination*, 2009, **249**, 786.
146. J. Beltrán-Heredia, J. Sánchez-Martín and M. Barrado-Moreno, *Chem. Eng. J.*, 2012, **180**, 128.
147. S. A. A. Jahn, *J. Am. Water Works Assoc.*, 1988, **80**, 43.
148. F. Anwar and M. I. Bhanger, *J. Agric. Food Chem.*, 2003, **51**, 6558.
149. A. Ndabigengesere, K. S. Narasiah and B. G. Talbot, *Water Res.*, 1995, **29**, 703.
150. S. Muller, Wirkstoffe zur trinkwasseraufbereitung aus samen von moringa oleifera, Diplomarbeit, University of Heidelberg, 1980.

151. S. A. A. Jahn, *Ambio.*, 1991, **20**(6), 244.
152. A. Ndabigengesere, K. S. Narasiah and B. G. Talbot, *Water Res.*, 1995, **29**, 703.
153. U. Gassenschmidt, K. K. Jany, B. Tauscher and H. Niebergall, *Biochem Biophys Acta*, 1995, **1243**, 477.
154. K. A. Ghebremichael, K. R. Gunaratna, H. Henriksson, H. Brumer and G. Dalhammar, *Water Res.*, 2005, **39**, 2338.
155. T. Okuda, A. U. Baes, W. Nishijima and M. Okada, *Water Res.*, 2001, **35**, 830.
156. J. Beltran-Heredia, J. Sanchez-Martın and A. Delgado-Regalado, *Ind. Eng. Chem. Res.*, 2009, **48**, 6512.
157. M. C. Menkiti, C. I. Nwoye, C. A. Onyechi and O. D. Onukwuli, *Adv. Chem. Eng. Sci.*, 2011, **1**, 125.
158. J. Beltran-Heredia, J. Sanchez-Martın and C. Solera-Hernandez, *Ind. Eng. Chem. Res.*, 2009, **48**, 5085.
159. M. S. Ozacar and I. A. Sengil, *Water Res.*, 2000, **34**, 1407.
160. M. S. Ozacar and I. A. Sengil, *Turk. J. Eng. Environ. Sci.*, 2002, **26**, 255.
161. M. S. Ozacar and I. A. Sengil, *Colloids Surf. A.*, 2003, **229**, 85.
162. M. A. Aboulhassan, S. Souabi, A. Yaacoubi and M. Baudu, *Environ. Technol.*, 2005, **26**, 705.
163. J. Sánchez-Martín, J. Beltrán-Heredia and C. Solera-Hernández, *J. Environ. Manage.*, 2010, **91**, 2051.
164. J. Beltrán-Heredia, J. Sánchez-Martín and C. Martín-Sánchez, *Ind. Eng. Chem. Res.*, 2011, **50**, 686.
165. J. Beltrán-Heredia, J. Sánchez-Martín and M. Jiménez-Giles, *Water Air Soil Pollut.*, 2011, **222**, 53.
166. R. Ibanez-Camacho, M. Meckes-Lozoya and V. Mellado-Campos, *J. Ethnopharm*, 1983, **7**, 175.
167. R. L. E. Kossori, C. Villaume, E. E. Boustani, Y. Sauvaire and L. Mejean, *Plant Food Hum. Nutr.*, 1998, **52**, 263.
168. C. Saenz, E. Sepulveda and B. Matsuhiro, *J. Arid Environ.*, 2004, **57**, 275.
169. S. Trachtenberg and A. M. Mayer, *Phytochemistry*, 1981, **20**, 2665.
170. B. Manunza, S. Deiana, M. Pintore and C. Gessa, *Carbohydr. Res.*, 1997, **300**, 85.
171. S. Carpinteyro-Urban, M. Vaca and L. G. Torres, *Water Air Soil Pollut.*, 2012, **223**, 4925.
172. J. Zhang, F. Zhang, Y. Luo and H. Yang, *Process Biochem.*, 2006, **41**, 730.
173. J. Idris, A. Md Som, M. Musa, K. Halim, K. Hamid, R. Husen, M. Najwa and M. Rodhi, *J. Chem.*, 2013.
174. A. K. Sen and K. R. Bulusu, *Indian J. Environ. Health*, 1962, **4**, 233.
175. P. N. Tripathi, M. Chaudhari and S. D. Bokil, *Indian J. Environ. Health*, 1976, **18**, 272.
176. M. Adinolfi, M. M. Corsaro, R. Lanzetta, M. Parrilli, G. Folkard, W. Grant and J. Sutherland, *Carbohydr. Res.*, 1994, **263**, 103.

177. M. G. Antov, M. B. Šćiban and J. M. Prodanović, *Ecol. Eng.*, 2012, **49**, 48.

178. A. Diaz, N. Rincon, A. Escorihuela, N. Fernandez, E. Chacin and C. F. Forster, *Process Biochem.*, 1999, **35**, 391.

179. T. Okuda, A. U. Baes, W. Nishijima and M. Okada., *Water Res.*, 2001, **35**, 405.

180. A. Ndabigengensere and K. S. Narasiah, *Water Res.*, 1998, **32**, 781.

181. J. Sanchez-Martin, M. Gonzalez-Velasco and J. Beltran-Heredia, *Chem. Eng. J.*, 2010, **165**, 851.

182. M. Šćiban, M. Klašnja, M. Antov and B. Škrbić, *Bioresour. Technol.*, 2009, **100**, 6639.

183. Y. A. J. Al-Hamadani, M. S. Yusoff, M. Umar, M. J. K. Bashir and M. N. Adlan, *J. Hazard. Mater.*, 2010, **190**, 582.

184. H. Patel and R. T. Vashi, *J. Saudi Chem. Soc.*, 2012, **16**, 131.

185. R. Sanghi, B. Bhatttacharyaa, A. Dixit and V. Singh, *J. Environ. Manage.*, 2006, **81**, 36.

186. R. S. Blackburn, *Environ. Sci. Technol.*, 2004, **38**, 4905.

187. S. A. Abo-El-Eneina, M. A. Eissab, A. A. Diafullahc, M. A. Rizkc and F. M. Mohamed, *J. Hazard. Mater.*, 2011, **186**, 1200.

188. Y. Fu, C. Lin, L. Chen, Y. Song and H. Ma, *Adv. Mater. Res.*, 2012, **415-417**, 1667.

189. A. Ikeda, A. Takemura and H. Ono, *Carbohydr. Polym.*, 2000, **42**, 421.

190. G. T. Grant, E. R. Morris, D. A. Rees, P. J. C. Smith and D. Thom, *FEBS Lett.*, 1973, **32**, 195.

191. S. Zhao, B. Gao, X. Li and M. Dong, *Eng. J.*, 2012, **200–202**, 569.

192. A. H. King, in *Food Hydrocolloids*, ed. M. Glicksman, CRC Press, Boca Raton, FL, vol. II, 1983, pp. 115–188.

193. H. A. Devrimci, A. M. Yuksel and F. D. Sanin, *Desalination*, 2012, **299**, 16.

194. H. Altaher, *J. Hazard. Mater.*, 2012, **233–234**, 97.

195. Q. Chang, M. Zhang and J. X. Wang, *J. Hazard. Mater.*, 2009, **169**, 621.

196. H. Salehizadeh and S. A. Shojaosadati, *Biotechnol. Adv.*, 2001, **19**, 371.

197. W. J. Liu, K. Wang, B. Z. Li, H. L. Yuan and J. S. Yang, *Bioresour. Technol.*, 2010, **101**, 1044.

198. N. He, Y. Li, J. Chen and S. Y. Lun, *Biochem. Eng. J.*, 2002, **11**, 137.

199. S. G. Wang, W. X. Gong, X. W. Liu, L. Tian, Q. Y. Yue and B. Y. Ga, *Biochem. Eng. J.*, 2007, **36**, 81.

200. Y. Wei, J. Mu, X. Zhu, Q. Gao and Y. Zhang, *J. Environ. Sci.*, 2011, **23**, S142.

201. J. S. Chang, C. Chou, Y. C. Lin, P. J. Lin, J. Y. Ho and T. L. Hu, *Water Res.*, 2001, **35**, 2841.

202. S. B. Deng, R. B. Bai, X. M. Hu and Q. Luo, *Appl. Microbiol. Biotechnol.*, 2003, **60**, 588.

203. Q. Gao, X.-H. Zhu, J. Mu, Y. Zhang and X.-W. Dong, *Bioresour. Technol.*, 2009, **100**, 4996.

204. Z. Li, S. Zhong, H.-Y Lei, R.-W Chen, Q. Yu and H.-L. Li, *Bioresour. Technol.*, 2009, **100**, 3650.
205. S. Deng, G. Yu and Y. P. Ting, *Colloids Surf. B.*, 2005, **44**, 179.
206. W.-X. Gong, S.-G. Wang, X.-F. Sun, X.-W. Liu, Q.-Y. Yue and B.-Y. Gao, *Bioresour. Technol.*, 2008, **99**, 4668.
207. R. Guo, L. Zhang, Z. Jiang, Y. Cao, Y. Ding and X. Jiang, *Biomacro-molecules*, 2007, **8**, 843.

CHAPTER 2

Activated Carbon from Biomass for Water Treatment

SARRA GASPARD,*[a] NADY PASSÉ-COUTRIN,[a]
AXELLE DURIMEL,[a] THIERRY CESAIRE[b] AND
VALÉRIE JEANNE-ROSE[a]

[a] Laboratoire COVACHIMM, EA 3592, Université des Antilles et de la Guyane, BP 250, 97157 Pointe à Pitre Cedex, Guadeloupe; [b] Laboration GTSI, EA 2432, Université des Antilles et de la Guyane, BP 250, 97157 Pointe à Pitre Cedex, Guadeloupe
*Email: sarra.gaspard@univ-ag.fr

2.1 Introduction

Among the traditional precursors, activated carbon has long been prepared using bituminous coals, coke and coconut shells. However, many other agricultural by-products[1–5] can be used as a feedstock for the preparation of activated carbons. They include agricultural wastes such as bagasse,[3,4] coir pith (a soft biomass obtained from the coconut husk during its preparation in the coconut industry),[1,2] banana pith,[3] sago waste,[3] silk cotton hull,[3] corn cob,[4] maize cob,[3] rice straw,[6] rice hulls,[6] fruit stones,[7] nutshells,[6,7] pinewood,[8] sawdust,[9] coconut tree sawdust,[3,10] rice husk,[9–11] bamboo[11] and cassava peel.[12] These precursors have the advantage of not increasing the accumulation of CO_2 in the atmosphere and of being highly lignocellulosic; therefore, they are excellent materials for producing activated carbon.

Activated carbon synthesis can be realised by two well-known processes: physical activation involving a carbonisation followed by an activation step by

RSC Green Chemistry No. 25
Biomass for Sustainable Applications: Pollution Remediation and Energy
Edited by Sarra Gaspard and Mohamed Chaker Ncibi

an oxidising gas; and chemical activation involving a single carbonisation step of the precursor in the presence of a chemical agent. More recently, many works have described the search for alternative synthesis methods with energy and chemical savings while producing high value materials from renewable resources such as biomass waste. Thus, non-conventional methods such as microwave heating and hydrothermal carbonisation treatment are being developed for activated carbon preparation.

Activated carbon yield and textural characteristics are highly dependent on both the botanical composition of the lignocellulosic feedstock and the preparation process. This chapter describes textural and chemical characteristics governing the adsorptive properties of activated carbons prepared from some vegetal biomasses (agro-industrial by-products, fruits and nut peel or seeds, marine plants). Methods and theories for determining the textural characteristics of activated carbons are also described.

Adsorption properties depend on surface chemistry that gives the surface its charge and hydrophobicity, and the electronic density of the graphene layers. The surface chemistry of the activated carbons essentially depends on their heteroatom content, mainly on their surface oxygen groups, *e.g.* carboxyls, phenols, lactones, carbonyls and quinones. Methods for characterisation of these chemical groups are described. They include spectrophotometric methods such as Fourier transform infra-red (FTIR) spectroscopy and X-ray photo-electron spectroscopy (XPS), and a thermal programmed desorption technique generally applied for the characterisation of the functional groups on the surface of the activated carbon surface. Adsorption of chemical substances from solution, which is essentially an exchange process between the sorbent surface and the liquid phase, is generally investigated by adsorption kinetic and isotherm studies. Focusing on dyes, chlorinated compounds and metals, some studies on pollutants adsorption by biomass derived activated carbons are discussed.

2.2   Activated Carbon Preparation

During the past 20 years various new renewable feedstock including wood, fruit stones or shells, and waste from agro-industries have been used as activated carbon precursors. The preparation of activated carbon by the classical experimental procedures, physical and chemical activation, but also using non-conventional methods such as microwave heating and hydrothermal carbonisation treatment are described below.

2.2.1   Physical Activation

For physical activation, the precursor is developed into activated carbon by carbonisation (or pyrolysis) of the lignocellulosic material followed by an activating step with an oxidising gas such as air, carbon dioxide, steam or their mixtures. During carbonisation, the carbonaceous starting material is converted to a fixed carbon mass that has a rudimentary pore structure. The precursor thermal behaviour and the carbon yield are related to the feedstock composition as carbonisation removes non-carbon elements, which are volatilised at low temperature.

Plant biomass[13] is mainly composed of cellulose (long polymers of glucose without branches), hemicellulose (various branched saccharides) and lignin (Figure 2.1), along with smaller amounts of pectin, protein, extractives and ash. The composition of these constituents can vary from one plant species to another. Cellulose is the main structural constituent of plant cell walls where it is found in an organised fibrous structure.[13]

Some authors were able to show a dependence between the char yields and the botanical composition of the biomass. In 1995, Raveendran *et al.*[14] compared the thermal decomposition of different biomasses (*e.g.* bagasse, coconut, cotton, rice stalk, rice straw, subabul wood) and studied the influence of demineralisation. They found that: (1) the char yield increased upon demineralisation of coir pith, groundnut shell and rice husk as observed by thermal analysis; (2) the increase in char yield upon demineralisation was much greater for rice husks than for coir pith or groundnut shells; and (3) there was a substantial increase in liquid yield for all five materials, ranging from 77% for wood to 13% for groundnut shell. González *et al.*[15] compared different biomass samples (almond shells, walnut shells, almond tree prunings and olive stones) in a thermoanalytical investigation to evaluate their thermal behaviour and to correlate these data with their lignocellulosic composition. They observed that the four materials presented similar char yields, decreasing in the following order: almond tree prunings > olive stones > almond shells > walnut shells, in agreement with previous studies by Antal,[16] who stated that carbon yield was higher for materials with greater lignin content. Bouchelta *et al.*[17] prepared activated carbon from wastes of Algerian date stones in a heated fixed-bed reactor by pyrolysis and physical activation in the presence of water vapour. The effect of the pyrolysis temperature and activation hold time on the textural and chemical surface properties and the resulting carbon materials were studied. A relationship between yield and temperature could be constructed

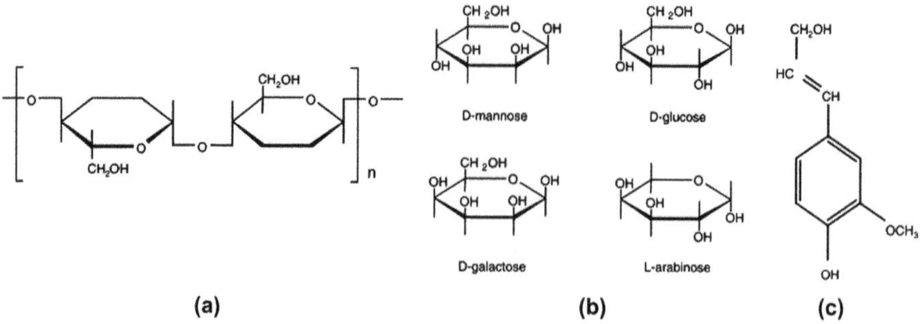

Figure 2.1 Structure of (a) cellulose molecule, (b) principle sugar residues of hemicelluloses, (c) phenylpropanoid units found in lignin (from ref. 13).

showing that the yield of the pyrolysed date stones decreased as the temperature increased from 500 to 700 °C. This phenomenon was attributed to the removal of volatile matter resulting from the decomposition of major compounds in the date pits, *i.e.* cellulose and hemicellulose. Above 700 °C, the yield remained constant because all the cellulose and hemicellulose decomposed. The only remaining component was lignin, the decomposition of which is more difficult. Later, Krzesinska and Zachariasz[18] investigated the effect of pyrolysis temperature on the physical properties of new monolithic porous carbon materials derived from exceptional types of bamboo characterised by solid, very strong stems, *i.e.* iron bamboo (*Dendrocalamus strictus*). Their results indicated that the temperature dependence of weight loss and carbon content tended to saturate above 600 °C, while the true density and elastic parameters continued to increase up to 900 °C. They explained that further heating above 600 °C did not remove any compounds from the bamboo stem, but that a reorganisation of the carbonised structure was likely resulting in a more compact matrix.

Ncibi *et al.*[19] investigated the pyrolysis of renewable, highly available and low-cost *Posidonia oceanica* marine fibres for preparation of carbon. The characterisation of the raw biomass showed that the major component was holocellulose (59% of the dry weight) with 27% lignin, and the content of extractives and ash was 12.2% and 1.8%, respectively. A thermogravimetric analysis of *P. oceanica* (L.) fibres showed a hemicellulose decomposition around 290 °C with a corresponding weight loss of 11.5% (initially representing 21% of the *P. oceanica* (L.) fibres). Cellulose degradation occurred around 330 °C (with a weight loss of 24.9%); this component representing 38% of the weight of the fibre showed the greater weight loss. Lignin degradation was shown at 400 °C with a corresponding weight loss of 9.7%. The degradation of the three component, cellulose, hemicellulose and lignin, occurred with an exothermic effect and the corresponding enthalpy values, $\Delta H = -24.63$ J g^{-1}, $\Delta H = -31.91$ J g^{-1} and $\Delta H = -11.41$ J g^{-1}, respectively. Several heating temperatures were experimented for studying the pyrolysis process of raw *P. oceanica* fibres, as well as two heating period (1 and 2 h). The increase in the pyrolysis temperature was proportional to an increase in the weight loss. From a temperature of 400 °C to 800 °C, the biomaterial loss in weight passed from 54.6% to 67.7%, for a heating exposure of 1 h. It was proposed that the increase in temperature quickens the volatilisation reactions of the amorphous components, which obstruct the pores. Also, the temperature increase would enhance the gaseous release of carbon monoxide and water within the marine biomass. Regarding the two heating times (1 and 2 h), no statistically significant difference was depicted between the weight losses for all produced chars.

Since 1970, attempts have been made to correlate mathematically the carbon yield to the precursor composition. Philpot[20] studied 20 lignocellulosic materials and found a correlation between char yield and the precursor inorganic content. Rothermel[21] recognised the need to also consider the organic fraction lignin, cellulose and other components present in a given precursor. In 2003, Ouensanga *et al.*[22] proposed an equation that mathematically described the dependence of the ideal char yield, Y_{ideal}, with the amounts of the components present in the carbons precursor prepared by carbonisation at high temperature (600–900 °C) of different tropical lignocellulosic wastes, guava (*Psidium guava*) seeds, dende (*Acrocomia karukerana*) shells, tropical almond (*Terminalia catappa*), jujube seeds (*Psidium guajava*) and cacao husks (*Theobroma cacao*):

$$Y(wt\%) = L[0.59 - 2.7 \times 10^{-4}(t\,^{\circ}C - 600) + 0.22C + A + yE] \qquad (2.1)$$

where L, C, A and E were, respectively, the percentage (wt%) of lignin, cellulose, ash and extractive in the dry precursor. The term, yE, accounted for the contribution of the extractives that partly volatilised at medium temperature (500–600 °C), whereas carbonaceous deposits formed *via* cracking and condensation reactions led to stable tars. The expression for y was given by $y = a + b$ (600 $- t\,^{\circ}C$), where a corresponded to the vaporised fraction of extractives and b to the carbonaceous deposit resulting from condensation and cracking reactions. The contribution yE was taken to account only for a precursor with a significant amount of extractive. Extractives represent a class of component that may vary considerably with the parent material, and therefore their volatilisation and the formation of secondary products may vary with the parent material. In a feasibility study of activated carbon production from vetiver roots, the yield of char produced after pyrolysis at different temperatures from 670 °C to 900 °C ranged from 25 to 28%. Those values were similar to those generally found for other lignocellulosic material.[24] The yields, Y, obtained for vetiver roots carbon production could be fitted using eqn (2.1).

This equation was latter modified[25] to introduce a multiplying coefficient $\cdot a_c$. The real char yield was written as:

$$Y_{real} = a_c Y_{ideal} \text{ with } a_c = (D_f / ln\varepsilon)^{1/2} \qquad (2.2)$$

where Df was the mean fractal dimension from mercury porosimetry data and $\varepsilon = [(1 - \rho_a/\rho_t) \cdot 100]$ was calculated from the apparent (ρ_a) and true (ρ_t) densities of the char measured by mercury intrusion.

The optimal pyrolysis temperature of lignocellulosic material could be determined using transmission electron microscopy (TEM).[26] Indeed, TEM is commonly used to characterise the carbon phase structure and composition; electron energy loss spectroscopy (EELS) used for chemical characterisation of such material reveals specific near edge structures at the level of the carbon K-edge appearing to depend on the state of matter organisation.[27,28] Carbonaceous compounds obtained by pyrolysis of lignocellulosic material often present an ill-organised structure whose middle–long range order depends on the pyrolysis temperature. EELS spectra at different temperatures of pyrolysed almond (*Terminalia Catappa*) shells, a tropical lignocellulosic precursor, are presented in Figure 2.2.

It was shown that, in the low loss energy range, the classical plasmon peak of carbonaceous material beyond 22 eV was still present, while specific features appeared about 6.5 eV. Two specific near edge structures increased with

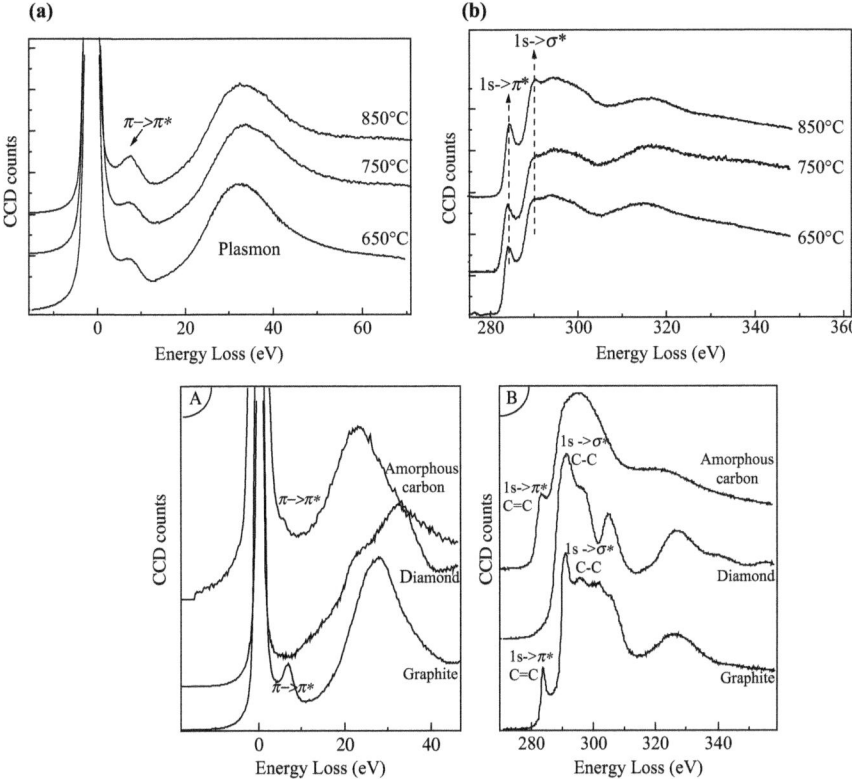

Figure 2.2 EELS Spectra for almond shell char respectively (a) Low Loss Spectra and (b) C K edge spectra Evolution of carbon K Near Edge and low loss structures for various carbon phases (CCD: Charge Coupled Device). (A) Low Loss spectra (B) C K-edge spectra (adapted from ref. 26).

the pyrolysis temperature, at the level of the carbon K-edge. These were characteristic of the chemical bonding of carbon atoms and of structure order.[29] Low loss and carbon-K edge spectra of amorphous carbon, diamond and graphite were used as reference for analysing the spectra of pyrolysed samples. In the low loss spectra and at the carbon K-edge, respectively, transition around 6.5 eV and a near feature at 284 eV are shown only in the amorphous carbon and graphite spectra. In agreement with the respective electronic band structures of diamond and graphite,[27–29] they could be attributed to electronic transition from π bonding levels into π^* anti-bonding level for the first one and from the 1s level to anti-bonding π^* states characteristic of a double bond (C=C) for the second one. The feature at 291 eV present in the graphite and diamond spectra, spread over a larger energy range in the case of amorphous carbon, could be attributed to electronic transition from the 1s level to anti-bonding σ^* states characteristic of a single bond (C–C). Evolution of pyrolysed tropical almond spectra shows a temperature dependence of the intensity of the peaks. These features were correlated to the C=C double bonds and C–C single bonds, and to the spatial extension of the organised structure. A quantitative measurement of the intensity of the 1s → π* and 1s → σ* transitions in the carbon K near edge structures and of the π → π* transition allowed the increase in concentration of the double bond and the short-range order to be followed. Thus, Jeanne-Rose *et al.*[26] showed that the amount of double bonds increased when the temperature rose from 600 °C to 700–750 °C, and then stabilised at higher temperatures. For temperatures >700 °C, structural ordering operated with the appearance of a well-defined 1s → σ* near edge structure, while intensity increased with increasing temperature (Figure 2.2). Finally, at the highest temperature (900 °C) the structure of the final product was less organised than graphitised carbon and the optimal temperature for pyrolysis was 800 °C.

The second step in the preparation of activated carbon by physical activation involves a controlled gasification of the char at high temperature with steam, carbon dioxide, air or a mixture of these. This gasification selectively eliminates most reactive carbon atoms from the sample, generating the porosity and the final carbon with the pore structure sought. Activation with CO_2 or H_2O broadens the carbon microporosity with a small shift to meso- and macro-porosity at high burn-offs. The final activated carbon has a well-developed microporosity with a very small contribution of mesoporosity but a well-developed macroporosity. The extent of macroporosity development is related to the cellular structure of the original material and to the activation holding time.

The chemical reactions involving H_2O and CO_2 as oxidising gas are as follows:[30]

$$C + H_2O \rightarrow CO + H_2 \ \Delta H_{298K} = +117 \, kj \cdot mol^{-1} \tag{2.3}$$

$$C + CO_2 \rightarrow 2\,CO \ \Delta H_{298K} = +159 \, kj \cdot mol^{-1} \tag{2.4}$$

Around 800 °C the following equilibrium involving steam may occur:

$$CO + H_2O \leftrightarrow CO_2 + H_2 \ \Delta H_{298K} = +41 \, kj \cdot mol^{-1} \tag{2.5}$$

Gasification reactions are endothermic and *in situ* combustion of the gaseous products, CO and H_2, are possible:

$$CO + \frac{1}{2}O_2 \rightarrow CO_2 \ \Delta H = -285 \, kj \cdot mol^{-1} \tag{2.6}$$

$$H_2 + \frac{1}{2}O_2 \rightarrow H_2O \ \Delta H = -238 \, kj \cdot mol^{-1} \tag{2.7}$$

Oxygen can also be used as activating gas.[31,32] The reaction being exo-thermic, the gas concentration and temperature of the following carbon–oxygen reactions must be strictly controlled:[31,32]

$$C + O_2 \rightarrow CO_2 \ \Delta H = -386 \, kj \cdot mol^{-1} \tag{2.8}$$

and

$$2C + O_2 \rightarrow 2CO \; \Delta H = -256 \, kj \cdot mol^{-1} \tag{2.9}$$

A study of the preparation of activated carbon by a one-step method combining carbonisation and activation of the raw material at high temperature in a gas (H_2O or CO_2) flow showed that this method was less expensive because the holding time at high temperatures was shorter than in the two-step method.[33]

2.2.2 Chemical Activation

For chemical activation, the carbonaceous material is generally impregnated with chemicals, such as KOH, NaOH, H$_3$PO4, $ZnCl_2$, H_2SO_4, $(NH_4)_2SO_4$, HCl, $MgCl_2$, HNO_3 or $CaCl_2$ followed by heating under a nitrogen flow at 450–900 °C, depending on the impregnant used. The chemical is introduced into the precursor, and produces physical and chemical changes modifying the thermal degradation process. It influences the pyrolytic decomposition of the starting materials, suppressing tar formation and lowering the pyrolysis temperature. The temperature of the process does not need to be high. During impregnation, there is a weakening of the precursor structure, hydrolysis reactions with loss of volatile matter, increase of elasticity, and swelling of the particle. The amount of chemical agent incorporated in the precursor and the impregnation duration govern the porosity of the carbon. This activation method is considered very flexible for the preparation of activated carbons with different pore size distributions.[25]

KOH activation is a well-known activating method. The main products formed during KOH activation are H_2, H_2O, CO, CO_2, potassium oxide (K_2O) and potassium carbonate (K_2CO_3).[34] The global proposed reaction occurring stoichiometrically between carbon and KOH is the following:

$$6KOH + 2C \rightarrow 3H_2 + 2K_2CO_3 + 2K \tag{2.10}$$

The standard Gibbs free energy change of this reaction is positive at room temperature and becomes negative at *ca.* 570 °C. The potassium compounds, K_2O and K_2CO_3, can also be reduced by carbon to produce metallic potassium at temperatures over 700 °C. The three main KOH activation mechanisms steps are:[30,35–38] (1) etching of the carbon framework by the redox reactions between various potassium compounds leading to pore network generation; (2) further development of the porosity through the gasification of carbon by *in situ* H_2O and CO_2 formed during chemical activation process; and (3) the as-prepared metallic potassium efficiently intercalating into the carbon lattices of the carbon matrix during the activation, results in the expansion of the carbon lattices. After washing of the intercalated potassium, the previous nonporous structure is not retrieved but high microporosity is created.

When phosphoric acid is used as the chemical activating agent, the reaction with the lignocellulosic precursor begins as soon as the components are mixed.

At low temperature, phosphoric acid attacks first hemicellulose and lignin, possibly because of easier access to these amorphous biopolymers than to the crystalline cellulose. Phosphoric acid hydrolyses glycosidic linkages in polysaccharides (hemicellulose and cellulose) and cleaves aryl ether bonds in lignin. These reactions are accompanied by further chemical reactions including dehydration, degradation and condensation releasing CO, CO_2 and CH_4, Phosphorus compounds can form ester linkages with –OH groups on cellulose at temperatures below 200 °C,[35] helping to crosslink the polymer chains. As the temperature increases, cyclisation and condensation reactions lead to an increase in aromaticity and in the size of the polyaromatic units, enabled by the scission of PPO–C bonds. Between 350 to 500 °C the char is considered to be stable, but at 430 °C and above, the continued cleavage of crosslinks leads to a very extensive growth in the size of the aromatic units. Above 450 °C the aromatic cluster size increases greatly. It is proposed that pores are formed due to acid that serves to 'swell the material structure' allowing the cellulose microfibrils to separate. Upon carbonisation, the altered microfibrils form an open porous structure.[35] At higher temperatures, cyclisation and condensation reactions form phosphate linkages, such as phosphate and polyphosphate esters, connecting and crosslinking biopolymer fragments. However, he activating agent is partly transformed into $H_4P_2O_7$ and continues to play its constructive role in activation progress. Typical yields obtained during phosphoric acid activation of lignocellulosic precursors are between 40 and 50%. Indeed, Altenor *et al.*[39] prepared activated carbon by phosphoric activation of vetiver roots, and found samples yields of 47–49%. When vetiver roots activated carbon was prepared by physical activation with steam, the carbon yield value was low (12.5%). Other precursors activated with phosphoric acid provide similar results: peach stones, 42–44%;[36] and coconut shells, 49–52%.[37] The higher yields obtained by chemical activation than physical activation were explained by the fact that the chemical agents used have dehydrogenation properties inhibiting formation of tar and reducing the production of other volatile products during pyrolysis.[38]

Activated carbons with a variety of pore size distributions can be prepared by either physical or chemical activation of lignocellulosic materials. Nevertheless, the activation mechanism and the flexibility to produce different pore size distributions are very different,[25] depending on the activating agent used. For physical activation with CO_2 or steam, a carbonisation step of the precursor at a temperature lower than the one used for activation is necessary, leading to lower activated carbon yields and higher energy consumption. Although chemical activation produced higher activated carbon yields[39,40] and remains a most economical method, it requires the use of chemical agents that produces effluent which must be recycled and/or treated, which may entail additional costs.

2.2.3 Microwave Heating

Microwave heating processing is attracting increasing attention for activated carbon preparation as its environmental impact is lower due to time and space saving.[41–43] Microwave energy is transformed into heat inside the particles by dipole rotation and ionic conduction. When high frequency voltages are applied to a material, the response of the molecules with a permanent dipole to the applied potential field is to change their orientation in the direction opposite to that of the applied field. The synchronised agitation of molecules then generates heat.[42,44,45] Therefore, the high temperature gradient from the interior of the char particle to its cool surface allows the microwave-induced reaction to proceed more quickly and effectively, resulting in energy savings and a shorter reaction time compared with the conventional process.[46,47] Microwave irradiation was previously demonstrated to interact with biomass at specific temperatures leading to a structural changes within the biomass due to both internal and volumetric heating.[43] To achieve activated carbon preparation, microwave heating is realised in the presence of an oxidising gas or a chemical activating agent similar to the classical physical and chemical activation processes, leading to activated carbon samples with various physical and chemical characteristics.

Guo and Lua[48] used a palm oil char obtained by a pyrolysis at 700 °C and observed no influence of microwave heating at 750 mW during 30 min, under a N_2 flow rate of 200 cm^3 min^{-1} char sample. Irrespective of the gas flow rates used, the yield after exposure to microwave heating was 100% (*i.e.* no weight loss). When CO_2 at a flow rate of 50 cm^3 min^{-1} was used instead of N_2, a weight loss of 50% was obtained indicating occurrence of a carbon-CO_2 reaction. They demonstrated that microwave power and exposure time are both important parameters for the process. For instance, proximate analyses of activated carbons prepared by microwave-induced CO_2 activation of oil palm stones[48] under different microwave power levels and times showed that the losses of volatile matter and fixed carbon were more significant at higher microwave power (750 W) than at lower power (450 W). At longer activation time (60 min), the losses were larger than at a shorter activation time (40 min). Indeed, at a fixed radiation time of 60 min, the volatile content weight losses (based on their respective individual starting weights in the char samples) at 450 W and 750 W increased by 60.3% and 87.2%, respectively. The fixed carbon decreased to 14.7% at 450 W and 46.5% at 750 W. At the same microwave power of 750 W, the weight loss of volatile content at 5 and 60 min was 38.5% and 87.2%, respectively.

Yang *et al.*[49] prepared activated carbon samples from coconut shell using a two-stage 1000 °C carbonisation and activation at 900 °C by different activating agents such as steam, CO_2 and a combination of steam–CO_2 with microwave heating. Samples with high surface area reaching 2000 m^2 g^{-1} were obtained and the activation time using microwave heating was very much shorter, while the yield of the activated carbon was similar to that obtained with the conventional heating methods.

Xin-Hui *et al.*[50] compared conventional and microwave heating systems for the preparation of activated carbon from Jatropha hull with different activation agents (steam and CO_2). They found that the yield of activated carbon did not vary significantly for steam activation, irrespective of the heating method, while it was found to double when CO_2 was used as activating gas with microwave heating compared with conventional heating.

When phosphoric acid was used as the activating agent for preparing activated carbons from different precursors such as waste tea, almond shells, tomato stems and leaves, at a microwave power of 900 W, and different treatment time of 1, 2, 3 and 4 minutes, the yields of activated carbons were in the range of 32–55%.[51] Similar to the values reported by Liu et al.[52] using bamboo as precursor, the carbon yield decreased from 61.2% at 200 W to 45.5% at 400 W.[52] At high microwave power level a greater weight loss of the carbon precursor was obtained. The carbon yields in this study were close to the results obtained in other studies dealing with classical phosphoric acid activation methods, ranging from 40 to 50%.[36,37,53]

2.2.4   Hydrothermal Treatment

The recent hydrothermal carbonisation (HTC) processing of biomass appears to be another low-cost, environmentally friendly route for activated carbon production. Hydrothermal carbonisation consists of heating raw material dispersed in an aqueous solution and autoclaving it at temperatures between 150 to 350 °C for about 2–24 h at saturated pressures. During this step, water-soluble organics and a carbon-rich, hydrophilic solid called 'hydrochar' are formed.[53] The HTC process is governed by dehydration and decarboxylation reactions which are exothermic and render the process self-sufficient after activation.[54] At subcritical conditions, biomass is first converted to monomers by hydrolysis, and then into soluble organics by dehydration and fragmentation.[55] A reduction in pH catalyses the initial hydrolysis step.[56] Polymerisation and condensation of organic solubles increase the solution concentration. Nucleation, and thus the growth, takes place. The final carbonaceous material has a core–shell structure with a hydrophobic core and a stabilising hydrophilic shell which contains oxygen functional groups such as hydroxyl, phenol, carbonyl or carboxyl.[54,57–59]

Using glucose as a carbohydrate model[60] and ^{13}C nuclear magnetic resonance (NMR) spectroscopy, the carbonaceous product structure and formation mechanism obtained upon hydrothermal carbonisation were proposed. Furan-rich structures occurring in the coalification process, leading to condensed polyaromatic structures, were proposed as intermediates. Under stronger hydrothermal carbonisation conditions (T > 200 °C), arene-rich chars, with a furan-to-arene ratio falling below 1, were obtained. Otherwise, pyrolysis does not provide furan-rich structures. Hydrothermal carbonisation process was stated as richer and more controllable in terms of carbon chemistry than the conventional pyrolysis.[60]

Sunflower stem, walnut shells and olive stone were carbonised hydrothermally at 220 °C for 20 hours.[61] The samples obtained were then activated using CO_2 and air respectively. The burn-off values calculated for each activated carbon prepared were higher than the values obtained for the corresponding activated carbons produced from chars resulting from classical pyrolysis. Previous studies involving traditional physical activation using the same precursors showed that longer periods of activation time were needed to achieve carbon conversions near 50%. It was then concluded that these activated carbons also showed a greater porosity development.[61]

Unur[62] compared the treatment of hazelnut shells by conventional and hydrothermal carbonisation. During classical pyrolysis, hazelnut shells started to decompose at 200 °C. At 400 °C, decomposition of the cellulosic fraction and a part of lignin took place with a weight loss of 57%. A 10% weight loss was then obtained at 600 °C, followed by an additional 10% weight loss, associated with the decomposition of aromatic lignin content when the temperature was maintained at 600 °C for 2 h. The final yield was 15% at 950 °C. During carbonisation of a sample previously hydrothermally carbonised at 250 °C in an autoclave for 7.5 h, sample decomposition also started at 200 °C and about 20% of the weight was lost as the temperature reached 400 °C and 20% at 600 °C. An additional 20% of the weight was lost during a 2 h holding time at 600 °C. The final yield of 28% at 950 °C was twice that obtained by classical carbonisation due to decomposition of cellulosic content in the earlier hydrothermal treatment.[62] Hydrothermal carbonisation at subcritical conditions is known to generate carbonaceous nanostructures featuring a hydrophilic surface. Hydrothermal carbonisation of hazelnut shell led to an increase of carbon content from 47.13 wt% to 65.78 wt%, and to slight changes in hydrogen and nitrogen contents. These results confirmed the formation of a condensed hydrocarbon due to oxygen and inorganic material loss. Subsequent pyrolysis of the hydrochar at 600 °C increased the carbon content to 78.38 wt% while the oxygen content fell from 23.32 wt% to 9.03 wt%. The drastic decrease in oxygen content indicates the disruption of surface functional groups with increasing temperature. However, the chemical activation of the hydrochar with KOH increased the oxygen content to 31.00 wt% due to oxidation, while the carbon C content remained at 64.5 wt%.

2.3   Activated Carbon Characterisation

2.3.1   Surface Chemistry of Activated Carbons

2.3.1.1   *Characterisation by Fourier Transform Infrared (FTIR) Spectroscopy*

Characterisation of the surface of activated carbons can be performed by infrared spectroscopy (IR) providing qualitative information of characteristic functional groups on the surface. The assignment of the bands is based on the data published by other authors (Table 2.1), among papers reporting the FTIR spectra of lignocellulosic biomass, coals or activated carbons.[63–66] Changes in the lignocellulosic structure during pyrolysis or activation can thus be observed.

Bilba and Ouensanga[63] studied the pyrolysis of sugarcane bagasse following the FTIR spectral changes and detected absorption bands characteristic of lignin for untreated bagasse. The broad band between 3100 and 3800 cm^{-1} was characteristic of O–H stretching, the bands at 2925 cm^{-1}, between

Table 2.1 Assignment of FTIR absorption bands of pyrolysed and activated lignocelulossic materials.

σ (cm^{-1})	Assignment	Designation
3700	v (OH)	Free OH group[63]
3400	v (OH)	Stretching in hydroxyl groups[66]
2926	v (C–H)	Stretching in alkyl groups[66]
2870	v (C–H)	Stretching in alkyl groups[66]
1745	v (C=O)	Stretching in aldehydes, ketones groups and esters[66]
1642	v (C=O)	Stretching in cyclic amide[64]
1640	v (C=C) v (C=O)	Stretching in olefins[66] carbonyl groups of carbon material highly conjugated in graphite layer[64]
1552	v (C=O)	carbonyl groups conjugated in aromatic[63,65]
1540	v (C=C)	Assigned to skeletal stretch in condensed aromatic system[66]
1520	v (C=C)	Aromatic skeletal stretching bands[66] (Bouchelta et al., 2008)
1462	v (C-H)	Stretching in aromatic ring[66]
1460	v (C=O)	Stretching in cyclic amide[65]
1444	v (C=C) and δ (CH)	Stretching in aromatic skeletal and ester[66]
1150–1200	v (C–O–C), v (C C), methoxyl–O–CH₃	Stretching vibration in pyranose ring skeletal or stretching in aromatic ring[63–66]
1137	v (C–O)	stretching in phenolic ring, carboxylic moiety[64]
1069	v (C–O-C)	Assymetrical stretch vibrations[65]
870, 690	γ (C–H)	Aromatic C–H out-of-plane bending vibrations[63,66]
610	v (O–H)	Stretching in OH groups[66]
450	v (C–C)	Stretching in C–C vibrations[66]

$1700–1730$ cm^{-1}, and between 700 and 900 cm^{-1} were assigned to $C–H_n$ stretching vibration, and C=O stretching of ketones and carbonyls (and aromatic hydrogens), respectively. They also observed signals due to the C=C stretching vibration of the benzene ring at 1632 cm^{-1}, characteristic of aromatic skeleton at 1608 and 1516 cm^{-1}, due to CH deformation (1402 and 835 cm^{-1}) and of CO deformation (1108 and 1060 cm^{-1}). According to the authors, during pyrolysis the most important structure modifications appeared between 300 and 400 °C and were characterised by the reduced intensities of the C–O and C=C bands and the appearance of alkyl bands.

FTIR spectra of raw vetiver roots, the corresponding chars prepared at different temperatures and activated carbons prepared by physical activation using steam and CO_2 as activation gas are presented in Figure 2.3.[23] The spectrum of the vetiver roots exhibited bands at 3354 cm^{-1} which were attributed to OH stretching vibrations in hydroxyl groups.[23] The band at 2919 cm^{-1} was attributed to C–H vibration of aliphatic groups, the bands between 1700 and 1400 cm^{-1} to C=C, and the bands between 1300 and 850 cm^{-1} to C–O vibrations.[55] The band at 1243 cm^{-1} could be due to esters and the strong one at 1037 cm^{-1} to C–O vibration in C–OH moiety. The pyrolysis of the vetiver roots led to a progressive decrease of the intensity of the latter band absorbing at 1037 cm^{-1} when the temperature increased. Only bands attributed to aromatic C=C and C–O

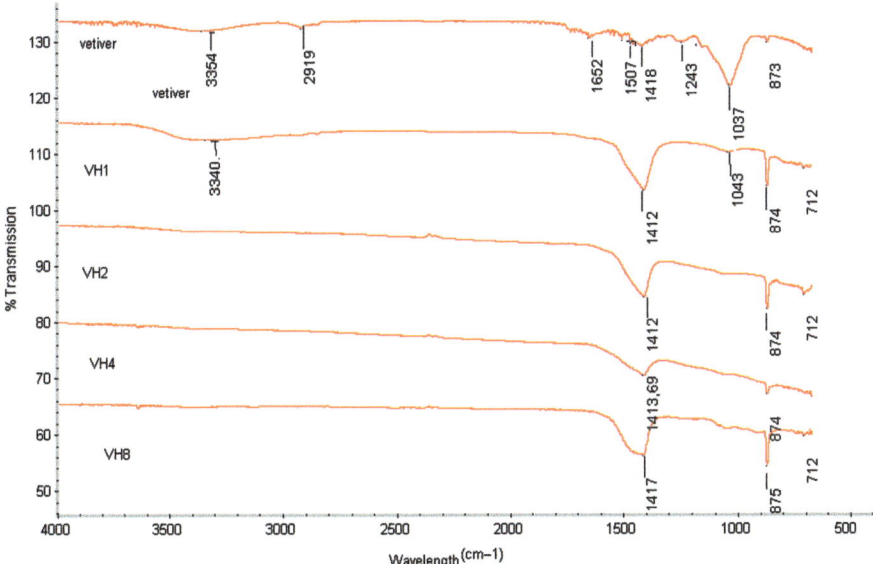

Figure 2.3 FTIR spectra of raw vetiver roots and activated carbons prepared by steam activation (adapted from ref. 67).

vibrations absorbing between 1400 and 850 cm^{-1} and a band at 873 cm^{-1} due to aromatic C–H vibrations were clearly displayed in the spectra of all chars prepared.[55] This showed that during pyrolysis the number of oxygen-containing groups decreased and that aromatisation of the samples occurred. The spectra of the activated carbons prepared by physical activation using steam and CO_2, respectively, contained principally the bands associated with the C=C, C=O and C–H aromatic vibrations between 1400 and 850 cm^{-1}.

Similarly, a study of the FTIR spectral evolution of jute stick during pyrolysis[68] demonstrated that the functional groups on the jute stick surface differed significantly from those of charcoal. Similarly to what was obtained for the vetiver roots spectra, the broad band at 3350 cm^{-1} in the spectrum of jute sticks was ascribed to the OH$^-$ stretching vibrations of hydroxyl groups in phenols. Two bands at 2925 and 2900 cm^{-1} were representative of asymmetric and symmetric stretching vibrations of methylene groups, while their bending vibrations were observed between 1430 and 1380 cm^{-1}. The strong band at 1740 cm^{-1} was attributed to carbonyls (C=O). The bands at 1590 and 1510 cm^{-1} corresponded to aromatic C=C bonds, while a band at 1460 cm^{-1} was ascribed to an aromatic methyl group (–CH$_3$). The C–O stretching or –OH deformation band in carboxylic acid was observed at 1430 cm^{-1}. The band at 1330 cm^{-1} was attributed to C–O vibrations in carboxylate groups, and the bands at 1240 and 1160 cm^{-1} were attributed to esters, ethers or phenols. A shoulder at 1110 cm^{-1} and a relatively intense band at 1060 cm^{-1} were assigned to alcohol (R–OH) groups. The C–H out of plane deformation and C–H out-of-plane bending in benzene derivatives caused the bands at 894 and

831 cm^{-1}, respectively. In the charcoal spectrum, some of the functional groups completely disappeared during the pyrolysis of jute sticks, while some were newly generated. According to the authors, the bands at 3040 and 876 cm^{-1} were newly generated and corresponded to =C–H and =CH$_2$ groups. The bands at 814 and 752 cm^{-1} represented the C–H out-of-plane bending in benzene derivatives and the C–H bending vibration in *cis*-RCH=CHR bonds.

Tsang *et al.*[69] compared the FTIR spectra of waste wood pallets and wood waste-derived activated carbon (at different activation temperatures). For waste wood pallets, the authors attributed the bands at 1300–1400 cm^{-1} to the aromatic CH and carboxyl-carbonate structures, and the peaks at 1020–1130 cm^{-1} to C–OH stretching and –OH deformation. After activation, they reported a decrease in the peak at 1022 cm^{-1} because phosphoric acid activation converted hydroxyl groups into acid oxygen-containing functional groups on the carbon surface. The spectra also showed definite bands at 1620–1700 cm^{-1} assigned to the C=O stretching vibrations of aldehydes, ketones, lactones or carboxyl groups; the stretching modes of hydrogen-bonded P=O; O–C stretching vibrations in P=O–C linkages; and P=OOH. The 1050–1290 cm^{-1} adsorption band was attributed to the vibration of the C=O group in lactones. The 1500–1560 cm^{-1} region showed considerable overlap of different absorption bands attributed to quinonic and carboxylate groups.

Liou and Wu[70] used FTIR spectroscopy to characterise the surface groups of raw rice husk and its derived activated carbon samples prepared by H$_3$PO$_4$ and ZnCl$_2$ activation. The spectrum of the raw material was found to have many similarities with pistachio nut shell,[60] another type of lignocellulosic material. A wide band located at 3400 cm^{-1} was attributed to O–H hydroxyl groups or adsorbed water. The band at approximately 2900 cm^{-1} corresponded to C–H vibrations. The band at 1692 cm^{-1} was attributed to carbonyl C=O groups, and the band at approximately 1615 cm^{-1} may have been due to aromatic C=C stretching vibrations. A very small peak near 1492 cm^{-1} was attributed to CH$_2$ vibrations. The band at 1158 cm^{-1} was likely to be C–O vibrations in phenols, ethers or esters. In addition, the peak shoulder at 1051 cm^{-1} could be attributed to alcohol R–OH groups. Finally, the bands at approximately 464, 735 and 1105 cm^{-1} were most likely from silicon atoms attached to the oxygen in the rice husk. After carbonising the rice husk in a N$_2$ atmosphere, Liou and Wu[70] observed the disappearance of many bands, indicating the vaporisation of organic matter. When the rice husks were activated with H$_3$PO$_4$ and ZnCl$_2$, the bands at 464, 735 and 1105 cm^{-1} disappeared completely, indicating the removal of ash. In addition, bands appeared at 1692, 1564 and 1158 cm^{-1} for the H$_3$PO$_4$ activated samples, and these bands were attributed to C=O, C=C and C–H vibrations, respectively. The shoulder at approximately 1000 cm^{-1} might be a chain of P–O–P vibrations for phosphorus compounds. For comparison, for the ZnCl$_2$ activated samples, the bands at 1590 and 1158 and a very small shoulder at 840 cm^{-1} were attributed to C=C, C=O and C–H vibrations, respectively. For both activation procedures, the intensity of these bands decreased as the activation temperature increased, indicating that the proportion of carbon content increased at high temperatures.

Deng *et al.*[71] found that the physically and KOH chemically activated cotton stalks samples had similar absorbance bands. From the literature, they suggested surface functionalities with C=O (carboxylic, anhydride, lactone and ketene groups) having IR bands at 1750–1630 cm^{-1}, C–O (lactonic, ether and phenol, among others), with a very intensive band at 1300–1000 cm^{-1} and C–C at 1640–1430 cm^{-1}. In addition, weak bands at 2649 and 1154 cm^{-1} in chemically activated carbons might be ascribed to traces of potassium carbonate and metallic potassium at 873 cm^{-1}. These signals indicated that, despite the extremely prolonged washing, a trace amount of potassium remained inside the pore structure. The most significant differences noticed by the authors in the spectra of the samples were sharp bands located between 1627 and 1604 cm^{-1} that disappeared in the activated carbon spectra samples. This behaviour suggested that the surface groups formed on the carbons could be influenced by the activation agent.

2.3.1.2 Characterisation by X-ray Photoelectron Spectroscopy (XPS)

XPS analysis can be used to determine the elemental surface composition obtained over a depth of about a few nanometres and atomic ratios obtained from the curve fitting of the XPS spectra. The changes in the chemical bonding states and concentrations of the surface functional groups of activated carbon samples can also be determined.

The C_{1s} spectrum can be de-convoluted into five components with chemical shifts corresponding to: (I) graphite type (284.1–284.4 eV); (II) amorphous carbon, hydroxyl groups, phenolic, alcohol or ether aromatic carbon (284.8–285.2 eV); (III) carbonyl groups (285.5–286.1 eV); (IV) carboxyl and ester groups (286.3–287.6 eV); and (V) a peak corresponding to π–π^* transitions in the aromatic carbon (289.5–290.0 eV).[38,72] The heteroatoms, such as nitrogen and oxygen, play an important role in surface chemistry. Thus, oxygen functional groups determine the surface properties such as hydrophobicity and surface polarity of the carbons and hence their quality as adsorbents. The O_{1s} spectrum can also be fitted to three components corresponding to: (I′) C=O groups (530–531.6 eV); (II′) C–OH or C–O–C groups (532.7–533.3 eV); and (III′) corresponding to chemisorbed oxygen (534.8–535.7 eV).[72] The N_{1s} spectra can be deconvoluted and fitted considering pyridinic, pyrrolic and quaternary nitrogens and nitrogen oxides. Peak I″ can be ascribed to N-6 or pyridine-like structures (398.0–398.1 eV), peak II″ to pyridon-N moieties, peak III″ to N-5, *i.e.* pyrrolic and/or quaternary nitrogen (401.460.5 eV) to N-Q (400.1–400.7 eV) and peak IV″ to N-oxides (402–405 eV), respectively.[73–77] According to the area simulating curve, the percentage of each component can be calculated.

As illustrated in Figure 2.4, Altenor *et al.*[53] observed from XPS analysis that vetiver root activated carbons obtained by steam activation with a burn-off of 50% had a higher content of graphitic carbon than activated carbons obtained by phosphoric acid activation. However, the content of carboxyl groups in steam activated carbon was lower than that obtained by chemical activation. The contribution of C=O groups to the O_{1s} profile increased with *Xp*, the

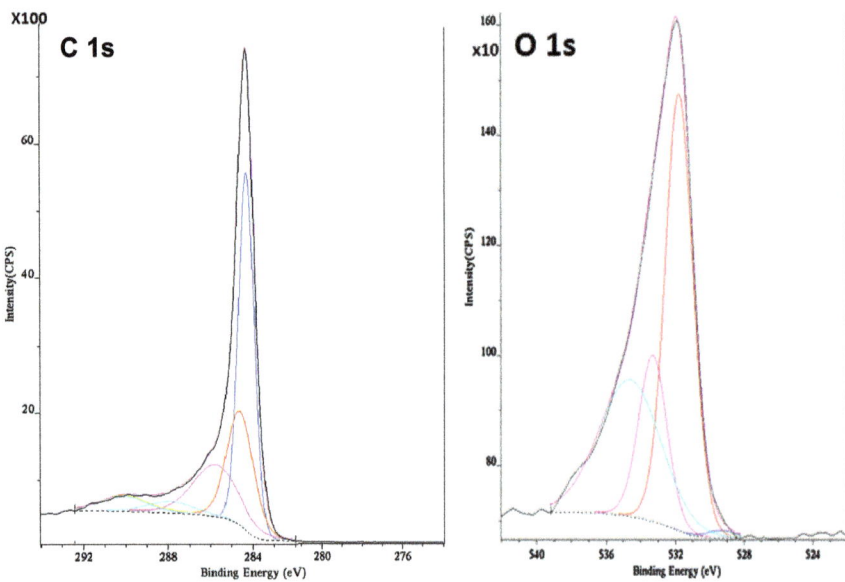

Figure 2.4   C1s and O1s spectra of vetiver roots activated carbon prepared by steam activation.

phosphoric acid/precursor ratio. For activated carbons prepared by steam activation of sugarcane bagasse, the oxygen content was shown to be higher. The ratio of oxygen to carbon content indicated the degree of surface oxidation, which was higher for the steam activated carbon sample than for the samples prepared by phosphoric acid activation. Nevertheless, the content of graphitic carbon increased with increasing Xp value. The hydroxyl and carboxyl groups were lower for the steam activated carbon than for the samples prepared by chemical activation, whereas the presence of C–OH and C–O–C bonds was remarkable for steam activated bagasse.

In another paper, Altenor *et al.*[40] compared the XPS spectra of raw *Turbinaria turbinata*, a brown alga, and its derived activated carbon prepared by steam activation and phosphoric acid activation, respectively. Raw turbinaria and steam activated turbinaria carbon showed the lowest graphitic carbon content (37.2% and 20.7%, respectively) compared with the phosphoric activated sample (55.2%). This difference in graphitic carbon content was directly related to the elemental carbon composition, which was lower for raw turbinaria and steam activated turbinaria (64.5% and 43%, respectively) than for activated carbon prepared by phosphoric acid activation (87%). Thus, among the tested sorbents, the lowest O/C was registered for this last sample (0.07). The graphitic content of activated carbon prepared by steam activation of turbinaria alga was significantly much lower than those of activated carbon samples obtained by steam activation of bagasse and vetiver roots (64.4 and 68%, respectively). For raw *T. turbinata* alga, the main surface chemical groups are carbonyls and hydroxyls (25.29 and 23.6%, respectively), followed by the carboxyl groups (13.6%). Regarding the pyrolysed and activated carbons derived from *T. turbinata*, the carbonyls and hydroxyls remain the main chemical groups, with a significant increase in the carbonyl groups for steam activated turbinaria (38%) and a decrease of the carboxyl groups for the phosphoric acid activated turbinaria (5.5%).

2.3.1.3 Characterisation by Temperature Programmed Desorption Technique

Temperature programmed desorption (TPD) coupled to mass spectroscopy (TPD-MS) can be used to determinate the nature of surface functional groups on activated carbons. Using this technique, gases produced from the decomposition of these groups are detected. The only gases evolved are carbon monoxide, carbon dioxide or water vapour and it is possible to recognise the presence of a functional group on the activated carbon surface considering its specific temperature of decomposition. Thus, surface oxygen groups on carbon materials decompose upon heating producing CO and CO_2 at different temperatures. TPD peaks can be tentatively assigned to the different functional groups by comparison with the data listed in the literature. The formation of water at high temperatures is related to dehydration reactions of neighbour carboxylic groups and phenol groups dehydration reactions to give anhydrides, lactones and ethers.[78,79] These groups then decompose to a mixture of CO and CO_2 above 600 °C.[66] Recently, a comparison between TPD-MS and TPD-XPS analysis allowed a more precise correlation between desorbed species and surface groups.[80]

Mestre *et al.*[81] used sisal waste as a precursor for the preparation of activated carbon by chemical activation with K_2CO_3. A ratio of sisal to chemical agent of 0.5 : 1 was used and the impregnated sample was carbonised at 700 and 800 °C, respectively. It was shown that the sample prepared at 700 °C had a pH_{pzc} of 7, while the one prepared at 800 °C had a pH_{pzc} of 8.2 in agreement with a higher content of carboxylic groups detected from TPD-MS measurements.

Bagasse activated carbon prepared by phosphoric acid activation were characterised using TPD-MS analysis.[82] Desorption spectra showed a peak at 230 °C that could correspond to the presence of carboxylic groups, and another one at 800 °C characteristic of anhydride or lactones. The TPD profile obtained for bagasse activated carbon prepared by steam activation presented a desorption peak at 580 °C, which was attributed to carbonates on the activated carbon surface. The curves obtained for all activated carbons showed a water desorption peak between 200 and 400 °C corresponding to water in microporosity, or hydrogen bonds with hydrophilic oxygenated surface groups such as carboxyls. An additional peak was observed at 600–800 °C for phosphoric acid activated carbons; the low temperature peak was related to water desorption. The relatively high temperature (above 200 °C) of this desorption showed that water was bonded to the carbon surface through relatively high energy interactions, *i.e.* water in microporosity, or hydrogen bonded with a hydrophilic oxygenated surface. The formation of water at high temperatures is related to the dehydration reactions of carboxylic and phenol groups to give anhydrides, lactones and ethers.[78,80] These groups then decompose to a mixture of CO and CO_2 above 600 °C.[80] These results were in agreement with XPS studies.

2.3.1.4 Characterisation by Boehm Titration

The Boehm titration[83] is a simple technique that quantifies acidic or basic functional groups on the surface of activated carbons. This titration is a selective neutralisation of surface acidic groups of various strengths using bases that have conjugate acids with a wide range of acid dissociation constants (Ka). For example, for quantification of acidic groups, basic aqueous solutions— sodium hydroxide (NaOH), sodium carbonate (Na_2CO_3) or sodium bicarbonate ($NaHCO_3$)—are used. $NaHCO_3$ ($pKa = 6.37$) uptake corresponds only to strong carboxylic acidity, while Na_2CO_3 ($pKa = 10.25$) further reacts with lactonic and carboxylic functionality, and NaOH ($pKa = 15.74$) reacts with phenolic, carboxylic and lactonic acidity. For the characterisation of weaker acidic groups, sodium ethoxide ($NaOC_2H_5$) is the other basic product used with absolute ethanol as solvent. However sodium ethoxide is not often used because of the need to perform the experiment in non-aqueous media and under oxygen-free conditions.[84] If the chosen concentrations of the used solutions decrease, the carbon dioxide effect can clearly appear on the titration curves. The relatively low concentration value of the reactant 0.01 mol L^{-1} used allows determination of small amounts of functional surface groups with an interesting precision. But it also reveals clearly the carbon dioxide effect due to the presence of atmospheric CO_2, which dissolves in the solutions.

The Boehm method turns out to be experimentally complex when it is necessary to prevent any dissolution of atmospheric CO_2 in the reacting media.[85,86] Kim et al.[87] proposed a protocol to realise Boehm titration without complicated devices to eliminate CO_2. The assumption, well demonstrated by the authors, was that the end point of titration of the leaving quantity of base was not modified by the presence of atmospheric carbon dioxide. With sodium hydroxide as the base, it was easy to show that the latest end point allows the determination of the total number of acidic functional groups. When using sodium carbonate and sodium bicarbonate as reactants, the atmospheric CO_2 concentration effect should be taken into account.[86] It turned out to be unnecessary to exclude CO_2 to determine the equivalence of the reaction base because the position of the second inflection point indicating the equivalence of the reaction base remained unchanged even though the reaction base interacts with CO_2.

Although it was demonstrated that the chemical groups at the surface play an important role in the adsorption of organics in aqueous solutions, there is very few studies that relate the composition of initial precursors to the nature and quantity of functional groups on the carbon. Toles et al.[87] found that oxidation of charcoals with oxygen at 400 °C produced oxygen groups thought to be carboxylic acid groups. It was suggested that oxidation of carbon dioxide activated peach stone and plum stone carbons in air at 300 °C would not produce significant carboxyl and lactonic groups due to instability of these groups at these temperatures. Toles *et al.*[87] studied the effect of the different activated carbon preparation strategies on the surface properties of various nutshell carbons. They showed that subsequent oxidation of phosphoric acid activated nutshell carbons converted carbonyl groups to carboxylic acid, phenolic and lactonic groups. Overall, carbons prepared by air activation contained the greatest total number of functional groups.

Valix *et al.*[88] attempted to correlate the heteroatom composition with the preparation process and the adsorptive properties of activated some bagasse activated carbons. Guo and Rockstraw[89] studied the influence of primary activation parameters (activation temperature, impregnation ratio between phosphoric acid and precursor, and soaking time at final activation temperature) on the formation of acidic surface groups during the phosphoric acid activation of pecan shell. Considering the effect of activation temperature and impregnation ratio on the formation of acidic surface groups, they proposed three mechanisms contributing to the formation of acidic surface groups on phosphoric acid activated pecan shell based carbons: (1) hydrolysis of starting material at acidic condition leading mainly to the formation of carboxylic groups of different strengths; (2) reaction between the activation agent and starting material and the hydrolysis products of starting material, mainly resulting in the formation of phosphorous-containing groups, which are stable under high temperatures; and (3) the reaction between the starting material and air leading to the formation of acidic surface groups. A fundamental remark is that one can distinguish different types of acidic groups, *i.e.* those that are temperature sensitive and phosphorous-containing acidic groups that are temperature insensitive. Carboxylic groups appear at low temperature and then disappear when the activation temperature is $>350\ °C$. At temperatures $\leq300\ °C$, the formation of carboxyl groups was favoured,[90] while the formation of phosphorus-containing groups was favoured at temperatures $>300\ °C$. Low impregnation ratio and activation temperature are favourable to the formation of acidic surface groups. Similarly, when microwave heating was used for the preparation of cotton stalk activated carbons at high temperature, an important decrease of in the number of carboxylic groups from 0.32 to 0.06 was observed (Table 2.2).[91]

It can be concluded that the nature of the functional groups created during activation depends on the activation agent and the activation temperature. The textural characteristics control the access of the solute molecules to the material surface within the pore network. The following sections present methods for studying the texture of carbon materials.

2.3.2 Textural Characterisation of Activated Carbons

The textural characteristics of the activated carbon including surface area and pore size distribution are the main information required for studying the porous texture formation during activated carbon synthesis and for determining the potential applications of the material.[92,93] Porous materials contain micropores (<2 nm), mesopores ($2–50$ nm) and macropores (>50 nm),

Table 2.2 Acido-basic properties of AC prepared from different precursors, the ratio gives the impregnant mass/precursor mass.

Precursor	Activation	Ratio (w/w)	Total acidic groups (meq g^{-1})				Basic groups (meq g^{-1})	pH$_{PZC}$
			Carboxylic	Lactonic	Phenolic	Total		
Artichoke AC[90]	H$_3$PO$_4$	2/1	0.46	0.68	0.38	1.52	0.03	6.0
Parkinsonia aculeata Wood sawdust[83]	H$_3$PO$_4$ 300/400 °C	2/1				1.9	0.1	5.1
	KOH 300/400 °C					1.2	0.7	6.8
Jackfruit peel waste[89]	H$_3$PO$_4$	4/1	0.7807	0.7395	0.4350	1.9552	0.0808	1.9
Lotus stalk[91]	H$_3$PO$_4$ 405 °C	1/2	0.32	0.15	1.11	1.58		4.09
Lotus stalk stalk[91]	H$_3$PO$_4$ Microwave	1/2	0.06	0.21	0.10	0.37		4.35

according to the classification adopted by the International Union of Pure and Applied Chemistry (IUPAC).[94,95]

Gas adsorption techniques allow determination of the textural parameters. During gas adsorption experiments, the amount adsorbed, a, expressed per unit of mass of solid (adsorbent) is dependent on the gas (adsorbate) pressure, p, temperature, T, and the adsorbent properties and gas–solid interactions.[96] Thus for a given adsorbate sorbed on a particular surface one can write:

$$a = f(p, T) \qquad (2.11)$$

and at a constant temperature:

$$a = f(p / p_0) \qquad (2.12)$$

with p_0 being the saturation pressure of the adsorbate at the temperature of measurement.

Physisorption occurs when the enhancement of the amount of gas molecules on the solid surface is due to van der Waals forces, and chemisorption when the enhancement is caused by covalent or ionic bonding. Physisorption isotherms can be measured by two types of experiment: volumetry and gravimetry. The solid sample is first heated under vacuum to empty the surface from any adsorbed species. The temperature at the gas–liquid equilibrium point of the gas is used. The pressure increases then, on reaching equilibrium, the amount of gas adsorbed for each value of pressure is measured, either gravimetrically or volumetrically. According to the Brunauer–Deming–Deming–Teller; (BDDT) classification, typical gas adsorption isotherms of categories I, II, III and IV can be obtained. Choma and Jaroniec[97] have given a comprehensive review description of the gas adsorption isotherms.

The mechanism of the process in which molecules are adsorbed into the space within the micropore is simply called micropore filling. The mechanism of pore filling depends on the pore size. The mesopores are supposed to be filled by capillary condensation linked to the multilayer formation. In the mesoporous range, the desorption isotherm curve is not superimposed to the adsorption one. This hysteresis phenomenon is explained either by the difference between rate of capillary emptying and filling, or by the interconnection of the meso-pores blocked by filled micropores. The filling of macropores is supposed to proceed *via* multilayer formation. Methods for the determination of textural parameters, such as specific surface area, micropore and mesopore volume, are presented in the following sections.

2.3.2.1 Specific Surface Area Determination

The specific surface area expresses the total surface of the cavities of a porous material. The surface area is generally determined from the equilibrium ad-sorption isotherm of a gas measured in a range of relative pressures from 0.01 to 0.3. Currently two major methods are generally used to evaluate specific surface area from gas adsorption data: the Brunauer–Emmett–Teller (BET) method[98–101] and the I-point method,[102] which was proposed to overcome a limitation of the BET method. The recently proposed Freundlich specific surface area[103] is an enlargement of the specific surface area determined by the BET method.

2.3.2.1.1 BET Specific Surface Area. The derivation of the BET equa-tion[98–101] involved the following major assumptions: (1) the surface is flat; (2) all adsorption sites exhibit the same adsorption energy; (3) there are no lateral interactions between adsorbed molecules; (4) the adsorption energy for all mol-ecules except those in the first layer is equal to the liquefaction energy; and (5) an infinite number of layers can be formed. In the case of adsorption on acti-vated carbons, some of these assumptions are often not valid. In particular, surfaces are geometrically and energetically heterogeneous, lateral interactions between adsorbed molecules do occur, and the interactions of adsorbed mol-ecules with the surface vary with the distance from the surface.[104]

The BET specific surface area is an evaluation of the monolayer capacity of an adsorbent surface by fitting experimental gas adsorption data to the BET equation:

$$a = \frac{a_m c \frac{p}{p_0}}{\left(1 - \frac{p}{p_0}\right)\left(1 + (c-1)\frac{p}{p_0}\right)} \tag{2.13}$$

where a is the total amount adsorbed, a_m is the monolayer capacity (*i.e.* the number of adsorbed molecules in the monolayer on the surface of a material), p/p_0 the relative pressure, and c is a constant related to the heat of the first adsorption layer:

$$c = exp\left(\frac{q_1 - q_L}{RT}\right) \tag{2.14}$$

where $q_1 - q_L$ is the difference between the heat of adsorption in the first layer and the heat of condensation, T is the absolute temperature and R is the universal gas constant. Eqn (2.13) was derived for an infinite number of adsorbed layers. If the cross-sectional area σ for a single molecule in the monolayer formed on a given surface is known, the surface area S_{BET} can be evaluated by using the following formula:

$$S_{BET} = a_m \, \sigma \, N_A \tag{2.15}$$

where N_A is the Avogadro constant. If diazote is the adsorbate, the cross-sectional area σ is equal to $0.162 \times 10^{-18} \ \text{m}^2$.

To obtain the a_m value, the BET expression is transformed into a linear form (eqn (2.16)):

$$\frac{\frac{p}{p_0}}{a_m\left(1 - \frac{p}{p_0}\right)} = \frac{1}{c\,a_m} + \frac{c-1}{c\,a_m}\frac{p}{p_0} \tag{2.16}$$

and $\dfrac{\frac{p}{p_0}}{a_m\left(1-\frac{p}{p_0}\right)}$ is plotted *versus* p/p_0. The linear part of this plot concerns the values of p/p_0 smaller than 0.3. The slope (s) and the intercept (i) depend on the a_m value and the latest can be calculated (eqn (2.17)):

$$a_m = \frac{1}{s+i} \tag{2.17}$$

The surface area is calculated with a_m being the number of corresponding moles on the monolayer giving, $S_{BET} = a_m \, \sigma \, N_A/22400$.

2.3.2.1.2 Freundlich Surface Area. The Freundlich surface area, S_f, was proposed from the observation that a linear correlation exists between the Freundlich constant K_f and the BET surface area obtained from the nitrogen adsorption isotherms analysis of a great number of carbon samples.[103] The Freundlich surface area S_f was proved to have a larger range of applications than the BET surface area: it allows BET energetic parameter c not to be constant (varying from the value c_2 up to infinite values), that is to say it takes into account for the energetic heterogeneities. It has been mathematically derived from the BET and Freundlich isotherms expressions[103] as:

$$S_f = \frac{1}{22400} K_f \, \sigma \, N_A \, \frac{\alpha^\alpha}{(\alpha+1)^{\alpha+1}} \tag{2.18}$$

with

$$c_2 = 2\frac{1+\sqrt{1-\alpha^2}}{\alpha^2} \tag{2.19}$$

σ is the cross-sectional area for a single molecule in the monolayer, alpha is obtained from the following Freundlich isotherm equation fitting:

$$a = K_f\left(\frac{p}{p_0}\right)^\alpha \tag{2.20}$$

where a is the total amount adsorbed. The standard gas molar volume, $22\,400\text{ cm}^3$, appears in eqn (2.18) because, traditionally, the amount adsorbed is counted in STP cm^3 of the concerned gas.

2.3.2.1.3 I-point Method. Pomonis *et al.*[102] proposed the I-point method for estimating the specific surface area of solid material by rearranging the BET equation in a Scatchard type form[105] (eqn (2.21)). The c and a_m parameters could be separated and c was assumed not to be constant:

$$\frac{a(1-x)}{x} = c\,a_m - (c-1)a(1-x) \tag{2.21}$$

The I-point is the inversion point of the plot $\dfrac{a(1-x)}{x}$ versus $a(1-x)$ and the maximum of the plot $a(1-x)$ *versus* x.[106] Assuming c is very large compared with one, the specific surface area can be calculated easily from the determination of the a_m parameter from the following relationship:

$$a_m = [a(1-x)]_{I-point} \tag{2.22}$$

2.3.2.2 Micropore and Mesopore Analysis

2.3.2.2.1 Comparative Plots for Evaluation of Micropore Volume and Mesopore Surface Area. The main idea behind comparative plots is to use a reference sample similar to the porous studied material, but being non-porous. The adsorption mechanism on the reference sample will be a multilayer formation. Two types of comparative plots currently used are the t-plot[107–110] and the α_s plot,[96,99,111] which differ only in the reference solid adsorption isotherm presentation. The thickness of the surface film on the reference solid t is used to construct the t-plot.[112,113]

For the α_s plot, the micropores are supposed to be completely filled at relative pressure $p/p_0 = 0.4$. The quantity adsorbed on the studied sample is the ordinate of the plot, the abscissa is α_s. 'α_s' is the ratio of the quantity adsorbed on the reference solid to the quantity adsorbed on the material for $p/p_0 = 0.4$. The slope of the quasi-linear segment for α_s values <1 is proportional to the total surface area, as it concerns the three different parts of the surface (micropores, mesopores and flat surface). The slope of the linear segment at $\alpha > 1$ is proportional to the external surface area of mesopores, whereas its intercept determines the maximum amount adsorbed in micropores, which can be converted to the micropore volume:

$$\alpha_s = \frac{a_s(p/p_0)}{a_s(0.4)} \tag{2.23}$$

and

$$a(p/p_0) = a_{mi}^0 + \eta\,\alpha_s \tag{2.24}$$

$$V_{mi} = a_{mi}^0 F \tag{2.25}$$

$$S_{me} = \frac{\eta}{a_s(0.4)} S_{S\,BET} \tag{2.26}$$

$$S_t = \frac{\eta_1}{a_s(0.4)} S_{S\,BET} \tag{2.27}$$

where η_1 is the slope of the first line (at the ratio less than one), and F is the ratio of the molar volume of liquid nitrogen at 77 K and atmospheric pressure to the molar volume of gaseous nitrogen at normal temperature and pressure conditions (273 K and atmospheric pressure).

2.3.2.2.2 Micropore Volume from the Adsorption Potential Distribution.
The adsorption potential, A, is defined as the Gibbs free enthalpy:

$$A = -\Delta G = RT\ln(p_0/p) \tag{2.28}$$

It can be seen as another way to quantify the pressure at a constant temperature from an isotherm plot. The amount of molecules adsorbed 'a' versus A is the representation of an adsorption isotherm allowing the variations at low pressure values to be seen. When the pressure increases, the inflection points of the adsorption isotherm curve are associated with a change in the adsorption mechanism. $da/d(-A)$ versus $-A$ was successively attributed to the end of micropores filling and then to the end of the adsorbed monolayer formation.[114] $da/d(-A)$ was proposed to represent the adsorption potential distribution when the local adsorption isotherm followed the condensation approximation model.[114,115] This assumption had a wide generality because it was admitted that adsorption phenomenon could change gaseous species to liquid type species, whatever the pore dimension.

The development of the theory of micropore volume filling by Dubinin,[116] around the thermodynamic concept of adsorption potential, was followed by the proposition of the Dubinin–Radushkevich (D–R) (eqn (2.29)) and Dubinin–Asthakov (D–A) (eqn (2.30)) equations, giving a very good fit of adsorption isotherms for microporous materials.[99,116–118] D–A represents a more general expression:

$$a_{mi} = a_{mi}^0 \exp\left[-\left(\frac{A}{\beta E_0}\right)^2\right] \tag{2.29}$$

$$a_{mi} = a_{mi}^0 \exp\left[-\left(\frac{A}{\beta E_0}\right)^n\right] \tag{2.30}$$

B is temperature independent and related to the structural parameters associated with the micropore sizes and β is a similarity coefficient.[116] E_0 is the characteristic energy of adsorption. a_{mi} and a°_{mi} are respectively the amount adsorbed in micropore at relative pressure p/p_0 and the maximum one, and $A = -RT\ln(p/p_0)$ is the adsorption potential.

2.3.2.2.3 Pore Size Analysis

2.3.2.2.3.1 Pore Size Distribution from Kelvin Theory. To describe pore size distribution from adsorption data analysis, it is necessary to find a relationship between the measured pressure and the minimal pore size concerned in the adsorption. For macropores and large mesopores, mercury intrusion porosimetry provides a link between the meniscus radius and the applied pressure leading to the Washburn relation, which gives a relationship between minimal pore radius and pressure.[119] For the whole range of activated carbon, the Kelvin equation can play a role similar to the Washburn that is applicable to gas adsorption.[120–126] Indeed, Barrett *et al.*[120] developed an experimental data analysis for pore size distribution determination from nitrogen adsorption isotherms. Adsorption in mesopores was depicted as a multilayer formation followed by capillary condensation, while desorption as an evaporation followed by thinning of the multilayer. The Kelvin type equation giving a relationship between pore size and capillary condensation and evaporation pressures, and as well between the pressure and the multilayer thickness, was established. This equation concerned the whole domain of mesopores and a great part of the larger micropores,[127] allowing a large applicability to activated carbons.

2.3.2.2.3.2 The Horvath Kawazoe Method. This method consisted of searching for a pore volume distribution function $J(x)$, x being the pore width for solving the problem of the lack of method dedicated to micropore size distribution. The amount adsorbed $a(A)$ is given by:

$$a(A) = a_t \; x_2 \; \theta_x(A, x) \; J(x) \; dx \qquad (2.31)$$

where a_t is the maximum amount adsorbed, $\theta_x(A, x)$ is the amount adsorbed for the pores of width x, with x_1 the smallest micropore width and x_2 the largest micropore width. A is the adsorption potential.

While the most popular solutions of eqn (2.32) concern capillary condensation and multilayer adsorption, Horvath and Kawazoe[128] proposed relating the average adsorption potential in a micropore of a given geometry with the pore of size x. Because of A being a function of x, the method can be considered as an extension of the condensation approximation method to the region of fine pores.[129]

2.3.2.2.3.3 Density Functional Theory. Density functional theory (DFT) is used to model the local interactions of an adsorbed fluid particle with the solid material porous structure and far, with the bulk fluid.[130] The Non Local Density Functional Theory was proven to give better results[131] and consisted of adding environmental interactions to the general ones. The general principle was to minimise the interactions to find the most probable configuration. These molecular modelling techniques were a major advance in the fitting of adsorption experimental data to the model.

2.3.2.3   Textural Characteristics of Biomass-based Activated Carbons

Many studies correlate the textural properties of activated carbons with the initial composition of the precursor. A widely accepted theory about the evolution of pore structure of chemically activated carbons prepared from lignocellulosic materials is that the chemicals left in the carbonised samples after being washed out lead to the formation of porosity, especially microporosity. For example, it was reported that during activation by phosphoric acid, the distribution of the chemicals in the precursor prior to the carbonisation process governed the pore size distribution of the final carbon products.[89,132] Textural parameters of different activated carbon samples evaluated from N_2 adsorption/desorption isotherms are presented in Table 2.3.

The texture of the raw material obviously affects the surface area of porous carbons, although the bioresources used contained similar components. In general, biomass materials with loose texture favour the formation of high surface area microporous carbon products, whereas only activated carbon with low surface area obtained from raw materials with a compact texture. Zhang *et al.*[133] showed it was possible to prepare activated carbon with a BET surface area >3000 m^2 g^{-1} depending on the raw material and the preparation conditions. In addition, the initial chemical content of the raw biomass was shown to play a role on the textural characteristics. For example, rice husk usually exhibited a low surface area due to high ash content. Daifullah *et al.*[134] proposed the use of the base-leaching or acid-washing process to enhance the adsorption capacities of activated carbons from rice husk. During the activated carbon preparation process, an increase in temperature generally improved textural properties until a maximal value of temperature for both physical and chemical activation However, reaching a higher final activation temperature required a longer heating time, which allowed the gas to enter deeper layers of the material, thus developing the internal porous structure and leading to higher specific surface area and porous volume.[135]

Nevertheless, the greatest effect on the surface area is the sample activation method. When a conventional heating procedure is used, activated carbon samples prepared by chemical activation generally have a much more developed surface area and greater pore volume than samples activated by the physical method.[135–137] This is attributed to the specific chemical reactions occurring inside the material volume during chemical activation. In addition, chemical activation with acidic agents led to activated carbons with a higher BET surface area and total pore volume than those prepared using a basic impregnant.[135] The ratio of microporous volume to the mesoporous one is related to the nature of the precursor and to its intrinsic porosity. The heating process may also strongly influence the textural properties of the activated carbon. Microwave heating gives higher yields and performance than conventional heating, particularly for physically activated samples (Table 2.3) due to the vibrational effect of microwaves creating internal heating that shorten the activation process duration. Xin-Hui *et al.*[50] compared the preparation of Jathropha hull activated carbon using

Table 2.3 Textural characteristics of activated carbons prepared from biomass.

Precursor	Activation	Specific surface area/ $m^2\ g^{-1}$	Total pore volume/ $cm^3\ g^{-1}$	Microporous volume/ $cm^3 g^{-1}$	Mesoporous volume/cm^3 g^{-1}
Artichoke[90]	H_3PO_4 at 110/300 °C	2106	2.47	0.61	1.80
Soybean oil cake[137]	Chemical (KOH) at 800 °C	618.5	0.291	0.143	0.148
	Chemical (K_2CO_3) at 800 °C	1352.9	0.680	0.400	0.280
Rice husk W-75[139]	Chemical (H_3PO_4)	376	0.431	0.0718	0.2921
Rice husk[53]	Chemical (H_3PO_4) at 400 °C	508	0.278	0.193	0.048
	Chemical (H_3PO_4) at up to 400 °C	1278	0.722	0.366	0.308
Parkinsonia aculeate wood sawdust[84]	Chemical (H_3PO_4)	968	0.70	0.18 (by DFT)	
	Chemical (KOH)	768	0.37	0.27 (by DFT)	
Vetiver roots[53]	Physical (H_2O)	558	0.40	0.24	0.16
	Chemical (H_3PO_4)	959	0.84	0.36	0.48
Commercial[53]		1226	0.50	0.46	0.04
Coconut shell[138]	Chemical ($ZnCl_2$)	1510	0.752	0.643	0.109
		2191	1.913	0.549	1.364
Coconut shell[138]	Chemical (KOH)	2309	1.584	0.903	0.681
Palm stone[138]	Chemical ($ZnCl_2$)	1291	0.785	0.463	0.322
		1267	1.263	0.075	1.188
Prunus domestica[141]	Chemical	217	1.09	1.057	
	Physical	470	0.23	0.221	
Pecan shell[143]		1017	0.537	0.284	
Walnut shell[140]	Chemical (KOH)	2263	1.10	1.07	
	Physical (CO_2)	697	0.37	0.34	
Jackfruit peel waste[141]	Chemical (H_3PO_4)	1260	0.733	0.471	
Corn stalks[142]	KOH (4–24 mol L^{-1})	3200	1.38		
Rice straws[142]	KOH (4–24 mol L^{-1})	3315			
Pine needles[142]	KOH (4–24 mol L^{-1})	3545	1.70		
Pine cone hulls[142]	KOH (4–24 mol L^{-1})	397			
Lotus stalk[91]	H_3PO_4, conventional heating at 450 °C	1220	1.191	0.2858	
Lotus stalk[91]	H_3PO_4, microwave heating at 450 °C	1431	1.337	0.3064	

Table 2.4 Textural parameters of activated carbons prepared from coconut shells[152–156] by microwave and conventional heating.

Heating	Activation	Duration/ min	Specific surface area/ $m^2 g^{-1}$	Total pore volume/ $cm^3 g^{-1}$	Microporous volume/ $cm^3 g^{-1}$	Yield/ %
Conventional[144]	Physical ($N_2 + H_2O$)	120		0.39	0.35	76.3
Conventional[145]	Physical	60	378	0.26	0.12	
Conventional[146]	Chemical and physical (KOH, CO_2)		1026	0.5768		
Conventional[147]	Physical	120	524	0.226	0.210	
Conventional[148]	Chemical (KOH)	120	2451	1.201		23.6
Conventional[149]	Physical (steam)	120	1926	1.260	0.931	39.1
Microwave[150]	Physical (steam)	75	2079	1.212	0.9735	42.2
Microwave[150]	Physical (CO_2)	210	2288	1.299	1.012	37.5
Microwave[150]	Physical ($CO_2 +$ steam)	75	2194	1.293	1.010	39.2

conventional and microwave heating. The pore volume and the surface area were found to double using microwave heating with steam, but to be of the same order of magnitude using CO_2 for both heating process. Although the carbon textural properties were similar for CO_2 activation using both heating process, the activation temperature, the activation time and CO_2 flow rate were significantly lower than those of conventional heating rendering the process more economical than the conventional heating (Table 2.4).

This observation was confirmed for chemical activation by Yagmur[51] who used phosphoric acid as the activating agent for the preparation of activated carbons from different precursors such as waste tea, almond shells, tomato stems and leaves. The soaked samples were heated in a microwave oven and samples with surface area reaching 1432 $m^2 g^{-1}$ were obtained using a microwave power of 350 W, a radiation time of 20 min, and a phosphoric acid to precursor ratio of 1 : 1. Conventional heating of the same precursor at 600 °C for 30 and 60 min gave surface areas of 1215 and 1416 $m^2 g^{-1}$, respectively. These results show that much lower processing time was necessary for producing activated carbon samples with similar textural properties by microwave radiation heating compared with conventional heating.

2.4 Adsorption of Pollutants by Activated Carbons from Various Biomass

2.4.1 Liquid Phase Adsorption Studies

Adsorption of chemical substances from solution is essentially an exchange process. The adsorption of a solute by activated carbons in water depends on the adsorbent textural properties and the adsorption mechanism may be governed by either electrostatic or/and non-electrostatic interactions between the solute and the carbon surface. Electrostatic interactions appear with electrolytes, essentially when they are ionised.[151] Non-electrostatic interactions are essentially due to dispersion and hydrophobic interactions. The surface chemistry of the carbon has a great influence on both types of interactions.

The surface chemistry of an activated carbon essentially depends on its heteroatom content which determines the charge of the surface, its hydrophobicity, and the electronic density of the graphene layers. It is well known that surface oxygenated groups that include carboxyls, phenols, lactones, carbonyls and as well quinones play a key role in the adsorption process.[151] The adsorption behaviour of solutes onto activated carbon is generally investigated using adsorption kinetics and isotherms studies. In order to predict the mechanism involved during the adsorption process, kinetic models such as the pseudo first-order or Lagergren,[152] pseudo second-order,[153] intraparticle[154] and Brouers–Weron–Sotolongo (BWS)[155–157] models are generally used to fit the experimental data. Adsorption equilibrium data are usually modelled by isotherms that relate the relative concentrations of solute adsorbed to the solid (q_e) and in solution (C_e) at equilibrium. To obtain a realistic estimation of the maximum capacity of adsorption, it is important to establish the most appropriate correlation with the equilibrium curves. Thus, the isotherms data can be analysed using five of the most commonly used equilibrium models: Langmuir,[158] Freundlich,[159] Redlich–Peterson,[160] Temkin[161] and isotherm models, and the Brouers–Sotolongo model.[162,163]

2.4.1.1 Adsorption Kinetic Theories

It has been shown by Brouers and Sotolono-Costa[155] and Figaro *et al.*[157] that a fractal kinetic equation associated with a complex adsorption mechanism, which has been extensively applied to biophysical problems, can also be a useful theoretical tool for the study of adsorption processes and also to describe the dynamics of adsorption phenomena on activated carbon:

$$q_{n,\alpha}(t) = q_e[1 - (1 + (n-1)(t/\tau_{q,\alpha})^\alpha)^{-1/(n-1)}] \tag{2.32}$$

This equation was developed from the general fractional differential equation:

$$-\frac{dq}{dt^\alpha} = K_{\alpha,n}q^n \tag{2.33}$$

where $q(t)$ is the mass of solute adsorbed per gram of activated carbon, q_e, the maximum adsorbed quantity, n, the reaction order and α a fractional time index. When n and α are different from 1, it is no longer possible to define a time-independent rate constant and the relevant quantity characterising the time evolution of the process is the characteristic time $\tau_{q,\alpha}$. One can define a 'half-reaction time' $\tau_{1/2}$, which is the time necessary to adsorb half of the equilibrium quantity by solving the equation:

$$(1 + (n-1)(t/\tau_{q,\alpha})^\alpha)^{-1/(n-1)} = 1/2 \tag{2.34}$$

Different cases can occur:

a. If $n=1, \alpha=1$, and, $-\dfrac{dq}{dt} = K_1 q$, the solution of this equation is

$$q_t = q_e \exp(-k_1 t) \tag{2.35}$$

which is a first-order kinetics[152] with k_1, the first order rate constant.

b. In the case where $n \to 1$ and $n \to 1$ and $\alpha \neq 1$, it can be written as a Weibull distribution:

$$q_\alpha(t) = q_e[1 - \exp(-t/\tau_\alpha)^\alpha] \tag{2.36}$$

with $\tau_{1/2} = (\ln(2))^{1/\alpha}\tau_\alpha$.

c. If $\alpha=1, n=2$, we have, $-\dfrac{dq}{dt} = K_2 q^2$ leading to:

$$\frac{1}{q_e - q_t} = K_2 t + \frac{1}{q_e} \tag{2.37}$$

with K_2, the second-order constant. This is second-order kinetics, corresponding to the very popular pseudo second-order kinetic.[153] This equation can be rearranged into the very convenient form:

$$\frac{t}{q_t} = \frac{1}{K_2 q_e^2} + \frac{t}{q_e} \tag{2.38}$$

If $\alpha \neq 1, n=2$, eqn (2.38) becomes:

$$q(t) = q_e\left[1 - \frac{1}{1 + (t/\tau_{n,\alpha})^\alpha}\right] \tag{2.39}$$

and $\tau_{1/2} = \tau_{n,\alpha}$.

The intraparticle model has the following expression:

$$q_t = k_{id} t^{1/2} + C \tag{2.40}$$

where k_{id} is the intraparticle rate constant and C a constant. The plot of q_t *versus* $t^{1/2}$ may present a multilinearity indicating the following three steps taking place:

(i) The first sharper portion is attributed to the diffusion of adsorbate through the solution to outer toward the interface space, so-called external diffusion.

(ii) The second portion describes the gradual adsorption stage, corresponding to diffusion of sorbate molecules inside the pore of the adsorbent, where the intraparticle diffusion is rate limiting.

(iii) The third portion is attributed to the final equilibrium stage for which the intraparticle diffusion starts to slow down due to the low adsorbate concentration left in the solution.

2.4.1.2 Adsorption Isotherm Theories

To evaluate the liquid phase adsorption ability of activated carbons, adsorption isotherms are performed using phenol or methylene blue as model molecules. Because solute molecules may be rejected by the solution or adsorbed by the solid, adsorption of pollutants is a complex interplay between electrostatic and non-electrostatic interactions between the solutes and the adsorbents. Electrostatic interactions appear with electrolytes, essentially when they are ionised.[151] Non-electrostatic interactions are essentially due to dispersion and hydrophobic interactions. The Langmuir[158] and Freundlich[158] equations are the most frequently used models to describe the experimental data of liquid phase adsorption isotherms. In this work, however, three other models are used: Redlich–Peterson,[160] Temkin[161] and the newly established model, the Brouers–Sotolongo isotherm.[162]

Langmuir's isotherm model suggests that uptake occurs on homogeneous surface by monolayer sorption without interaction between sorbed molecules.[158] Its well-known expression is given by:

$$Qe = (Q^o \cdot K_L \cdot Ce)/(1 + K_L \cdot Ce) \qquad (2.41)$$

where Qe (mg g^{-1}) and Ce (mg L^{-1}) are the amount of adsorbed solute molecules per unit weight of AC and the residual concentration in solution at equilibrium, respectively. Q_0 is the maximum amount of the solute per unit weight of AC to form a complete monolayer on the surface bound and K_L (L mg^{-1}) is a constant related to the affinity of the binding sites.

The empirical Freundlich equation[159] based on sorption onto a heterogeneous surface is given by:

$$Q_e = K_F \times C_e^{1/n} \qquad (2.42)$$

where K_F and n are the Freundlich constants characteristic of the system. K_F (mg$^{1-1/n}$ L$^{1/n}$g^{-1}) and n are indicators of adsorption capacity and adsorption intensity, respectively.

The Redlich–Peterson model[160] incorporates the features of the Langmuir and Freundlich isotherms into a single equation and presents a general isotherm equation as follows:

$$Qe = (A_{RP} \cdot Ce)/(1 + K_{RP} \cdot Ce^{\beta}) \qquad (2.43)$$

where A_{RP} [(L mg^{-1})$^{\beta}$] and K_{RP} (L g^{-1}) are the Redlich–Peterson isotherm constants. The exponent, β, generally lying between 0 and 1, has two limiting behaviours: Langmuir form for $\beta = 1$ and Henry's law form for $\beta = 0$.

The Brouers–Sotolongo isotherm (BSI)[162] is given by a deformed exponential (Weibull) function. Its expression is given by:

$$Qe = Q_{max}(1 - \exp(-K_w \cdot Ce^{\alpha})) \qquad (2.44)$$

The parameters Q_{max}, K_w and α, which can be determined by a non-linear curve fitting procedure, have a clear physical meaning: Q_{max} is the saturation value, $K_w = K_F/Q_{max}$ where K_F is the Freundlich constant for a given temperature, and the exponent α is a measure of the width of the sorption energy distribution and therefore of the energetic heterogeneity of the sorbent surface.

Considering the effects of some indirect adsorbate–adsorbate interactions on adsorption isotherms and a linear decrease of the heat of adsorption of the molecules in the adsorbed layer with the coverage, Temkin and Pyzhev[161] developed the following Temkin isotherm equation:

$$Q_e = \frac{RT}{b}\ln(K_T C_e) \tag{2.45}$$

where BT is the Temkin constant related to the heat of the adsorption, KT (L g^{-1}) is the equilibrium binding constant corresponding to the maximum binding energy, R (8.314 J mol^{-1} K^{-1}) is the universal gas constant and T is the absolute solution temperature.

2.4.2 Adsorption of Pollutants by Biomass-based Activated Carbons

Because of their abundance, vegetal wastes from agriculture and wood industry are considered products with a low economic value that may cause environmental problems. Many activated carbons have been prepared from different biological waste materials such as conventional wastes from agro-industry, agriculture or the wood industry; non-conventional wastes from municipal residues have also been used.

The preparation of activated carbons from wastes is an example of how high value products are obtained from low cost materials, simultaneously bringing solutions to the problem of wastes and water pollution. Most studies focus on the use of rigid biological waste materials such as wood, shells and/or stones of fruits like nuts, peanuts, olives, dates, almonds, apricots and cherries. However, the by-products of agro-industries such as cereals, rice, coffee, soybean, maize and corn as well as olive cakes, sugar cane, coir pith, palm oil shell (from palm oil processing mills) and various seed wastes are already used as are sewage sludge by-products. Some examples of activated carbons prepared from various waste materials and their application for the removal from water of pollutants such as dyes, chlorinated compounds and metals are presented below.

2.4.2.1 Treatment of Dye Contaminated Water

Dyes containing wastewaters are produced in many industries including paint manufacture, dyeing, textiles and paper. According to Chakrabarti et al.[164] there are nearly 40 000 dyes and pigments with over 7000 different chemical structures. Most of them are not biodegradable.[165] About 10 000 different commercial dyes and pigments exist and are produced worldwide each year.[166] Some 20–27% of the dye is lost in the effluent during the dyeing process, manufacturing and processing operations, and approximately 20% of this lost dye enters the industrial wastewaters.[167] Dyes containing wastewaters need to be treated properly because these contaminants may have carcinogenic, teratogenic and mutagenic effects on both humans and aquatic life.[168] Therefore, the treatment of these dye-contaminated effluents with cost-effective and efficient technologies such as adsorption on activated carbons is of high scientific and public interest.[169]

2.4.2.1.1 Treatment of Methylene Blue Contaminated Waters. Baçaoui *et al.*[31,170] determined the optimal conditions for producing activated carbons from olive waste cakes using a procedure in which carbonisation and steam activation were performed in one step. When the preparation conditions were optimised, the sample obtained had a surface area around $1100 m^2 g^{-1}$ and a pore volume of $0.4 cm^3 g^{-1}$. At an intermediate activation temperature, *e.g.* 800 °C and an activation time of 25–55 min, the capacity of the activated carbons to adsorb methylene blue was constant ($240 mg g^{-1}$) and increased to $345 mg g^{-1}$ at 80 min of activation.

Oh and Park[171] prepared activated carbons from rice straw using an original procedure. The raw material was first carbonised at temperatures from 700 to 1000 °C, then impregnated with KOH and activated by heat treatment. The porous carbon with a surface area of $2410 m^2 g^{-1}$ had a total pore volume of $1.4 cm^3 g^{-1}$. The resulting carbon had high surface area (up to $2410 m^2 g^{-1}$) and high adsorption capacity for methylene blue of $800 mg g^{-1}$.

Palm kernel shell activated carbons (PKSAC) were produced by steam activation[172] at a temperature of approximately 1000 °C. The BET surface area was between 1146 and $1599 m^2 g^{-1}$. All activated carbons consisted of micropores with almost similar pore diameter, which fell within the narrow range of 5.5 to 6.4 Å. PKSAC were predominantly microporous with micropores accounting for 68–72% of total porosity. The sorption kinetics and isotherms of methylene blue onto PKSAC were studied. The adsorption isotherm data fitted well to the Langmuir isotherm. The kinetic data were better fitted by the pseudo second-order than the pseudo first-order and intraparticle diffusion equations. The adsorption data of methylene blue on the activated carbon sample was well fitted with the Langmuir adsorption equation ($R^2 > 0.994$), but not well with Freundlich equation. The Langmuir monolayer capacity values were between 172 and $263 mg g^{-1}$.

Altenor *et al.*[53] utilised vetiver roots for the preparation of activated carbons by chemical activation with different impregnation ratios of phosphoric acid (g $H_3PO_4 g^{-1}$ precursor: 0.5:1, 1:1 and 1.5:1). Textural characterisation showed mixed microporous and mesoporous structures with high surface areas ($>1000 m^2 g^{-1}$) and high pore volumes (up to $1.19 cm^3 g^{-1}$). Activation of vetiver roots by phosphoric acid leads to acidic carbons with various oxygen containing species on their surface and a higher percentage of mesopores than the sample obtained by steam activation. Among different kinetic equations used, the BWS equation was the best kinetic model to describe adsorption of methylene blue on the vetiver roots activated carbons, indicating a complex mechanism of adsorption. The Brouers–Sotolongo and Redlich–Peterson equations were the most suitable isotherm models for describing methylene blue adsorption equilibrium on vetiver roots activated carbons. Acidic groups such as carboxylic groups were shown to favour methylene blue adsorption due to electrostatic interactions between the methylene blue molecules and the deprotonated carboxylic groups. A high adsorption capacity value of methylene blue $Q_L = 423$ mg g^{-1} was found for the samples prepared with the higher weight H_3PO_4 to precursor ratio of 1.5.

Powder activated carbons were prepared from peanut hulls[173] by chemical activation with H_3PO_4, $ZnCl_2$ and KOH, respectively, and physical activation with water. Steam pyrolysis gave a low surface area activated carbon. Contrary to that observed for other precursors, the carbon prepared by impregnation with KOH had a low surface area and a developed mesoporosity. The H_3PO_4 treatment created abundant microporosity with an extended surface area up to 1177 m^2 g^{-1} and a pore volume of 0.597 cm^3 g^{-1}. The activated carbon prepared by phosphoric acid activation could highly adsorb methylene blue, as a monolayer adsorption capacity up to of 388 mg g^{-1} was obtained.

Attia *et al.*[174] used pistachio shells to obtain activated carbons. The raw biomass was soaked in a phosphoric acid solution for 1 and 3 days respectively, and then pyrolysed at 500 °C. The higher surface area was obtained for the activated carbon prepared (up to 1436 m^2 g^{-1}) by impregnation with H_3PO_4 for 3 days; its total pore volume was 0.542 m^2 g^{-1}. This sample could adsorb methylene blue with an adsorption capacity of 129 mg g^{-1}.

Artichoke leaves were selected as a suitable precursor for activated carbon preparation[175] due to their abundance in Algeria and their disposal in the environment without treatment. Artichoke leaves were impregnated for 2 h with phosphoric acid solutions, with impregnation ratio (*Xp*) of 100, 200 and 300 wt%, giving samples AC1/1, AC2/1 and AC3/1, respectively. The activation was performed at 500 °C. The BET surface areas were between 793 and 2160 m^2 g^{-1}. Total pore volume was between 1.1 and 2.46 cm^3 g^{-1}. The sample AC2/1, exhibiting the higher surface area and pore volume, showed the highest adsorption uptake of methylene blue (~ 780 mg $\cdot$ g^{-1}) at pH $= 9$ compared with AC3/1 (~ 639 mg $\cdot$ g^{-1}) and AC1/1 (~ 587 mg $\cdot$ g^{-1}). It was shown that phenolic, carboxylic, phosphorous acidic groups and carbonyl group contents decreased as the impregnation ratio increased from 100% to 300%. Methylene blue adsorption was proposed to occur into the carbon supermicropores and to be controlled by dispersive interactions at pH $= 3$ and electrostatic interactions at pH $= 9$. The nature of the adsorption phenomena was dependent on the oxygenated groups; oxygen content at the activated carbon surface decreased with increasing impregnation ratio (from 100% to 300%).

In 2011 Foo and Hameed[176] prepared activated carbons by microwave-induced KOH activation of date stones and tested their potential for methylene blue adsorption. The precursor was first impregnated with KOH at a char : KOH ratio of 1 : 1.75 (wt%). The microwave input power was 600 W and irradiation time 8 min. The BET surface area and total pore volume of the activated carbon obtained were 856 m^2 g^{-1} and 0.47 cm^3 g^{-1}, respectively. The methylene blue adsorption isotherm was fitted by the Freundlich, Langmuir and Temkin isotherm models. The best fit was obtained with the Langmuir isotherm model, suggesting a homogeneous adsorption process and a mono-layer coverage of dye molecules at the outer surface of activated carbon sample. The monolayer adsorption capacity of the activated carbon for methylene blue was 316.11 mg g^{-1}. When the pH increased from 2 to 12, an enhancement of the adsorption capacity from 237.79 to 275.89 mg g^{-1} was observed. This was explained by methylene blue protonation in the acidic medium; it was also associated with the presence of excess H^{+} ions at low pH competing with dye cations for the adsorption sites. These authors also used pomelo skin[177] from a fruit juice processing plant for activated carbon preparation using microwave heating. The precursor was mixed with sodium hydroxide solution with a NaOH : char ratio of 1 : 1.25 (wt%). The input power was 800 W and ir-radiation time 5 min. The BET surface area and total pore volume were 1335 m^2 g^{-1}, 0.77 cm^3 g^{-1}, respectively. The adsorption behaviour was well described by the Langmuir isotherm model, and the monolayer adsorption capacity for methylene blue and acid blue were 501.10 mg g^{-1} and 444.45 mg g^{-1}, respectively. In another work,[178] the same authors used pistachio nut shell, an abundant biomass available from the pistachio nut processing industry, as a precursor for activated carbon preparation (PSAC) *via* microwave-assisted KOH activation. The precursor was impregnated with an impregnation ratio (char: KOH) of 1 : 1.75 (wt%). The activation was performed at the microwave input power of 600 W and irradiation time of 7 min. The sample obtained had a BET surface area and total pore volume of 700.53 m^2 g^{-1} and 0.375 cm^3 g^{-1}, respectively. Methylene blue equilibrium adsorption data were best fitted by the Langmuir isotherm model, showing a monolayer adsorption capacity of 296.57 mg g^{-1}.

Activated carbons were prepared by chemical activation of cotton stalk with KOH and K$_2$CO$_3$ under microwave radiation.[179] Cotton stalk was mixed with KOH or K$_2$CO$_3$ solution at various concentrations for 24 h at ambient tem-perature. After mixing, the slurries were heated in a microwave oven. Activated carbons obtained from chemical activation with K$_2$CO$_3$ and KOH were de-noted as KCAC and KAC, respectively. KCAC and KAC had a surface area of 621 and 729 m^2 g^{-1}, respectively. The total pore volume was 0.38 cm^3 g^{-1} for both samples, while the micropore volume was 0.11 and 0.26 cm^3 g^{-1}, re-spectively, corresponding to 29% and 68% of the total pore volume. Chemical activation could develop both microporosity and mesoporosity; KOH gener-ated a greater micropore volume than K$_2$CO$_3$. The equilibrium data of meth-ylene blue adsorption were well fitted to the Langmuir isotherm for both activated carbons, giving monolayer adsorption capacities of 285 and 294 mg/g for KCAC and KAC, respectively.

Unur[180] pre-carbonised hazelnut shells (HS), an abundant agricultural biomass under hydrothermal conditions, to a more stable hydrochar (HH). Successive calcination of HH under inert atmosphere under a flow of argon at 600 °C for 2 h yielded to a sample named HH-T, with a hydrophobic nanoarchitectured carbon material and a specific surface area of 250 m^2 g^{-1} and pore volume of 0.17 cm^3/g (HH-T). This improvement in textural prop-erties can be attributed to the formation of micropores by the dissociated leaving groups during pyrolysis of HH. HH was also mixed with KOH in a weight ratio of 1 : 4, heated to 600 °C under a flow of nitrogen at a rate of 10 °C min^{-1}, and kept at that temperature for 2 h. A highly porous hydrophilic carbon material designed by HH-KT was obtained, with a specific surface area and pore volume of 1700 m^2 g^{-1} and 0.79 cm^3 g^{-1}, respectively. HH-KT had a honeycomb-like structure dominated by micropores and mesopores. The monolayer adsorption capacity was as high as 524 mg g^{-1}. The adsorption isotherms for the three samples fitted the Langmuir model better, suggesting a monolayer limited adsorption. The adsorption mechanism was shown to be dominated by electrostatic attractions arising from the presence of oxygen functionalities on the adsorbent surface.

Brown algae, *Turbinaria turbinata* (turb-raw), was used as an activated carbon precursor by Altenor *et al.*[40] Pyrolysed carbon was prepared by carbonisation at 800 °C for 1 h; physically activated carbon by steam at 800 °C gave turb-H$_2$O and turb-P1 was obtained by impregnation by phosphoric acid with a ratio of 1 : 1 by weight followed by a carbonisation at a temperature of 600 °C for 1 h. The total pore volume increased drastically from 0.001 cm^3 g^{-1} for the raw algal precursor to 0.242 and 1.316 cm^3 g^{-1} for turb-H$_2$O and turb-P1, respectively. The BET surface area was 328 and 1307 m^2 g^{-1} for turb-H$_2$O and turb-P1, respectively. The sorption results revealed that the sorption capacities were average for untreated turbinaria biomass (63 mg g^{-1}), and the produced pyrolysed sample and activated carbons showed a notable increase in methylene blue removal efficiency. The best performing activated carbon was 'turb-H$_2$O' with a sorption capacity of 411 mg g^{-1} (Langmuir's $Q°$), while values of 345 and 163 mg g^{-1} were found for turb-P1 and turb-pyr, respectively.

2.4.3.1.2 Treatment of Water Contaminated with Other Dyes. Mahogany sawdust was used by Malik[181] to produce activated carbons by steam activation. The raw material was first pyrolysed at 500 °C for 1 h and then activated by steam for 1 h at 800 °C. This adsorbent had a BET surface area of 516 m^2 g^{-1}; the number of acidic and basic groups were 0.02 and 2.34 meq g^{-1}, respectively. The kinetics of direct dye adsorption on sawdust carbon followed the pseudo second-order model. The equilibrium data fitted well the Langmuir adsorption model, showing monolayer coverage of dye molecules at the outer surface of sawdust carbon. The sample showed Langmuir adsorption capacities of 518 and 327 mg g^{-1} for Direct Blue 2B and Direct Green B, respectively.

Activated carbon was prepared from 'waste' bamboo culms (BAC) and tested for the removal of Disperse Red 167 (DR167), an azo disperse dye.[182] The bamboo culms waste was impregnated in a H$_3$PO$_4$ solution at 253 K for 2 days and carbonised at 500 °C for 2 h. The S_{BET} value was 747 m^2 g^{-1} and the total pore volume of BAC was 0.451 cm^3 g^{-1}, with total acidic groups 0.625 mmol g^{-1} and total basic groups 0.267 mmol g^{-1}. Low initial pH or concentration of dye solution favoured the adsorption process; adsorption of Disperse Red 167 azo dye on the activated carbon sample was endothermic. Kinetic data showed good correlation with both the intraparticle diffusion and pseudo first-order models. The equilibrium data fitted both the Temkin and Freundlich isotherms well. A low Langmuir monolayer adsorption capacity of 1.89 mg g^{-1} was obtained.

Several studies have reported the preparation of activated carbons from sugar cane bagasse.[183–188] In 2004, Valix *et al.*[183] prepared activated carbon by carbonisation of bagasse at 160 °C followed by a gasification with carbon dioxide at 900 °C. The activated carbon samples obtained had a surface area between 403 and 1433 m^2 g^{-1} and pore volumes between 0.013 and 0.91 cm^3 g^{-1}. The acid blue dye adsorption capacity, expressed as the Langmuir constant, was between 59.9 and 384 mg g^{-1}. In another study,[188] activated carbons were prepared from bagasse by physical activation with CO_2 and also using three different chemical activating agents, $ZnCl_2$, $MgCl_2$ and $CaCl_2$. High surface area activated carbons (1353 m^2 g^{-1}) could be obtained using $ZnCl_2$, whereas low surface area values of 115 and 202 m^2 g^{-1} were obtained for the samples prepared with $CaCl_2$ and $MgCl_2$, respectively. The adsorption capacities of acid blue dye on these carbons were in the range of 15 to 358 mg g^{-1}. Positive charges on the carbon surface were shown to increase anionic AB80 dye adsorption, indicating an electrosorption mechanism.

Namane *et al.*[189] prepared activated carbons by impregnation of coffee grounds with an equimolar mixture of $ZnCl_2$ and H_3PO_4 for 24 h and then carbonisation at 600 °C for 45 min. A sample with a specific surface area of 650 m^2 g^{-1} and a pore volume of 0.95 cm^3 g^{-1} was obtained. A 90% removal of acid blue and basic blue was found with monolayer adsorption capacity values of 3.57 and 10 mg g^{-1}, respectively. The previously described pistachio shells activated carbons[175] could adsorb rhodamine with a monolayer adsorption capacity of 95 mg g^{-1}.

Palm flower activated carbon[190] was prepared by treatment of the precursor with a concentrated H_2SO_4 solution at a 1 : 1 ratio, for 48 h followed by carbonisation. The dried palm flower activated carbon (LCBPF) had a surface area of 9.57 m^2 g^{-1}. Total pore volume was 0.073 cm^3 g^{-1} and the pore width was 246.56 nm. The surface acidity of LCBPF was determined by the Boehm titration method and the total surface acidity was calculated as 0.593 meq g^{-1} including phenolic groups (0.487 meq g^{-1}) and traces of lactonic (0.041 meq g^{-1}) and carboxylic (0.064 meq g^{-1}) groups. The zero point charge was at pH 2.5 and the maximum adsorption occurred at pH 2.3. Amido Black 10B adsorption data were analysed by different adsorption isotherm models such as Langmuir, Freundlich and Temkin equations, and it was found that the Freundlich isotherm model best fitted the adsorption data. The Langmuir monolayer adsorption capacity was between 2.42 and 3.68 mg g^{-1} depending on the carbon particle size.

Grapevine rhytidome, the outer layer of bark on the trunk, is an abundant and low-cost by-product of the grapevine industry.[191] Granular activated carbons were prepared by impregnation of the precursor with different phosphoric acid concentrations (45, 70 and 85 wt%) with acid to precursor weight ratios from 2 : 1 to 30 : 1 for 24 or 48 h at room temperature. The impregnated samples were heated in a microwave oven at different powers (120, 400 and

700 W), and radiation times (1–10 min). The optimal conditions were: acid concentration 85%; acid/precursor weight ratio 5 : 1; impregnation time 24 h; microwave power 400 W; microwave radiation time 2 min; oven temperature 110 °C; and heating duration of 7 h. The activated carbon sample obtained had an amorphous feature, its BET surface area, total pore volume and yield were 1607 m^2 g^{-1}, 1.42 cm^3 g^{-1} and 25%, respectively. Adsorption isotherm data of methyl violet (MV) on this sample were best described by the Redlich–Peterson model, suggesting that this adsorbent provided an heterogeneous surface. The percentage of MV removal increased with increase of pH from 3 to about 7. Thus, the number of positively charged sites on the adsorbents may decrease, favouring the cationic dye (MV) adsorption because of electrostatic repulsion reduction between the surface and cationic dye.

Salima et al.[192] investigated marine algae Ulva lactuca (ULV-activated carbon) and Systoceira stricta (SYS-activated carbon) based activated carbon adsorbents for the removal of hazardous cationic dyes, Malachite Green and Safranin O. The activated carbon was prepared by chemical activation involving an impregnation step with phosphoric acid (20%) at 170 °C for 90 min, followed by pyrolysis at 600 °C for 3 h. The ULV-activated carbon and SYS-activated carbon had specific surface area values of 883 and 526 m^2 and a total pore volume of 1.47 and 0.57 cm^3, respectively. At the optimum pH of 4, the Langmuir adsorption capacities obtained for ULV-activated carbon and SYS-activated carbon were 400 and 172 mg g^{-1} for Malachite Green and 384 and 526 mg g^{-1} for Safranin O, respectively. The adsorption kinetics for both dyes tested were best described using the pseudo second-order model. The calculated thermodynamic parameters showed that Malachite Green and Safranin O dye adsorption from aqueous solution was spontaneous and endothermic.

Sludge-based activated carbon (SAC) was prepared from sewage sludge obtained from a paper mill sewage treatment plant.[193] Dehydrated sewage sludge was carbonised at 300 °C in a muffle furnace for 60 min followed by activation at 850 °C for 40 min in a steam atmosphere. Specific surface areas of SAC and raw sludge were about 130–140 m^2 g^{-1}. The equilibrium adsorption data were better represented by the Langmuir isotherm equation than the Freundlich model, indicating a monolayer coverage of methylene blue and Reactive Red 24 (RR24) onto the SAC surface. The adsorption monolayer capacity values calculated from the Langmuir model at 20 °C were 263.64 and 34.49 mg g^{-1} for methylene blue and RR24 respectively.

2.4.2.2   Treatment of Water Contaminated with Chlorinated Compounds

There are only very few studies reporting the removal of organochlorinated compounds from contaminated water by biomass-based activated carbon. Hameed et al.[194] attempted to use date palm (Phoenix dactylifera) stones (DSAC), a low-cost, abundantly available and renewable precursor, for the production of activated carbons. They evaluated the adsorption potential of date stone based activated carbon for 2,4-dichlorophenoxyacetic acid (2,4 D), which is commonly used in the agriculture sector for the control of a wide range of broad leaf weeds and grasses in plantation crops such as sugar cane, oil palm, cocoa and rubber. 2,4-D is a low-cost herbicide with good selectivity that is moderately toxic but is poorly biodegradable. DSAC was very effective for 2,4-D adsorption and a maximum adsorption capacity of 238.10 mg g^{-1} was obtained. The authors studied 2,4-D adsorption with respect to pH, and their results indicated an increase in 2,4-D adsorption when increasing pH from 2 to 11. This behaviour suggested that the adsorption was dominated by interactions between the pesticide and the chemical groups at the adsorbent surface.

To prevent the propagation of the banana weevil (*Cosmopolite sordidus*), which attacks the roots of banana plants, chlorinated pesticides such as chlordecone, hexachlorocyclohexane (HCH) and dieldrine were used extensively in the French West Indies islands of Guadeloupe and Martinique until the beginning of the 1990s, resulting in the contamination of soils and surface waters. Chlordecone was used from 1972 and banned in early 1993. Technical HCH, the mixture of the six HCH isomers, was used from 1951; its use was banned in France in 1972. More than 20 years after HCH was banned, however, the molecule can still be found in soils, water and the food chain in Guadeloupe and Martinique. Both compounds, chlordecone and HCH, are included in the Stockholm Convention on Persistent Organic Pollutants, which bans their production and use worldwide.

Durimel[195] used bagasse activated carbons to study chlordecone and HCH adsorption. One activated carbon sample was prepared by steam activation of bagasse and had a BET surface area of 1242 m^2 g^{-1}, a pore volume of 0.69 cm^3 g^{-1} and a percentage of micropores of 60%. Three other samples were prepared by phosphoric acid activation with a different phosphoric acid impregnation ratio, X_P (g H$_3$PO$_4$ g^{-1} precursor: 0.5 : 1, 1 : 1, 1.5 : 1). They had BET surface areas between 1262 and 1502 m^2 g^{-1}, a pore volume between 0.78 and 1.69 cm^3 g^{-1}, and a percentage of micropores between 8.5 and 65%. The isotherm adsorption data correlated with the surface functional groups composition. β-HCH adsorption was favoured by the presence of high amounts of acidic groups at the activated carbon surface, suggesting that hydrogen bounds may be involved in the adsorption mechanism. The adsorption capacity for HCH was between, 94 and 115 mg g^{-1} when calculated with the Langmuir model. A recent paper describes a study of the adsorption of chlordecone on bagasse activated carbons.[82] It was shown that chlordecone adsorption capacity increased with the amount of carbon atoms and acidic groups at the activated carbon surface whereas basic groups, hydroxyl and ether groups were detrimental to chlordecone adsorption. The adsorption capacity was maximised at a solution pH level equal to the pH$_{pzc}$ of the considered activated carbon. From temperature programmed desorption studies,[82] it was proposed that chlordecone adsorption mechanism onto activated carbons was mainly governed by interactions with carboxylic groups. Q_m was between 51 and 91 mg g^{-1} at pH 7.

Amendment with activated carbon is a promising *in situ* technique for the treatment of porewater of sediments contaminated by hydrophobic organic compounds such as polychlorinated biphenyls (PCBs).[196] Amstaetter *et al.*[197] compared biomass-based activated carbon fabricated from coconut shells (CP1) and anthracite-based activated carbon (BP2) for PCB immobilisation in the presence and absence of sediment. CP1 and BP2 had similar surface area values of 1158 and 1199 m^2 g^{-1}, respectively, while their pore volume values were 0.539 and 0.966 cm^3 g^{-1}, respectively. In the absence of sediment, similar adsorption capacities of PCB were found for CP1 and BP2 (log KF, 9.91 ± 0.13 and 9.70 ± 0.36, respectively). The slightly higher sorption by CP1 than BP2 was attributed to its higher pore volume and pore surface area in the interval 3.5–15 Å; It was however shown that through the presence of sediment, PCB adsorption on anthracite-based activated carbon was reduced by a factor of five and by almost a factor of 100 for the biomass-based activated carbon. This was explained by the presence of narrower pores on the biomass-based activated carbon.

2.4.2.3   Treatment of Metal Contaminated Water

The main sources of wastewaters containing heavy metals are industrial activities such as mining, painting, car manufacturing, metal plating, and tanneries, nuclear power plants aerospace industries and agricultural activities that use metal-containing fertilisers and fungicides. The main heavy metals which cause metal ion pollution are Cd, Pb, Cr, As, Hg, Cu, Th and U. Heavy metals are considered to be one of the most hazardous water contaminants, according to the World Health Organization (WHO).[197,198] Due to their large micropore and mesopore volumes and the resulting high surface area, activated carbons are widely used for the removal of heavy metal contaminants from wastewaters.[200] Many studies report the conversion of lignocellulosic materials into activated carbon for the treatment of heavy metal containing wastewaters as an economically favourable and technically easy process.[199,201] Data on the adsorption of heavy metals such as lead, chromium, copper, arsenic and uranium by biomass-based activated carbons are reviewed in the following section.

2.4.2.3.1   Treatment of Lead Contaminated Water.   Lead is an important metal used as an intermediate in processing industries such as plating, paint and dyes, glass operations, and lead batteries. The compound is a general metabolic poison and enzyme inhibitor,[202] toxic to the neuronal system, and affecting the function of brain cells.[197] The presence of excess lead in drinking water causes diseases such as anaemia, encephalopathy and hepatitis.[203]

In a recent study, *Enteromorpha prolifera* was used as activated carbon precursor for studying lead adsorption.[204] The activated carbon sample was prepared by impregnation of *E. prolifera* with zinc chloride using a 1 : 1 weight ratio for 12 h.[205] The mixture was dehydrated at 100 °C for 24 h and then pyrolysed at 500 °C for 1 h. The activated carbon prepared had a high surface area of 1688 m^2 g^{-1} and a pore volume of 1.04 cm^3 g^{-1}. Kinetic studies showed that adsorption data of lead on the activated carbon sample followed a pseudo second-order kinetic model and the adsorption isotherm data was well represented by the Freundlich model. At 25 °C, the Langmuir monolayer adsorption capacity was 146.8 mg g^{-1}. Thermodynamic studies indicated that the adsorption reaction was a spontaneous and endothermic process. The influence of the pH on lead adsorption on *E. prolifera* activated carbon was studied and showed lower Pb(II) ion adsorption capacity at pH 2.0. This was explained by a competition between H$^+$ and Pb(II) ions for the same adsorption active sites. The activated carbon sample had a isoelectric point of 2.71, and consequently a positive surface charge at pH 2.0 that repelled cations due to electrostatic repulsion. At pH >2.71, however, the amount of Pb^{2+} adsorbed increased rapidly due to ionisation of COOH groups and other negative charges on the activated carbon surface.[204] At pH 5.0, the adsorption capacity reached a maximum. At pH >6.0, species such as [Pb(OH)]$^+$, [Pb$_3$(OH)$_4$]$^+$, and [Pb(OH)$_2$] precipitates were produced according to the lead speciation diagram.[205]

European black pine (*Pinus nigra* Arn.) is a widespread conifer in southern and eastern Europe, growing up to 50 m, which contains green needles and 5–10 cm long cones with rounded scales. Large quantities of cones are produced annually throughout the world, especially in pine plantations grown for the pulp and paper industry.[206] Activated carbons prepared from the European black pine cones was used as adsorbent for lead removal. The cones were impregnated with H$_3$PO$_4$ in a weight ratio of 1:3.[207] The mixture was first heated at 170 °C for 60 min and then carbonised at 500 °C for 60 min. The activated carbon sample obtained had a BET surface area of 1094 m^2 g^{-1} and a pore volume of 1.095 cm^3 g^{-1}. The amount of acidic and basic groups at the activated carbon surface were 2.96 and 1.35 mmol g^{-1}, respectively It was shown that most of the acidic and oxygen containing functional groups were carboxylic groups (1.742 mmol g^{-1}), followed by lactonic and phenolic groups. The Langmuir isotherm provided the best fit to the equilibrium data with a maximum adsorption capacity of 27.53 mg g^{-1}. Removal of lead(II) ions was pH dependent and higher in basic medium, though, at pH values higher than 6.7 it was ascribed to lead precipitation.

Activated carbons were prepared from cotton stalk by impregnation with a 50% (w/v) phosphoric acid solution with a mass ratio of 3 : 2 and pyrolysis at 500 °C for 60 min.[207] The cotton stalk activated carbon had a BET surface area of 1570 m^2 g^{-1} and a total pore volume of 0.904 cm^3 g^{-1}. The amount of acidic surface groups on the activated carbon surface was 1.43 mmol/g^{-1}, which included a high amount of carboxylic groups. The Langmuir maximum adsorption capacity of lead(II) was more than 119 mg g^{-1}.

Singh *et al.*[208] prepared activated carbon from tamarind wood material by chemical activation with sulphuric acid. The drywoods were soaked in concentrated sulphuric acid (H$_2$SO$_4$). The H$_2$SO$_4$ treated wood was then heated at 200 °C. The BET surface area of this material was found to be 612 m^2 g^{-1} and its total pore volume was 0.508 cm^3 g^{-1}. The maximum removal rate of Pb(II) from dilute aqueous solution was 97.95% and the Langmuir monolayer adsorption capacity was 134.22 mg/g at a Pb(II) initial concentration of 40 mg L^{-1}, adsorbent dose 3 g L^{-1} and pH 6.5.

Tamarind wood activated carbon was also prepared by chemical activation with zinc chloride.[209] The chemical ratio (activating agent/precursor) was 296% in this case; the activation time was 40 min and the carbonisation temperature was 439 °C. These optimised conditions were determined using response surface methodology. The BET surface area was 1322 m^2 g^{-1}. The average pore diameter was 5.3 Å and the pore volume, 1.042 cm^3 g^{-1}. Adsorption of lead(II) obeyed the pseudo second-order equation with good correlation. The Langmuir monolayer adsorption capacity determined at pH 5.68 and 30 °C was 43.85 mg g^{-1}.

2.4.2.3.2 Treatment of Chromium Contaminated Water. Chromium can be released into the environment through human activities such as leather tanning, electroplating, manufacturing of dye, paint and paper production,[210–212] but also by natural processes like volcanic activity and weathering of rocks.[213] Chromium exists in six oxidation states, with the most common and stable states being Cr(0), Cr(III) and Cr(VI).[214] Due to its low solubility, Cr(III) hydroxide is relatively immobile.[215] However, Cr(VI) is 500 times more toxic than Cr(III).[215] Hexavalent chromium has been classified as a group A carcinogen by the US Environmental Protection Agency (USEPA) based on its chronic effects[216] which modify DNA transcription process causing important chromosome aberration. Exposure of Cr(VI) causes digestive tract and lung cancer, epigastric pain, nausea, vomiting, severe diarrhoea and hemorrhage.[217]

The tamarind wood activated carbon sample prepared by chemical activation with zinc chloride[209] described in the previous section also demonstrated a high removal rate of Cr(VI) up to >89% at pH 5.41. Equilibrium data were well fitted with the Langmuir model with a maximum adsorption capacity of 28.02 mg g^{-1}. Adsorption was feasible, spontaneous and endothermic.

Giri *et al.*[218] used as activated carbon precursor, roots of *Eichhornia crassipes* (family Pontederiaceae) commonly known as water hyacinth, which is an aquatic weed found abundantly. The roots were impregnated with a concentrated H$_2$SO$_4$ solution at a 1 : 1 ratio, at 150 °C for 24 h and then pyrolysed at 600 °C. The activated carbon sample obtained had a low surface area of 109 m^2 g^{-1}. The activated carbon maximum removal capacity of Cr was found to be 36.34 mg g^{-1} at optimum conditions of pH 4.5, contact time of 30 min, biomass dosage of 7 g L^{-1} and temperature of 25 °C. The adsorption isotherm data were well fitted with the D–R model which indicated a chemisorption process. The pseudo second-order equation described the Cr(VI) adsorption kinetics well, and thermodynamic parameters revealed that the adsorption process was endothermic, spontaneous and feasible in nature.

Hevea brasilinesis (rubber wood) sawdust activated carbons[219] were prepared by mixing the precursor with phosphoric acid in a 1 : 2 weight ratio and soaking for 24 h. The mixture was dried at 110 °C for 1.5 h and then carbonised at

400 °C for 1 h. The BET surface area of the activated carbon sample obtained was 1673.86 $m^2 g^{-1}$. Cr(VI) removal by activated carbon was pH dependent and found to be maximum at pH 2.0. Adsorption kinetics of Cr(VI) ions onto rubber wood sawdust activated carbon was best described by the pseudo second-order model. The Langmuir isotherm described the adsorption data better than the Freundlich or Temkin isotherms. At 30 °C, the Langmuir monolayer adsorption capacity was 44 $mg g^{-1}$. Adsorption of Cr(VI) was found to be effective in the lower pH range and at higher temperatures.

Mohanty et al.[220] prepared different structured activated carbons by chemical activation of *Terminalia arjuna* nuts. The precursor was impregnated with $ZnCl_2$ for 1 h at 50 °C with a different chemical ratio (activating agent/precursor) ranging from 100 to 500%. Samples were then held at different final temperatures (300, 400, 500 or 600 °C) for different carbonisation times of 1, 2 and 3 h. The porosity of the resulting activated carbon increased with the carbonisation temperature and duration. The optimal conditions for the production of high surface area activated carbon from *T. arjuna* nuts by chemical activation were: chemical ratio of 300%; carbonisation duration 1h; and temperature of 500 °C. At these optimal conditions, the BET surface area of the sample obtained was 1260 $m^2 g^{-1}$. Adsorption kinetic data of Cr on the activated carbon were best fitted with the Lagergren pseudo first-order model. The isotherm equilibrium data were well fitted by the Langmuir and Freundlich models. The maximum removal of chromium was obtained at pH 1.0, with a value of 28.4 $mg g^{-1}$.

Valix et al.[221] prepared activated carbons from sugar cane bagasse by impregnation of bagasse with sulfuric in a 3 : 4 ratio (by weight) and heating at 160 °C for two hours in air. The carbonised chars were subsequently gasified with carbon dioxide concentrations of 10, 50 and 100% (v/v). Other activated carbon samples were prepared by chemical activation with $ZnCl_2$, $MgCl_2$ and $CaCl_2$ in a 10 : 7.5 fibre to activating agent ratio by weight and a carbonisation temperature between 400 and 500 °C under nitrogen for 2 h. The authors showed that increasing both total surface area and pore size promoted Cr(VI) adsorption. However, reduction of chromium to Cr(III) also increased. High sulfur and nitrogen content and minimal hydrogen and oxygen content favoured chromium removal. This can be explained by the basic properties of sulfur and nitrogen. Hydrogen and oxygen containing functional groups with an acidic property were detrimental to both the adsorption and reduction effects. The authors concluded that optimising the adsorption of chromium required carbon with a basic surface, a high nitrogen and sulfur content, and minimal hydrogen and oxygen content.

Demiral et al.[222] prepared activated carbons from olive bagasse. The precursor was first carbonised for 500 °C for 1 h. The sample was then treated by steam activation for 0.5 h. The activated carbon sample had a surface area of 718 $m^2 g^{-1}$ and its pore volume was 0.37 $cm^3 g^{-1}$. Cr(VI) removal from aqueous solutions by adsorption showed a maximum adsorption yield at an initial pH of 2. The adsorption increased from 11.2 $mg g^{-1}$ to 48.5 $mg g^{-1}$ when the pH decreased from 10.4 to 2. Hexavalent chromium exists in different forms in aqueous solution such as H_2CrO_4, $HCrO_4^-$ and CrO_4^{2-} with stability depending on the pH.[223] The higher adsorption at low pH by the activated carbon was attributed to the large number of H^+ ions present at low pH values which neutralised the negative charges on the adsorbent surface, thereby reducing hindrance to the diffusion of chromate ions.[224] The dominant form of Cr(vi) between pH 1.0 and 4.0 is $HCrO_4^-$,[223,224] so $HCrO_4^-$ was adsorbed preferentially on the carbon. The decrease in removal at higher pH was explained by the abundance of OH^- ions which caused increased hindrance to diffusion of dichromate ions. Increasing the pH would shift the concentration of $HCrO_4^-$.[224] It was reported that Cr(vi) could be reduced to Cr(iii) in the presence of activated carbon under highly acidic conditions (pH <2). At low Cr(iii) concentrations and at high pH, the Cr(v) ions could precipitate as hydroxide.[225,226] The adsorption process of chromium was spontaneous in nature and more favourable at higher temperature. The pseudo second-order rate expression fitted the adsorption kinetic data well. The Langmuir equation was found to fit well the equilibrium data for Cr(vi) adsorption.

Xie *et al.*[227] developed activated carbons carbons from walnut shell. The powdered precursor was impregnated with H_3PO_4 in a shell to H_3PO_4 weight ratio of 1 : 2 (w/w). The mixture was then irradiated with ultrasound for 30 min (frequency 40 KHz, power 200 W and temperature 258 °C) and dried in an oven at 105 °C for at least 8 h. The impregnated samples were then heated at 500 °C in a N_2 atmosphere for 60 min. The sample had a BET surface area of 1097 m^2 g^{-1} and a total pore volume of 1.21 cm^3 g^{-1}.

At solution pH <2, the main chromium compounds are H_2CrO_4 and $HCrO_4^-$. $HCrO_4^-$ had a low adsorption free energy, while H_2CrO_4 cannot undergo electrostatic interaction with the adsorbent surface. At pH value between 2 and 4.5, $HCrO_4^-$ was the predominant species. Xie *et al.*[227] observed that maximum adsorption occurred at pH 3.0. As shown by Demiral *et al.*[222] because of protonation the carbon surface became more positive leading to an electrostatic attraction between the positively charged surface and the negative complex. In acidic medium, reduction of hexavalent chromium to trivalent chromium may increase the chromium adsorption capacity due to its smaller size.[227] As the pH increased, $HCrO_4^-$ was converted to CrO_4^{2-}, while the increasing amount of OH^- competed more efficiently with CrO_4^{2-} resulting in a decrease of the adsorption capacity. The Cr(vi) adsorption capacity of the activated carbon reached a maximum value of 59.38 mg g^{-1} at pH 3.0, and then decreased to its lowest value of 24.90 mg g^{-1} at pH 6.0. The kinetics data agreed well with the pseudo second-order model. The Langmuir model was shown to give a better description of the adsorption data than the Freundlich and D–R models, implying a monolayer adsorption. The authors concluded that Cr uptake was directly related to the solution pH value and to the carbon textural properties.

2.4.2.3.3   Treatment of Arsenic Contaminated Water.   Arsenic can be spread in the environment[228,229] by agricultural activities that use arsenic as pesticides, herbicides and fertilisers, through discharges from coal-fired thermal power plants, the petroleum refining industry, ceramic industries, but also due to natural geochemical reactions.[228,230] Arsenic in natural waters is a worldwide problem—arsenic pollution has been reported in many countries such as USA, China, Chile, Bangladesh, Taiwan, Mexico, Argentina, Poland, Canada, Hungary, New Zealand, Japan and India.[228] Arsenic is one of the most toxic pollutants.[228,229] Long term As exposure causes skin, lung, bladder and kidney cancer as well as pigmentation changes, skin thickening (hyperkeratosis) neurological disorders, muscular weakness, loss of appetite and nausea. Acute poisoning typically causes vomiting, oesophageal and abdominal pain, and bloody 'rice water' diarrhea.[228,229]

Activated carbon were produced from oat hulls and its efficiency for As(v) adsorption was determined.[230,231] The precursor was submitted to a fast pyrolysis at 500 °C in an inert nitrogen atmosphere. The char was then activated with steam at 800 °C. The activated carbon sample had a BET surface area of 522 mg g^{-1}.[230] The adsorption capacity decreased from 3.09 to 1.57 mg As g^{-1} when the initial pH increased from 5 to 8. Chuang *et al.*[230] used a modified linear driving force model (LDF) coupled with the Langmuir isotherm to describe simultaneous rapid and slow kinetic process. In the LDF model the uptake was assumed to be linearly proportional to a driving force, defined as the difference between the surface concentration and the average adsorbed phase concentration. The adsorption process was well described by this modified LDF model. It was proposed that the adsorbate could be easily and rapidly adsorbed on the activated carbon surface. In the interior (or micropore surface) of activated carbon, the adsorbate (pore) and/or surface diffusion mechanism would occur, resulting in a slower adsorption.

Budinova *et al.*[229] prepared low-cost activated carbon from bean pod waste. Conventional physical (water vapour) activation was used to synthesise the adsorbent. Bean pods were carbonised at 600 °C. The solid char was then activated at 700 °C with water vapour for 1 h. The obtained activated carbon had a BET surface area of 258 m^2 g^{-1} and a total pore volume of 0.206 cm^3 g^{-1}. Hydroxyl and carbonyl groups accounted for 0.21 and 1.52 meq g^{-1}, respectively, and total basic groups for 7.40 meq g^{-1}. Adsorption of As(III) followed a Langmuir-type isotherm with a maximum loading capacity of 1.01 mg g^{-1}. The adsorption capacity on the obtained activated carbon was found to be similar to that of more expensive commercial carbons showing high adsorption capability. The experimental data enabled the authors to propose that the high content of basic groups favoured arsenic adsorption. The removal of As(III) increased with the pH of the system, reaching a maximum around 7.7. Above this value, a sharp decrease was obtained in the uptake of As(III). This behaviour was attributed to the changes in the carbon surface. The carbon had a pH$_{PZC}$ value of 11.9. At solution pH lower than 11.2, the carbon surface was predominantly positively charged, whereas at values above the pH$_{PZC}$ value, negative charges appear on the surface. Meanwhile, below pH 7, the arsenic neutral form, H_3AsO_3, was predominant in the solution and the carbon surface was positively charged. This corresponded to the maximum uptake. When the pH increased, the amount of negative arsenic species, $H_2AsO_3^-$ and $HAsO_3^{2-}$, was higher while the number of positive surface charges decreased. Therefore, at pH above the pH$_{PZC}$, the negative charges on the surface appeared and there were also electrostatic repulsive interactions with the anionic forms of As(III), H$_2$AsO$_3^-$ and HAsO$_3^{2-}$, leading to a decrease in arsenic uptake. The authors concluded that, in an acidic pH range, van der Waals forces between H$_3$AsO$_3$ and activated carbon were most likely the driving forces for the adsorption which resulted in lower adsorption. At values near neutral pH (7–9), along with physical adsorption, the low dissociation of weak arsenic acid (H$_3$AsO$_3$) produced arsenite ions, H$_2$AsO$_3^-$, which were electrostatically attracted to the positive charges on the carbon surface. This enhanced the overall uptake. Oxygen functional groups were proposed to play a key role in arsenic adsorption due to their involvement in the carbon basic functionalities and/or to the formation of complexes with As(III) ions.

Budinova and collaborators[232] prepared activated carbons from olive stones and solvent-extracted olive pulp by physical and chemical activation. The raw material was heated with a heating rate of 60 °C min^{-1} and maintained at a final temperature of 800 °C for 10 min. After carbonisation, the carbon obtained from extracted olive pulp was activated in steam at 800 °C for 2 h. The duration of this treatment at the final temperature was 2 h giving Carbon A. For chemical activation, solvent-extracted olive pulp and olive stones were mixed with K$_2$CO$_3$ (in the proportion 1 : 1) for 12 h. The sample was then carbonised at 950 °C under a N$_2$ flow for 10 min. The sample obtained by chemical activation of olive pulp and olive stones were named Carbon B and Carbon C, respectively. Carbon D was obtained by Carbon A oxidation with nitric acid. The surface areas of Carbons A, B, C and D were 1030, 1850, 1610 and 732 m^2 g^{-1}, respectively, and the total pore volumes were 0.665, 0.898, 0.843 and 0.645 cm^3 g^{-1}, respectively. The authors showed that the removal of arsenic followed the order A (water vapour activation, $Q° = 1.393$ mg g^{-1}) > B (K$_2$CO$_3$ activation of extracted olive pulp, $Q° = 0.855$ mg g^{-1}) > C (K$_2$CO$_3$ activation of olive stones, $Q° = 0.738$ mg g^{-1}) > D (HNO$_3$ activation, $Q° = 0.210$ mg g^{-1}). The oxygen group content and surface characteristics influenced arsenic adsorption on the carbons. As(III) adsorption was optimum in the pH range 7–9. The removal of As(III) increased as the pH increased and reached a maximum at pH 7.9, followed by a sharp decrease up to pH 10. In highly acidic medium, the adsorbent surface was highly protonated and adsorption of neutral H$_3$AsO$_3$ species was not favoured. At pH 9.2, dissociation of H$_3$AsO$_3$ (pKa $= 9.2$) began, resulting in less adsorption. At near neutral pH values (7–9), slow dissociation of weak arsenic (H$_3$AsO$_3$) produced arsenite ion. This partially neutral and partially negatively charged arsenite ion was attracted to the positively charged (below 8.2) surface, resulting in high As(III) removal in this range to form H$_2$AsO$_3$. When the pH was greater than 8.2, the carbon surface became negatively charged. The decrease in As(III) removal by adsorbent in the pH range of 9–10 may be due the adsorbent's negative charge. The authors concluded that the oxygen functional groups may play a major role during the adsorption process of As(III) as Carbon A, with a lower surface area than Carbons B and C, showed a higher As(III) adsorption capacity. The different oxygen functionalities might form complexes with the As(III) ion.

2.4.2.3.4   Treatment of Uranium Contaminated Water.   Uranium is usually found in the hexavalent form in the soil and rock. The average uranium concentration in earth crust is about 3 mg kg^{-1}.[233–235] Uranium is mainly used as fuel for the generation of electricity in nuclear reactors as well as a catalyst and high temperature ceramic.[236] Excessive releases of uranium into the environment are associated with activities of the nuclear fuel cycle as well as a current mining operations[233,235] and processing. Uranium concentration is a health hazard due to its natural radioactivity and chemical toxicity.[235] Uranium may accumulate in bones and cause damage to kidneys. It has been reported that the radioactivity affects living organisms and humans negatively, with problems such as skin erythema, trichomadesis, cataract, leukaemia, cancer and genital system injury.[237–239] Due to its industrial importance and toxicity to living beings, and the expected shortage of uranium in the near future, uranium removal and recovery is of great interest. Adsorption has been proven to be an effective and convenient process for this purpose because of its cost-effective treatment and easy operation.[234] Various types of adsorbents such as biosorbent, organic and inorganic sorbents and carbon-based materials have been developed for uranium recovery.[234]

In aqueous solutions, uranium can be present in both tetravalent and hexavalent stable forms, uranous U$^+$ and uranyl UO$_2^{2+}$ ions.[240] For adsorption studies, uranyl nitrate hexahydrate is always used to prepare the uranium-containing solution, meaning that the UO$_2^{2+}$ ion constitutes the main adsorbate.[240]

Yi and collaborators[241] used nut shell activated carbon with a specific surface area of 1000 m^2 g^{-1} to study the adsorption of uranyl cations UO$_2^{2+}$. They studied the effects of pH, contact time, temperature and adsorbent dosage on the adsorption kinetics and equilibrium adsorption isotherms of U(VI). The maximum adsorption was obtained at an initial solution pH of 6.0; temperatures over the range of 25 to 45 °C had little effect on U(VI) adsorption. The adsorption process followed both the Langmuir and Freundlich isotherms. On the basis of the Langmuir model, the maximum adsorption capacity was found to be 59.17 mg U(VI) g^{-1} adsorbent. The adsorption kinetics was very well defined by the pseudo first-order rate model.

An activated carbon prepared by physical activation of milled coconut shells with a high density and a total surface area of 1150 m^2 g^{-1} was used to study uranium adsorption.[240] The adsorption kinetics followed the pseudo first-order kinetic model and the equilibrium adsorption data were well fitted by the Langmuir and Freundlich isotherm models, giving a maximum monolayer adsorption capacity of 55.32 mg g^{-1}.

Activated carbon prepared by the chemical activation of olive stones was used to study uranium adsorption from aqueous solutions.[242] The activated carbon was prepared by impregnation with ZnCl$_2$ with a precursor/activating agent (ZnCl$_2$) ratio (1 : 2). The sample obtained was carbonised at 500 °C.

The BET surface area of the activated carbon was 464.68 m^2 g^{-1}. The Barrett–Joyner–Halenda (BJH) adsorption cumulative volume of pores was 0.020 cm^3 g^{-1}. The total adsorption capacity was found to be 0.171 mmol g^{-1}. The adsorption kinetic data followed the pseudo second-order model. The process was found to be endothermic in nature.

Pine needles were used as activated carbon precursor;[243] 3.0 g pine needle powder was added into a 50 mL autoclave containing 0.1 g citric acid and 30 mL of deionised water. The autoclave was sealed and tempered at 200 °C for 16 h before being cooled to room temperature. The products were filtered off and washed several times with deionised water and ethanol, and finally dried at 60 °C in a vacuum oven. The saturated monolayer sorption capacity of uranium by the biochar obtained was 62.7 mg g^{-1} at pH 6.0, a contact time of 50 min and temperature of 298 K. U(VI) sorption on the activated carbon was well fitted to the pseudo second-order kinetics model. The thermodynamic parameters clearly indicated that the adsorption process was feasible, spontaneous and endothermic in nature.

Adsorption of metallic ions[199] from aqueous solution is far from being a straightforward process. The predominant interactions involved in the metallic species adsorption mechanism[78] on activated carbon are electrostatic. The adsorption process on activated carbon is mainly controlled by: (1) the speciation of the metal ion or metal ion complex; (2) the solution pH and the point of zero charge of the surface; (3) the surface area and porous distribution; (4) the surface chemistry; and (5) the size of the metallic species.

2.5   Conclusions

With the aim of developing environmentally friendly technologies and achieving the status of green environmental policy, various researchers have advocated using biological waste materials as activated carbon precursors and the development of fabrication processes which have limited environmental impact. Other factors that have to be taken in account in the choice of a new precursor include its abundance and local availability for activated carbon production at industrial scale at affordable cost. Since research is ongoing on the theoretical basis of the factors governing the textural and chemical characteristics of the activated carbons, it is also necessary to determine the chemical, physical and adsorptive characteristics of the activated carbon produced from each new precursor tested. The properties of a resulting activated carbon depend on the precursor and type of activation, *i.e.* physical, chemical, conventional, microwave or hydrothermal heating. For a given precursor, the optimal parameters such as temperature, retention time, microwave power, radiation time have to be determined in addition to the type of activating agent and the chemical agent ratio for chemical process.

Biomass can be a source of activated carbon products with high surface areas comparable with those of commercial activated carbons. Although the use of activated carbons has been common for the adsorption of pollutants for a long time, continuous research efforts on studying textural and chemical properties are made due to their versatile properties depending of the nature of precursor and of the preparation mode. Current ongoing works deal with the search for new precursors providing activated carbon with higher pollutant adsorption capacity or adsorption rates, and elucidation of the adsorption principle underlying its isotherms or kinetics study.

Surface area is not always related to the adsorption capacity of activated carbon. Other parameters that affect the adsorption capacity are the microporosity, the chemical used for activation and the properties of the adsorbate. Parameters such as the extent of graphitisation, chemical groups at the surface, surface pH and solution pH are quite well correlated to adsorption data of molecules. Physico-chemical properties such as hydrophobicity, size of molecules, charge and dipolar moment are also of great help in predicting the extent of adsorption of a given molecule on an activated carbon.

The adsorbing capacities for both organic and inorganic materials of activated carbons produced from biomass are generally comparable with those of commercially available carbons. Nevertheless, the overwhelming majority of the publications in the field of activated carbons have dealt with the use of synthetic wastewaters or polluted waters containing a single pollutant, and show generally that microporous activated carbon can adsorb a single pollutant efficiently. More detailed research with real industrial effluents, multi-component pollutants and continuous flow studies should also be conducted to identify practical applications of these modified adsorbents in the water industry. Size exclusion effects may occur when microporous activated carbons are considered for the adsorption of complex effluents that may contain natural organic matter. Activated carbons with the largest volumes of mesopores may show higher adsorption capacities. Moreover, total cost evaluation is required to make full-scale implementations possible.

Activated carbons produced by different activation processes from various precursors can also be used for various applications such as adsorption air conditioning systems, adsorption refrigeration systems, solar adsorption refrigeration and adsorption technologies in automobiles[244] and widely as electrodes for electrical double layers capacitors[245] or supercapacitors. The application of biomass-based activated carbon for supercapacitors is developed in Chapter 9.

References

1. C. Namasivayam, M. D. Kumar, K. Selvi, R. A. Begum, T. Vanathi and R. T. Yamuna, *Biomass Bioenergy*, 2001, **21**, 477.
2. C. Namasivayam and D. Kavitha, *Dyes Pigm.*, 2002, **54**, 47–58.
3. K. Kadirvelu, M. Kavipriya, C. Karthika, M. Radhika, N. Vennilamani and S. Pattabhi, *Bioresour. Technol.*, 2003, **87**, 129–132.
4. R.-L. Tseng and S.-K. Tseng, *J. Colloid Interface Sci.*, 2005, **287**, 428–437.
5. N. Kannan and M. M. Sundaram, *Dyes Pigm.*, 2001, **51**, 25–40.

6. M. Ahmedna, W. E. Marshall and R. M. Rao, *Bioresour. Technol.*, 2000, **71**, 113–123.
7. A. Aygun, S. Y. Karakas and I. Duman, *Microporous Mesoporous Mater.*, 2003, **66**, 189–195.
8. R. L. Tseng, F. C. Wu and R. S. Juang, *Carbon*, 2003, **41**, 487–495.
9. P. K. Malik, *Dyes Pigm.*, 2003, **56**, 239–249.
10. K. Kadirvelu, M. Palanival, R. Kalpana and S. Rajeswari, *Bioresour. Technol.*, 2000, **74**, 263–265.
11. F. Wu, R. Tsen and R. Juang, *J. Environ. Sci Health, Part A:* 1999, **34**, 1753–1775.
12. S. Rajeshwarisivaraj, P. Sivakumar, P. Senthilkumar and V. Subburam, *Bioresour. Technol.*, 2001, **80**, 233–235.
13. A. A. El-Hendawy, *J. Anal. Appl. Pyrolysis*, 2006, **75**, 159–166.
14. K. Raveendran, A. Ganesh and K. C. Khilart, *Fuel*, 1995, **74**, 1812–1822.
15. J. F. González, S. Román, J. M. Ehncinar and G. Martínez, *J. Anal. Appl. Pyrolysis*, 2009, **85**, 134–141.
16. M. J. Antal, *Ind. Eng. Chem. Res.*, 1983, **22**, 366–375.
17. C. Bouchelta, S. Medjram, O. Bertrand and J. P. Bellat, *J. Anal. Appl. Pyrolysis*, 2008, **82**, 70.
18. M. Krzesinska and J. Zachariasz, *J. Anal. Appl. Pyrolysis*, 2007, **80**, 209–215.
19. M. C. Ncibi, V. Jeanne-Rose, B. Mahjoub, C. Jean-Marius, J. Lambert, J. J. Ehrhardt, Y. Bercion, M. Seffen and S. Gaspard, *J. Hazard. Mater.*, 2009, **165**, 240–249.
20. C. Philpot, *Forest Sci.*, 1970, **16**, 461–471.
21. R. C. Rothermel, *Thermal Uses and Properties of Carbohydrates and Lignins*, Academic Press, New York, 1976.
22. A. Ouensanga and L. M.-A. Arsene, *Microporous Mesoporous Mater*, 2003, **59**, 85–91.
23. S. Gaspard, S. Altenor, P. A. Dawson and A. Ouensanga, *J. Hazard. Mater.*, 2007, **144**, 73–81.
24. H. Wang, Q. Gao and J. Hu, *J. Am. Chem. Soc.*, 2009, **131**, 7016–7022.
25. N. Passé-Coutrin, V. Jeanne-Rose and A. Ouensanga, *Fuel*, 2005, **84**, 2131–2134.
26. V. Jeanne-Rose, V. Golabkan, J.-L. Mansot, L. Largitte, T. Césaire and A. Ouensanga, *J. Microsc.*, 2003, **210**, 53–59.
27. R. F. Egerton, *Electron Energy Loss Spectroscopy in the Electron Microscope*, Plenum Press, New York, 1996.
28. J.-L. M. Hallouis and J. M. Martin, *Colloids Surf. A*, 1993, **71**, 123–134.
29. J. Wery and J.-L. Mansot, *Microsc., Microanal., Microstruct.*, 1993, **4**, 87–100.
30. R. C. Bansal, J.-B. Donnet and F. Stoeckli, *Active Carbon*, Marcel Dekker, New York, 1998.
31. A. Baçaoui, A. Yaacoubi, C. Bennouna, F. J. Maldonado-Hdar, J. Rivera-Utrilla, F. Carrasco-Marin and C. Moreno-Castilla, *Environ. Sci. Technol.*, 2002, **36**, 3844–3849.

32. E. A. Dawson, G. M. B. Parkes, P. A. Barnes and M. J. Chinn, *Carbon*, 2003, **41**, 571–578.
33. K. Gergova, A. Galushko, N. Petrov and V. Minkova, *Carbon*, 1992, **30**, 721–727.
34. T. Otowa, R. Tanibata and M. Itoh, *Gas Sep. Purif.*, 1993, **7**, 241–245.
35. M. Jagtoyen and F. Derbyshire, *Carbon*, 1998, **36**, 1085–1097.
36. M. Molina-Sabio, F. Rodriguez-Reinoso, F. Caturla and M. J. Selles, *Carbon*, 1995, **33**, 1105–1113.
37. M. K. B. Gratuito, T. Panyathanmaporn, R. A. Chumnanklang, N. Sirinuntawittaya and A. Dutta, *Bioresour. Technol.*, 2008, **99**, 4887–4895.
38. S. Biniak, G. Szymanski, J. Siedlewski and A. Swiatkoski, *Carbon*, 1997, **35**, 1799–1810.
39. S. Altenor, M. C. Ncibi, N. Brehm, E. Emmanuel and S. Gaspard, *Waste Biomass Valorization*, 2012.
40. S. Altenor, M. C. Ncibi, E. Emmanuel and S. Gaspard, *BioChem. Eng. J.*, 2012, **67**, 35–44.
41. E. T. Thostenson and T. W. Chou, *Composites, Part A*, 1999, **30**, 1055–1071.
42. D. A. Jones, T. P. Lelyveld, S. D. Mavrofidis, S. W. Kingman and N. J. Miles, *Resour. Conserv. Recycl.*, 2002, **34**, 75–90.
43. T. J. Appleton, R. I. Colder, S. W. Kingman, I. S. Lowndes and A. G. Read, *Appl. Energy*, 2005, **81**, 85.
44. C. O. Ania, J. B. Parra, J. A. Menéndez and J. J. Pis, *Microporous Mesoporous Mater.*, 2005, **85**, 7–15.
45. J. M. V. Nabais, P. J. M. Carrott, M. M. L. Ribeiro Carrott and J. A. Menéndez, *Carbon*, 2004, **42**, 1315–1320.
46. R. Hoseinzadeh Hesas, W. M. A. Wan Daud, J. N. Sahu and A. Arami-Niya, *J. Anal. Appl. Pyrolysis*, 2013, **100**, 1–11.
47. L. M. Norman and C. Y. Cha, *Chem. Eng. Commun.*, 1996, **140**, 87–110.
48. J. Guo and A. C. Lua, *Carbon*, 2000, **38**, 1985–1193.
49. K. Yang, J. Peng, C. Srinivasakannan, L. Zhang, H. Xia and X. Duan, *Bioresour. Technol.*, 2010, **101**, 6163–6169.
50. D. Xin-Hui, C. Srinivasakannan, P. Jin-hui, Z. Li-Bo and Z. Zheng-Yong, *Biomass Bioenergy*, 2011, **35**, 3920–3926.
51. E. Yagmur, *J. Porous Mater.*, 2012, **19**, 995–1002.
52. Q. S. Liu, T. Zheng, P. Wang and L. Guo, *Ind. Crops Prod.*, 2010, **31**, 233–238.
53. S. Altenor, B. Carene, E. Emmanuel, J. Lambert, J. J. Ehrhardt and S Gaspard, *J. Hazard. Mater.*, 2009, **165**, 1029–1039.
54. M. Sevilla and A. B. Fuertes, *Carbon*, 2009, **47**, 2281–2289.
55. A. Funke and F. Ziegler, *Biofuels, Bioprod. Biorefin.*, 2010, **4**, 160–177.
56. L. Zhao, N. Baccile, S. Gross, Y. Zhang, W. Wei, Y. Sun, M. Antonietti and M. M. Titirici, *Carbon*, 2010, **48**, 3778–3787.

57. A. Thomas, P. Kuhn, J. Weber, M. Titirici and M. Antonietti, *Macromol. Rapid Commun.*, 2009, **30**, 221–236.
58. B. Hu, K. Wang, L. Wu, S. H. Yu, M. Antonietti and M. Titirici, *Adv. Mater.*, 2010, **22**, 813–828.
59. X. Sun and Y. Li, *Angew. Chem., Int. Ed.*, 2004, **43**, 597–601.
60. C. Falco, F. Perez Caballero, F. Babonneau, C. Gervais, G. Laurent, M. M. Titirici and N. Baccile, *Langmuir*, 2011, **27**, 14460–14471.
61. S. Román, J. M. Valente Nabais, B. Ledesma, J. F. González, C. Laginhas and M. M. Titirici, *Microporous Mesoporous Mater.*, 2013, **165**, 127–133.
62. E. Unur, *Microporous Mesoporous Mater.*, 2013, **168**, 92–101.
63. K. Bilba and A. Ouensanga, *J. Anal. Appl. Pyrolysis*, 1996, **38**, 61–73.
64. Z. Jiang, Y. Liu, X. Sun, F. Tian, F. Sun, C. Liang, W. You, C. Han and C. Li, *Langmuir*, 2003, **19**, 731–736.
65. A. P. Terzyk, *Colloids Surf. A*, 2001, **177**, 23–24.
66. C. Bouchelta, M. S. Medjram, O. Bertrand and J.-P. Bellat, *J. Anal. Appl. Pyrolysis*, 2008, **82**, 70–77.
67. T. Yang and A. C. Lua, *J. Colloid Interface Sci.*, 2003, **267**, 408–417.
68. M. Asadullah, M. Asaduzzaman, M. Shajahan Kabir, M. Golam Mostofa and T. Miyazawa, *J. Hazard. Mater.*, 2010, **174**, 437–443.
69. D. C. W. Tsang, J. Hu, M. Y. Liu, W. Zhang, K. C. K. Lai and I. M. C. Lo, *Water Air Soil Pollut.*, 2007, **184**, 141–155.
70. T. -H. Liou and S.-J. Wu, *J. Hazard. Mater.*, 2009, **171**, 693–703.
71. H. Deng, G. Li, H. Yang, J. Tang and J. Tang, *Chem. Eng. J.*, 2010, **163**, 373–381.
72. Z. Hu and M. P. Srinivasan, *Microporous Mesoporous Mater.*, 2001, **43**, 267–275.
73. K Laszlo and A Szucs, *Carbon*, 2001, **39**, 1945–1953.
74. J. R. Pels, F. Kapteijn, J. A. Moulijn, Q. Zhu and K. M. Thomas, *Carbon*, 1995, **33**, 1641–1653.
75. Q. Zhu, S. Money, A. E. Russell and K. M. Thomas, *Langmuir*, 1997, **13**, 2149–2157.
76. K. Laszlo, E. Tombac and K. Josepovits, *Carbon*, 2001, **39**, 1217–1228.
77. F. Kapteijn, J. A. Moulijn, S. Matzner and H. P. Boehm, *Carbon*, 1999, **37**, 1143–1150.
78. S. Haydar, C. Moreno-Castilla, M. A. Ferro-García, F. Carrasco-Marín, J. Rivera-Utrilla, A. Perrard and J. P. Joly, *Carbon*, 2000, **38**, 1297–1308.
79. G. S. Szymanski, Z. Karpinski, S. Biniak and A. Sowiatkowski, *Carbon*, 2002, **40**, 2627–2639.
80. P. Brender, R. Gadiou, J. C. Rietsch, P. Fioux, J. Dentzer, A. Ponche and C. Vix-Guterl, *Anal. Chem.*, 2012, **84**, 2147–2153.
81. A. S. Mestre, A. S. Bexiga, M. Proença, M. Andrade, M. L. Pinto, I. Matos, I. M. Fonseca and A. P. Carvalho, *Bioresour. Technol.*, 2011, **102**, 8253–8260.

82. A. Durimel, S. Altenor, R. Miranda-Quintana, P. Cousepel Du Mesnil, U. Jauregui-Haza, R. Gadiou and S. Gaspard, *Chem. Eng. J.*, 2013, **229**, 239–249.
83. H. P. Boehm, *Carbon*, 1994, **32**, 759–769.
84. G. V. Nunell, M. E. Fernandez, P. R. Bonelli and A. L. Cukierman, *Biomass Bioenergy*, 2012, **44**, 87–95.
85. P. Nowicki, R. Pietrzak and H. Wachowska, *Catal. Today*, 2010, **150**, 107–114.
86. Y. S. Kim, S. J. Yang, H. J. Lim, T. Kim and C. R. Park, *Carbon*, 2012, **50**, 3315–3323.
87. C. A. Toles, W. E. Marshall and M. M. Johns, *Carbon*, 1999, **37**, 1207–1214.
88. M. Valix, W. H. Cheung and G. McKay, *Langmuir*, 2006, **22**, 4574–4582.
89. Y. Guo and D. A. Rockstraw, *Bioresour. Technol.*, 2007, **98**, 1513–1521.
90. A. Farooq, L. Reinert, J. M. Levêque, N. Papaiconomou, N. Irfan and L. Duclaux, *Microporous Mesoporous Mater.*, 2012, **158**, 55–63.
91. L. Huang, Y. Sun, W. Wang, Q. Yue and T. Yang, *Chem. Eng. J.*, 2011, **171**, 1446–1453.
92. J. L. Figueiredo, M. F. R. Pereira, M. M. A. Freitas and J. J. M. Orfao, *Carbon*, 1999, **37**, 1379–1389.
93. F. Rodríguez-Reinoso and M. Molina-Sabio, *Adv. Colloid Interface Sci.*, 1998, **76–77**, 271–294.
94. J. Guo and A. C. Lua, *Carbon*, 2000, **38**, 1985–1193.
95. F. Rouquerol, J. Rouquerol and K. Sing, *Adsorption by Powders and Porous Solids*, 1999.
96. K. S. W. Sing, *Porosity in Carbons*, Edward Arnold, London, 1995.
97. J. Choma and M. Jaroniec, *Activated Carbon Surfaces in Environmental Remediation*, Elsevier, New York, 2006.
98. S. Brunauer, P. H. Emmett and E. J. Teller, *Am. Chem. Soc.*, 1938, **60**, 309–319.
99. S. J. Gregg and, K. S. Sing, *Adsorption, Surface Area and Porosity*, Academic Press, London, 1982.
100. H. Jankowska, A. Siatkowski and J. Choma, *Active Carbon*, Ellis Horwood, Chichester, UK, 1991.
101. K. S. W. Sing, D. H. Everett, R. A. W. Haul, L. Moscou, R. A. Pierotti, J. Rouquerol and T. Siemieniewska, *Pure Appl. Chem.*, 1985, **57**, 603–619.
102. P. J. Pomonis, D. E. Petrakis, A. K. Ladavos, K. M. Kolonia, G. S. Armatas, S. D. Sklari, P. C. Dragani, A. Zarlaha, V. N. Stathopoulo and A. T. Sdoukos, *Microporous Mesoporous Mater.*, 2004, **69**, 97–107.
103. N. Passe-Coutrin, S. Altenor, D. Cossement, C. Jean-Marius and S. Gaspard, *Microporous Mesoporous Mater.*, 2008, **111**, 517–522.
104. M. Kruk and M. Jaroniec, *Chem. Mater.*, 2001, **13**, 3169–3183.
105. G. Scatchard, *Ann. NY Acad. Sci.*, 1949, **51**, 660–672.

106. J. Rouquerol, P. Llewellyn and F. Rouquerol, *Stud. Surf. Sci. Catal.*, 2007, **160**, 49–56.
107. B. C. Lippens and J. H. de Boer, *J. Catal.s*, 1965, **4**, 319–323.
108. M. Ternan, *J. Colloid Interface Sci.*, 1973, **45**, 270–279.
109. C. P. Broekhoff and B. G. Linsen, *Physical and Chemical Aspects of Adsorbents and Catalysts*, Academic Press, London and New York, 1970.
110. K. S. W. Sing, *Chem. Ind.*, 1967, **20**, 829–830.
111. J. Rouquerol, D. Avnir, C. W. Fairbridge, D. H. Everett, J. M. Haynes, N. Pernicone, J. D. F. Ramsay, K. S. W. Sing and K. K. Unger, *Pure Appl. Chem.*, 1994, **66**, 1739–1758.
112. M. Jaroniec and K. Kaneko, *Langmuir*, 1997, **13**, 6589–6596.
113. M. Kruk, M. Jaroniec and K. P. Gadkaree, *J. Colloid Interface Sci.*, 1997, **192**, 250–256.
114. M. Jaroniec and J. Choma, *Stud. Surf. Sci. Catal.*, 1997, **104**, 715–744.
115. M. Jaroniec and R. Madey, *Physical Adsorption on Heterogeneous Solid Surfaces*, Elsevier, Amsterdam, 1988.
116. M. M. Dubinin, *Progress Surf. Membrane Sci.*, 1975, **9**, 1–70.
117. F. Rodriguez-Reinoso and A. Linares-Solano, *Chemistry and Physics of Carbon*, Marcel Dekker, New York, 1988.
118. M. M. Dubinin and V. A. Astakhov, *Bull. Acad. Sci. USSR Div. Chem. Sci.*, 1971, **20**, 3–7.
119. E. W Washburn, *Phys. Rev.*, 1921, **17**, 273–283.
120. E. P. Barret, P. B. Joyner and P. Halenda, *J. Am. Chem. Soc*, 1951, **73**, 373–380.
121. J. C. P. Broekhoff and J. H. de Boer, *J. Catal.*, 1967, **9**, 8–14.
122. J. C. P. Broekhoff and J. H. de Boer, *J. Catal.*, 1967, **9**, 15–27.
123. J. C. P. Broekhoff and J. H. de Boer, *J. Catal.*, 1968, **10**, 377–390.
124. J. C. P. Broekhoff and J. H. de Boer, *J. Catal.*, 1968, **10**, 368–376.
125. R. W. Cranston and F. A. Inkley, *Adv. Catal.*, 1957, **9**, 143–154.
126. D. Dollimore and G. R. Heal, *J. Appl. Chem.*, 1964, **14**, 109.
127. J. Choma, M. Jaroniec and M. Kloske, *Adsorption Sci. Technol.*, 2002, **20**, 307–316.
128. G. Horvath and K. Kawazoe, *J. Chem. Eng. Jpn.*, 1983, **16**, 470–475.
129. M. Jaroniec, *Access in Nanoporous Materials*, Plenum Press, New York, 1995.
130. N. A. Seaton, J. P. R. B. Walton and N. Quirke, *Carbon*, 1989, **27**, 853–861.
131. P. I. Ravikovitch, A. Vishnyakov, R. Russo and A. V. Neimark, *Langmuir*, 2000, **16**, 2311–2320.
132. B. S. Girgis and A.-A. El-Hendawy, *Microporous Mesoporous Mater.*, 2002, **52**, 105–117.
133. F. Zhang, G.-D. Li and J.-S. Chen, *J. Colloid Interface Sci.*, 2008, **327**, 108–114.
134. A. A. M. Daifullah, B. S. Girgis and H. M. H. Gad, *Colloids Surf. A*, 2004, **235**, 1–10.

135. P. Nowicki, M. Skrzypczak and R. Pietrzak, *Chem. Eng. J.*, 2010, **162**, 723–729.
136. S. Altenor, B. Carene-Melane and S. Gaspard, *Int. J. Environ. Technol. Manage.*, 2009, **10**, 308–326.
137. T. Taya, S. Ucarb and S. Karagözb, *J. Hazard. Mater.*, 2009, **165**, 481–485.
138. Z. Hu, H. Guo, M. P. Srinivasan and Ni Yaming, *Sep. Purif. Technol.*, 2003, **31**, 47–52.
139. A. S. Dastgheib and D. A. Rockstraw, *Carbon*, 2001, **39**, 1849–1855.
140. P. Nowicki, R. Pietrzak and H. Wachowska, *Catal. Today*, 2010, **150**, 107–114.
141. D. Prahas, Y. Kartika, N. Indraswati and S. Ismadji, *Chem. Eng. J.*, 2008, **140**, 32–42.
142. F. Zhang, G.-D. Li and J.-S. Chen, *J. Colloid Interface Sci.*, 2008, **327**, 108–114.
143. H. Trevino-Cordero, L. G. Juárez-Aguilar, D. I. Mendoza-Castillo, V. Hernández-Montoya, A. Bonilla-Petriciolet and M. A. Montes-Morio, *Ind. Crops Prod.*, 2013, **42**, 315–323.
144. B. Cagnon, X. Py, A. Guillot, F. Stoeckli and G. Chambat, *Bioresour. Technol.*, 2008, **100**, 292–298.
145. S. Kunwar, P. M. Amrita, S. Sarita and O. Priyanka, *J. Hazard. Mater.*, 2008, **150**, 626–641.
146. A. T. M. Din, B. H. Hameed and A. L. Ahmad, *J. Hazard. Mater.*, 2009, **161**, 1522–1529.
147. O.-W. Achaw and G. Afrane, *Microporous Mesoporous Mater.*, 2008, **112**, 284–290.
148. Z. Hu and M. P. Srinivasan, *Microporous Mesoporous Mater.*, 1999, **27**, 11–18.
149. L. Wei, Y. K. P. Jinhui, Z. Libo, G. Shenghui and X. Hongying, *Ind. Crops Prod.*, 2008, **28**, 190–198.
150. K. Yang, J. Peng, C. Srinivasakannan, L. Zhang, H. Xia and D. Xin-Hui, *Bioresour. Technol.*, 2010, **101**, 6163–6169.
151. C. Moreno-Castilla, *Carbon*, 2004, **42**, 83–94.
152. S. Lagergren, *Handlinger*, 1898, **24**, 1–39.
153. Y. S. Ho and G. McKay, *Process Biochem.*, 1999, **34**, 451–465.
154. W. J. Weber and J. C. Morris, in *Proceedings of the International Conference on Water Pollution Symposium*, Pergamon Press, Oxford, 1962, vol. 2, pp. 231–266.
155. F. Brouers and O. Sotolongo-Costa., *Physica A*, 2006, **368**, 165–175.
156. S. Gaspard, S. Altenor, N. Passe-Coutrin, A. Ouensanga and F. Brouers, *Water Res.*, 2006, **40**, 3467–3477.
157. S. Figaro, J. P. Avril, F. Brouers, A. Ouensanga and S. Gaspard, *J. Hazard. Mater.*, 2009, **161**, 649–656.
158. I. Langmuir, *J. Am. Chem. Soc.*, 1918, **40**, 1361–1403.
159. H. Freundlich, *J. Phys. Chem.*, 1906, **57**, 385–470.
160. O. Redlich and D. L. Peterson, *J. Phys. Chem.*, 1959, **63**, 1024.

161. M. J. Temkin and V. Pyzhev, *Acta Physiochim. USSR*, 1940, **12**, 217–222.
162. F. Brouers, O. Sotolongo, F. Marquez and J. P. Pirard, *Physica A*, 2005, **349**, 271–282.
163. C. Ncibi, S. Altenor, S. Mongi, F. Brouers and S. Gaspard, *Chem. Eng. J.*, 2008, **145**(2), 196–202.
164. T. Chakrabarti, O. V. R. Subrahmanyan and B. B. Sundaresan, Biodegradation of recalcitrant industrial wastes, in: *Biotreatment Systems*, vol. II, CRC Press, Boca Raton, FL, 1988, pp. 171–234.
165. P. Pitter and J. Chudoba, *Biodegrability of Organic Substances in the Aquatic Environment*, CRC Press, Boca Raton, FL, 1990.
166. V. K. Grag, R. Kumar and R. Gupta, *Dyes Pigm.*, 2004, **62**, 1–10.
167. A. Demirbas, *J. Hazard. Mater.*, 2009, **167**, 1–9.
168. Y. Feng, H. Zhou, G. Liu, J. Qiao, J. Wang, H. Lu, L. Yang and Y. Wu, *Bioresour. Technol.*, 2012, **125**, 138–144.
169. E. Mitter, G. dos Santos, É de Almeida, L. Morão, H. Rodrigues and C. Corso, *Water Air Soil Pollut.*, 2012, **223**, 765–770.
170. A. Baçaoui, A. Yaacoubi, A. Dahbi, C. Bennouna, R. Phan Tan Luu, F. J. Maldonado-Hodar, J. Rivera-Utrilla and C. Moreno-Castilla, *Carbon*, 2001, **39**, 425–432.
171. G. H. Oh and C. R. Park, *Fuel*, 2002, **81**, 327–336.
172. S. M. Mak, B. T. Tey, K. Y. Cheah, W. L. Siew and K. K. Tan, *Adsorption*, 2009, **15**, 507–519.
173. B. S. Girgis, S. S. Yunis and A. M. Soliman, *Mater. Lett.*, 2002, **57**, 164–172.
174. A. A. Attia, B. S. Girgis and S. A. Khedr, *J. Chem. Technol. BioTechnol.*, 2003, **78**, 611–619.
175. M. Benadjemia, L. Millière, L. Reinert, N. Benderdouche and L. Duclaux, *Fuel Process. Technol.*, 2011, **92**, 1203–1212.
176. K. Y. Foo and B. H. Hameed, *Chem. Eng. J.*, 2011, **170**, 338–341.
177. K. Y. Foo and B. H. Hameed, *Chem. Eng. J.*, 2011, **173**, 385–390.
178. K. Y. Foo and B. H. Hameed, *Biomass Bioenergy*, 35, 2011, **32**, 57–61.
179. M. Hejazifara, S. Aziziana, H. Sarikhanib, Q. Li and D. Zhao, *J. Anal. Appl. Pyrolysis*, 2011, **92**, 258–266.
180. E. Unur, *Microporous Mesoporous Mater.*, 2013, **168**, 92–101.
181. P. K. Malik, *Dyes Pigm.*, 2003, **56**, 239–249.
182. L. Wang, *Environ. Sci. Pollut. Res.*, 2013, **20**, 4635–4646.
183. M. Valix, W. H. Cheung and G. McKay, *Chemosphere*, 2004, **56**, 493–501.
184. R. S. Juang, F. C. Wu and R. L. Tseng, *Colloids Surf. A.*, 2002, **201**, 191–199.
185. W. T. Tsai, C. Y. Chang, M. C. Lin, S. F. Chien, H. F. Sun and M. F. Hsieh, *Chemosphere*, 2001, **45**, 51–58.
186. I. D. Mall, V. C. Srivastava, N. K. Agarwal and I. M. Mishra, *Chemosphere*, 2005, **61**, 492–501.
187. K. Zhang, W. H. Cheung and M. Valix, *Chemosphere*, 2005, **60**, 1129–1140.

188. M. Valix, W. H. Cheung and G. McKay, *Langmuir*, 2006, **22**, 4574–4582.
189. A. Namane, A. Mekarzia, K. Benrachedi, N. Belhaneche-Bensemra and A. Hellal, *J. Hazard. Mater.*, 2005, **119**, 189–194.
190. S. Nethaji and A. Sivasamy, *Chemosphere*, 2011, **82**, 1367–1372.
191. M. Hejazifar, S. Azizian, H. Sarikhani, Q. Li and D. Zhao, *J. Anal. Appl. Pyrolysis*, 2011, **92**, 258–266.
192. A. Salima, B. Benaouda, B. Noureddine and L. Duclaux, *Water Res.*, 2013, **47**, 3375–3378.
193. W.-H. Li, Q.-Y. Yue, B.-Y. Gao, Z. -H. Ma, Y.-J. Li and H.-X. Zhao, *Chem. Eng. J.*, 2011, **171**, 320–327.
194. B. H. Hameed, J. M. Salman and A. L. Ahmad, 2009, **163**, 121–126.
195. A. Durimel, PhD thesis, Université des Antilles et de la Guyane, 2008.
196. K. Amstaetter, E. Eek and G. Cornelissen, *Chemosphere*, 2012, **87**, 573–578.
197. World Health Organization, *Guidelines for Drinking-water Quality*, Chemical Fact Sheets, World Health Organization, Geneva, 2004.
198. World Health Organization, Guidelines for Drinking-water Quality, First Addendum, World Health Organization, Geneva, 2006.
199. J. M. Diasa, M. C. M. Alvim-Ferraza, M. F. Almeidaa, J. Rivera-Utrilla and M. Sanchez-Polo, *J. Environ. Manage.*, 2007, **85**, 833–846.
200. F. Fu and Q. Wang, *J. Environ. Manage.*, 2011, **92**, 407–418.
201. Y. H. Li, S. Wang, J. Wei, X. Zhang, C. Xu, Z. Luan, D. Wu and B. Wei, *Chem. Phys. Lett.*, 2002, **357**, 263–266.
202. Y. S. Ho, J. C. Y. Ng and G. McKay, *Sep. Sci. Technol.*, 2001, **36**, 241–261.
203. B. L. Martins, C. C. V. Cruz and A. S. Luna, *Biochem. Eng. J.*, 2006, **27**, 310–314.
204. Y. Li, Q. Dua, X. Wang, P. Zhang, D. Wang, Z. Wang and Y. Xi, *J. Hazard. Mater.*, 2010, **183**, 583–589.
205. M. Machida, Y. Kikuchi, M. Aikawa and H. Tatsumoto, *Colloids Surf. A*, 2004, **240**, 179–186.
206. M. Momčilović, M. Purenović, A. Bojić, A. Zarubica and M. Ranđelović, *Desalination*, 2011, **276**, 53–59.
207. K. Li, Z. Zheng and Y. Li, *J. Hazard. Mater.*, 2010, **181**, 440–447.
208. C. K. Singh, J. N. Sahu, K. K. Mahalik, C. R. Mohanty, B. Raj Mohanc and B. C. Meikap, *J. Hazard. Mater.*, 2008, **153**, 221–228.
209. J. Acharyaa, J. N. Sahub, C. R. Mohantyc and B. C. Meikap, *Chem. Eng. J.*, 2009, **149**, 249–262.
210. M. J. Udy, *Chromium*, Reinhold Publishing Corporation, New York, 1956.
211. L. J. Casarett and J. Doul, *Toxicology: The Basic Science of Poisons*, Macmillan, New York, 1980.
212. J. O. Nriagu and E. Nieboer, *Chromium in the Natural and Human Environment*, Wiley, New York, 1988.

213. N. Serpone, E. Borgarello, E. Pelizzeti and E. Schiavello (ed.), *Photocatalysis and Environment*, Kluwer Academic, Dordrecht, The Netherlands, 1988.
214. K. J. Cronje, K. Chetty, M. Carsky, J. N. Sahu and B. C. Meikap, *Desalination*, 2011, **275**, 276–284.
215. Z. Kowalski, *J. Hazard. Mater.*, 1994, **4937**, 137–144.
216. D. B. Kaufaman, *Am. J. Dis. Children*, 1970, **119**, 374–381.
217. E. Browning, in *Chromium in Toxicity of Industrial Metals*, Butterworths, London, 2nd edn, 1969, pp. 76–96.
218. A. K. Giri, R. Patel and S. Mandal, *Chem. Eng. J.*, 2012, **185–186**, 71–81.
219. T. Karthikeyan, S. Rajgopal and L. R. Miranda, *J. Hazard. Mater.*, 2005, **B124**, 192–199.
220. K. Mohanty, M. Jha, B. C. Meikap and M. N. Biswas, *Chem. Eng. Sci.*, 2005, **60**, 3049–3059.
221. M. Valix, W. H. Cheung and K. Zhang, *Adsorption*, 2008, **14**, 711–718.
222. H. Demiral, I. Demiral, F. Tumsek and B. Karabacakoglu, *Chem. Eng. J.*, 2008, **144**, 188–196.
223. V. K. Singh and P. N. Tiwari, *J. Chem. Technol. Biotechnol.*, 1997, **69**, 376–382.
224. M. Rao, A. V. Parwate and A. G. Bhole, *Waste Manage.*, 2002, **22**, 821–830.
225. A. Baran, E. Biçak, S. Hamarat-Baysal and S. Önal, *Bioresour. Technol.*, 2006, **98**, 661–665.
226. M. Kobya, *Bioresour. Technol.*, 2004, **91**, 317–321.
227. R. Xie, H. Wang, Y. Chen and W. Jiang, *Environ. Progress Sustainable Energy*, 2013, **32**, 688–696.
228. D. Mohan and C. U. Pittman Jr, *J. Hazard. Mater.*, 2007, **142**, 1–53.
229. T. Budinova, D. Savova, B. Tsyntsarski, C. O. Ania, B. Cabal, J. B. Parra and N. Petrov, *Appl. Surf. Sci.*, 2009, **255**, 4650–4657.
230. C. L. Chuang, M. Fan, M. Xu, R. Brown, S. Sung, B. Saha and C. P. Huang, *Chemosphere*, 2005, **61**, 478–483.
231. M. Fan, W. Marshall, D. Daugaard and R. C. Brown, *Bioresour. Technol.*, 2004, **93**, 103–107.
232. T. Budinova, N. Petrov, M. Razvigorova, J. Parra and P. Galiatsatou, *Ind. Eng. Chem. Res.*, 2006, **45**, 1896–1901.
233. A. Abdelouas, W. Lutze and E. H. Nuttall, *Rev. Mineral. Geochem.*, 1999, **38**, 433–473.
234. P. Ilaiyaraja, Ashish Kumar Singh Deb, K. Sivasubramaniana, D. Ponraju, B. Venkatraman and B. Venkatraman, *J. Hazard. Mater.*, 2013, **250–251**, 155–166.
235. Z. Wang, S.-W. Lee, J. G. Catalano, J. S. Lezama-Pacheco, J. R. Bargar, B. M. Tebo and D. E. Giammar, *Environ. Sci. Technol.*, 2013, **47**, 850–858.
236. M. Betti, *J. Environ. Radioactivity*, 2003, **64**, 113–119.
237. J. Wang and C. Chen, *BioTechnol. Adv.*, 2006, **24**, 427–451.
238. Z. Salem and K. Allia, *Int. J. Chem. Reactor Eng*, 2008, **6**, A10.

239. X. L. Shi and G. W. Qian, *Ind. Saf. Environ. Prot*, 2004, **30**, 69.
240. M. Caccin, F. Giacobbo, M. Da Ros, L. Besozzi and M. Mariani, *J. Radioanal. Nucl. Chem.*, 2013, **297**, 9–18.
241. Z. Yi, J. Yao, F. Wang, H. Chen, H. Liu and C. Yu, *J. Radioanal. Nucl. Chem.*, 2013, **295**, 2029–2034.
242. C. Kutahyalı and M. Eral, *J. Nuclear Mater.*, 2010, **396**, 251–256.
243. Z. Zhang, X. Cao, P. Liang and Y. Liu, *J. Radioanal. Nucl. Chem.*, 2013, **295**, 1201–1208.
244. T. H. C. Yeo, I. A. W. Tan and M. O. Abdullah, *Renewable Sustainable Energy Rev.*, 2012, **16**, 3355–3363.
245. L. Wei and G. Yushin, *Renewable Sustainable Energy Rev.*, 2012, **16**, 3355–3363.

CHAPTER 3

Plants for Soil Remediation

BORHANE MAHJOUB

High Institute of Agronomic Sciences, BP 47, 4042, Chatt Meriem,
University of Sousse, Sousse, Tunisia
Email: mahjoub.borhane@gmail.com

3.1   Introduction

Phytoremediation is a term applied to a group of technologies that use plants to reduce, remove, degrade or immobilise environmental toxins, primarily those of anthropogenic origin, with the aim of restoring area sites to a condition useable for private or public applications. The concept of using plants to clean up contaminated environments is not new. About 300 years ago, plants were proposed for use in the treatment of wastewater. The use of plants in environmental remediation has been called 'green remediation', 'botanical bioremediation', *etc.*

In essence, phytoremediation employs human initiative to enhance the natural attenuation of contaminated sites and, as such, is a process that is intermediate between engineering and natural attenuation. Because phytore-mediation depends on natural, synergistic relationships among plants, micro-organisms and the environment, it does not require intensive engineering techniques or excavation. Human intervention may, however, be required to establish an appropriate plant–microbe community at the site or apply agro-nomic techniques (such as tillage and fertiliser application) to enhance natural degradation or containment processes.

This technology has been receiving attention as being innovative, cost-effective and aesthetically pleasing to the public compared with alternate remediation strategies involving the excavation/removal or chemical *in situ*

RSC Green Chemistry No. 25
Biomass for Sustainable Applications: Pollution Remediation and Energy
Edited by Sarra Gaspard and Mohamed Chaker Ncibi
© The Royal Society of Chemistry 2014
Published by the Royal Society of Chemistry, www.rsc.org

stabilisation/conversion used at hazardous waste sites. As shown in Figure 3.1, the general strategies for phytoremediation of soil are to: phytostabilise soil contaminants into persistently non bioavailable forms in the soil; phytoextract and accumulate the soil elements into the plant shoots for recycling or less expensive disposal; rhizodegradate persistent organic pollutants in the root system zone; and phytodegradate these latter pollutants through metabolism within the plant after they have been absorbed, or phytovolatilise the soil trace elements through the plant leaves.

Phytoremediation strategies have been used effectively to remediate inorganic and organic contaminants in soil and groundwater. Some common examples of phytoremediation applications are shown in Table 3.1. Various plants, including *Thlaspi caerulescens, Viola calaminaria, Brassica napus* L., *Avena sativa, Hordeum vulgare*, tolerate and accumulate metals such as selenium, copper, nickel, cadmium and zinc.[1,2] Alamo switch grass (*Panicum virginatum*) accumulates the radionuclides Cs-137 and Sr-90, compounds present in nuclear fallout from

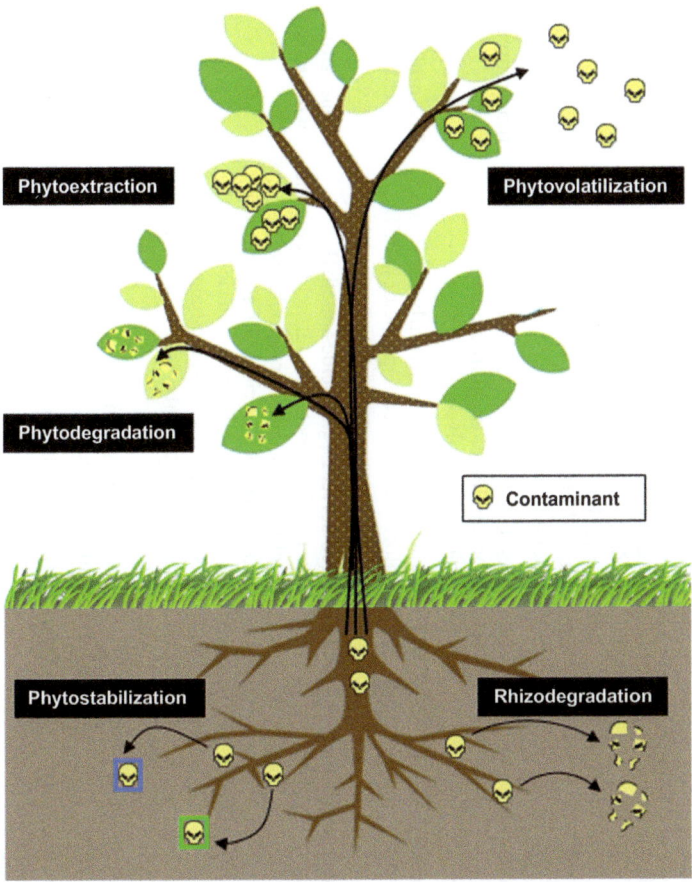

Figure 3.1   The main natural processes underlying the phytoremediation.

Table 3.1 Overview of some phytoremediation applications.

Mechanism	Contaminants	Main plants used	Status
Phytoextraction	Metals (Ag, Cd, Co, Cr, Cu, Hg, Mn, Mo, Ni, Pb, Se, Zn), metalloids, radionuclides, perchlorate, BTEX, PCP, short-chained aliphatic and other organic compounds not tightly bound to soils	Various: Indian mustard, pennycress, alyssum, sunflowers, hybrid poplars	Laboratory pilot and field applications
Phytostabilisation	Metals (As, Cd, Cr, Cu, Hg, Pb, Zn), phenols, tetrachloromethane, trichloromethane and other chlorinated solvents	Indian mustard, hybrid poplars, grasses	Field applications
Rhizodegradation	BTEX, other petroleum hydrocarbons, PAHs, PCP, chlorinated solvents, pesticides, PCBs, and other organic compounds	Red mulberry, grasses, hybrid poplars, catcall, rice	Field applications
Phytodegradation	Chlorinated solvents, herbicides, methyl bromide, tetrabromoethane, tetrachloroethane, dichloroethene, atrazine, organochlorine insecticide P based pesticides, PCBs, phenols, anilines, nitriles, nitrobenzene, picric acid, the nitro based explosives, nitromethane, nitroethane and nutrients, and other organic compounds.	Algae, stonewort, hybrid poplars, black willow, bald cypress	Field demonstrations
Phytovolatilisation	Chlorinated solvents, some inorganics (Se, Hg, As) tritium, m-xylene	Poplars, alfalfa, black locust, Indian mustard	Laboratory and field applications
Vegetative cover (evapotranspiration cover)	Organic and inorganic compounds	Poplars, grasses	Field applications

BTEX = benzene, toluene, ethylbenzene and xylenes; PAHs = polycyclic aromatic hydrocarbons; PCBs = polychlorinated biphenyls; PCP = pentachlorophenol.

weapons testing and reactor accidents.[3] Hybrid poplar trees (*e.g. Populus deltoides* × nigra) reduce the concentration of nitrate in superficial groundwater[4] and degrade the herbicide atrazine from contaminated soils.[5]

Forage grasses inoculated with bacteria degrade individual chlorinated benzoic acids as well as mixtures of these compounds;[6] chlorinated benzoic acids arise out of the degradation of polychlorinated biphenyls (PCBs) and chlorinated herbicides. Various plants have also been found to increase the removal of petroleum hydrocarbons from contaminated soil.[7,8]

3.2 Phytoextraction

3.2.1 Contaminant Hyperaccumulation and Enhanced Phytoextraction

Where some plant species are shown to be endemic to areas of well-defined soil composition, they have been used as geobotanical indicators in mineral exploration. However, some metal-tolerant plant species are also characterised by the ability to accumulate very high concentrations of certain elements, far in excess of normal physiological requirements, and far in excess of the levels found in the majority of other species tolerant of metalliferous soils. These species are known to accumulate levels at least as high as 1% of the plant shoot dry matter for the elements Zn, Ni, Se, Cu, Co or Mn, and over 0.1% Cd. Accumulation of other elements may also occur depending on the degree of soil contamination. These 'special' plants are an important resource for researchers investigating various strategies of soil remediation. The idea of using plants to extract pollutants from contaminated soil was re-introduced and developed by researchers such as Chaney,[9] and the first field trial on Zn and Cd phytoextraction was conducted in 1991.[10] In the last decade, extensive research has been conducted to investigate the biology of metal phytoextraction.

Phytoextraction refers to the use of accumulating plants, which are able to transport and concentrate contaminants from the soil into their harvestable, aboveground parts. Contaminant-enriched plant biomass can be safely disposed of as a hazardous material or eventually used for product recovery. The plant biomass containing the extracted contaminant can also be a resource. For example, biomass that contains selenium, an essential nutrient, has been transported to areas that are deficient in Se and used for animal feed.[11]

Phytoextraction is generally limited to the immediate zone of influence of the roots; thus, root depth determines the depth of effective phytoextraction. The root zones of most metal accumulators are limited to the top foot of soil. To make phytoextraction possible, the contaminants must be bioavailable for plants and plants must have a genetic predisposition for compartmentalisation of extracted pollutants. Phytoextraction is mainly used for accumulation of heavy metals. There are two general approaches to phytoextraction: continuous and chemically enhanced phytoextraction.[12] The first approach uses naturally hyperaccumulating plants with the ability to accumulate an exceptionally high metal content in the shoots. Plant populations colonising metalliferous soils have developed a series of physiological and biochemical adaptations in order to overcome metal toxicities and the physical and chemical stresses frequently associated with such soils. Baker *et al.*[13] proposed two basic strategies by which plants respond to large concentrations of heavy metals in their environment: metal exclusion, whereby uptake and transport of the metal is restricted; and metal accumulation, when metals are accumulated to a high degree in the upper plant parts. There is an intermediate type, that of indicator plants, in which the concentration of heavy metals mirrors that in the soils in a linear relationship. The main attraction of using hyperaccumulators for phytoremediation is the possibility of employing species that remove and concentrate large amounts of pollutants from the soil without significant chemical intervention (except the eventual application of fertilisers). Hyperaccumulating plants are also mostly slow growing with a small biomass and shallow root systems, and lacking in good agronomic characteristics.[14] In non-hyperaccumulating plants, factors limiting their potential for phytoextraction include small root uptake and little root-to-shoot translocation of heavy metals.

The induced phytoextraction approaches enhance contaminant accumulation by addition of accelerants or chelators to the soil. Indeed, common crop plants with high biomass can be triggered to accumulate high amounts of low bioavailable metals, when their mobility in the soil and translocation from the roots to the green part of plants is enhanced by the addition of mobilising agents. A chelating agent is a multidentate ligand which forms complexes with contaminants. In the case of heavy metals, chelators like ethylenediaminetetraacetic acid (EDTA) assist in the mobilisation and subsequent accumulation of soil contaminants such as Pb, Cd, Cr, Cu, Ni and Zn in *Brassica juncea* (Indian mustard) and *Helianthus anuus* (sunflower).[15] The ability of other metal chelators such as *trans*-1,2-cyclohexanediaminetetraacetic acid (CDTA), diethylenetriaminepentaacetic acid (DTPA), ethyleneglycoltetraacetic acid (EGTA), ethylenediamine-N,N'-bis(2-hydroxyphenylacetic acid) (EDDHA) and nitrilotriacetic acid (NTA) to enhance metal accumulation has also been assessed in various plant species.[16,17]

The chemical forms of contaminants in soils affect their solubility, which directly influences their availability to plants. Indeed, plant uptake of metals shows marked dependence on the chemical speciation and soil fractionation of the contaminants (metals particularly). The fraction of metals soluble in the soil solution consists of free hydrated ions, water-soluble organic and inorganic complexes, and metals sorbed on dissolved organic matter. They are the most mobile forms of metals in soil and directly available to plants.

The fractionation of contaminants into soluble compounds in soil solution and exchangeable from soil colloids (both bioavailable to plants) and contaminants strongly bound into the soil solid phase is probably controlled by many types of reactions, mainly: adsorption/desorption reactions; precipitation; penetration into the crystal structure of minerals; and biological mobilisation and immobilisation. These reactions are presumably determined and constrained by soil properties: soil texture; content of organic matter; content and type of clay minerals and Al, Fe and Mn oxides; and prevailing physicochemical conditions in the soil including soil saturation, soil aeration, pH and redox potential.[18]

In addition to the inherent metal characteristics and soil factors which determine fractionation and, consequently, the bioavailability of metals, plants can themselves change metal mobility. Plants can regulate metal solubility by acidification of the rhizosphere due to the extrusion of H^+ from roots and by exuding their own chelating agents, *e.g.* malic and citric acids. The localised excretion of plant chelating agents mobilises nutrients such as Zn, Fe, Mn and other metals. Water-soluble chelating agents increase and maintain metal concentration in the soil aqueous phase. These plant mechanisms are sufficient to desorb readily bioavailable polluting metals, such as Cd, which predominantly form easily hydrolysable, low-energy bonds to the soil solid phase.[19]

To make plants take up Pb and other low bioavailable metals, which are much more strongly bound to the soil solid phase than Cd, strong chelating agents have been used to bring metals into the soil solution. Chelating agents, mostly EDTA, are widely used in agriculture, as ingredients of mineral fertilisers to increase the phytoavailability of Fe and other soil micronutrients, and to maintain the solubility of micronutrients in hydroponic solutions. Chelating agents are also used for *ex situ* chemical extraction of polluting metals from contaminated soils. The literature to date reports a number of chelating agents that have been tested for enhanced heavy metal phytoextraction (Table 3.2).

Several plants have been used in combination with chelating agents for enhanced phytoextraction. The ideal plant should be fast growing and produce a large biomass while accumulating high concentrations of polluting metals. It should also be tolerant enough to grow in contaminated soils and be resistant to the chelating agent. It should have a known agronomics practice and produce usable fruit or biomass to generate some financial income after harvesting.

Table 3.2 Most important chelating agents tested for enhanced phytoextraction of metals from the soil and their stability constant of complex formation (log K at 20–25 °C and ionic strength 0.1–1.0) for low bioavailable Pb, Cr and U in the soil, and readily available Fe and Ca.[20]

	Stability constants of complex formation (log K)				
Chelating agent	Pb^{2+}	Cr^{3+}	U^{4+}	Fe^{3+}	Ca^{2+}
EDTA	18.0	23.4	25.8	25.1	10.6
CDTA	20.2	23.0	27.6	30.1	13.1
DTPA	18.8	–	28.8	27.8	10.7
EDDS	12.7	–	–	–	4.7
NTA	11.5	6.2	9.6[a]	15.9	6.3
HEDTA	15.6	6.1	–	19.7	8.1
EGTA	14.6	–	9.4[a]	20.5	10.8
Citric acid	4.4	–	7.4[a]	11.2	3.5

[a]UO_2^+

EDDS = ethylenediamine-N,N′-disuccinic acid; HEDTA = *N*-(2-hydroxyethyl)ethylenediamine-N, N′, N′-triacetic acid

B. juncea possesses several of these characteristics and is the most commonly used plant species in this remedial approach.

However, there may be risks associated with the use of certain chelators, considering the high water solubility of some chelator–toxin complexes which could result in movement of the complexes to deeper soil layers and potential groundwater and estuarian contamination.[21]

3.2.2   Hyperaccumulating Plants (Hyperaccumulators)

In 1935, Beath *et al.*[22] reported that plants of the genus *Astragalus* were capable of accumulating up to 0.6% selenium in dry shoot biomass. A decade later, Minguzzi and Vergnano[23] identified plants able to accumulate up to 1% Ni in shoots. There are records of strikingly high plant levels of cobalt and copper from the 1960s, and cadmium, arsenic and manganese from the 1970s.

Plants capable of accumulating exceptionally large concentrations of metals such as Zn, Cd, Ni and Pb have been termed 'hyperaccumulators'.[24] Hyperaccumulator plants typically contain > 100 mg kg^{-1} Cd, > 1000 mg kg^{-1} Ni or $> 10\,00$ mg kg^{-1} Zn in their leaf tissue (dry weight). Most plants suffer toxicity and experience yield reduction when leaves contain about 400–00 mg kg^{-1} Zn or 50–100 mg kg^{-1} Ni.

Few hyperaccumulator species have so far been studied agronomically, and it is not known what yields of plant dry matter might be achieved under optimum conditions of climate, nutrition and plant density, particularly since in their natural habitats they are usually growing under adverse physical and nutritional conditions. The inter-element selectivity of the metal uptake process is also well established in only a few cases. This is particularly relevant when a plant species is proposed for use in phytoremediation of a substrate that is chemically quite different from the normal habitat of the plants.

Hyperaccumulator plants are generally found in the Brassicaceae, Euphorbiaceae, Asteraceae, Lamiaceae or Scrophulariaceae plant families. For instance, *Brassica juncea* can accumulate Pb, Cr(VI), Cd, Cu, Ni, Zn, ^{90}Sr, B and Se.[25–27] It has over 20 times the biomass yield of *Thlaspi caerulescens*.[25] Among the different plant species screened, *B. juncea* had the best ability to transport lead to the shoots, accumulating $> 1.8\%$ lead in the shoots (dry weight). With the exception of sunflower (*Helianthus annuus*) and tobacco (*Nicotiana tabacum*), other non-brassica plants had phytoextraction coefficients < 1. A total of 106 *B. juncea* cultivars varied widely in their ability to accumulate Pb, with different cultivars ranging from 0.04% to 3.5% Pb accumulation in the shoots and 7–19% in the roots.[26]

Baker *et al.*[13] found 80 species of nickel-accumulating plants in the Buxaceae (including boxwood) and Euphoribiaceae (including cactus-like succulents) families. Some euphorbs can accumulate up to 5% of their dry weight in nickel. *Brassica napus* and *B. canola* have been shown to accumulate Se. Kenaf (*Hibiscus cannabinus* L. cv. Indian) and tall fescue (*Festuca arundinacea* Schreb cv. Alta) also take up Se, but to a lesser degree than canola.[1] Hybrid poplar trees were also used in a field study in mine tailing wastes contaminated with

As and Cd.[28] Lambsquarter leaves had relatively higher As concentrations (14 mg kg^{-1} As) than other native plant or poplar leaves (8 mg kg^{-1}) in mine tailing wastes.[28] Sunflowers took up Cs and Sr, with Cs remaining in the roots and Sr moving into the shoots.[29]

Notable metal accumulators include several members of the Cruciferae. For example, *Alyssum* and *Thlaspi* species native to serpentine soils can accumulate concentrations of Ni in excess of 2% on a dry matter basis,[30] while species of *Thlaspi* from Zn/Pb mineralised soils can accumulate Zn to more than 2%, Cd up to 0.1%, and Pb up to 0.8%.[30,31]

It is important to notice that other metal accumulator plants such as the crop plants corn, sorghum and alfalfa may be more effective than hyperaccumulators and remove a greater mass of metals due to their faster growth rate and larger biomass. Additional study is needed to quantify their contaminant removal efficiency. Besides, the cultivation of fast-growing tree species for remediation purposes and the production of renewable energy from contaminated biomass is an approach to the use of post-mining polluted areas which offers an alternative to more traditional types of land use.[32,34] Compared with herbaceous species, fast-growing trees have several advantageous characteristics such as a deeper root system, a high productivity and transpiration activity.

The number of taxonomic groups (taxa) of hyperaccumulators varies according to which metal is hyperaccumulated. To date, about 400 taxa have been identified as hyperaccumulators.[35] The majority of them are Ni hyperaccumulators (*cf.* Table 3.3).

Possibly, the best-known metal hyperaccumulator is *Thlaspi caerulescens* (alpine pennycress). *Thlaspi caerulescens* is primarily a Zn and Cd hyperaccumulator and is the most extensively studied. It actually requires abnormal amounts of Zn to be able to grow normally.[30,36] *Thlaspi caerulescens* is not uniform in metal accumulation capability.[37] The accumulation rates vary among populations and are influenced by the physical and chemical characteristics of soils.[38] Some populations of *T. caerulescens* from the south of France exhibited extraordinary Cd hyperaccumulating ability where foliar Cd concentrations could reach 3000 mg kg^{-1}.[39-41] These southern France populations appear to have the great potential to make Cd phytoextraction a reality. The nomenclature of this species has a complex history, but it is now generally

Table 3.3 Number of metal hyperaccumulator plants.

Metal	Criterion/% in leaf dry matter	Number of taxa	Number of families
Cadmium	>0.01	1	1
Cobalt	>0.1	28	11
Copper	>0.1	37	15
Lead	>0.1	14	6
Manganese	>1.0	9	5
Nickel	>0.1	317	37
Zinc	>1.0	11	5

accepted that *T. caerulescens* is the legitimate name for the species that occurs sporadically throughout central and western Europe, often at sites of Zn and Pb mining and/or smelting.[42] The name *T. alpestre*, formerly widely used for this species, properly applies to a different plant.[43] *Thlaspi caerulescens* is remarkable in its ability to show extreme Zn accumulation even from soils of normal Zn status.[44] This species also has wider multi-element accumulating abilities. Indeed, Ni, Co, Mn and Cd could all be accumulated to high concentrations in the leaves. Other elements such as Pb, Cu, Fe, Al and Cr were accumulated strongly in the roots, but not readily translocated to the shoots. Several findings have drawn attention to other *Thlaspi* species as metal accumulators. Rascio[45] showed that the species now listed as *T. cepaeifolium* subsp. rotundifolium from zinc-polluted soils near the border of Italy and Austria was also a hyperaccumulator of zinc. A more complete study of the genus identified high concentrations of Ni and/or Zn in *Thlaspi* species from the Pacific Northwest of the United States, and from Turkey, Cyprus and Japan. In Turkey, in particular, there are several serpentine-endemic Ni-accumulating *Thlaspi* species, including relatively recent discoveries such as *T. cariense*.[46,47]

Reeves and Baker[48] demonstrated that the ability of the Austrian species *T. goesingense* to accumulate nickel and zinc was an innate property, unrelated to the geochemistry of the area from which the seed originated, giving rise to the notion of a 'constitutional' metal-tolerance and metal-uptake ability. Although much of the focus of recent work has been on the properties of *T. caerulescens*, more detailed study of the closely related *T. brachypetalum* and of other species such as *T. goesingense, T. cepaeifolium, T. alpinum* and *T. ochroleucum* is certainly justified. The North American *T. montanum* also deserves further investigation.

As shown in Table 3.4, hyperaccumulating plants usually accumulate only a specific metal. However, some plant species has been found that demonstrates a wide spectrum of metal hyperaccumulation.

Because there can be significant variation in metal-accumulating properties among isolated populations of sporadically distributed species, and intra-population variability as well, it is important to preserve all populations of such species and to see that they are maintained in a 'healthy' state. Metal-tolerant species, hyperaccumulators in particular, are a valuable and potentially useful biological resource, and concerted efforts are now needed to improve our field knowledge and distribution records of these plants in many parts of the world. This information will then be valuable in the formulation of strategies for conserving rare species and threatened populations. These issues have been discussed recently in some detail by Whiting *et al.*[49]

It is clear that the hyperaccumulating plants represent an unusual and valuable biological resource with great potential for use in a variety of strategies for soil remediation. There is an urgent need for more exploration of the world's contaminated soils, so that more accumulator species can be recognised, and the distribution and rarity of these species become better defined. Conservation measures need to be put in place to ensure that rare hyperaccumulator species are not lost through any of the above threats. In addition,

Table 3.4 Plants able to accumulate several metals.[30]

Plant name	Origin	Accumulated elements
Azolla filiculoides	Africa, floating	Cu, Ni, Pb, Mn
Bacopa monnieri	India, emergent species	Hg, Cu, Cr, Pb, Cd
Eichhornia crassipes	Pantropical/subtropical, troublesome weed	Cd, Cr, Zn, Hg, Pb, Cu
Hydrilla verticillata	Southern Asia but introduced and speading as troublesome weed in the warmer states of USA	Cd, Cr, Hg, Pb
Lemna minor	Native to North America and widespread	Pb, Cd, Cu, Zn
Pistia stratiotes	Pantropical, native to southern USA	Cu, Cd, Hg, Cr
Salvinia molesta	India	Cr, Ni, Pb, Zn
Spirodela polyrhiza	Native to North America	Cd, Ni, Cr, Pb, Zn
Vallisneria americana	Native to Europe and North Africa but widely cultivated	Cu, Cd, Cr, Pb
Brassica juncea	Cultivated	Pb, Zn, Ni, Cu, Cr, Cd, U
Helianthus annuus	Cultivated	Pb, U, Sr, Cs, Cr, Cd, Cu, Mn, Ni, Zn
Agrostis castellana	Portugal	As, Pb, Zn, Mn, Al
Thlaspi caerulescens	Europe	Zn, Cd, Co, Cu, Ni, Pb, Cr
Athyrium yokoscense	Japan	Cu, Cd, Zn, Pb

there are issues to be resolved about the way in which both scientific and commercially-oriented work can proceed, while promising benefits to the people in whose territory the resource is found in nature.

3.3 Phytostabilisation

Plants can be used to stabilise the land and to reduce or eliminate the movement of toxic elements from the contaminated soil to the general environment.[50] This special form of stabilisation, mediated by plants, is usually called 'phytostabilisation'. It is particularly applicable for elemental contaminants in soils at waste sites for which the best alternative is often to hold these pollutants in place to: (i) prevent greater bioavailability during removal; (ii) avoid more worker exposure; and (iii) minimise disruption of critical ecosystems. In general, phytostabilisation can be temporarily effective for metals tightly bound to soils. The immobilisation of the contaminants and the reduction of their migration through the soil is generally due to their absorbing and binding to the plant structure. This process effectively reduces the bioavailability of the harmful contaminants. The value of the phytoremediated-land may also be increased if previously derelict land could be used for some economically or socially beneficial purpose. Examples include growing biofuel crops or forestry on the land, or turning the land into a park or nature reserve. This technique is based on a sound fundamental knowledge of soil chemistry, agricultural practices, experience with the reclamation and revegetation of contaminated sites and industrial plant ecology.

Phytostabilisation occurs through root-zone microbiology and chemistry, and/or alteration of the soil environment or contaminant chemistry. This process takes advantage of plant roots ability to alter soil environment conditions, such as pH and soil moisture content. Soil pH may be changed by plant root exudates or through the production of CO_2. Phytostabilisation can change metal solubility and mobility, or impact on the dissociation of organic compounds. The plant-affected soil environment can convert metals from a soluble to an insoluble oxidation state.[25] Phytostabilisation can occur through sorption, precipitation, complexation or metal valence reduction.

Phytostabilisation of Pb, Cd, Zn, As, Cr, Mo, Cs, Sr, U and other elements is thereby theoretically possible. Many divalent metals (*e.g.* Pb, Cd, Zn, Ni, Cu, Co and Hg) precipitate as sulfide minerals under anaerobic conditions. Thus, occasionally the strategy may be to lower redox conditions in the soil environment and to reduce sulfate to sulfide. In addition, some relative stability is possible for redox metals that have less mobile valence states in natural settings. Chromium, for example, is much less mobile and less toxic in the reduced state [Cr(III)] than in the oxidised state [Cr(VI)]. A self-sustaining or low-maintenance application of phytostabilisation may be possible for elements that are more stable in an oxidised state. Arsenic and molybdenum are less mobile in the oxidised valence states [arsenate As(v) and molybdate Mo(v)], and thus vigorously growing, deep rooting or deep planted trees, shrubs or wetland plants might keep these pollutants oxidised and immobile. Sorbents such as Fe(III) oxides may be necessary during seasonal lulls in growth to buffer the lack of continuous pumping of oxygen by plants into the soil. Finally, root control will be difficult, but exudation may sustain reducing conditions below healthy oxidised roots.[51]

This immobilisation requires heavy metal-tolerant living plants that reduce the mobility of heavy metals in soil by uptake and storage in the roots.[25] The plants growing on these soils will increase the stability of the heavy metals and prevent wind or water erosion.[52] By choosing and/or maintaining an appropriate cover of plant species, coupled with appropriate soil amendments, it may be possible to stabilise certain contaminants (particularly metals) in the soil and reduce the interaction of these contaminants with associated biota.[53] In some cases, contaminant stabilisation might be due primarily to the effects of soil amendments, with plants only contributing to stabilisation by decreasing the amount of water moving through the soil and by physically stabilising the soil against erosion.

As an example, *in situ* inactivation stabilises soil Pb both chemically and physically through the use of soil amendments and a vegetative cover. Soil amendments alter the existing Pb chemistry in the soil and reduce the biological availability of Pb by inducing the formation of very insoluble Pb species. Plants with dense canopies and rooting systems, such as grasses, physically stabilise the soil against rain impact and erosion or leaching, as well as restrict off-site migration of the contaminants.

The use of metals by plants is very limited, so that stabilising these elements *in situ* is sometimes the best alternative for sites with low contaminant levels (*i.e.* near risk thresholds) and other sites where vast areas are contaminated (*i.e.* large-scale removal action).

Phytostabilisation has several advantages. It has a lower cost and is less disruptive than other more vigorous soil remedial technologies. Other advantages of this strategy is that disposal of the hazardous plant material is not required and soil removal is unnecessary. This technique can also be used to re-establish a vegetative cover in sites where natural vegetation is lacking due to high metal concentrations in surface soils or physical disturbances to surficial materials. Metal-tolerant species can be used to restore vegetation to the sites, thereby decreasing the potential migration of contamination through transport of exposed surface soils and leaching of soil contamination to groundwater.

Plants used for phytostabilisation had to be tolerant to the contaminants present in the particular site, but the accumulation of pollutants in their aerial parts may be disadvantageous. Unlike phytoextraction, the ideal plants for phytostabilisation of metals in soils are those that do not take up or that limit uptake to the roots, thus avoiding translocation to the shoots. In addition, the root zone, root exudates, contaminants and soil amendments must be monitored to prevent an increase in metal solubility and leaching. The vegetation and soil may then require long-term maintenance to prevent re-release of the contaminants. As well, aboveground translocation and, to a lesser degree, belowground root accumulation, expose the food chain to the potential toxicity associated with bioaccumulation of metals. Hence, immobilisation of toxic elements such as Pb, Cd, Zn and As by revegetation may require soil amendments to increase soil organic matter, raise the pH (using lime), or bind some constituents with phosphate or carbonate.[4] Without amendments, cadmium readily translocates to the leaves of many plants, which represents a risk of food chain bioaccumulation. Such concern may limit the consideration of phytostabilisation to some waste sites. For these reasons, phytostabilisation is often considered to be only an interim measure.

Concerning soil contaminated with xenobiotic organic pollutants, phytostabilisation is based on sequestration processes (thus the term 'phytosequestration'), which may be better classified as 'phytotransformation' according to the 'green liver' concept.[54] These processes include humification, covalent or irreversible binding and lignification. The term 'phytolignification' has been used to refer to a form of phytostabilisation in which organic compounds are incorporated into plant lignin.[53] Humification reduces bioavailability and occurs from both microbial and plant mediated metabolism or transformation. The key to the phytostabilisation of organic compounds is the benefit from plants and fungi in achieving greater humification, irreversible binding, and lignification. Specifically, phytostabilisation pertains to increased humification and binding of contaminants by soil bacteria whose numbers and activity are enhanced by the presence of plants, and humification and binding by fungi and by external plant enzymes.[4] Lignification by fungi, which does not occur within bacteria, would appear to be important in both phytostabilisation and phytotransformation. Thus, only increased humification and binding by microorganisms enhanced by plants would seem to constitute phytostabilisation of organic xenobiotic compounds in the rhizosphere.

3.4   Rhizodegradation

3.4.1   Rhizosphere Biodegradation

In most remedial strategies for organic pollutants, the complete destruction of the pollutant molecule is highly desirable. Due to the persistence, toxicity and widespread abundance of recalcitrant organic pollutants in the terrestrial environment, efficient removal of the contaminant can be a result of the combined influence of both plants and root-associated microbes through rhizodegradation.

The growth of plant roots in the soil has profound effects. The rhizosphere is the enlarged zone of soil under the direct influence of a plant root. The abiotic and biotic characteristics in the rhizosphere differ greatly from the rootless, bulk soil. The rhizosphere provides a unique terrestrial environment that harbours a remarkable diversity of soil microorganisms. Within the rhizosphere, consortia of organisms mineralise simple and complex naturally occurring organic compounds. The presence of plant roots will often increase the size and variety of microbial populations in the soil surrounding roots (the rhizosphere) or in mycorrhizae (associations of fungi and plant roots). Significantly higher populations of total heterotrophs, denitrifiers, pseudomonads, BTX (benzene, toluene, xylenes) degraders, and atrazine degraders were found in rhizosphere soil around hybrid poplar trees in a field plot than in non-rhizosphere soil. The increased microbial populations are due to stimulation by plant exudates.[55]

Plant exudates are compounds produced by plants and released from plant roots. They include sugars, amino acids, organic acids, fatty acids, sterols, growth factors, nucleotides, flavanones, enzymes and other compounds.[4,56] The microbial populations and activity in the rhizosphere can be increased due to the presence of these exudates and can result in increased organic contaminant biodegradation in the soil. Additionally, the rhizosphere substantially increases the surface area where active microbial degradation can be stimulated. Degradation of the exudates can lead to co-metabolism of contaminants in the rhizosphere.[57]

Stimulation of soil microbes by plant root exudates can also result in alteration of the geochemical conditions in the soil, such as pH, which may result in changes in the transport of inorganic contaminants.

The increased microbial populations and activity in the rhizosphere can result in increased contaminant biodegradation in the soil, and degradation of the exudates can stimulate co-metabolism of contaminants in the rhizosphere. Increased biodegradation could occur even in the absence of root exudates. Indeed, plants and plant roots can affect the water content, water and nutrient transport, aeration, structure, temperature, pH, or other parameters in the soil, often creating more favourable environments for soil microorganisms.[57]

Because the rhizosphere extends only about 1 mm from the root and initially the volume of soil occupied by roots is a small fraction of the total soil volume, the soil volume initially affected by the rhizosphere is limited. With time, however, new roots will penetrate more of the soil volume and other roots will decompose, resulting in additional exudates to the rhizosphere. Thus, the extent of rhizodegradation will increase with time and with additional root growth. The effect of rhizodegradation might extend slightly deeper than the root zone. If the exudates are water-soluble, not strongly sorbed and not quickly degraded, they may move deeper into the soil.[57] As a result, researchers have focused on the rhizosphere as a zone of enhanced biodegradation[58,59] and on the plant–microbe interactions that occur there.[60]

Rhizodegradation is then the enhancement of naturally occurring biodegradation in soil through the influence of plant roots, and ideally will lead to destruction or detoxification of organic contaminants. Rhizodegradation is also known as plant-assisted degradation, plant-assisted bioremediation, plant-aided *in situ* biodegradation or enhanced rhizosphere biodegradation. For many organic compounds, the rhizodegradation is a dominant fate and overall the *in situ* destruction is a desired fate, limiting transfer into aboveground tissues and to the atmosphere.

Rhizosphere microorganisms are actively involved in the decomposition and cycling of organic carbon in the environment. Organic contaminants in soil can often be broken down into daughter products or completely mineralised to inorganic products such as carbon dioxide and water by naturally occurring bacteria, fungi and actinomycetes.

In a number of cases the aerobic conditions can lead to enhanced hydrocarbon degradation. Degradation of hydrocarbon fuel spills has been noted[61,62] and can decrease the role of processes that are directly plant mediated, such as uptake and translocation. While numerous applications of phytoremediation for petroleum are in place and volatile organic compound (VOC) fractions are likely to be present, little published work exists on VOCs in particular. Increased benzene degradation in planted reactors was clearly noted in a laboratory study, although the artefacts associated with the laboratory study probably increased the extent of degradation enhancement.[63] To evaluate the impacts of terrestrial phytoremediation on hydrocarbon VOCs, more research is needed.

Rhizosphere bioremediation can capitalise on the innate qualities of vegetation and the potential for the mineralisation of organic pollutants in the rhizosphere. Understanding the dynamic nature and ecological processes within the rhizosphere is essential to employing rhizodegradation as a remedial option.

3.4.2 Benefits and Impediments of Rhizodegradation

Appealing features of rhizodegradation include destruction of the contaminant *in situ*, the potential complete mineralisation of organic contaminants, and that translocation of the compound to the plant or atmosphere is less likely than with other phytoremediation technologies since degradation occurs at the source of the contamination. Harvesting of the vegetation is not necessary since there is contaminant degradation within the soil, rather than contaminant accumulation within the plant. Root penetration throughout the soil may allow a significant percentage of the soil to be contacted. However, at a given time only a small percentage of the total soil volume is in contact with living roots. Development of an extensive root zone is likely to require substantial time. Also, root penetration into soil can be limited due to the physical structure or moisture conditions of the soil, leading to some portions of the soil never being contacted by roots.

Possibly, the most serious impediment to successful rhizodegradation is its limitation to the depth of the root zone. Many plants have relatively shallow root zones and the depth of root penetration can also be limited by soil moisture conditions or by soil structures such as hard pans or clay pans that are impenetrable by roots. However, in some cases roots may extend relatively deep (110 cm) and extend into soil with high contaminant concentrations.[64]

Other potential impediments to successful rhizodegradation include the often substantial time that may be required to develop an extensive root zone. Indeed, time is needed to allow new roots to penetrate more of the soil volume and other roots decompose. But this root turnover, although time-consuming, adds exudates to the rhizosphere.[65]

Stimulation of rhizosphere organisms does not always lead to increased contaminant degradation, as populations of microorganisms that are not degraders might be increased at the expense of degraders. Competition between the plants and the microorganisms can also impact the amount of biodegradation.[66]

Plant uptake can occur for many of the contaminants that have been studied. Laboratory and field studies need to account for other loss and phytoremediation mechanisms that might complicate the interpretation of rhizodegradation. For instance, if plant uptake occurs, phytodegradation or phytovolatilisation could occur in addition to rhizodegradation.

3.4.3 Contaminants and Applicable Plants

A wide range of organic contaminants are candidates for rhizodegradation, such as petroleum hydrocarbons, polyaromatic hydrocarbons (PAHs), pesticides, chlorinated solvents, pentachlorophenol (PCP), polychlorinated biphenyls (PCBs) and surfactants. Higher populations of benzene-, toluene- and *o*-xylene-degrading bacteria were found in soil from the rhizosphere of poplar trees than in non-rhizosphere soil, although it was not clear that the populations were truly statistically different. Schwab and Banks[67] investigated total petroleum hydrocarbon (TPH) disappearance at several field sites contaminated with crude oil, diesel fuel or petroleum refinery wastes, at initial petroleum hydrocarbon contents of 1700 to 16 000 mg kg^{-1} TPH. The presence of some species led to greater TPH disappearance compared with other species or in unvegetated soil. At the crude oil contaminated field site near the Gulf of Mexico, an annual rye–soybean rotation plot and a St Augustine grass–cowpea rotation plot had significantly ($P < 0.05$) greater TPH disappearance than did sorghum–Sudan grass or unvegetated plots, in 21 months. In addition, at the diesel fuel-contaminated Craney Island field in Norfolk, Virginia, the fescue plot had significantly ($P < 0.10$) greater TPH disappearance than did an unvegetated plot.

The role of vegetation in the enhanced rhizodegradation of PAHs in the rhizosphere has been actively studied over the past decade. In a seminal paper on the subject, a mixture of eight common prairie grasses was evaluated in the greenhouse for enhanced disappearance of four- and five-ringed PAHs in the rhizosphere.[68] This study used a mix of prairie grasses: big bluestem (*Andropogon gerardi*); little bluestem (*Schizachyrium scoparius*); indiangrass (*Sorghastrum nutans*); switchgrass (*Panicum virgatum*); Canada wild rye (*Elymus canadensis*); western wheatgrass (*Agropyron smithii*); side oats grama (*Bouteloua curtipendula*); and blue grama (*Bouteloua gracilis*). The presence of the prairie grasses, each having a deep, fibrous root system, significantly lowered the concentration of high molecular weight PAHs in the rhizosphere after 219 days compared with the unplanted controls.

In a study conducted in a greenhouse, statistically greater loss of fluoranthene, pyrene and chrysene occurred in soil planted with perennial ryegrass (*Lolium perenne*) than in unplanted soil.[69] In another landmark paper, pyrene was significantly dissipated in the rhizosphere of fescue *(Festuca arundinacea)*, Sudan grass *(Sorghum vulgare)*, switch grass *(Panicum virgatum)* and alfalfa *(Medicago saliva)* after 24 weeks compared with the unplanted control.[70] Pyrene removal was further stimulated in the rhizosphere by the addition of simple organic acids *(e.g.* formic and succinic), suggesting the importance of root-released compounds and rhizosphere microorganisms in the biodegradation of PAHs.

The concentrations of several recalcitrant hydrocarbons were significantly reduced in the rhizosphere of ryegrass *(Lolium perenne)* by a consortium of microorganisms.[71]

Rhizodegradation of more recalcitrant PAHs has also been observed. For example, the enhanced degradation of [^{14}C]-benzo(a)pyrene was demonstrated in the rhizosphere of fescue *(Festuca arundinaced)* even though the pollutant was highly sorbed to the organic material present in the soil.[72]

In addition, a most-probable-number analysis that quantified bacteria growing on a mixture of phenanthrene, chrysene, pyrene and benzo(a)pyrene showed that the number of PAH degraders increased over a one-year period in PAH-contaminated soil planted with perennial ryegrass (*Lolium perenne*), fescue (*Festuca arundinaced*), kleingrass (*Panicum coloratum*) and (annual) Indian mustard (*Brassica juncea*).[73] The results of this experiment illustrate the role of plants in the rhizodegradation of PAHs, and suggest an important role for species-specific root-released compounds.

Fungi inhabiting the rhizosphere of select plant species may play an important role in the rhizodegradation of PAHs. The establishment and maintenance of plant species growing on PAH-contaminated soils may be greatly improved by symbiotic relationships with arbuscular mycorrhizal fungus.[74] White-rot fungi *(Phanerochaete chysosporiuni)* have been shown to partially oxidise PAHs to intermediates having higher water solubility, leading to increased bioavailability and biodegradation of the metabolised PAH.[75] The combined influence of fungi and bacteria may greatly enhance the degradation of PAHs, and such microbial interactions may be stimulated in the rhizosphere of select plant species.[76]

Pesticide biodegradation was found to be influenced by plants. *Kochia* sp. rhizosphere soil increased the degradation of herbicides (0.3 g g^{-1} trifluralin, 0.5 g g^{-1} atrazine and 9.6 g g^{-1} metolachlor) relative to non-rhizosphere soil. These laboratory experiments used rhizosphere soil but were conducted in the absence of plants to minimise any effects of root uptake.[77] In a laboratory study, bush bean (*Phaseolus vulgaris* cv. 'Tender Green') rhizosphere soil had higher mineralisation rates for 5 g g^{-1} of the organophosphate insecticides parathion and diazinon than non-rhizosphere soil. Diazinon mineralisation in soil without roots did not increase when an exudate solution was added, but parathion mineralisation did increase.[78]

A greenhouse study indicated that rice (*Oryza sativa* L.) rhizosphere soil with 3 g g^{-1} propanil herbicide had increased numbers of Gram-negative bacteria that could rapidly transform the propanil. It was hypothesised that the best propanil degraders would benefit from the proximity to plant roots and exudates.[79] Microorganisms capable of degrading 2,4-dichlorophenoxyacetic acid (2,4-D) occurred in elevated numbers in the rhizosphere of sugar cane, compared to non-rhizosphere soil.[80] As well, the rate constants for 2,4-D and 2,4,5-trichlorophenoxyacetic acid (2,4,5-T) herbicide biodegradation in a laboratory evaluation were higher in field-collected rhizosphere soil than in non-rhizosphere soil.[81]

Chlorinated solvents may be subject to rhizodegradation. In a growth chamber study, trichloroethene (TCE) mineralisation was increased in soil planted with a legume (*Lespedeza cuneata*), Loblolly pine (*Pinus taeda* (L.)), and soybean (*Glycine max* (L.) Merr., cv. Davis), compared with non-vegetated soil. In another study, the presence of alfalfa possibly contributed to the dissipation of 100 and 200 L L^{-1} TCE and 50 and 100 L L^{-1} 1,1,1-trichloroethane (TCA) in groundwater, through the effect of root exudates on soil bacteria.[82] Newman *et al.*[83] did not find any rhizodegradation of TCE in a two-week long laboratory experiment using hybrid poplars; however, they could not conclusively rule out the occurrence of microbial degradation of TCE in the soil.

Other contaminants are also candidates for rhizodegradation, as indicated by a variety of greenhouse, laboratory and growth chamber studies. Mineralisation rates of 100 mg kg^{-1} PCP were greater in soil planted with crested wheatgrass than in unplanted controls.[84] Proso millet (*Panicum miliaceum* L.) seeds treated with a PCP-degrading bacterium germinated and grew well in soil containing 175 mg L^{-1} PCP compared with untreated seeds.[85]

The capability of vegetation to stimulate the biodegradation of PCBs by root-associated fungi has been demonstrated. Ectomycorrhizal fungi, which can exist in the soil as free-living organisms or in plant-specific associations, have been shown to metabolise a variety of PCB congeners in the laboratory.[86] Aerobic metabolism of PCBs by bacteria generally takes place only in the presence of a co-metabolite that resembles the biphenyl structure of the PCB backbone. A variety of secondary plant compounds including catechin, coumarin and flavonoids, resemble the biphenyl structure. Research into the phytoremediation of PCBs has shown that these plant compounds support the growth and degradative activity of known PCB degrading bacteria.[87] Spearmint (*Mentha spicata*) extracts contained a compound that induced co-metabolism of a PCB.[88] Red mulberry (*Morus rubra* L.), crab apple [*Malus fusca* (Raf.) Schneid] and osage orange [*Maclura pomifera* (Raf.) Schneid] produced exudates with relatively high levels of phenolic compounds, at concentrations capable of supporting growth of PCB-degrading bacteria.[89]

The biodegradation of PCBs from the co-metabolism of carvone can be further enhanced by the addition of surfactants.[90] By the way, the surfactants linear alkylbenzene sulfonate (LAS) and linear alcohol ethoxylate (LAE) at 1 mg L^{-1} had greater mineralisation rates in the presence of cattail (*Typha latifolia*) root microorganisms than in non-rhizosphere sediments.[91]

In summary, the overall mechanisms responsible for enhanced removal of recalcitrant organic pollutants in the rhizosphere are complex and not well understood. Plants are capable of the direct uptake, accumulation and transformation of organic compounds,[91] but given the low aqueous water solubility and significant amount of sorption, plant uptake of PCBs and PAHs is unlikely to be a significant factor in biodegradation. However, because root-associated microorganisms play a major role in the decomposition of recalcitrant and naturally occurring organic compounds, it is reasonable to assume that enhanced rhizodegradation of organic pollutants should proceed at appreciable rates within the rhizosphere of select vegetation. Compelling evidence exists regarding the beneficial use of phytoremediation for the treatment of recalcitrant organic pollutants, but the underlying mechanisms are not readily understood.

In addition, the choice of plants suitable for the rhizodegradation of organic pollutants has received little attention, as easy-to-grow grasses and agronomic plants are usually employed for greenhouse-based experiments and field demonstration sites. From an ecological background, plant selection should be directed toward vegetation that can not only tolerate the contaminated environment at a given locale (including climatic and soil conditions), but also has the potential to form rhizosphere-specific associations between plant roots and pollutant-degrading microorganisms. As a result, more attention should be paid to the ecological and environmental factors that are likely to be quite important for the successful development of a rhizosphere-based bioremediation technology of recalcitrant organic pollutants.

Examinations of naturally revegetated terrestrial sites contaminated with recalcitrant organic pollutants provide compelling evidence to support the long-term efficacy of rhizodegradation.[92] The effectiveness of rhizodegradation may be site-specific and not universal, but applying sound ecological principles and appropriate planting and management practices at contaminated sites can accelerate, maximise and sustain important natural processes that will lead to the rhizodegradation of recalcitrant organic pollutants and remediation of polluted sites.

3.5 Phytodegradation

3.5.1 Plant Uptake and Metabolism

A contaminant can be directly degraded and completely mineralised *via* phytodegradation, also known as phytotransformation.[93] Phytodegradation is the breakdown of contaminants taken up by plants through metabolic processes within the plant, or the breakdown of contaminants external to the plant through the effect of compounds produced by the plants. The main mechanism is plant uptake and metabolism. Phytodegradation is then a contaminant destruction process. It is not dependent on microorganisms associated with the rhizosphere. Any degradation caused by microorganisms associated with the plant root is considered rhizodegradation.

Phytodegradation that occurs outside the plant is mainly due to the release of enzymes that cause transformation. Plant-formed enzymes have been identified for their potential use in degrading contaminants such as munitions, herbicides and chlorinated solvents. Enzymes of particular interest for phytodegradation include: (i) dehalogenase (transformation of chlorinated compounds); (ii) peroxidase (transformation of phenolic compounds); (iii) nitroreductase (transformation of explosives and other nitrated compounds); (iv) nitrilase (transformation of cyanated aromatic compounds); and (v) phosphatase (transformation of organophosphate pesticides).

So that the phytodegradation occurs within the plant, the plant must be able to take up the compound. Uptake of contaminants is dependent on hydrophobicity, solubility and polarity. Moderately hydrophobic organic compounds are most readily taken up by and translocated within plants. Very soluble compounds will not be sorbed onto roots or translocated within the plant. Hydrophobic compounds can be bound to root surfaces or partitioned into roots, but cannot be further translocated within the plant.[4,14] Nonpolar molecules with molecular weights <500 will sorb to the root surfaces, whereas polar molecules will enter the root and be translocated.[94] For these reasons, short chain halogenated aliphatic compounds could be taken up by plants.[95] Plant uptake of organic compounds can also depend on type of plant, age of contaminant, and many other physical and chemical characteristics of the soil.

3.5.2 Concerned Contaminants and Applicable Plants

Organic compounds are the main category of contaminants subject to phytodegradation. They include organic contaminants such as munitions, chlorinated solvents, herbicides and insecticides. Inorganic nutrients can also be removed through plant uptake and metabolism.

Plants can metabolise a variety of organic compounds, including the chlorinated solvent (TCE),[96] the explosive trinitrotoluene (TNT)[97] and the herbicide atrazine.[92] Partial metabolism by wheat and soybean plant cell cultures was found for a variety of compounds including 2,4-D), 2,4,5-T, 4-chloroaniline, 3,4-dichloroaniline, PCP, diethylhexylphthalate (DEHP), perylene, benzo(a)pyrene, hexachlorobenzene, dichlorodiphenyltrichloroethane (DDT) and PCBs.[98–100] In phytodegradation applications, transformation of a contaminant within the plant to a more toxic form, with subsequent release to the atmosphere through transpiration, is undesirable. The formation and release of vinyl chloride resulting from the uptake and phytodegradation of TCE has been a concern. However, although low levels of TCE metabolites have been found in plant tissue, vinyl chloride has not been reported.[96] Plant-produced enzymes that metabolise contaminants may be released into the rhizosphere, where they can remain active in contaminant transformation. These enzymes are associated with transformations of chlorinated compounds, munitions, phenols, the oxidative step in munitions, and herbicides, respectively. Some examples of plant-produced enzyme with their targeted contaminants are shown in Table 3.5.

In one week, the dissolved TNT concentrations in flooded soil decreased from 128 ppm to 10 ppm in the presence of the aquatic plant parrot feather *(Myriophyllum aquaticum)*, which produces nitroreductase enzyme that can partially degrade TNT.[4] The nitroreductase enzyme has also been identified in a variety of algae, aquatic plants, and trees.[4] Hybrid poplar trees metabolised TNT to 4-amino-2,6-dinitrotoluene (4-ADNT), 2-amino-4,6-dinitrotoluene (2-ADNT), and other unidentified compounds in laboratory hydroponic and soil experiments.[97]

Poplar trees (*Populus* spp.) are capable of transforming TCE in soil. Uptake and degradation of TCE has also been confirmed in poplar cell cultures and in hybrid poplars.[96] Laboratory studies have demonstrated the metabolism of methyl *tert*-butyl ether (MTBE) by poplar cell cultures, and provided some indication of MTBE uptake by eucalyptus trees.[95] Atrazine degradation has occurred in hybrid poplars *(Populus deltoides* × nigra DN34, Imperial

Table 3.5 Plant-formed enzymes with known phytodegradation activity.

Enzyme	Plant known to produce enzymatic activity	Targeted contaminants
Dehalogenase	Hybrid poplars (*Populus* spp.), algae (various spp.)	Chlorinated solvents
Laccase	Parrot feather (*Myriophyllum aquaticum*), stonewort (*Nitella* spp.)	TNT, triaminotoluene
Nitrilase	Willow (*Salix* spp.)	Cyanide groups from aromatic rings
Nitroreductase	Hybrid poplars (*Populus* spp.), parrot feather (*Myriophyllum aquaticum*), stonewort (*Nitella* spp.)	Explosives and other nitroaromatic compounds
Peroxidase	Horseradish (*Armoracia rusticana*)	Phenols
Phosphatase	Giant duckweed (*Spirodela polyrhiza*)	Phosphates, organophosphate pesticides

Carolina). Atrazine in soil was taken up by trees and then hydrolysed and de-alkylated within the roots, stems, and leaves. Metabolites were identified within the plant tissue, and a review of atrazine metabolite toxicity studies indicated that the metabolites were less toxic than atrazine.[5,101]

The herbicide bentazon was degraded within black willow (*Salix nigra*) trees, as indicated by loss during a nursery study and by identification of metabolites within the tree. Bentazon was phytotoxic to six tree species at concentrations of 1000 and 2000 mg L^{-1}, but allowed growth at 150 mg L^{-1}. At this concentration, bentazon metabolites were detected within tree trunk and canopy tissue samples. Black willow, yellow poplar (*Liriodendron tulipifera*), bald cypress (*Taxodium distichum*), river birch (*Betula nigra*), cherry bark oak (*Quercus falcata*) and live oak (*Quercus viginiana*) were all able to support some degradation of benta-zon.[101,102] Nitrate can be taken up by plants and incorporated into proteins or other nitrogen-containing compounds, or transformed into nitrogen gas. Deep-rooting techniques can increase the effective depth of this application.[101]

Dec and Bollag[93] have described plants that can degrade aromatic rings in the absence of microorganisms. PCBs have been metabolised by sterile plant tissues. Phenols have been degraded by plants such as horseradish, potato (*Solanum tuberosum*) and white radish (*Raphanus sativus*) that contains peroxidase.[93]

The main advantage of phytodegradation is that contaminant degradation can occur in an environment free of microorganisms (microorganisms may be killed by high contaminant levels). Plants are able to grow in sterile soil and also in soil that has concentration levels that are toxic to microorganisms. Thus, phytodegradation potentially could occur in soils where biodegradation cannot.

The main inconvenient is the possible production of toxic intermediates or phytodegradation products. In a study unrelated to phytoremediation research, PCP was metabolised to the potential mutagen tetrachlorocatechol in wheat plants and cell cultures.[103]

3.6   Phytovolatilisation

Phytovolatilisation, or 'biovolatilisation' as used by Cunningham and Ow,[104] is a mechanism by which plants convert a contaminant into a volatile form, thereby removing the contaminant from the soil at a contaminated site.[105] For effective phytoremediation, the degradation product or modified volatile form should be less toxic than the initial contaminant. Phytovolatilisation is pri-marily a contaminant removal process, transferring the contaminant from the original medium (soil water) to the atmosphere. This process can only be used if risk assessment and follow-up monitoring establish that sufficient risk re-duction occurs. For elemental contaminants, phytovolatilisation involves up-take and transformation or re-speciation to volatile forms that are released through the roots, stems or leaves. Volatile organic compounds are taken up and transpired with water vapour or diffused out of roots, stems and leaves. In consequence, for significant transpiration to occur, the soil must be able to transmit sufficient water to the plant.

Metabolic processes within the plant might alter the form of the contaminant, and in some cases, transform it to less toxic forms. Examples include the reduction of highly toxic mercury species to less toxic elemental mercury, or transformation of toxic selenium (as selenate) to the less toxic dimethyl selenide gas.[26] In some cases, contaminant transfer to the atmosphere allows much more effective or rapid natural degradation processes to occur, such as photodegradation.

Phytovolatilisation can occur with soluble inorganic contaminants, particularly elements of sub-groups II, V and VI of the periodic table like mercury, selenium and arsenic. However, other elements do not readily form volatile chemical species in the soil environment or in plant shoots, so phytovolatilisation cannot be applied to these elements. The inorganic contaminants Se and Hg, along with As, can form volatile methylated species.[28] For example, plants, possibly in association with microorganisms, can convert selenium to dimethyl selenide. Dimethyl selenide is a less toxic, volatile form of selenium. The importance of plants in the removal of Se through volatilisation has been established by several researchers who showed that the addition of plants to soil increased the rate of Se volatilisation above that found for soil alone.[106] Experiments conducted by Zayed and Terry showed that the addition of barley to Se-contaminated soil increased significantly the volatilisation rate.[107] Root and rhizosphere transformation of selenium to volatile dimethyl selenide avoids accumulation of toxic levels of selenium in plants and the food chain, especially in birds and fish. Air transport of dimethyl selenide is not known to represent a health risk; in fact, many soils are selenium poor, and plants with limited accumulation may represent a good supplement to cattle feed.[108]

Zayed and Terry[107] found that some commercial vegetable crops are quite effective in phytovolatilisation; broccoli annual Se removal was promising, although sulfate strongly inhibited Se emission. Phytovolatilisation offers significant opportunity to alleviate soil Se contamination while redistributing the Se to a much larger land area where the concentration would not comprise risk.

Plant species differ substantially in their ability to take up and volatilise Se.[108] Rice, broccoli, cabbage, cauliflower, Indian mustard and Chinese mustard volatilised Se at rates exceeding 1500 µg Se day^{-1} kg^{-1} plant dry weight when supplied with 20 µm Se as sodium selenate. Sugar beet, bean, lettuce and onion exhibited very low rates of Se volatilisation, below 250 µg Se day^{-1} kg^{-1} plant dry weight. Other plant species tested, including carrot, barley, alfalfa, tomato, cucumber, cotton, eggplant and maize, showed intermediate values of 300 to 750 µg Se day^{-1} kg^{-1} plant dry weight. Plant species from the Brassicaceae family were particularly effective Se volatilisers. There was evidence that the ability to volatilise Se was associated with the ability to accumulate Se in plant tissues—the rate of Se volatilisation by different plant species was strongly correlated with the plant tissue Se concentration.[108] Hyperaccumulator plants appear to have a significant role to play in phytoextraction of Se as well because of their ability to accumulate and phytovolatilise Se even in the presence of high levels of sulfate compared to crop plants.[108] For example, Indian mustard and canola (*Brassica napus*) may be effective for phytovolatilisation of selenium and, in addition, accumulate the selenium.[1]

Phytovolatilisation may then be a useful, inexpensive means of removing selenium from sites contaminated with high concentration selenium wastes. This is because selenium is volatilised by roots and not appreciably accumulated in aboveground vegetation in forms available to wildlife. Recent research showed that Se volatilisation by plant roots may require rhizosphere microorganisms. Other methylated metalloids such as mercury and arsenic may also be good candidates for volatilisation by plants, depending on local risk assessments of the volatilised material.

Some transgenic plants have converted organic and inorganic mercury salts to the volatile, elemental form. For example, *Arabidopsis thaliana* can be genetically engineered with bacterial genes to resist mercury poisoning. In addition, the same genetic engineering provides genes to demethylate organic complexes of mercury and transpire elemental mercury.[109] Testing the concept of transplanting bacterial genes into plants to increase tolerance, transformation and accumulation is a remarkable precedent.

Organic Hg compounds are the principle source of environmental Hg poisoning because these compounds are bioaccumulated in aquatic food chains, and both predator birds and mammals are poisoned. Any potential for adverse effects of Hg(0) phytovolatilised from contaminated soils is very small compared with the reduction in risk of adverse effects by hydrolysing any methyl mercury in soils. In laboratory experiments, tobacco (*Nicotiana tabacum*) converted ionic mercury [Hg(II)] to the less toxic metallic mercury [Hg(0)] and volatilised it.[110] Similarly transformed yellow poplar (*Liriodendron tulipifera*) plantlets had resistance to, and grew well in, normally toxic concentrations of ionic mercury. The transformed plantlets volatilised about 10 times more elemental mercury than did untransformed plantlets.[109]

Phytovolatilisation can also occur with organic contaminants, such as TCE, generally in conjunction with other phytoremediation processes. Uptake of TCE by the hybrid poplar trees occurred, with unaltered TCE being found within the trees. Oxidation of TCE also occurred within the trees, indicated by the presence of TCE oxidative metabolites. Analysis of entrapped air in bags placed around leaves indicated that about 9% of the applied TCE was transpired from the trees during the second year of growth, but no TCE was detected during the third year.[83]

It is not clear to what degree phytovolatilisation of TCE occurs under different conditions and with different plants, since some other studies have not detected transpiration of TCE. However, measurement of transpired TCE can be difficult, and measurements must differentiate between volatilisation from the plant and volatilisation from the soil. In addition, it is almost certain that several phytoremediation processes (rhizodegradation, phytodegradation and phytovolatilisation) occur concurrently in varying proportions, depending on the site conditions and on the plant. Questions remain as to chlorinated solvent metabolism within plants and transpiration from the plants.

Wiltse *et al.*[111] observed leaf burn in alfalfa plants growing in crude oil-contaminated soil. The authors suggested that an unidentified compound from the contaminated soil was being translocated through the plant and then transpired. The leaf burn gradually disappeared as the experiment progressed, indicating that the contaminants responsible for this effect had dissipated. Watkins *et al.*[112] found that the volatilisation of [^{14}C] naphthalene was enhanced in sandy loam soil planted with Bell rhodesgrass compared with unplanted soil. The results of the study suggest that naphthalene was taken up by the roots of the grass, translocated within the plant, and transpired through the stems and leaves. The authors noted that this mechanism of removal would reduce the amount of naphthalene available in soil, but may have implications regarding subsequent contamination of the atmosphere and, consequently, regulatory compliance with air quality guidelines. In a study of poplar cuttings in hydroponic solution,[92] about 20% of the benzene and TCE in the initial solution was volatilised from the leaves, with little remaining within the plant. About 10% of toluene, ethylbenzene and *m*-xylene was volatilised. There was little volatilisation of nitrobenzene and no volatilisation of 1,2,4-trichlorobenzene, aniline, phenol, PCP or atrazine. The percentage of applied compound taken up into the plant was 17.3% for 1,2,4-trichlorobenzene, 40.5% for aniline, 20.0% for phenol, 29.0% for PCP and 53.3% for atrazine. For 1,2,4-trichlorobenzene, aniline, phenol, and PCP, the largest percentage of compound taken up was found in the bottom stem, as opposed to the root, upper stem or leaves. For atrazine, the largest percentage of compound taken up was found in the leaves. Of the 11 compounds tested, nine had 2.4% or less of the applied compound in the leaves, but aniline had 11.4% and atrazine had 33.6% in the leaves. All compounds had 3.8% or less in the upper stem.[92] However, the chemical fate and translocation is most likely concentration-dependent and other concentrations may give different results.

Phytovolatilisation of organic chemicals is then a two-edged sword, and being similar to that of mercury transpiration, will be very site-dependent. Even though the half-life of TCE in the atmosphere is on the order of hours to days and atmospheric dispersion significant, the lack of general and local risk assessments have slowed application of trees for phytohydraulic control of TCE and MTBE plumes. However, the scientific consensus is that phytovolatilisation of TCE is acceptably small (immeasurable to 9%) and that continued monitoring is sufficient until a general risk assessment can be undertaken.[113] However, further unpublished studies associated with Hong *et al.*[114] have shown that very little MTBE is volatilised under field conditions, suggesting storage or metabolism by poplar *(Populus* spp.) trees in the field.

Research on phytovolatilisation has successfully identified superior volatilising plant species, characterised the optimum environmental conditions for maximum rates of some contaminants volatilisation, and begun to unravel, using modern molecular biology techniques, the mechanisms by which compounds are evolved by plants. In addition, recent findings on the microbial involvement in plant uptake and volatilisation of some pollutants like Se or Hg has provided another tool to accelerate the phytovolatilisation process through the creation of superior plant/microbe associations. However, because phytovolatilisation involves the transfer of contaminants to the atmosphere, the impact of this contaminant transfer on the ecosystem and on human health needs to be addressed.

3.7   Vegetative Cover Systems

Vegetative cover systems are used at some contaminated sites, landfills and other types of waste disposal sites to control moisture and percolation, promote surface water runoff, minimise erosion, prevent direct exposure to the waste, control gas emissions and odours, prevent occurrence of disease vectors and other nuisances, and meet aesthetic and other end-use purposes. Vegetative cover systems are intended to remain in place and maintain their functions for an extended period of time. In addition, cover systems are also used in the remediation of hazardous waste sites, such as source areas contaminated at or near the ground surface or at abandoned dumps. In addition to being called vegetative cover systems, these types of covers have also been referred in the literature as evapotranspiration covers, water balance covers, alternative earthen final covers, soil–plant covers, store-and-release covers, vegetative soil caps, phytoremediation cover systems or evapotranspiration cover systems.

3.7.1   Conventional Capping and Evapotranspiration Cover Systems

Capping is a process used to cover contaminated soils to prevent the migration of the pollutants. This migration can be caused by rainwater or surface water moving over or vertically through the site. Thus, the primary purpose of a cap is to minimise contact between rain or surface water and the contaminated soil. The two main types of capping system are shown in Figure 3.2.

   To minimise percolation, conventional cover systems use low-permeability barrier layers. These barrier layers are often constructed of compacted clay, geomembranes, geosynthetic clay liners, concrete or combinations of these materials. Depending on the material type and construction method, the saturated hydraulic conductivities for these barrier layers are typically between 10^{-5} and 10^{-9} cm s^{-1}. In addition, conventional cover systems generally include: additional layers, such as surface layers to prevent erosion; protection layers to minimise freeze/thaw damage; internal drainage layers; and gas collection layers.[51,115] The design of cover systems is site-specific and depends on the intended function of the final cover; components can range from a single-layer system to a complex multilayer system. The type of cap that is generally acceptable is the multi-layer cap. This type of cap generally has four layers: vegetation, drainage, water-resistant and foundation. The vegetation layer prevents erosion of the soils of the cap. The drainage layer channels rainwater away from the cap and keeps water from collecting on the water-resistant layer which covers the waste. The foundation layer is composed of soil materials that are structurally capable of supporting the weight of the finished cap. Structural stability tests should be run on each increment to assure uniformity. The vegetation layer should be at least two feet (61 cm) thick to accommodate root penetration. It should be spread evenly and not overly compacted. The vegetation should be non-woody plants, preferable grasses, which will require low maintenance and do not have deep roots.

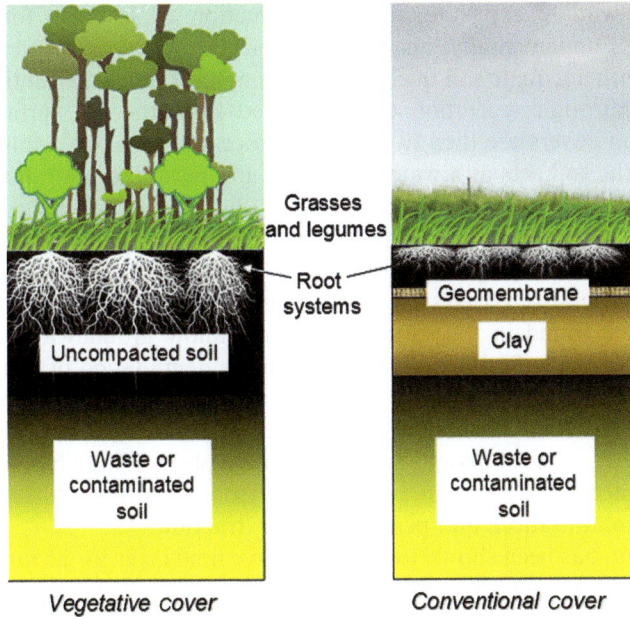

Figure 3.2 A vegetative cover compared with a conventional cover.

In many cases, conventional covers have proven to be expensive, especially if the building materials are not available on-site.

One type of 'alternative earthen final cover' exploits the water storage capacity of finer textured soils and the water removal capability of vegetation.[116] This type of cover is referred to as an evapotranspiration cover. Evapotranspiration covers are caps composed of soil and plants engineered to maximise the available storage capacity of soil, evaporation rates and transpiration processes of plants to minimise water infiltration. Generally, they do not contain barrier layers. Evapotranspiration covers work as follows: water that infiltrates into the cover due to precipitation is stored in the soil (sponge) and subsequently removed by transpiration and surface evaporation, and returned to the atmosphere (pump). A complete discussion of the 'sponge and pump' mechanism can be found in Licht *et al.*[117] Evapotranspiration is then the sum of transpiration by the plant and evaporation from the soil surface. Controlling water is often the key to controlling and remediating contaminated sites because many contaminants are either soluble in or moved by water and the evapotranspiration cap is a form of hydraulic control by plants. It differs from more conventional cover designs in that they rely on obtaining an appropriate water storage capacity in the soil rather than an initial as-built engineered hydraulic conductivity.

Evapotranspiration cover system designs are based on using the hydrological processes (water balance components) at a site, which include the water storage capacity of the soil, precipitation, surface runoff, evapotranspiration and infiltration. Risk reduction relies on the isolation of contaminants to prevent human or wildlife exposure, and the reduction of leachate formation or movement. Fundamentally, an evapotranspiration cover is a layer of mono-lithic soil with adequate soil thickness to retain infiltrated water until it is either transpired through vegetation or evaporated from the soil surface. Evapo-transpiration covers use then two natural processes to control infiltration into the waste: the soil acts as a water reservoir, and natural evaporation from the soil plus plant transpiration empties the soil water reservoir before it can in-filtrate into the contaminated zone or into the waste to generate leachate.

In addition to moving large amounts of water from the soil to the atmos-phere, the vegetative cover systems may also accumulate, transfer or destroy contaminants found in the vadose zone or shallow groundwater underlying waste or pollutants. Risk reduction relies on the degradation of contaminants, the isolation of contaminants to prevent human or wildlife exposure, and the reduction of leachate formation or movement. Mechanisms include water up-take, root-zone microbiology, and plant metabolism. This phytoremediating cover can incorporate certain aspects of phytodegradation, rhizodegradation and phytovolatilisation, and perhaps phytoextraction.[101]

Vegetation has been shown to be an effective final layer for hazardous waste site covers.[118] At many landfill sites, evapotranspiration covers can meet the requirements for landfill covers by (i) controlling infiltration into the polluted zone and (ii) isolating the contaminants and preventing contact with recep-tors. The concentration of the contaminants in the underlying material is not a concern, as long as the plants are not in contact with materials having phytotoxic concentrations. Evapotranspiration landfill covers are inexpen-sive, practical, and easily maintained biological systems that will remain effective over extended periods of time at low cost. Indeed, evapotranspira-tion landfill covers typically cost about half as much to build as conventional barrier-type landfill covers and have low maintenance costs over time. Vegetation may encourage aerobic microbial activity in the root zone; such activity could discourage formation of anaerobic landfill gases or degrade them. However, vegetative covers are not appropriate for certain landfill units that generate gas in chronic, large, or uncontrolled amounts. Indeed, landfill gases can be toxic to plants and therefore must be considered. To date, vegetative cover systems have not been shown to prevent the diffusion of gases from landfills. Gas control may then be incorporated into evapotranspiration covers where needed.

Evapotranspiration is often the largest controlling factor in the use of plants in remediation. It is also a primary factor that should be carefully considered before spending money on the design and installation of remediation systems that use plants. Evapotranspiration may control plant effectiveness in at least two ways:

(i) Limited water supply may reduce evapotranspiration, and consequently plant growth, thus limiting remediation effectiveness for some plants.
(ii) Limited evapotranspiration may not remove enough water from the vadose zone to perform the desired remediation function.

Both potential (PET) and actual (AET) evapotranspiration are important design criteria because they determine the effectiveness of the evapotranspiration landfill cover. The PET is the maximum amount of water that plants can remove, while the AET estimate indicates how plants may actually perform in the environment of the site. AET by a plant system is almost always less than PET, and is reduced by factors that limit plant growth. These factors include water supply, incident solar radiation, humidity, air temperature, wind, dormant seasons, immaturity of the plants, dry soil layers, plant type, plant disease, insect attack, soil fertility and soil physical properties. Climatic factors from this group are collectively the largest potential limitation for plant growth and they control PET. Hydrologic factors that control the amount of water actually removed from a contaminated site by evapotranspiration include surface runoff, plant vigour and area of soil surface available for plant production.

3.7.2 Factors to be Considered

Capping is a reliable technology for sealing off contamination from the aboveground environment and significantly reducing underground migration of pollutants away from the site. Hauser *et al.*[115] reviewed extensive and long-term field measurements which demonstrated that the evapotranspiration cover concept is sound and can control water movement. Recent performance data from large-scale demonstration indicates that evapotranspiration covers have percolation rates comparable with conventional cover designs.[119] Licht *et al.*[120] also suggested that evapotranspiration covers percolate much less precipitation than conventional covers. Caps can be constructed over virtually any site and can be completed relatively quickly if the ground is not frozen or saturated with water. Compared with conventional membrane or compacted clay cover systems, evapotranspiration cover systems are expected to cost less to construct. They can reduce construction costs compared to more complex, multi-layered cover designs and reduce maintenance needs once the vegetative system is established, such as minimising surface erosion by establishing a self-sustainable ecosystem, and may require fewer repairs than a barrier system. Additionally, they are often aesthetic because they employ naturalised vegetation.

There has been renewed interest in vegetated soil caps as an alternative process, particularly in arid and semi-arid locations. However, they are less effective in achieving low percolation in humid climates. Licht *et al.*[120] suggest that evapotranspiration covers tend to leak more in wet climates. Albright *et al.*[121] reported high percolation rates at sites in humid climates.

Monitoring the performance of evapotranspiration covers has been generally focused on evaluating the ability of these covers to minimise water drainage into the waste or contaminated site. In most cases, percolation monitoring for evapotranspiration covers is measured directly using lysimeters or estimated indirectly using soil moisture content and calculating a percolation rate.

Vegetative cover systems are increasingly being used at municipal solid waste (MSW) landfills, hazardous waste landfills, industrial monofills and mine sites. All cover systems should consider the goals of the cover in terms of protectiveness, including the pathways of risk from contained material, and the lifecycle of the containment system. The containment system needs to be protective of direct contact of people and animals with the waste, prevent surface and groundwater water pollution, and minimise release of airborne contaminants. A vegetative cover must demonstrate equivalent performance with generic cover designs. Although the actual designs tend to be site-specific and vary widely between regions, evapotranspiration cover designs tend to all have locally available soils that have a relatively high water storage capacity and vegetation to increase evapotranspiration.[115,122] The vegetation should be ecologically sustainable for the site, taking into consideration the conditions that sustain climax ecosystems.

To use plants successfully, it is necessary to understand the requirements for plant growth and to estimate the probable performance of plants at a particular site. Poplar trees and grasses have been used commercially to construct vegetative covers. Ideally, the vegetation selected for the system should be a mixture of native plants and consist of warm- and cool-season species. Robust plant growth most completely achieves remediation goals. Many factors may limit plant growth, which in turn may limit the effectiveness of plants in environmental clean-up systems. Most plant-based cover designs will be effective only in a specific climate. Universally applicable designs may not be possible.

While most containment strategies have been based on the dry tomb strategy of keeping waste dry, there are some sites where adding or allowing moisture to help decompose organic waste is the current plan. Evapotranspiration covers may work well in places where complete exclusion of precipitation is not needed. Vegetative covers may be appropriate to address contaminated surface soil or sludge, certain waste disposal units, waste piles and surface impoundments. These covers have the potential to enhance the biodegradation of contaminants in soils, sludges and sediments, but contaminants should not be at phytotoxic levels since the plant roots need to be in contact with the contaminated waste.

3.8   Conclusions

Phytoremediation capitalises on the innate abilities of vegetation to remediate and restore sites contaminated with a variety of pollutants. Although the concept of phytoremediation has been in existence for many years, only in the past decade has interest in this technology accelerated to the point of commercial application.

3.8.1   Advantages and Constraints of Phytoremediation

Despite the diversity of potential options, phytoremediation is in its initial stages. The majority of the research has been conducted in laboratories under relatively controlled conditions for short periods of time. More extensive research under field conditions for longer durations is required for a better understanding of the potential role of phytoremediation. A limiting factor of phytoremediation treatment includes the fact that a specific phytoremediation 'prescription' cannot be applied to every site with a certain chemical contaminant because different site-specific conditions (*e.g.* soil and climate) may not be suitable for the target plant. Plants also interact with and are affected by other living organisms such as insects, pests and pathogens, and exposure of plants to contaminants and related stresses can make the phytotreatment more susceptible to these other agents, ultimately influencing the outcome of phytoremediation attempts. Additionally, phytoremediation generally is restricted to sites where the concentration levels of contaminants are not toxic to the plants proposed for remediation. The concentrations of soil pollutants should be considered, as in some contaminated sites, high levels of contaminants are toxic to plants and prevents successful phytoremediation.

Among the limitations of phytoremediation strategies has been that *in situ* remediation often takes many years to accomplish compared with traditional decontamination approaches to substantially restore a polluted area.[4] Finally, the contaminants must be accessible to the tissue responsible for uptake (*e.g.* root system) in plants. Another limitation to consider is the availability of the contaminants in question to the plants; Schnoor *et al.*[4] noted that areas where contamination is less than 5 m in depth are the best suited sites for phytoremediation. Furthermore, the solubility of the contaminant in the soil determines whether phytoextraction is possible; in the case of metals, only metals found as free metal ions, soil soluble metal complexes, or metals adsorbed to inorganic soil constituents at ion exchange sites are readily available for uptake by the plants. Metals that are bound to soil organic matter, precipitated (oxides, hydroxides, carbonates), or embedded in the structure of silicate minerals are not available to the plants. It has been suggested that phytoremediation is best suited for removing: moderately hydrophobic pollutants such as BTEX compounds (benzene, toluene, ethylbenzene and xylenes), chlorinated solvents, or nitrotoluene ammunition wastes; or excess nutrients (nitrate, ammonium and phosphate).[4] As a result, *in situ* phytoremediation using live plants is restricted to sites conducive to growth of the selected plant with the contaminant located within the potential root zone of the selected plant.

Despite the constraints outlined above, plants have many features that result in a high potential for environmental clean-up. Energy costs and expenses are reduced and natural resources are conserved because plants use solar energy. Phytoremediation costs at least 10 times less than traditional methods of excavation and removal,[51] and if an additional economic incentive were present (not only an environmental benefit) such as phytomining or forestry, then phytoremediation would be viewed as economically viable. Plants are adapted to a wide range of environmental conditions and are capable of modifying conditions of the environment to some extent. The unique enzyme and protein systems of some plant species appear to be beneficial for phytoremediation. Additionally, since plants lack the ability to move, many plants have developed unique biochemical systems for nutrient acquisition, detoxification, and controlling local geochemical conditions. Some plants that grow well in nutrient poor soils may have useful mechanisms for removing and transforming contaminants that resemble certain nutrients. Infiltration is a primary pathway in contaminant migration to groundwater and plants play an important role in regulating water content in soil. Plant roots aerate the soil, which may stimulate microbial activity in the soil. Root exudates may be a nutrient source for microorganisms, since the rhizosphere generally contains significantly higher numbers and more active microorganisms than similar soils without plants.

Thus, plants can contribute in many ways to enhance biodegradation in the soil. Furthermore, phytoremediation provides an aesthetically pleasing alternative to structural remediation and decontamination technologies. Ecologists have recognised the importance of vegetation in restoring disturbed environments. Through succession, vegetation has the remarkable capacity to restore disturbed sites to conditions resembling those prior to the disturbance. Thus, phytoremediation should not be considered a 'new technology', but a technology that can build on a long history of observations and research. As a result of these advantages, phytoremediation has considerable potential for environmental restoration of contaminated sites.

3.8.2   Future Outlook

Compared with physico-chemical methods, phytoremediation shows great advantage. Studies in this field are increasing, its mechanism is more and more clear, and there are lots of approaches of phytoremediation available. However for the best choice, it is essential to first investigate the conditions of the soil and the contaminants. Managers must know the types of contaminants at the site, together with any metals, salts and/or pesticides associated with the target contaminants. Following site and contaminant characterisation, the manager should screen for appropriate plants and microorganisms, identifying whether there are native species that could be used in the phytoremediation process. Further assessment of these plants in field situation would be useful. Finally, it is important to keep in mind that a variety of remediation approaches may be required to accomplish all recovery goals at a contaminated site.

There is a need to optimise the agronomic practices to maximise the clean-up potential of remediative plants. Since in many instances metal absorption in roots is limited by low solubility in soil solution, it is important to further investigate the use of chemical amendments to induce metal bioavailability. Significant results have been obtained in this area. However, there is a need to find cheaper, environmentally benign chemical compounds with metal-chelating properties. Research is also needed to identify phytoremediating species capable of being rotated to sustain the rate of contaminant removal.

Organic and inorganic contaminants are under the biotic influence of both the plant growing in the site and the microorganisms inhabiting the root zone. It is important to stress that optimising the dynamic interactions between plants and microbes will largely determine the success of phytoremediation strategies such as volatilisation, extraction, stabilisation, transformation and degradation of the contaminant in the plant (*viz.* phytodegradation) or rhizosphere (*viz.* rhizodegradation).

As many sites can present several contaminants, it is interesting to combine some remediation techniques in order to have a more efficient decontamination process. In many cases, phytoremediation involves combinations of the mechanisms described in this chapter, as one can be a complement to the other. For example, phytoextraction and phytovolatilisation have been credited with the removal of excess selenium in soil.[123] It is likely that both processes occur simultaneously. The treatment of TCE in groundwater using poplar trees requires extraction of the groundwater by the plant (phytoextraction) that will also degrade TCE (phytodegradation) within the plant. Another example is degradation of PCBs by plant cells, as well as by microorganisms stimulated by plants,[124,125] creating the opportunity to combine phytodegradation and rhizodegradation. Combining technologies described in this chapter offers the greatest potential to efficiently phytoremediate contaminated sites.

Generally, phytoremediation is employed using single plant species or, occasionally, a few related species grown together. These management practices are at odds with the natural ecological recovery of disturbed sites, which involves multiple, diverse plant species in an evolving plant community. Plant selection for enhanced phytoremediation should be patterned after native vegetation of the given climatic area and plants that have been demonstrated to grow on previously contaminated soils. However, it is important to apply grasses with fast growth first, until the levels of the soil contaminants are lower, followed by the removal of such vegetation, incineration and recovery of the metals. This measure makes possible the use of tree species to then phytostabilise the contaminants. It is also essential to attend to acceptable reference values given by the governmental institutes and to check beforehand the tolerance levels of the candidate species. Furthermore, the contaminated site can first be stabilised with a variety of pioneer plants capable of withstanding severe environments. As site conditions improve, additional native plants would be planted at the site and ultimately the site would be revegetated with a variety of annual and perennial grasses and forbs, as well as shrubs and trees. Appropriate plant management strategies, including watering, fertilisation, and plant clipping could be devised to assure the success of this engineered form of natural succession.

In conclusion, phytoremediation is emerging as a technology with great potential. However, prior to becoming available for use on a large scale, it is necessary to integrate research in many disciplines, including molecular biology, plant physiology, biochemistry and microbiology (at molecular, cell and plant levels), as well as ecology, agronomy, hydrology, soil science, agricultural engineering and entomology (at field level). For example, using transgenic technology is a tendency for the future to create a purposely ideal species. The creation of an optimum plant–soil–microbes combination would be a promising way for future development. However, genetic pollution must be taken into consideration. Hence, overall, these studies must be considered in relation to economics, environmental impact assessment and government regulatory policies.

References

1. G. S. Banuelos, H. A. Ajwa, B. Mackey, L. Wu, C. Cook, S. Akohoue and S. Zambrzuski, *J. Environ. Qual.*, 1997, **26**, 639.
2. S. B. Ebbs and L. V. Kochian, *J. Environ. Qual.*, 1997, **26**, 776.
3. J. A. Eutry, L. C. Watrud and M. Relves, *Environ. Pollut.*, 1999, **104**, 449.
4. J. L. Schnoor, L. A. Licht, S. C. McCutcheon, N. L. Wolfe and L. H. Carreira, *Environ. Sci. Technol.*, 1995, **29**, 7.
5. J. G. Burken and J. L. Schnoor, *Environ. Sci. Technol.*, 1997, **31**, 1399.
6. S. D. Siciliano and J. J. Germida, *Environ. Toxicol. Chem.*, 1998, **17**, 728.
7. K. A. Reilley, M. K. Banks and A. P. Schwab, *J. Environ. Qual.*, 1996, **25**, 212.
8. C. M. Reynolds, D. C. Wolf, T. J. Gentry, L. B. Perry, C. S. Pidgeon, B. A. Koenen, H. B. Rogers and C. A. Beyrouty, *Polar Record*, 1999, **35**, 33.
9. R. L. Chaney, in *Land Treatment of Hazardous Wastes*, ed. J. F. Parr, P. B. Marsh and J. M. Kla, Noyes Data Corp., Park Ridge, NJ, 1983, pp. 50–76.
10. A. J. Baker, R. D. Reeves and S. P. McGrath, in *In Situ Bioreclamation*, ed. R. E. Hinchee and R. F Olfenbuttel, Butterworth Heinemann, Boston, MA, 1991, pp. 539–544.
11. G. S. Banuelos, H. A. Ajwa, L. Wu and S. Zambrzuski, *J. Contamin. Soil*, 1998, **7**, 481.
12. D. E. Salt, R. D. Smith and I. Raskin, *Annu. Rev. Plant Physiol. Plant Mol. Biol.*, 1998, **49**, 643.
13. A. J. Baker, S. P. McGrath, C. M. Sidoli and R. D. Reeves, *Resour. Cons. Recycl.*, 1994, **11**, 41.
14. S. C. Cunningham, W. R. Berti and J. W. Huang, *Trends Biotechnol.*, 1995, **13**, 393.
15. M. J. Blaylock, D. E. Salt, S. Dushenkov, O. Zakharova, C. Gussman, Y. Kapulnik, B. D. Ensley and I. Raskin, *Environ. Sci. Technol.*, 1997, **31**, 860.
16. J. W. Huang, J. Chen, W. R. Berti and S. D. Cunningham, *Environ. Sci. Technol.*, 1997, **31**, 800.
17. E. Lombi, F. J. Zhao, S. J. Dunham and S. P. McGrath, *J. Environ. Qual.*, 2001, **20**, 1919.
18. J. S. Rieuwerts, I. Thornton, M. E. Farago and M. R. Ashmore, *Chem. Spec. Bioavail.*, 1998, **10**, 61.
19. B. J. Alloway, in Heavy Metals in Soils, ed. B. J. *Alloway*, Blackie Academic and Professional, Glasgow, 1995, pp. 122–147.
20. A. E. Martell and R. M. Smith, *NIST Critically Selected Stability Constants of Metal Complexes: Version 7.0*, National Institute of Standards and Technology, Gaithersburg, MD, 2003.
21. J. Wu, F. C. Hsu and S. D. Cunningham, *Environ. Sci. Technol.*, 1999, **33**, 1898.
22. O. A. Beath, H. F. Eppson and C. S. Gilbert, *Wyo. Agric. Stand. Bull.*, 1935, **206**, 1.

23. C. Minguzzi and O. Vergnano, *Atti Soc. Toscana Sci. Nat.*, 1948, **55**, 49.
24. R. R. Brooks, J. Lee, R. D. Reeves and T. Jaffre, *J. Geochem. Explor.*, 1977, **7**, 49.
25. D. E. Salt, M. J. Blaylock, D. B. Kumar, V. Dushenkov, B. D. Ensley, I. Chet and I. Raskin, *Environ. Sci. Technol.*, 1995, **13**, 468.
26. P. B. Kumar, V. Dushenkov, H. Motto and I. Raskin, *Environ. Sci. Technol.*, 1995, **29**, 1232.
27. I. Raskin, P. B. Kumar, S. Dushenkov, M. J. Blaylock and D. Salt, in *Emerging Technologies in Hazardous Waste Management VI*, ACS Industrial & Engineering Chemistry Division Special Symposium, Volume I, Atlanta, GA, 1994, pp. 19–21.
28. G. M. Pierzynski, J. L. Schnoor, M. K. Banks, J. C. Tracy, L. A. Licht and L. E. Erickson, *Environ. Sci. Technol.*, 1994, **1**, 49.
29. T. Adler, *Sci. News*, 1996, **150**, 42.
30. R. D. Reeves, in *Phytoremediation of Metal-Contaminated Soils*, ed. J.-L. Morel, G. Echevarria and N. Goncharova, Springer, Dordrecht, The Netherlands, 2002, pp. 25–52.
31. D. W. Shimwell and A. E. Laurie, *Environ. Pollut.*, 1972, **3**, 291.
32. R. Bungart and R. F. Huttl, *Eur. J. Forest Res.*, 2004, **123**, 105.
33. L. Sebastiani, F. Scebba and R. Tognetti, *Environ. Exp. Bot.*, 2004, **52**, 79.
34. P. J. Tharakan, T. A. Volk, C. A. Nowak and L. P. Abrahamson, *Can. J. For. Res.*, 2005, **35**, 421.
35. A. J. Baker, R. D. Reeves and A. S. Hajar, *New Phytol.*, 1994, **127**, 61.
36. Z. G. Shen, F. J. Zhao and S. P. McGrath, *Plant Cell Environ.*, 1997, **20**, 898.
37. A. J. Pollard, K. D. Harper and J. A. Smith, *Crit. Rev. Plant Sci.*, 2002, **21**, 539.
38. M. Molitor, C. Dechamps, W. Gruber and P. Meerts, *New Phytol.*, 2005, **165**, 503.
39. E. Gerard, G. Echevarria, T. Sterckeman and J. L. Morel, *J. Environ. Qual.*, 2000, **29**, 1117.
40. E. Lombi, F. J. Zhao, S. J. Dunham and S. P. McGrath, *New Phytol.*, 2000, **145**, 11.
41. R. D. Reeves, C. Schwartz, J. L. Morel and J. Edmondson, *Int. J. Phytorem.*, 2001, **3**, 145.
42. R. D. Reeves and A. J. Baker, in *Phytoremediation of Toxic Metals, Using Plants to Clean Up the Environment*, ed. I. Raskin and B.D. Ensley, Wiley, New York, 2000, pp. 193–229.
43. J. R. Akeroyd, in *Flora Europaea*, ed. T. G. Tutin *et al.*, (eds) assisted by J. R. Akeroyd and M. E. Newton, Cambridge University Press, Cambridge, 2nd edn, vol. 1, 1993, pp. 287–292.
44. J. Escarre, C. Lefebvre, W. Gruber, M. Leblanc, J. Lepart, Y. Rivier and B. Delay, *New Phytol.*, 2000, **145**, 429.
45. N. Rascio, *Oikos*, 1977, **29**, 250.
46. R. D. Reeves, *Taxon*, 1988, **37**, 309.

47. R. D. Reeves, A. R. Kruckeberg, N. Adiguzel and U. Kramer, *S. Afr. J. Sci.*, 2001, **97**, 513.
48. R. D. Reeves and A. J. Baker, *New Phytol.*, 1984, **98**, 191.
49. S. N. Whiting, R. D. Reeves and A. J. Baker, *Mining Environ. Manage.*, 2002, **10**(2), 11.
50. N. Marmiroli, in *Phytoremediation. Transformation and Control of Contaminants*, ed. S. C. McCutcheon and J. L. Schnoor, John Wiley, Hoboken, NJ, 2003, pp. 85–119.
51. US Environmental Protection Agency, *Introduction to Phytoremediation*, EPA/600/R-99/107, US ENvirohnmental Protection Agency, Washington DC, 2000.
52. J. Vangronsveld, F. Van Assche and H. Clijsters, *Environ. Pollut.*, 1995, **87**, 51.
53. S. D. Cunningham, W. R. Berti and J. W. Huang, *Trends Biotechnol.*, 1995, **13**, 393.
54. H. Sandermann, *Pharmacogenetics*, 1994, **4**, 225.
55. J. L. Jordahl, L. Foster, J. L. Schnoor and P. J. Alvarez, *Environ. Toxicol. Chem.*, 1997, **16**(6), 1318.
56. J. F. Shimp, J. C. Tracy, L. C. Davis, E. Lee, W. Huang, L. E. Erickson and J. L. Schnoor, *Environ. Sci. Technol.*, 1993, **23**, 41.
57. M. Mackova in *Phytoremediation and Rhizoremediation: Theoretical Background*, ed. M. Mackova, D. Dowling and T. Macek, Springer, Dordrecht, The Netherlands, 2006, pp. 143–167.
58. T. A. Anderson, E. A. Guthrie and B. T. Walton, *Environ. Sci. Technol.*, 1993, **27**, 2630.
59. L. E. Erickson, M. K. Banks, L. C. Davis, A. P. Schwab, N. Muralidharan, K. Reilley and J. C. Tracy, *Environ. Prog.*, 1994, **13**(4), 226.
60. S. D. Siciliano and J. J. Germida, *Environ. Rev.*, 1998, **6**, 65.
61. E. Carman, T. Crossman and E. Gatliff, *J. Soil Contam.*, 1998, **7**, 455.
62. C. Wiltse, W. Rooney, A. Schwab and M. Banks, *J. Environ. Qual.*, 1998, **27**, 169.
63. J. G. Burken, C. Ross, L. M. Harrison, A. Marsh, L. Zetterstrom and J. S. Gibbons, *Pract. Period. Hazard. Toxic, Radioact. Waste Manage*, 2001, **5**, 161.
64. P. E. Olson and J. S. Fletcher, *Bioremed. J.*, 1999, **3**, 27.
65. P. E. Olson and J. S. Fletcher, *Environ. Sci. Pollut. Res.*, 2000, **7**, 1.
66. M. Molina, R. Araujo and J. R. Bond, *Symposium on Bioremediation of Hazardous Wastes, Research, Development, and Field Evaluations*, EPA/540/R-95/532, US Environmental Protection Agency, Office of Research and Development, Washington DC, 1995, pp. 32–34.
67. A. P. Schwab and M. K. Banks, in *Bioremediation of Contaminated Soils*, ed. D. C. Adriano and J. M. Bollag, W. T. Frankenburger and R. C. Sims, Agronomy Monograph 37, American Society of Agronomy, Madison, WI, 1999, pp. 381–454.
68. W. Aprill and R. C. Sims, *Chemosphere*, 1990, **20**, 253.

69. A. M. Ferro, S. A. Rock, J. Kennedy, J. J. Herrick and D. L. Turner, *Int. J. Phytoremed.*, 1999, **1**, 289.

70. K. A. Reilley, M. K. Banks and A. P. Schwab, *J. Environ. Qual.*, 1996, **25**, 212.

71. T. Gunther, U. Dornberger and W. Fritsche, *Chemosphere*, 1996, **33**, 203.

72. M. K. Banks, E. Lee and A. P. Schwab, *J. Environ. Qual.*, 1999, **28**, 294.

73. B. A. Wrenn and A. D. Venosa, *Can. J. Microbiol.*, 1996, **42**, 252.

74. C. Leyval and P. Binet, *J. Environ. Qual.*, 1998, **27**, 402.

75. R. Meulenberg, H. H. Rijnaarts, H. J. Doddema and J. A. Field, *FEMS Microbiol. Lett.*, 1997, **152**, 45.

76. R. Canet, J. G. Birnstingl, D. G. Malcolm, J. M. Lopez-Real and A. J. Beck, *Bioresour. Technol.*, 2001, **76**, 113.

77. T. A. Anderson, E. L. Kruger and J. R. Coats, *Chemosphere*, 1994, **28**, 1551.

78. T. S. Hsu and R. Bartha, *Appl. Environ. Microbiol.*, 1979, **37**, 36.

79. R. E. Hoagland, R. M. Zablotowicz and M. A. Locke, in Bioremediation through Rhizosphere Technology, ed. T. A. Anderson and J. R. Coats, *ACS Symposium Series*, American Chemical Society, Washington DC, 1994, vol. 563, pp. 160–183.

80. E. Sandmann and M. A. Loos, *Chemosphere*, 1984, **13**, 1073.

81. J. J. Boyle and J. R. Shann, *J. Environ. Qual.*, 1995, **24**, 782.

82. M. Narayanan, L. C. Davis and L. E. Erickson, *Environ. Sci. Technol.*, 1995, **29**, 2437.

83. L. A. Newman, X. Wang, I. A. Muiznieks, G. Ekuan, M. Ruszaj, R. Cortellucci, D. Domroes, G. Karscig, T. Newman, R. S. Crampton, R. A. Hashmonay, M. G. Yost, P. E. Heilman, J. Duffy, M. P. Gordon and S. E. Strand, *Environ. Sci. Technol.*, 1999, **33**, 2257.

84. A. M. Ferro, R. C. Sims and B. Bugbee, *J. Environ. Qual.*, 1994, **23**, 272.

85. W. F. Pfender, *J. Environ. Qual.*, 1996, **25**, 1256.

86. P. K. Donnelly and J. S. Fletcher, in *Bioremediation through Rhizosphere Technology*, ed. T. A. Anderson and J. R. Coats, American Chemical Society, Washington DC, vol. 563, 1994, pp. 93–99.

87. P. K. Donnelly, R. S. Hegde and J. S. Fletcher, *Chemosphere*, 1994, **28**, 981.

88. E. S. Gilbert and D. E. Crowley, *Applied Environ. Microbiol.*, 1997, **63**, 1933.

89. J. S. Fletcher and R. S. Hegde, *Chemosphere*, 1995, **31**, 3009.

90. A. C. Singer, E. S. Gilbert, E. Luepromchai and D. E. Crowley, *Appl. Microbiol. Biotechnol.*, 2000, **54**, 838.

91. T. W. Federle and B. S. Schwab, *Appl. Environ. Microbiol.*, 1989, **55**, 2092.

92. J. G. Burken and J. L. Schnoor, *Environ. Sci. Technol.*, 1998, **32**, 3379.

93. J. Dec and J. M. Bollag, *Biotechnol. Bioeng.*, 1994, **44**, 1132.

94. R. M. Bell, *Higher Plant Accumulation of Organic Pollutants from Soils*, EPA/600/R-92/138, Risk Reduction Engineering Laboratory, Cincinnati, OH, 1992.

95. L. A. Newman, S. L. Doty, K. L. Gery, P. E. Heilman, I. Muiznieks, T. Q. Shang, S. T. Siemieniec, S. E. Strand, X. Wang, A. M. Wilson and M. P. Gordon, *J. Soil Contam.*, 1998, **7**, 531.

96. L. A. Newman, S. E. Strand, N. Choe, J. Duffy, G. Ekuan, M. Ruszaj, B. B. Shurtleff, J. Wilmoth, P. Heilman and M. P. Gordon, *Environ. Sci. Technol.*, 1997, **31**, 1062.

97. P. L. Thompson, L. A. Ramer and J. L. Schnoor, *Environ. Sci. Technol.*, 1998, **32**, 975.

98. H. Sandermann, D. Scheel and T. V. Trenck, *Ecotoxicol. Environ. Safety*, 1984, **8**, 167.

99. H. Harms and C. Langebartels, *Plant Sci.*, 1986, **45**, 157.

100. A. Wilken, C. Bock, M. Bokern and H. Harms, *Environ. Toxicol. Chem.*, 1995, **14**, 2017.

101. B. E. Pivetz, Phytoremediation of Contaminated Soil and Ground Water at Hazardous Waste Sites, *EPA-540-S-01-500*, US Environmental Protection Agency, Washington, DC, 2001.

102. R. M. Conger and R. Portier, *Remediation*, 1997, **7**, 19.

103. D. Komossa, C. Langebartels and H. Sandermann, in *Plant Contamination, Modeling and Simulation of Organic Chemical Processes.*, ed. S. Trapp and J. C. McFarlane, Lewis Publishers, Boca Raton, FL, 1995, pp. 69–103.

104. S. D. Cunningham and D. W. Ow, *Plant Physiol.*, 1996, **110**, 715.

105. N. Terry and A. M. Zayed, in *Environmental Chemistry of Selenium*, ed. W. T. Frankenberger and R. A. Engberg, Marcel Dekker, New York, 1998, pp. 633–657.

106. J. W. Biggar and G. R. Jayaweera, *Soil Sci.*, 1993, **155**, 31.

107. A. Zayed and N. Terry, *J. Plant Physiol.*, 1994, **143**, 8.

108. A. Zayed, E. Pilon-Smits, M. deSouza, Z. Q. Lin and N. Terry, in *Phytoremediation of Contaminated Soils and Water*, ed. N. Terry and G. Banuelos, Lewis Publishers, Boca Raton, FL, 2000, pp. 71–93.

109. C. L. Rugh, G. M. Gragson, R. B. Meagher and S. A. Merkle, *Hortscience*, 1998, **33**, 618.

110. R. B. Meagher, C. L. Rugh, M. K. Kandasamy, G. Gragson and N. J. Wang, in *Phytoremediation of Contaminated Soil and Water*, ed. N. Terry and G. Banuelos, Lewis Publishers, Boca Raton, FL, 2000, pp. 201–219.

111. C. C. Wiltse, W. L. Rooney, Z. Chen, A. P. Schwab and M. K. Banks, *J. Environ. Qual.*, 1998, **27**, 169.

112. J. W. Watkins, D. L. Sorensen and R. C. Sims, in *Bioremediation through Rhizosphere Technology*, ed. T. A. Anderson and J. R. Coats, American Chemical Society, Washington DC, vol. 563, 1994, pp. 123–131.

113. S. C. McCutcheon and S. Rock, *Int. J. Phytoremed.*, 2001, **3**, 1.

114. M. S. Hong, W. F. Farmayan, I. J. Dortch, C. Y. Chiang, S. K. McMillan and J. L. Schnoor, *Environ. Sci. Technol.*, 2001, **35**, 1231.

115. V. L. Hauser, B. L. Weand and M. D. Gill, *J. Environ. Eng.*, 2001, **127**, 768.

116. J. C. Stormont and C. E. Morris, *J. Geotech. Geoenviron. Engi.*, 1998, **124**, 297.
117. L. Licht, E. Aitchison, W. Schnabel, M. English and M. Kaempf, *Pract. Period. Hazard. Toxic Radioact. Waste Manage.*, 2001, **5**, 175.
118. C. C. McAneny, P. G. Tucker, J. M. Morgan, C. R. Lee, M. F. Kelley and R. C. Horz, Covers for Uncontrolled Hazardous Waste Sites, *EPA/540/2-85/002*, US Environmental Protection Agency, Washington DC, 1985.
119. K. Madalinski, D. Gratton and R. Wessman, *Remediation J.*, 2003, **14**, 55.
120. L. Licht, E. Aitchison and S. Rock, in Proceedings of 9th Annual SWANA Landfill Symposium, 21–25 June 2004, Monterey, CA, Solid Waste Association of North America, Silver Spring, MD, 2004.
121. W. Albright, C. Benson, G. Gee, A. Roesler, T. Abichou, P. Apiwantragoon, B. Lyles and S. Rock, *J. Environ. Qual.*, 2004, **33**, 2317.
122. T. E. Hakonson, K. V. Bostic, G. Trujillo, K. L. Maines, R. W. Warren, L. J. Lane, J. S. Kent and W. Wilson, *Hydrologic Evaluation of Four Landfill Cover Designs at Hill Air Force Base, Utah*, LAUR-93-4469, Los Alamos National Laboratory, Los Alamos, NM, 1994.
123. J. E. Cornish, W. C. Goldberg, R. S. Levine and J. R. Benemann, in *Bioremediation of Inorganics*, ed. R. E. Hinchee, J. L. Means and D.R. Burris, Battelle Press, Columbus, OH, 1995, pp. 55–63.
124. J. S. Fletcher, A. Groeger, J. McCrady and J. McFarlane, *Biotech. Lett.*, 1987, **9**, 817.
125. J. S. Fletcher, P. K. Donnelly and R. S. Hegde in *Bioremediation of Recalcitrant Organics*. ed. R. E. Hinchee, D. B. Anderson and R. E. Hoeppel, Battelle Press, Columbus, OH, 1995, pp. 131–136.

CHAPTER 4

Microorganisms for Soil Treatment

ALFREDO PÉREZ-DE-MORA,*[a,b]
LAURENT LAQUITAINE[c] AND SARRA GASPARD[c]

[a] Department of Chemical Engineering and Applied Chemistry, University of Toronto, 200 College Street, Toronto, ON, Canada; [b] Research Unit Biogeochemistry and Analytics, Helmholtz Centre Munich, Ingolstaedterlandstrasse 1, 85764 Neuherberg, Germany; [c] Department of Chemistry, COVACHIM-M2E Lab. University of Antilles and Guyane, UFR Exact and Natural Sciences, Pointe à Pitre 97159, Guadeloupe
*Email: perezdemora@gmail.com

4.1 Introduction

Every year millions of tons of inorganic and organic substances from a variety of sources—industrial, domestic and agricultural—enter the world's soils, becoming part of the biogeochemical cycles that affect all forms of life. While soils act as mitigation systems for wastes and chemical substances, their capacity to accommodate those is limited. Contamination of soils negatively affects life in our planet as soils perform a variety of essential functions including: (a) medium to support growth of food and biomass; (b) controlling the fate of water in the hydrological system; (c) recycling and buffering of elements as well as natural and man-made chemicals; (d) habitat for living organisms (*e.g.* microorganisms); (e) engineering medium (*e.g.* building materials, the foundation of any construction); and (f) physical and cultural environment for mankind.[1,2]

RSC Green Chemistry No. 25
Biomass for Sustainable Applications: Pollution Remediation and Energy
Edited by Sarra Gaspard and Mohamed Chaker Ncibi
© The Royal Society of Chemistry 2014
Published by the Royal Society of Chemistry, www.rsc.org

Contamination occurs when the concentration of a particular element (*e.g.* metal, metalloid, trace element), radionuclide, ion (*e.g.* nitrate) or organic compound is above its background or baseline concentration for that particular soil.[3] When contamination implies a direct hazard for living organisms, the term 'pollution' is preferred. About two decades ago efforts were initiated to identify and remediate contaminated soils and groundwater. In Europe alone, the European Environment Agency (EEA) has identified *ca.* 250 000 contaminated sites requiring immediate clean-up actions, whereas the total number of polluting sites is three million.[4] The cost of the remediation of the total soil and groundwater contaminated sites in the first 25 Member States of the European Union has been estimated at €119 billion, including feasibility studies and the remediation costs of both small- and large-scale sites.[4] In the USA, the US Environmental Protection Agency (USEPA) has estimated a total of $209 billion remediation costs for approximately 294 000 contaminated sites.[5] From these numbers it seems clear that developing sound technologies that can effectively aid the remediation of soil and groundwater at reasonable costs and timescales, and in such ways that these natural bodies also remain functional following treatment, is of paramount importance.

Microorganisms, referred herein as bacteria and archaea (prokaryotes) and fungi (eukaryotes), possess an enormous metabolic diversity and versatility; all naturally produced compounds, and most anthropogenic compounds, are substrates for microbial growth.[6] In addition, microorganisms have the ability to affect and/or mediate the mobilization or immobilization processes that finally determine the balance of inorganic species between soluble and insoluble phases.[7] Given that (a) soils naturally harbour millions of microbial cells (*e.g.* bacteria and archaea usually range from 10^7 to 10^{10} cells g^{-1} dry soil), (b) the majority of microbial processes in soil are carried out by microorganisms and (c) microorganisms entail numerous catalytic activities, it seems quite obvious that microorganisms offer unique and enormous potential to treat contaminated soils.

The purpose of this chapter is to provide a general description of the types of contaminants (both inorganic and organic) typically found in soils, the influence and transformations that microorganisms have on such contaminants as well as an overview of the different remediation practices to treat contaminated soil in which microorganisms are involved. The next section focuses on the role of bacteria in soil remediation. Fungi are the subject of section 4.3.

4.2 Bacterial Bioremediation

4.2.1 Heavy Metals, Metalloids and Radionuclides

4.2.1.1 Definition, Origin and Toxicity

By definition heavy metals are those elements with an atomic mass ≥Fe (55.8 g mol^{-1}) or an apparent density ≥5.0 g cm^{-3} (*e.g.* Cd, Cr, Cu, Fe, Mn, Pb and Zn). For metalloids there is no such rigorous definition; metalloids are elements that have properties in-between those of metals and non-metals or a mixture of the two (As, B, Ge, Po, Sb, Si and Te). Heavy metals and metalloids in the environment can originate from a variety of natural sources (*e.g.* volcanic eruptions, serpentine soils, forest fires) and anthropogenic activities (mining, fossil fuel burning, plastics, textile and microelectronics industry, chemicals used in agriculture and waste disposal),[8] although it is anthropogenic activities that generally account for the contamination of soil and groundwater.

The majority of metals and metalloids are naturally found in soils at low or trace concentrations (less than 1000 mg kg^{-1}); thus they are considered trace elements.[3] It should be noted that various heavy metals are essential for living organisms, that is, they are micronutrients (*e.g.* Cu, Fe, Mn or Zn) performing vital functions: (a) catalysts for biochemical reactions; (b) stabilizers of DNA, protein structures and bacterial cell walls; (c) redox processes; (d) part of complex biomolecules with specific function (*e.g.* vitamins); and (e) osmotic balance. Above a certain threshold, however, they become toxic. Toxicity at the cell level occurs through displacement of essential metals from their native binding sites or through ligand interactions. This results in alterations in the conformational structure of nucleic acids and proteins, osmotic balance and oxidative phosphorylation.[9] Acute symptoms of metal and metalloid poisoning include vomiting, diarrhoea or haemorrhage, whereas chronic effects can result in organ dysfunction, pneumoconiosis, neuropathies, hepatorenal degeneration, pulmonary fibrosis (Cr and Ni) and a variety of cancers.[10] Exposure to metals and metalloids might occur through the diet or from medications, but in most cases it occurs due to occupational exposure to metal dusts and fumes which are inhaled.

Radionuclides are elements capable of emitting ionizing radiation. Radioactive isotopes are of particular concern since the ions produced, even at low radiation power, can cause significant DNA damage. Radionuclides are known to enhance the risk of neurological disorders, birth defects, infertility and various types of cancer.[11] Unstable nuclides of all elements can exist; however, radionuclides of environmental concern are those that are routinely produced during the combustion of coal in thermal plants and nuclear power generation, or those released by mining activities, exploration of oil and gas, weapon manufacture and military operations, as well as those present in radioactive waste from nuclear medicine, research and other commercial purposes.[12] The USEPA lists 12 radionuclides as the most commonly found at Superfund sites: Am-241, Cs-137, Co-60, I-129 and I-131, Pu, Ra, Sr-90, Tc-99, ^{3}H, Th and U.[13] Some of these elements can emit α particles (two protons and two neutrons), β particles (electrons) and γ particles (photons with a frequency higher than 10^9 Hz) (*e.g.* U), whereas others only emit two (*e.g.* Cs-137 and Co-60) or one type of particle (*e.g.* Sr-90). Radionuclides are typically non-volatile and less soluble in water than other contaminants. Due to the cation exchange properties of soils, the amount of radionuclides transferred to the plants and thus the food chain is usually minimal.[14] Nonetheless, the mobility of radionuclides in soil is strongly related to their oxidation state. In the reduced valence state, radionuclides are found to be insoluble and geochemically inert; in the oxidized valence state, however, radionuclides become highly mobile migrating into groundwater.

In the environment, metals and radionuclides can be found in one of the following forms: (a) dissolved in the soil solution as ions or as part of soluble in/organic complexes; (b) interchangeable (adsorbed in the exchange complex, that is, clay minerals, Fe and Mn oxides, organic matter); (c) precipitated as carbonates or phosphates; (d) complexed with organic matter; (e) as salts of sulfides, sulfates or chlorides; (f) absorbed or trapped by microbial biomass, plant roots or the soil fauna; (g) occluded by Fe, Mn and Al oxides, oxy-hydroxides and hydroxides; and (h) as part of the crystalline structure of the clay minerals. Metals and radionuclides in forms (a) and (b) are considered as bioavailable, forms (c), (d), (e) and (f) as potentially bioavailable, and (g) and (h) as residual or non-bioavailable.[3] The distribution of metals and radio-nuclides among these compartments is greatly affected by the properties of the soil (pH, redox state, quantity and quality of organic matter, texture, mineral composition, amounts and forms of oxides and carbonates, *etc.*) and the element (abundance, characteristics, competition with other ions).[15] Both physico-chemical and biological mechanisms control the availability and mobility of trace elements in soil. Physico-chemical mechanisms include: (a) adsorption-desorption; (b) precipitation–dissolution; and (c) complexation–chelation. Biological processes related to microorganisms are described in more detail below.

The environmental challenge posed by trace element contamination is that, unlike organic compounds, they cannot be degraded or destroyed. However, they may be redistributed or interconverted among forms of varying chemical and physical properties. Here lies the key to many of the problems but also of the solutions to remediate metal and radionuclide contamination.[12] As fundamental drivers of the biogeochemical cycles of both trace elements and major elements such as C, N, P and S, microorganisms can effect and/or mediate the mobilization and immobilization processes of both metals and radionuclides (see next section) influencing their equilibria between soluble and insoluble phases (Figure 4.1).[16,17]

4.2.1.2 Interactions of Microorganisms with Metals, Metalloids and Radionuclides

Microorganisms can acidify their environment by proton efflux *via* plasma membrane H^+—ATPase protons compete with metals for sites in the exchange complex—or by release of low molecular mass organic acids which can also release H^+ in the environment.[18] Organic acids can also supply complexing organic anions (*e.g.* citrate and oxalate) which can form stable soluble complexes with numerous cationic metal species. Siderophores are another type of low molecular mass molecules which microorganisms excrete in order to assimilate Fe(III); siderophores can form complexes with other metals [Mn, Mg, Cr(III), Ga(III)] and radionuclides such as Pu(IV).[19] Sulfur-oxidizing bacteria

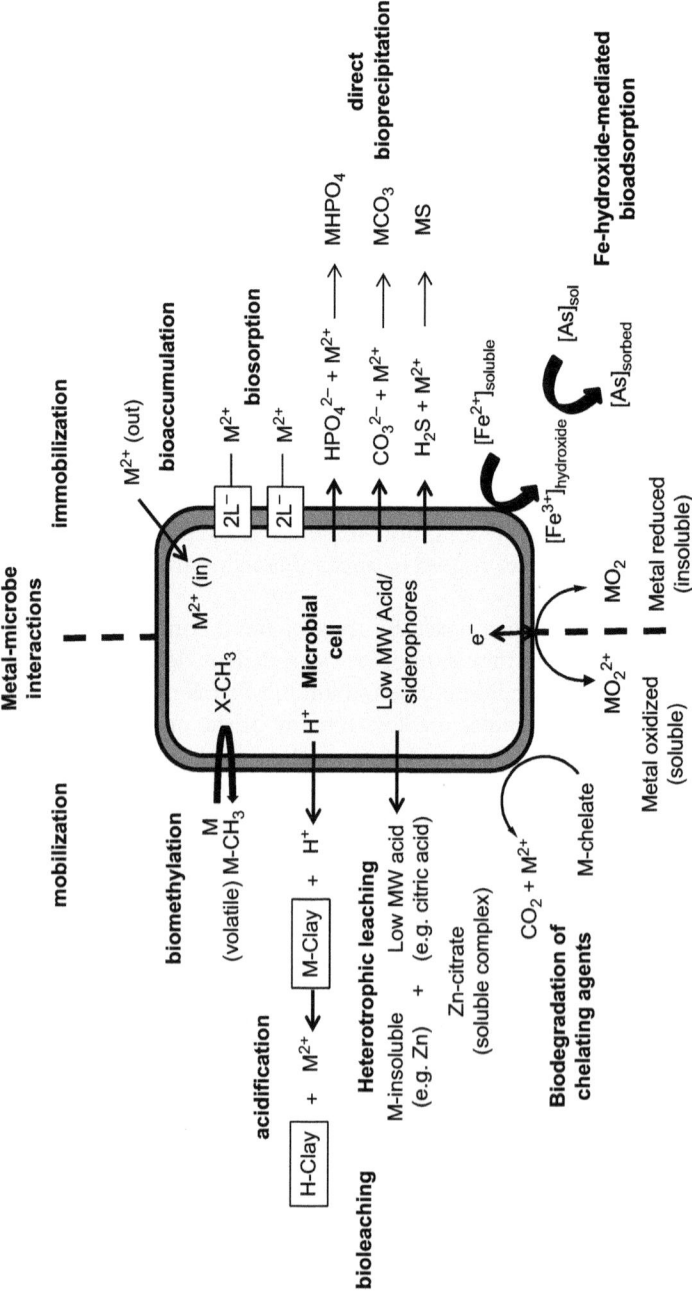

Figure 4.1 Metal/trace element–microbe interactions contributing to bioremediation. Adapted from refs. 16 and 17.

(*e.g. Thiobacillus thiooxidans*) can directly solubilize metals by oxidation of metal sulfides. In addition, the oxidation of sulfur and iron (by iron-oxidizing bacteria) acidifies the immediate environment potentially resulting in enhanced metal desorption from the exchange complex.[20] Fungi and bacteria can bio-methylate various metals and metalloids (As, Hg, Pb, Se, Sn and Te) as a means of detoxification. The methylated compounds are often volatile and lost from the soil, *e.g.* $(CH_3)_2Se$; $(CH_3)_3As$. Redox transformations by micro-organisms can also result in the mobilization of metals and metalloids. For instance, under anaerobic conditions microorganism can use Fe(III) and Mn(IV) as terminal electron acceptors.[21] In their reduced state, Fe and Mn oxides dissolve; metals precipitated on their surface might increase their mobility as well.

Microbial cells are sinks of metals. These can be sorbed onto cell walls including those of dead cells by different radicals present in cell wall peptidoglycans and other polymers; they can also accumulate intracellularly either sequestered by cysteine-rich metallothioneins or compartmentalized in vacuoles in the case of fungi. One important characteristic of cell walls as metal sinks is that bacterial cells have high surface to volume ratios. Extracellular polymeric substances (EPS), which are excreted by microbial biofilms and mats in the soil, also act as metal and metalloid traps. Redox transformations can also result in trace element immobilization. Microbial reduction of metals and radionuclides to a lower oxidation state decreases the mobility and by extent the toxicity of elements such as U(VI) to U(IV) and Cr(VI) to Cr(III).[22] Some sulfate-reducing bacteria such as *Desulfotomaculum reducens* can use H_2 or organic compounds as electron donors and various oxidized metal species such as Cr(VI), Pd(II), Tc(VII) and U(VI) as sole terminal electron acceptors to gain energy, a process known as dissimilatory metal reduction. The reduced forms are less mobile. Indirect reduction of metals to less mobile forms can be achieved with microbially reduced Fe(II), Mn(III) and S(-II) as intermediaries.[23] Microbial phosphatase activity can enhance the precipitation of metals in the form of phosphates. Fungi excrete oxalate to capture Ca from the environment, which precipitates forming crystals on the surface of the hyphae. Oxalate can also bind and precipitate other metals such as Cd, Co, Cu, Mn, Sr and Zn.[24]

4.2.1.3 Microbial Bioremediation of Metal, Metalloid and Radionuclide Contaminated Soils

Traditionally, treatment of soils contaminated with metals and/or radionuclides is achieved by landfilling, stabilization/solidification of the soil with cement or other material or vitrification—heating the soil up to 1200 °C in the presence of borosilicate or soda lime to form a glass-like product that incorporates the metals and/or radionuclides in the crystal structure-. These methods are expensive and render the soil inert. When metals and radionuclides are present in the subsurface and groundwater contamination, technology is generally based on pump-and-treat strategies coupled to some on-site physico-chemical treatment such as precipitation/flocculation, filtration or ion exchange using synthetic organic resins. Cheaper and more environmentally friendly technologies are desirable; however, bioremediation is not yet a mature technique for the treatment of metals and radionuclides. Many microbial metal and mineral transformations may have potential applications for remediation, although most processes are still at the laboratory scale and yet to be tested in full-scale applications.[25] In addition, most of the potential technologies are more likely to succeed as *ex situ* technologies, for instance in bioreactors, rather than *in situ*.

Bioleaching, a well-established process in the mining industry, can be used to recover metals from wastes such as sludge or fly ash coal, which may be later applied in the field.[26] Bioprecipitation by sulfides and phosphates has been tested in lab-based experiments to precipitate metals in contaminated waters and leachates from soils.[27] Indirect bioprecipitation can be achieved by ferrous iron-oxidizing bacteria that precipitate Fe hydroxyl-oxides; these minerals have excellent arsenic-sorbing capacity (1000–10 000 mg^{-1}).[28] Bioaccumulation and biosorption aims to exploit the high surface to volume ratio of microbial cells or the high sorption capacity of fungal biomass in an analogous manner to conventional ion exchange technology; nonetheless, there has been little exploitation by industry.[29]

Biotransformation based on dissimilatory reduction is one of the few bioremediation technologies that has been used successfully at field scale to treat metals and radionuclides in the subsurface.[30,31] This process couples the reduction of radionuclides such as U(VI), Tc(VII), Pu(V/VI) and Np(V) to growth by using electron donors such as H_2 or organic compounds (acetate, lactate and formate). The reduction might be catalysed directly by a reductase enzyme or indirectly by other inorganic species such as Fe^{2+} or S^{2-}.[32] Species of the genera *Chlostridium*, *Desulfovibrio*, *Geobacter* and *Shewanella* have been used in pilot experiments at US Department of Energy (DOE) sites.[30] In one of these experiments (at Rifle in Colorado), acetate was injected in the subsurface as electron donor. Soluble U(VI) concentrations decreased concomitantly with an increase in Fe(II) concentrations in groundwater and *Geobacter* species. However, after 39 days, sulfate-reducing organisms thrived and U(VI) in groundwater increased with a concomitant decrease in Fe(II) and acetate accumulation.[31] At the field research centre in Oak Ridge (TN), a series of single pull–push tests were used to study potential dissimilatory reduction of U(VI) and Tc(VII) in the presence of high (120 mM) nitrate concentrations.[33] Following addition of electron donor (ethanol, glucose or acetate) nitrate and Tc(VII) reduction was observed. Ur(VI) reduction did not occur until Fe(III)-reducing conditions were reached.

The most widespread bioremediation technology with implementation at full scale for the treatment of metals and metalloids is phytostabilization.[34] (See the phytoremediation chapter of this book for details). Here, microorganisms can play a significant role in improving plant survival in metal- and radionuclide-contaminated soils.

4.2.2 Organic Compounds

4.2.2.1 *Organic Compounds in Soils and their Interactions with Microorganisms*

Modern industrialized societies have developed thousands of synthetic organic compounds for multiple uses such as plastics, lubricants, refrigerants, fuels, solvents, pesticides, preservatives and munitions.[35] These contaminants enter the natural environment through accidental leakage and spills, such as leaking underground storage tanks, or through planned spraying (*e.g.* pesticides) or other activities (*e.g.* burning of fossil fuels and explosives). Despite the large number (around 70 000[36]) of synthetic organic chemicals in use or that have been used in the past century, natural microbial communities possess an immense metabolic diversity capable of transforming many of these compounds to less toxic or non-toxic products. The main reason for such great potential is that many synthetic chemicals possess similar structures to naturally occurring compounds or might even occurred naturally, *e.g.* polychlorinated biphenyls (PCBs), perchloroethene (PCE) and chloroform, and transformation of such compounds results in energy gain for microbial growth and/or metabolism.

Organic compounds that can be used by microorganisms to obtain energy for growth are the ones most susceptible to biotransformation in both naturally and engineered systems.[35] The best example is when the C, N, P and S present in the original organic compounds are converted to inorganic products (*e.g.* CO_2, H_2O and/or inorganic salts). This is known as mineralization. The term 'biodegradation' is a bit more ample in the sense that it involves a biologically catalysed transformation of chemicals to simpler compounds, but not necessarily to their inorganic products.[37] In many biodegradative transformations, organic compounds act as electron donors in energy metabolism, that is, they are oxidized by microorganisms; however, this is not always the case. Fermentations are microbial-driven transformations in which one part of an organic compound is oxidized while another part is reduced, in other words, the fermented compound acts both as electron donor and acceptor (*e.g.* fermentation of glucose to CO_2 and ethanol). Another common example is dehalorespiration, a process in which chlorinated chemicals serve as electron acceptors in energy metabolism.[38] The dehalogenated compound does not serve necessarily as the ultimate source of H_2 (electron power), but rather another compound (*e.g.* H_2 formed during fermentation of other organic compounds such as lactate or methanol) supplies the H_2 necessary for dehalogenation. It should be noted that hydrogen serves as electron donor for a variety of microbial transformation involving the reduction of various electron acceptors such as sulfate, Fe(III), Mn(IV), chlorinated solvents and CO_2. On some occasions, microbial activities bring about the biodegradation or biotransformation of an organic compound without using it as substrate for energy or growth. This is known as cometabolism. Usually, cometabolism depends on the presence of a second substrate that can support the growth or provide the energy for the organism carrying out the cometabolic reaction.

Despite the metabolic versatility of microorganisms the conversion of organic chemicals in soil to simpler products is not always possible. Several factors are critical for biodegradation: (a) the concentration of the target compound should not be too high so that it is toxic or so low that it cannot be used to support growth (threshold); (b) the compound should be bioavailable; (c) the appropriate microorganisms must be present and active; (d) the presence of nutrients (*e.g.* N, P, C) and appropriate electron donors; (e) suitable medium conditions (pH, redox and temperature and presence or absence of O_2; and (f) lack of inhibiting substances. Figure 4.2 shows examples of the various groups of organic contaminants considered in this chapter.

Bioavailability can be defined as the fraction of the total chemical that can interact with a biological target.[39] Determining the bioavailable fraction is a major challenge as there is no universal approach. Limited bioavailability of organic chemicals in soils can be attributed to one of the following causes:[40] (a) the target compound is strongly sorbed onto clay minerals (*e.g.* the herbicide paraquat which is positively charged) or organic matter (*e.g.* hydrophobic and non-polar compounds such as polycyclic aromatic hydrocarbons); (b) the target compound is located in micro- or nanopores inaccessible by microorganisms; (c) the target compound is present in a NAPL (non-aqueous phase liquid) and thus has very limited solubility (high K_{ow}); and (d) the target compound complexes with natural organic matter or other environmental constituents to form molecular species that, although containing the parent molecule, are in fact new molecular species non-amenable to microbial transformation.

The target substrate can be acted on outside the cell by extracellular or cell wall bound hydrolytic enzymes as in fungal biodegradation by ligninolytic enzymes. Alternatively, it might require to be taken up inside the cell (*e.g.* aliphatic hydrocarbons), or it may need close contact between the cell and the substrate (*e.g.* chlorinated solvents are reduced by reductive dehalogenases located on the periplasmic space). Microorganisms must rely, at least, on one of the following three mechanisms to access a particular substrate:[37,40] (a) cells use molecules that enter the aqueous phase by spontaneous partitioning; (b) microorganisms excrete biomolecules (biosurfactants) capable of entrapping the insoluble compound into micelles (emulsification) allowing the microorganisms to access the substrate; and (c) cells attach directly to the NAPL, frequently as biofilms. The lipophilic cell surface may facilitate absorption into the cell, or an emulsifier might be present as a thin layer on the cell surface assisting absorption.

For the degradation or transformation of organic compounds, microorganisms require enzymes and regulatory elements that control the expression of the genes coding for those enzymes or the catabolic operons in which those genes are integrated. Catabolic genes might be plasmid- or chromosome-regulated. Regulation implies a more efficient response to presence/absence of the target compound or other environmental signals. It most commonly occurs at the transcriptional level.[41] It should be noted at this stage that, apart from microbial degradation/transformation, there are also physico-chemical processes that influence the fate of organic chemicals in soils such as photo and

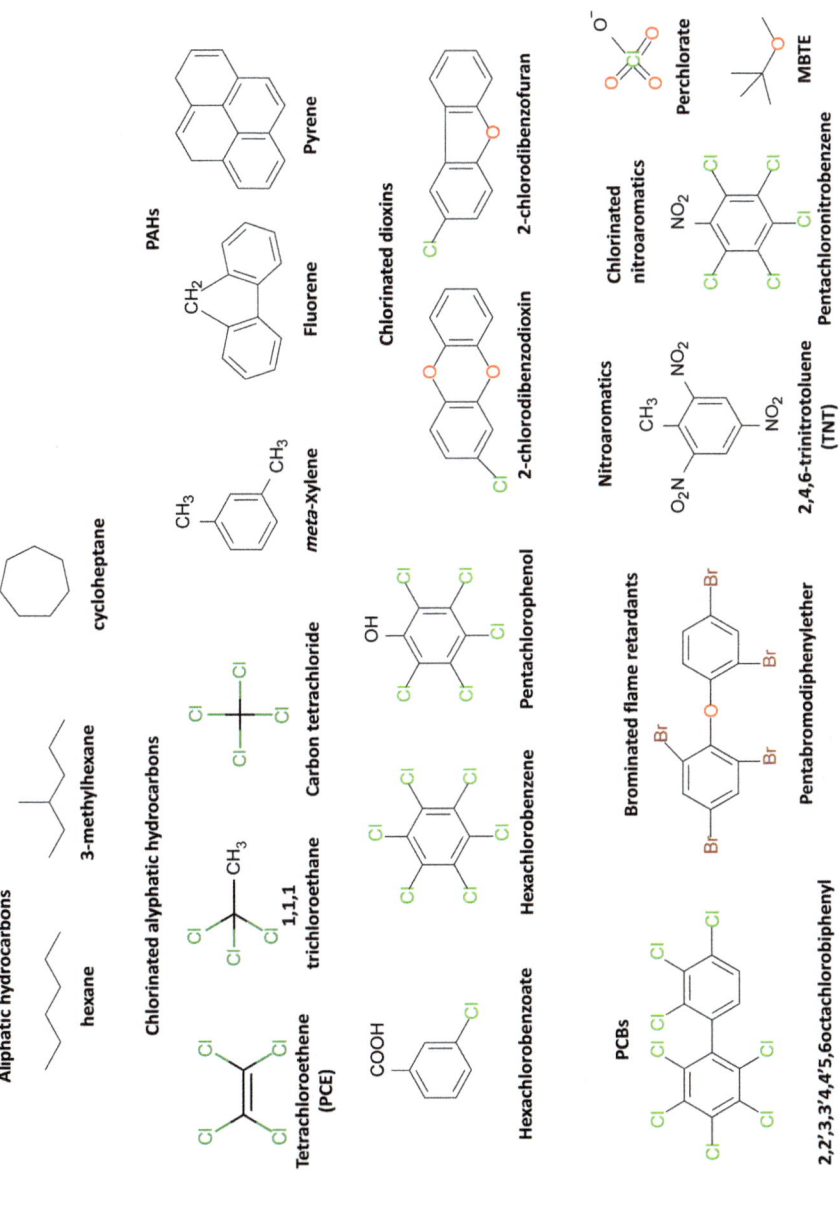

Figure 4.2 Overview of the various groups of organic contaminants (and perchlorate) relevant to this chapter.

chemical decomposition, volatilization, adsorption by clay minerals, oxides or organic matter, absorption by both plants and microorganisms, exudation by plant roots, solubilization and transportation through run-off and/or leaching.[1]

4.2.2.2   *Biotransformation of Organic Compounds*

The most widespread organic contaminants are hydrocarbons (HCs). These are compounds consisting entirely of H and C. Anthropogenic activities related to the production, transportation, processing, handling and disposal of oil have resulted in the contamination of numerous sites with different types of hydrocarbons in the last century. HCs may consist of aromatic rings (aromatic HCs) or not (aliphatic HCs). Aliphatic HCs might be entirely composed of single bonds (saturated HCs) or contain at least one double or triple bond (unsaturated HCs). Depending on their arrangement, aliphatic HCs can be found as: linear chains (*n*-alkanes), branched (*iso*-alkanes) or cyclic (at least one carbon ring). HCs are non-polar compounds with limited solubility in water and high octanol–water coefficients. They are highly hydrophobic and adsorb well onto organic matter. Due to their non-polar nature, HCs are not affected by the charges of the clay minerals.[42] Substitutions by halogens and other groups (*e.g.* nitro group, hydroxide) introduce new properties and lead to a variety of compounds that are also of environmental concern (see next sections for details).

4.2.2.2.1   Aliphatic Hydrocarbons.   Aliphatic hydrocarbons are typical components of mineral oils and gasoline; they may account for up to 75% of the total HCs present.[43] Those having between one to four carbon atoms are gaseous at ambient temperatures, whereas HCs with more than four carbon atoms are liquid or solid. Due to their apolar nature they are chemically very inert, have low solubility in water and tend to accumulate in cell membranes.[44] Microorganisms are capable of completely mineralizing alkanes of various chain lengths all the way to CO_2 and H_2O (*e.g. Acinetobacter* sp. M1, *Pseudomonas putida* GOP1, *Mycobacterium vanbaalenii*).[45]

The first step in the aerobic degradation of alkanes is carried out by alkane monooxygenases/hydroxylases. These enzymes usually act on alkanes of different chain length (*e.g.* alkane moonoxygenase from *P. putida* GOP1 can act on *n*-alkanes from C_3 to C_{15}). Relevant alkane hydroxylases include:[46] (a) soluble (sMMO) and particulate (pMMO) methane monooxygenases; (b) the alkB-type alkane hydroxylases with typical substrate range of C_3–C_{15}; (c) cytochrome P450 alkane hydroxylases, which are present in both bacteria and fungi; and (d) alkane hydroxylases, which act on longer alkanes (C_{15} and longer). The general pathway for *n*-alkanes of two or more carbon atoms is oxidation of a terminal methyl group to render a primary alcohol, which is first transformed to aldehyde and then to the corresponding fatty acid (Figure 4.3(a)). The fatty acid is combined with CoA to form acyl-CoA, which enters the β-oxidation pathway. Subterminal oxidation is also possible and coexists in some microorganisms.[47] In this case a secondary alcohol is formed which is subsequently converted to the corresponding ketone and then oxidized to an ester *via* a Baeyer-Villiger monooxygenase.[48] The ester is hydrolysed by an esterase generating an alcohol and a fatty acid (Figure 4.3(a)). Anaerobic transformation of *n*-alkanes has been reported under sulfate conditions by bacteria in the *Desulfosarcina/Desulfococcus* cluster and under nitrate-reducing conditions by *Azoarcus* sp HxN.[48] This process involves the addition of a fumarate molecule either at a subterminal or terminal C yielding an alkyl succinate (Figure 4.3(b); only the subterminal addition is shown). Subsequently CoA is linked and acyl-CoA is formed, which enters the β-oxidation pathway.[49]

4.2.2.2.2 Unsaturated Hydrocarbons. Unsaturated hydrocarbons have similar properties to those of their saturated counterparts with the difference that the double or triple bond is more reactive than the single bond. They follow a similar degradation path to that of saturated HCs under aerobic conditions. The first step is an oxidation and the resulting product either an alcohol (when acting on terminal or subterminal methyl groups) or an epoxide (if acting on the double bond). These intermediates are further processed to fatty acids which enter the β-oxidation pathway. Under anaerobic conditions, *n*-alcohols, *n*-ketones and linear (3-OH-) ketones have been recently shown as products of the anaerobic oxidation of *n*-alkenes by strains from the genus *Desulfatiferula*.[50]

4.2.2.2.3 Cycloalkanes. Cycloalkanes are more recalcitrant in the environment than most hydrocarbons including *iso*-alkanes, monoaromatic and polycyclic aromatics. Two different aerobic degradation pathways of cyclic alkanes are known: a lactone formation pathway and aromatization. In the well-defined lactone formation pathway, cyclohexane is oxidized to cyclohexanol and cyclohexanol is dehydrogenated to cyclohexanone from which epsilon-caprolactone is produced.[51] Epsilon-caprolactone is further oxidized to adipic acid, and ultimately to CO_2 as in *Acinetobacter* NCIB 9871 (Figure 4.3(c)).[52] In the aromatization pathway, cyclohexanone is desaturated to 2-cyclohexen-1-one which is aromatized to phenol as in *Rhodococcus* sp. NDKKK[53] phenol can be transformed to catechol, entering the regular pathway for degradation of aromatics.[54] Anaerobic degradation of cycloalkanes has been only observed in mixed enrichment cultures under sulfate, methanogenic and nitrate reducing conditions. The postulated activation mechanism of the cycloalkane involves the addition of fumarate analogous to that of n-alkanes.[49,55]

4.2.2.2.4 Chlorinated Aliphatic Hydrocarbons. Chlorinated aliphatic hydrocarbons (CAHs) possess at least one Cl atom covalently bound (Figure 4.2). The simple chlorinated HCs with one carbon atom (chlorinated methanes) or two carbon atoms (chlorinated ethanes and ethenes) are the most widespread contaminants in the subsurface and groundwater. Many of these compounds are known or suspected carcinogens, mutagenic agents and/or neurotoxins. Chlorinated solvents, including carbon tetrachloride (CT),

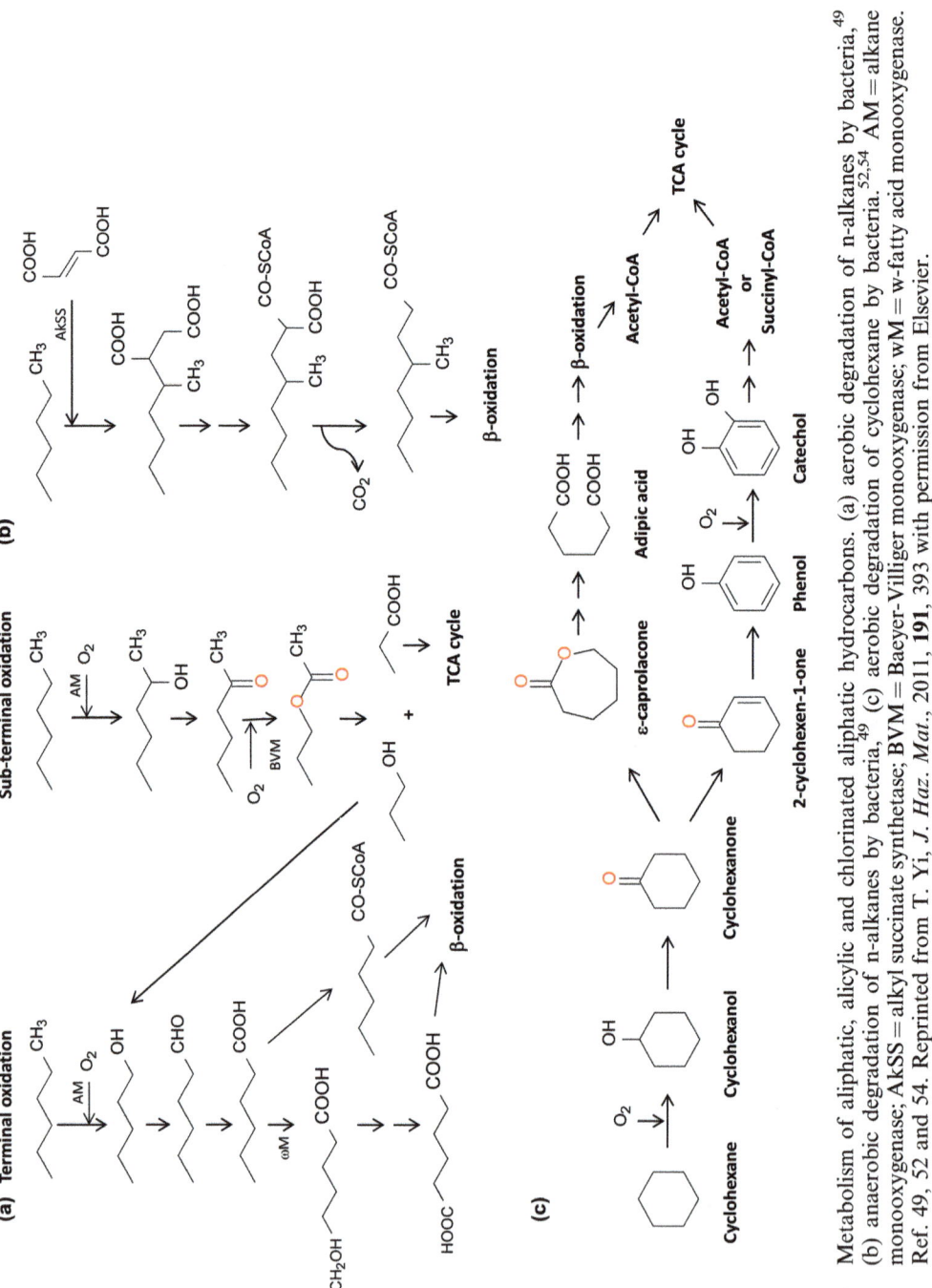

Figure 4.3 Metabolism of aliphatic, alicylic and chlorinated aliphatic hydrocarbons. (a) aerobic degradation of n-alkanes by bacteria,[49] (b) anaerobic degradation of n-alkanes by bacteria,[49] (c) aerobic degradation of cyclohexane by bacteria.[52,54] AM = alkane monooxygenase; AkSS = alkyl succinate synthetase; BVM = Baeyer-Villiger monooxygenase; wM = w-fatty acid monooxygenase. Ref. 49, 52 and 54. Reprinted from T. Yi, *J. Haz. Mat.*, 2011, **191**, 393 with permission from Elsevier.

perchloroethene/tetrachloroethene (PCE), trichloroethene (TCE) and 1,1,1-trichloroethane (TCA), have been extensively used as cleaning and degreasing agents. Other applications include adhesives, pharmaceuticals, textile processing, coating solvents and feed stocks for other chemicals.[56]

Chlorinated HCs usually have densities >1 g cm^{-3} and thus are denser than water. As a result, chlorinated HCs tend to migrate down the soil profile reaching the vadose and the saturated zone as dense non-aqueous phase liquids (dNAPLs). Here, they can dissolve slowly into the groundwater and move along with the flow resulting in a contamination plume.[57] However, short-chained CAHs are lost as gaseous emissions from contaminated soils and groundwater due to their volatility potentially affecting overlying buildings. CAHs can undergo both aerobic and anaerobic degradation (Figure 4.4). The tendency for the chlorinated ethenes and ethanes to undergo oxidation increases with decreasing number of chlorine substituents. Nonetheless; compounds with a highly oxidized character such as TCE or 1,1,1-TCA can be oxidized *via* cometabolism; for example, methanotrophic bacteria through methane monooxygenase can fortuitously oxidize TCE to TCE epoxide, an unstable compound that can be degraded to a variety of compounds which can be mineralized by other microorganisms.[58] In contrast, aerobic microbial degradation of dichloroethene (DCE) and vinyl chloride (VC) as primary substrates is possible.[59] For chlorinated methanes, microbial aerobic mineralization to CO_2 has been proven for dichloromethane and chloromethane. Chloroform can be mineralized to CO_2 by cometabolism, whereas mineralization of CT under aerobic conditions has not yet been reported.[60]

Highly chlorinated CAHs are subject to reductive dechlorination under anaerobic conditions when Cl atoms are replaced with H atoms. Reductive dechlorination may be cometabolic or the chlorinated HCs may act as terminal electron acceptors in energy metabolism, a process known as dehalorespiration. There are various organisms capable of reductive dechlorination of chlorinated ethenes (*e.g. Geobacter, Desulfitobacterium and Dehalobacter*), but only *Dehalococcoides* is known to reductively dechlorinate cis-1,2-dichloroethene (cDCE) and VC to ethene. In the subsurface organic substrates (*e.g.* acetate, butyrate, lactate, benzoate, proprionate, methanol, ethanol and formate) are fermented, generating H_2 which ultimately serves as electron donor.[61] In this fashion CAH substrates can be dechlorinated to non-toxic products such as ethene, ethane or methane. The enzymes responsible for dechlorination are reductive dehalogenases. Reductive dehalogenation is a well-exploited bioremediation technology for numerous contaminated subsurface systems. In this regard, both biostimulation of native communities through addition of electron donor (*e.g.* ethanol) or bioaugmentation using mixed cultures including *Dehalococcoides* have been shown successfully at field sites.[62,63]

4.2.2.2.5 Monoaromatic Hydrocarbons. Toluene, ethylbenzene and the three different isomers of xylene comprise the group known as BTEX (Figure 4.5(a)). These compounds are of particular concern because of their high solubility in water relative to other HCs and their toxicity; benzene is a

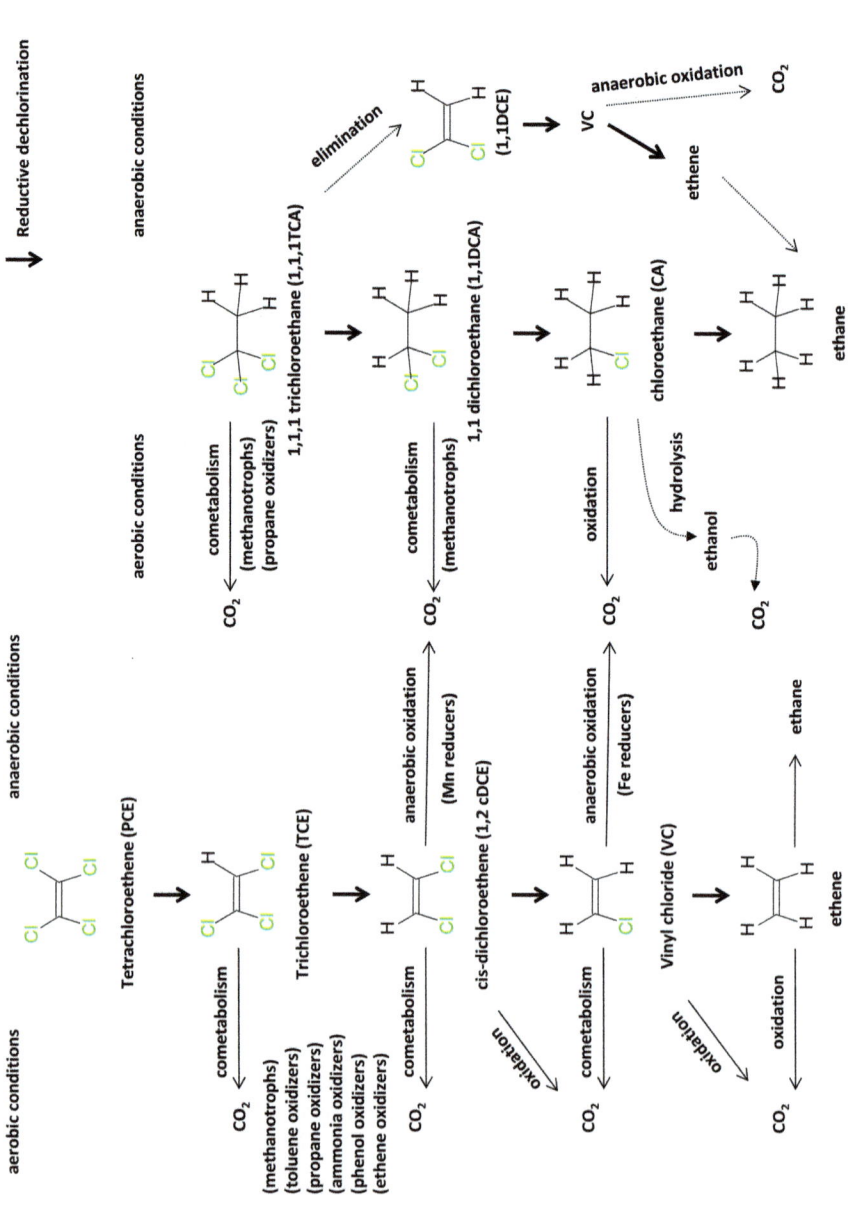

Figure 4.4 Chlorinated aliphatic hydrocarbons: biotransformations of chlorinated ethenes and ethanes under aerobic and anaerobic conditions. Adapted from ref. 141.

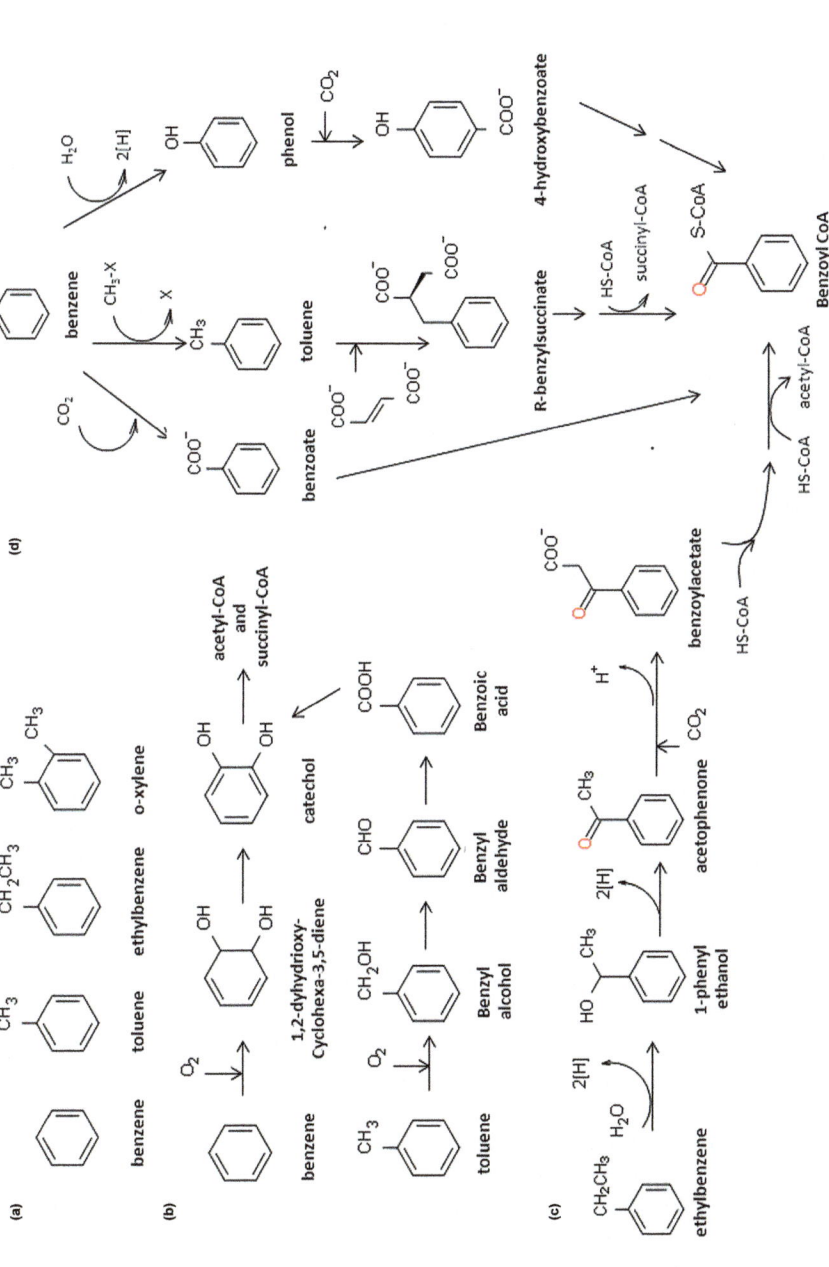

Figure 4.5 Monoaromatic hydrocarbons: (a) BTEX components (only the ortho form of xylene is shown); (b) bacterial aerobic degradation of benzene and toluene; (c) bacterial anaerobic degradation of ethylbenzene under nitrate conditions; and (d) proposed pathways for benzene degradation under anaerobic conditions. From Weelink *et al.*, *Rev. Environ. Sci. Biotechnol.*, 2010, **9**, 359–385 with permission from Springer Science and Business Media. Adapted from ref. 165 and 166.

potent carcinogen and the rest of the group are known to be harmful to the central nervous system.[64] The BTEX compounds are important components of gasoline; they are also employed in the production of pharmaceuticals, agrochemicals, polymers and explosives. Natural aromatic compounds include aromatic amino acids, phenols and quinones.

As for other HCs, monooxygenases/dioxygenases are the enzymes involved in the first activation mechanism in the aerobic degradation of the BTEX compounds (*e.g.* as in toluene-degrading *Pseudomonas* strains).[65] A variety of aromatic compounds (not only BTEX) can be transformed to a restricted number of intermediates (*e.g.* catechol, protocatechuate, phenylacetic acid, benzoate) which enter central aromatic pathways.[66] Ultimately, these intermediate metabolites are transformed to acetyl-CoA and succinyl-CoA, which can then enter the tricarboxylic acid cycle (Figure 4.5(b)). In anaerobic metabolism, all BTEX compounds are known to, or hypothesized to, be transformed to a central intermediate (benzoyl-CoA), which can be further oxidized *via* reductive ring cleavage to CO_2.[67]

The anaerobic degradation pathways for toluene, ethylene and xylene have been established. In many cases, addition of fumarate is responsible for the first activation step (Figure 4.5(c)). The initial activation and degradation steps of benzene remain unknown. Various mechanisms have been proposed including hydroxylation, methylation and carboxylation (Figure 4.5(d)).[68] Examples of organisms involved in benzene degradation under anaerobic conditions include *Dechloromonas* strains and bacteria similar to *Azoarcus* under NO_3^- reducing conditions) and bacteria related to the sulfate-reducing genus *Desulfosporinus*.[69,70]

4.2.2.2.6 Polycyclic Aromatic Hydrocarbons. Polycyclic aromatic hydrocarbons (PAHs) consist of two or more fused benzene rings combined in linear, angular or cluster arrangements (Figure 4.2). They are formed during the incomplete thermal decomposition of organic molecules and their subsequent recombination. Natural sources of PAHs include forest and rangeland fires, oil seeps, volcanic eruptions and exudates from trees.[71] Anthropogenic sources are related to the burning of fossil fuels, coal tar, wood and garbage. PAHs are of particular concern due to their toxicity, mutagenicity and carcinogenicity.

The main process of PAH degradation in the environment is microbial transformation. However, PAHs are among the most recalcitrant compounds in nature due to their high hydrophobicity—PAHs have low aqueous solubility and tend to adsorb strongly with solid particles in the soil, particularly with the humin fraction of the organic matter.[72] Both bacteria and fungi have been shown to degrade PAHs under aerobic conditions. The initial step requires the activity of a dioxygenase and O_2 which yields vicinal *cis*-dihydriols (Figure 4.6). The dihydroxylated intermediates may then be cleaved by intradiol or extradiol dioxygenases through *ortho*- or *meta*-cleavage, leading to central intermediates such as protocatechuates and catechols that are further converted to intermediates of the TCA cycle (Figure 4.6).[73,74]

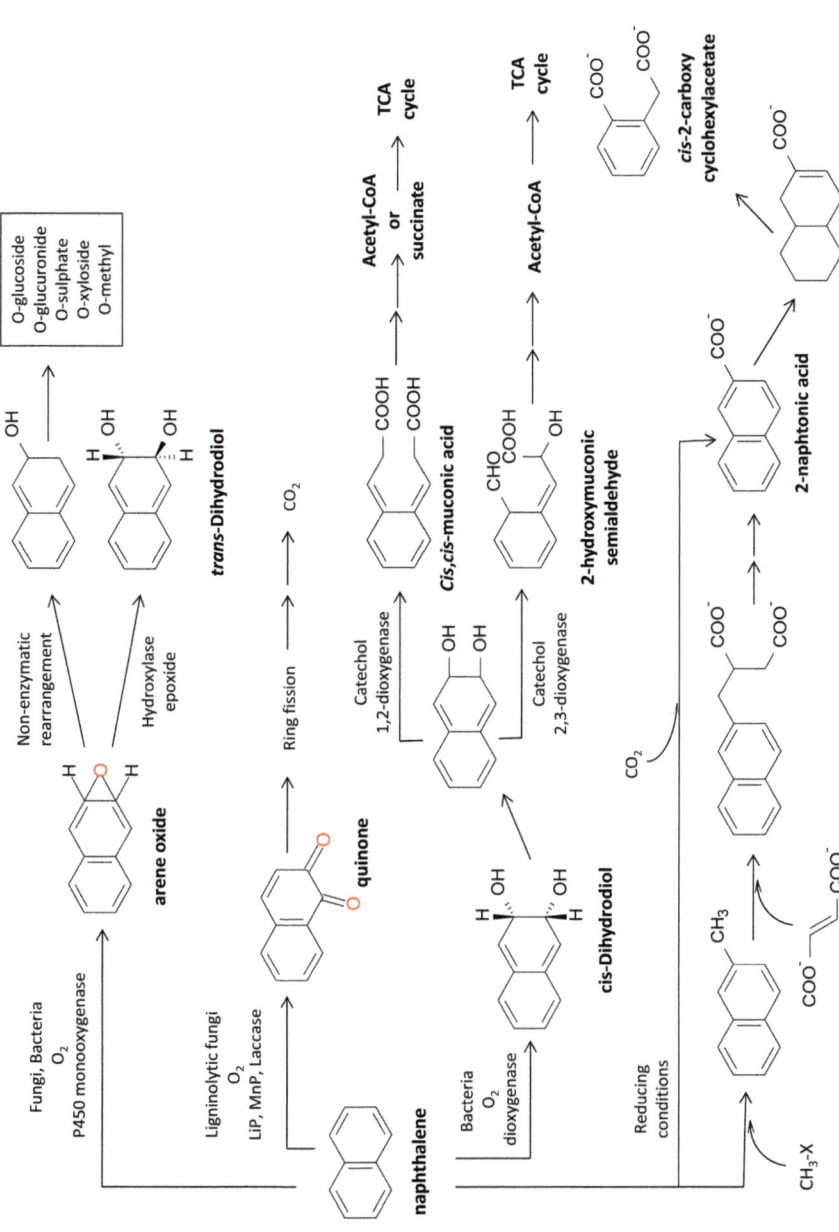

Figure 4.6 Schematic representation of aerobic and anaerobic degradation of PAHs (naphthalene as example) by both bacteria and fungi. Adapted from ref. 73 and 167.

Multiple groups of bacteria have been reported to oxidize both low and high molecular weight PAHs (*e.g.* strains of *Acinetobacter*, *Pseudomonas*, *Rhodococcus*, *Mycobacterium*).[74] While there are numerous reports on bacteria that can grow on low molecular weight PAHs (2–3 rings) as the sole carbon source, there are only a few studies that document extensive mineralization of PAHs that have more than four rings.[75] For instance, oxidation of benzo[a]pyrene (five rings) has been only reported for bacteria growing on other sources of carbon. Detailed information on the anaerobic degradation of PAHs is scarce and it is not clear yet whether PAHs having three or more rings can support growth or whether they are only oxidized through cometabolism.[76] Two pathways have been proposed for naphthalene and phenanthrene: carboxylation *versus* methylation and subsequent fumarate addition. Further metabolism of these PAH results either in dead-end metabolites or ring cleavage amenable to further reactions which presumably lead to a central metabolite such as benzoate (Figure 4.6). There are very few organisms isolated to date capable of PAH degradation under anaerobic conditions, namely strains belonging to the *Deltaproteobacteria* with affiliation to the *Desulfobacteriaceae*, strains with less than 91% affinity to other genera and Gram-positive microorganisms such as *Desulfotomaculum* strains.[77]

4.2.2.2.7 Chlorinated Aromatic Hydrocarbons. Chlorinated benzenes or chlorobenzenes (CBs) are isomeric chlorinated aromatic compounds having a benzene ring that is substituted with 1–6 chlorine atoms (Figure 4.2). There is limited information on the acute (short-term) effects of chlorobenzene on humans, whereas chronic (long-term) exposure is known to affect the central nervous system. So far there is no information on the carcinogenicity of chlorobenzene. Environmental contamination with chlorobenzenes is widespread due to their relevance as industrial intermediates (rubber chemicals, antioxidants, dye and pigments, agriculture products and pharmaceuticals, pesticides and solvents).[78] The degree of dechlorination is important for the chemical properties of chlorobenzenes: the greater the number of chlorine substituents the lower the solubility and the vapour pressure and the greater the hydrophobicity. Based on this the lower chlorinated benzenes are more prone to volatilization, whereas the higher chlorinated benzenes tend to sorb on the organic matter of soils and sediments.[79]

Various strains are known to grow on the lower chlorobenzenes as sole sources of carbon and energy—mostly strains of *Pseudomonas* and *Burkholderia*. The attack of CBs under aerobic conditions is carried out by dioxygenases producing chlorinated dihydrodiol intermediates that are subsequently re-aromatized by dihydrodiol dehydrogenases to chlorocatechols.[80] The chlorocatechols are further hydroxylated to chloromuconic acids by dioxygenases, which can be further metabolized into intermediates of the tricarboxylic acid cycle. Aerobic cometabolism of CBs has been shown by various methane-oxidizing bacteria and benzene-grown mixed cultures.[81] Under anaerobic conditions, CBs can be reductively dechlorinated and organisms (*e.g. Dehalococcoides* and *Dehalobacter*) can gain energy for growth from this process. Recently, reductive dechlorination of mono- and di-CBs all the way to benzene was reported in mixed methanogenic cultures.[82]

Chlorinated benzoates or chlorobenzoates (CBAs) include isomers of chlorobenzoate, dichlorobenzoate and trichlorobenzoate (Figure 4.2). Contamination of soils with CBAs is mainly due to their utilization as herbicides (*e.g.* 2,3,6-trichlorobenzoate) or as metabolites of other halogenated compounds such as PCBs, which are also employed as herbicides/fungicides.[83] CBAs can be also produced naturally by plants, insects, bacteria and fungi as intermediate metabolites or deliberately as defence mechanism.

Aerobic growth on mono- and di-chlorinated benzoates has been shown for several bacterial strains, many of them related to *Pseudomonas* and *Alcaligenes*. Field and Sierra-Alvarez[84] provide an overview on the reported aerobic pathways for 3-and 4-chlorobenzoate. Ring attack by dioxygenases and the subsequent formation of diols, which are further transform to chlorocatechols, is the most commonly reported pathway. Other pathways involve the conversion of 3-chlorobenzoate to protocatechuate or the replacement of the chlorine by a hydroxyl group. Aerobic bacterial cometabolism has been reported in the presence of benzoate and other aromatic substrates including benzene.[85] Under anaerobic conditions, CBAs can be reductively dechlorinated to benzoate which can be further metabolized to CO_2 and acetate as described by Dolfing and Tiedje[86] in a three-member system including: a halorespirer (*Desulfomonile tiedji*) that reductively dechlorinates 3-chlorobenzoate to benzoate; a heteroacetogen which converts the benzoate to H_2, CO_2 and acetate; and a H_2-consuming organism (the methanogen *Methanospirillum*) which yields the reaction thermodynamically favourable.

Chlorinated phenols or chlorophenols (CPs) have been introduced into the environment through their use as antiseptics (4-CP or 2,4-D), herbicides (2,4-D) or fungicides in wood preservative applications, *e.g.* pentachlorophenol (PCP). CPs can cause harmful effects on various organs (*e.g.* liver, kidneys, lungs and nervous system) and direct contact can cause irritation. PCP is considered a potential carcinogen.[87] Phenols with three or more chlorine atoms are significantly ionized at neutral pH. Being ionizable, the chlorinated phenols are quite soluble in water and thus are commonly found as groundwater contaminants.[35] However, they also have a high octanol–water partition coefficient (*e.g.* PCP $>10^5$) and partition quite strongly into high organic content soils and sediments.

There are two main strategies for the degradation of CPs under aerobic conditions: (a) the chlorocatechol pathway for lower chlorinates phenol (1–2 Cl atoms); and (b) the hydroquinone pathway for the higher chlorinated phenols (3–5 Cl atoms).[88] In the chlorocatechol pathway, the CPs are initially hydroxylated by monooxygenases yielding chlorocatechols as the first intermediates which are, in turn, subject to ring cleavage prior to dechlorination. In the hydroquinone pathway, the CPs are initially transformed to chlorohydroquinones; subsequent reactions remove the Cl atoms prior to ring cleavage. In this second pathway, the Cl atoms are removed through a sequence of hydrolytic and reductive dechlorinating steps.[89] Glutathione S-transferase is involved in the reductive dechlorinating steps. Some monooxygenases also possess hydrolytic activity and hence can dechlorinate CPs as reported for the monooxygenase of *Ralstonia eutropha* strain JMP134(pJP4).[90] CPs can be used as electron acceptors in reductive dechlorination to yield energy for microbial growth under anaerobic conditions.[91]

4.2.2.2.8 Polychlorinated Dibenzodioxins. Polychlorinated dibenzodioxins (PCDDs) refer to two families of tricyclic, planar aromatic compounds highly toxic including 2,3,7,8-tetrachlorodibenzo-*p*-dioxin (2378-TeCDD), classified as a Group 1 carcinogen (Figure 4.2). Teratogenic, mutagenic, carcinogenic, immunotoxic and hepatotoxic effects of PCDDs have also been described in animals. One of these families is the polychlorinated dibenzo-*p*-dioxins (PCDDs) with 75 possible congeners and the other is the polychlorinated dibenzofurans (PCDFs) with 135 different congeners.[92]

Dioxins are formed as unwanted by-products of pesticide manufacturing, combustion and incineration, chlorine bleaching and disinfection. Forest fires are also an important natural input of dioxins to the environment. Dioxins are highly hydrophobic and thus are tightly adsorbed by soil and sediments. There are no documented pure cultures capable of complete mineralization of PCDD/Fs under aerobic conditions. However, co-cultures containing a CDF-degrader (*Sphingomonas*) and an organism (*Pseudomonas* or *Burkholderia*) capable of degrading chlorosalicylic acid, an intermediate which would otherwise accumulate as a dead-end product, were able to completely mineralize monochlorodibenzofurans.[93] In most cases, degradation of PCDD/Fs under aerobic conditions has been reported as a cometabolic process, with dibenzofuran, biphenyl or carbazole as examples of primary substrates.[92] The best known degradation pathways involve the initial action of angular dioxygenases which attack a ring adjacent to the ether oxygen bridging the two rings (Figure 4.7(a),(b)).[94] The resulting diols spontaneously convert to trihidroxy-diphenyl ethers in the case of PCDD and trihydroxybiphenyl in the case of PCDF (Figure 4.7(a),(b)). The latter are subsequently oxidized by dioxygenases causing ring opening. The ring-opened products can be further metabolized to chlorinated catechols from PCDDs or chlorinated salicylates from PCDFs. Chlorinated dioxins also undergo reductive dechlorination under anaerobic conditions. *Dehalococcoides* strains have been shown to effectively dechlorinate various PCDD congeners with initial dechlorination either in the lateral or *peri* position (Figure 4.7(c)).[95]

4.2.2.2.9 Polychlorinated Biphenyls. Polychlorinated biphenyls (PCBs) are a large group of synthetic chemicals with 209 possible congeners (Figure 4.2). PCBs came into extensive use in the 1950s in electrical capacitors and transformers, hydraulic fluids and pump oils, and adhesives, dyes and inks. In practice, PCB contamination results from commercial preparations composed of complex mixtures containing 60–90 congeners.[96] PCBs are considered carcinogens. It has also been shown that PCBs cause reproductive, neurological and endocrine disruptions in both animals and humans.[97] PCBs

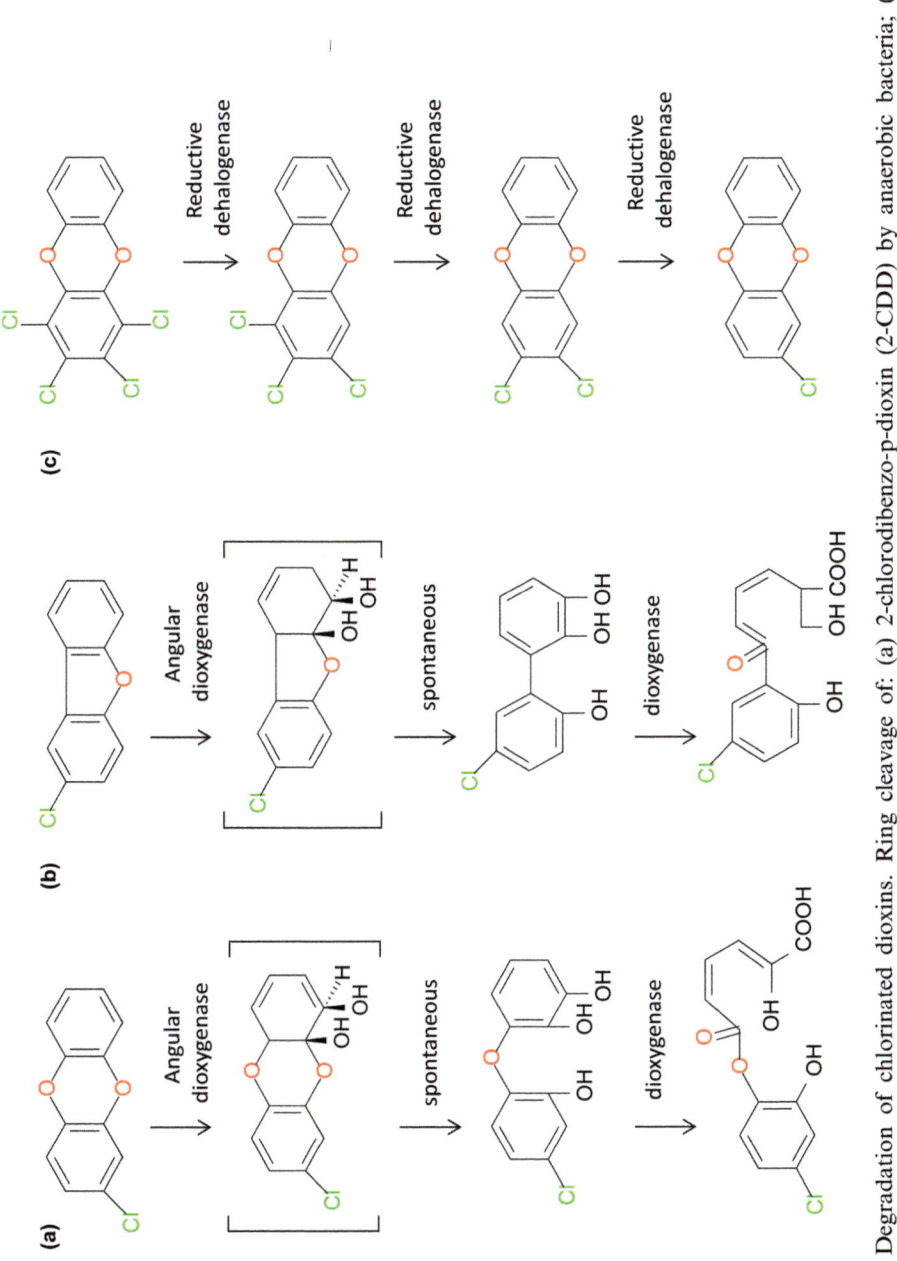

Figure 4.7 Degradation of chlorinated dioxins. Ring cleavage of: (a) 2-chlorodibenzo-p-furan (2-CDF) by aerobic bacteria; (b) 2-chlorodibenzo-p-dioxin (2-CDD) by anaerobic bacteria; (b) 2-chlorodibenzo-p-dioxin (2-CDD) by anaerobic bacteria; and (c) reductive dechlorination of 1234-tetrachlrodibenzo-p-dioxin (1234-TeCDD) by *Dehalococcoides* strain CBDB1. Adapted from ref. 96.

are considered persistent organic pollutants (POPs); due to their hydrophobic properties and their resistance to both chemical and biological transformations, they tend to become adsorbed by natural organic matter in soil, sediments and sludges. Their destruction by chemical thermal, and biochemical processes is extremely difficult, and presents the risk of generating extremely toxic dibenzodioxins and dibenzofurans through partial oxidation.

Some bacteria have the ability to grow on PCBs congeners of one or two Cl atoms (*e.g.* 4-chlorobiphenyl) as sole sources of energy and carbon. However, growth on some congeners (*e.g.* 2- and 3-chlorobiphenyl) is limited due to the accumulation of toxic intermediates.[98] The aerobic metabolism of PCBs starts with the oxidation of the PCB by biphenyl-2,3dioxygenase to the corresponding *cis*-dihydriol. This is subjected to further oxidation and hydrolytic steps which yield chlorobenzoic acid and an aliphatic acid.[99] The latter can be metabolized *via* the tricarboxylic acid cycle to CO_2, whereas the chlorobenzoic acid requires the contribution of specialized bacteria. Thus, the best degradation results can be achieved in co-cultures or mixed cultures. *Dehalococcoides* type organisms preferentially dechlorinate higher-chlorinated PCBs (four and five Cl atoms), while dechlorination of the lower chlorinated compounds is limited to cometabolic activity in the presence of highly chlorinated parent compounds.[100]

4.2.2.2.10 Brominated Flame Retardants. Brominated flame retardants (BFRs) are organobromide compounds that have an inhibitory effect on the ignition of combustible organic materials. They are the largest market group of flame retardants because of their low cost and high performance efficiency. Other flame retardants include chlorine-, phosphorus-, nitrogen-containing and inorganic flame retardants. There are five major classes of BFRs: brominated bis-phenols, diphenyl ethers, cyclododecanes, phenols and phthalic acid derivatives.[101] Concerns have arisen regarding the use of BFRs as immunotoxic, cytotoxic, neurotoxic and carcinogenic affects have been reported for some compounds.[102] As a result, production of polybrominated biphenyls (PBBs) was banned (Figure 4.2).

Degradation of BFRs, at least to some extent, has been observed under both aerobic and anaerobic conditions. Recently, An *et al.*[103] reported debromination and mineralization of tetrabromobisphenol-A (TBBPA) by a novel *Ochrobactrum* sp. grown aerobically. In a later study, Zu *et al.*[104] isolated a *Bacillus* strain capable of debrominating and mineralizing 2,4,6-tribromophenol (TBP). Based on the identification of various downstream metabolites degradation routes were proposed, although the full pathways still need to be completed and the corresponding enzymes identified. Recently, Lee *et al.*[105] described a co-culture of *Dehalococcoides* and *Desulfuvibrio* grown on acetate as carbon source and H_2 as electron donor that could reductively debrominate various tetrabrominated diphenyl ether congeners completely. Growth of *Dehalococcoides* was coupled to debromination whereas that of *Desulfovibrio* was not. In another study a sequential approach using nZVI (nano Zero Valence Iron) to reductively debrominate polybrominated diphenyl ethers (PBDEs) followed by mineralization of the low-brominated diphenyl ethers (tri- and mono-BDEs) by *Sphingomonas* sp. PH-07 was proposed as a means with potential for treatment of PBDEs.[106] As the time of writing, no literature was found to report treatment of soil contaminated with BFRs or PBDEs at large scale.

4.2.2.2.11 Nitroaromatic Compounds. Nitroaromatic compounds consist of at least one nitro group ($-NO_2$) attached to an aromatic ring (Figure 4.2). They are of particular concern due to their acute toxicity, mutagenicity or carcinogenicity. Soil contamination with nitroaromatics is strongly associated with the manufacture, storage and handling of munitions (*e.g.* TNT, RDX and HMX) and chemical manufacturing plants. Atmospheric contamination is also important as a result of the nitration of hydrocarbons and PAHs from coal plants and combustion engines.[107] Microorganisms produce nitroaromatic compounds with a variety of functions including antibiotics, siderophores and signalling molecules.[108]

Degradation of nitroaromatic compounds under aerobic conditions yields substituted phenols, quinones or catechols, which are then metabolized to intermediates of the tricarboxylic acid cycle. The elimination of the nitro group generally occurs by: (a) reduction of the nitro group to amino group by a nitroreductase; or (b) oxidative removal of the nitro group by mono or diooxygenase.[109] Oxygenolytic metabolism for aromatic compounds is limited to mononitro- and dinitroaromatics, that is, for trinitrotoluene (TNT) reductive metabolism is required even in the presence of O_2.[110] Reduction of TNT to tinitroaminetoluene (TAT) under anaerobic conditions by *Chlostridium* and *Desulfovibrio* has been reported in various studies; a *Desulfovibrio* strain can use TNT as its sole source of N with accumulation of toluene in the culture medium.[111]

Chlorinated nitroaromatics (CNAs) form an important group of organic contaminants highly toxic to humans and animals. They find multiple industrial applications including the manufacturing of drugs, herbicides, pesticides, dyes, antioxidants and gasoline additives.[112] Chlorinated nitroaromatics are particularly recalcitrant to microbial degradation due to the presence of the chloro and nitro groups which act as electron-withdrawing groups (Figure 4.2). The relative position of the nitro and chloro groups is important to degradation as relevant biodegradative enzymes are quite specific. The position of the nitro group is more critical than that of the chloro group. Recalcitrance to degradation increases with increasing number of nitro and chloro groups. Microbial degradation of CNAs can be initiated by one of the following mechanisms: (a) reduction of the nitro group to amino *via* a reductase followed by ring cleavage as in *Comanomonas* strain CNB-1[113] and *Pseudomonas putida* ZWL73[114] (Figure 4.8(a)); (b) oxidative removal of the nitro group by mono or dioxygenase as in *P. putida* OCNB-1,[115] the chloro group can be removed before or after ring cleavage; and (c) oxidative removal of both the chloro and nitro groups prior to ring cleavage (Figure 4.8(b)).[115] Initial transformations lead to some central intermediate of aromatic metabolism such as catechols which undergo ring cleavage. The metabolic products formed can then enter the

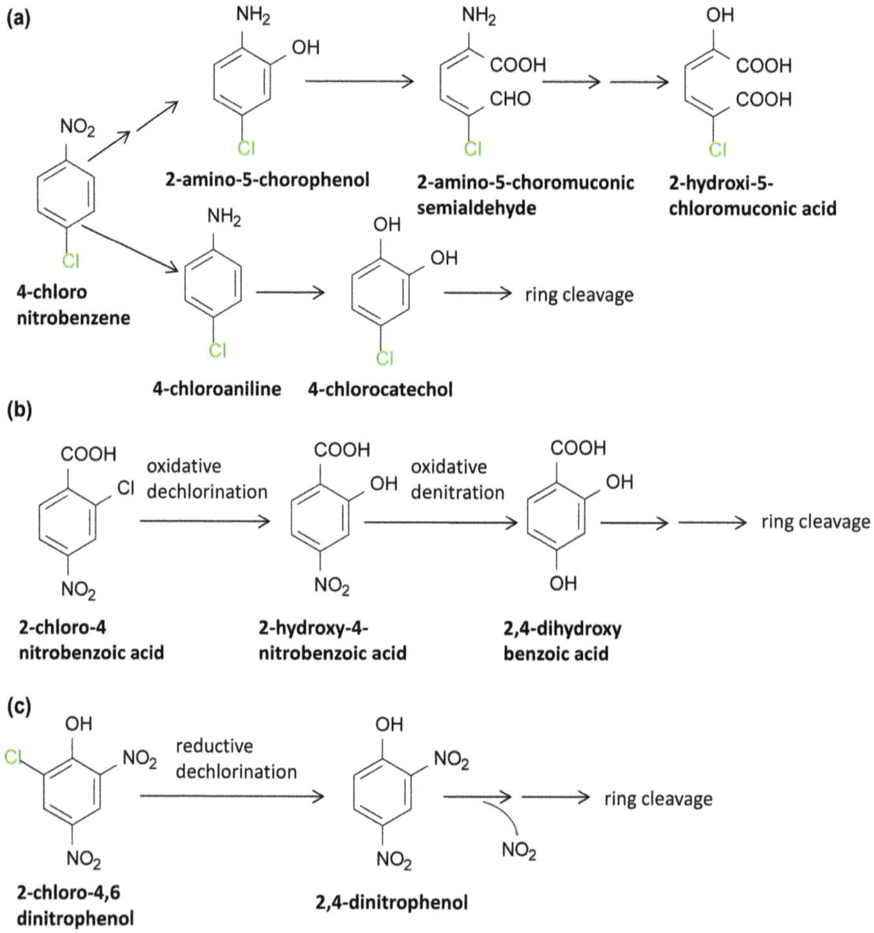

Figure 4.8   Degradation of chlorinated nitroaromatic compounds: (a) degradation of
4-chlorobenzene in *Comanomonas* sp. CNB-1 (upper pathway) and *Pseudomonas acidovorans* CA50 (lower pathway); (b) reductive dechlorination
of 2-chloro-4,6 dinitrophenol by *Rhodococcus erythropolis* HL 24-1; and
(c) oxidative dehalogenation by of 2-chloro-4-nitrobenzoic acid in *Acinetobacter* sp. RKJ12. Adapted from ref. 116.

tricarboxylic acid cycle. Dehalogenation can occur either before or after removal of the nitro group by: (a) hydrolysis (the O from the hydroxyl group
derives from water); (b) oxygenolysis (the O from the hydroxyl group comes
from O_2); and (c) reductive dechlorination (under anaerobic conditions)
(Figure 4.8(c)).[116] *Acinetobacter* sp. RKJ12 is capable of complete mineralization of CNAs such as 2-, 3- or 4-chloronitrobenzene as sole source of carbon, nitrogen and energy.[113] For more complex CNAs such as the commonly
used pesticide pentachloronitrobenzene, the complete degradation pathway is
still unknown, despite the fact that complete mineralization has been observed
in pure cultures of *Nocardioides* sp. Strain PD653.[117]

4.2.2.3   Biotransformation of Other Contaminants

4.2.2.3.1   Perchlorate. Perchlorate (ClO_4^-) is both a naturally occurring and man-made chemical that is used to produce rocket fuel, fireworks, flares and explosives (Figure 4.2). Perchlorate contamination of drinking water is of concern because it can disrupt the thyroid's ability to produce hormones needed for normal growth and development.[118] Due to its negative character, perchlorate is poorly adsorbed in soils and is highly soluble in water and organic solvents, moving into groundwater easily. It is also resistant to aerobic biodegradation. However, it can be completely reduced under anaerobic conditions to Cl^- and O_2.[119]

Most of the perchlorate-reducing bacteria found in the environment are members of the *Dechloromonas* and *Azospira* genus of the β subclass of the Proteobacteria, but isolates belonging to the α, γ and ε subclasses have been also identified.[120] Recently, two perchlorate-reducing bacteria, namely *Moorella perchloratireducens*[121] and *Sporomousa* sp.[122] belonging to the Gram-positive Firmicutes have been identified. It should be noted that not all chlorate-degrading bacteria can degrade perchlorate, although the converse is true. Perchlorate acts as electron acceptor, while a variety of gases (*e.g.* H_2) and carbon sources can act as electron donors (*e.g.* sugars, alcohols, volatile organic acids and organic wastes). Various approaches have been used to treat both soil and groundwater contaminated with perchlorate such as *in situ* bioremediation of groundwater, soil composting, and phytoremediation of surface water using wetland plants.[123]

4.2.2.3.2   Ether Oxygenates. Methyl *tert*-butyl ether (MBTE) and other ether oxygenates such as ethyl tert-butyl *ether* (EBTE), *tert*-butyl alcohol (TBA) and *tert*-amyl methyl ether (TAME) are added to gasoline to enhance octane and improve combustion efficiency (Figure 4.2). MBTE is highly soluble in water (48 g L^{-1}) and has lower log K_{oc} (1.0–1.1 *vs.* >1.5 for all BTEX) and Henry's constant (0.023–0.12 *vs.* >0.22 for all BTEX) compared with the BTEX group.[124] Thus, when MBTE is spilled it can rapidly spread into the saturated zone leading to wide plumes. Although health effects of MBTE at low-level contamination are still under debate, it has a potent impact on water taste at 10–30 parts per billion (ppb).[125]

Various bacteria capable of growing on MBTE as the sole carbon and energy source have been isolated including *Rubrivivax gelatinosus* PM1,[126] *Hydrogenophaga* gelatinosus[127] and *Mycobacterium austroafricanum* IFP2012.[128] Aerobic degradation starts with a monooxygenase attack and the subsequent release of a C_1-unit as formaldehyde or formate yielding TBA; the latter can be converted to 2-hydroxyisobutiric acid which, in turn, can be further metabolized. Complete degradation of MBTE under anaerobic conditions in the presence of different terminal electron acceptors has been demonstrated with TBA frequently found as an intermediate metabolite. Nonetheless, the mechanism is still unknown and the organisms involved from degrading-consortia have yet to be identified.[129] For treatment of MBTE impacted sites, it is common that biological *in situ* and *ex situ* (bioreactors) are combined, as well as with other non-biological methods such as *in situ* thermal desorption, soil vapour extraction, thermal oxidation and removal of hot spots of contaminated soil.[130]

4.2.2.3.3 Emerging Contaminants. Emerging contaminants are natural/synthetic chemicals or microorganisms that are not commonly monitored in the environment but cause known or suspected adverse ecological and/or human health effects.[131] They might originate from: (a) the synthesis of new chemicals; (b) changes in use or disposal of existing chemicals; and (c) chemicals/ microbial contaminants which have occurred for a long time, but may have not been recognized until new detection methods were developed.

Some major groups of emerging contaminants include:[132] (a) pharmaceuticals such as hormones (*e.g.* natural and synthetic estrogens) and antibiotics (*e.g.* penicillins, sulfonamides, *etc.*); (b) organophosphorus and brominated flame retardants (see above); (c) drinking/recreational water disinfection by-products which formed following reaction of chlorine, chloroamines, chlorine dioxide and ozone (disinfectants) with organic compounds; (d) nitrosamines, which are carcinogenic chemicals employed in the manufacture of cosmetics, pesticides and rubber products; (e) surfactants and their degradation products; (f) per- and polyfluorinated compounds, which are extremely difficult to biodegrade due to the strong C–F bond; (g) the corrosion inhibitors benzotriazoles and benzothiazoles; and (h) pathogens.[133]

Bacterial degradation for some of these contaminants is documented. For instance, various bacteria present in wastewater have been related to the estrogen-degrading capacity in municipal sludge including strains of *Rhodococcus*,[134] *E. coli*, *P. fluorescens* and *Bacillus thuriginensis*.[135] Organisms capable of utilizing fluorinated biphenyls as sole source of carbon and energy, namely *Pseudomonas pseudoalcaligenes* KF707 and *Burkholderia xenovorans* LB400 under aerobic conditions[136] or dehalogenating chlorofluorohydrocarbons under anaerobic conditions[137] such as *Desulfuromonas michiganensis* strain BB1, *Sulfurospirillum multivorans* and *Geobacter* sp. strain SZ have been reported. However, the toxicity of the fluorometabolites resulting from such activity can be of greater concern than the parent compound (*e.g.* more mobile in the environment). Thus, biological treatment of soils contaminated with emerging contaminants is not established yet as a mature full-scale technique for any of these groups.

4.2.2.4   Microbial Bioremediation of Organic Compounds in Soil

The clean-up of contaminated soil involves a variety of techniques ranging from biological processes to simple or complex engineering technologies. The initial steps consist of a thorough characterization of the site and an evaluation of the extent of contamination. If remediation is deemed necessary, choosing the adequate technology for a particular site requires careful consideration of various factors which can be grouped as follows: (a) technical (*e.g.* type and extent of contamination, soil characteristics, monitoring); (b) health- and environment-related (*e.g.* regulatory needs, hazardous by-products, odours, worker health and safety); and (c) economic (labour, monitoring, fuel, equipment, operating and maintenance).[138]

Depending on where the contaminated soil is treated and whether it needs to be excavated or not, technologies can be applied in three different ways:[139]

- *in situ:* contaminants are treated at the original location without excavating the soil
- on-site: contaminated soil is excavated, treated on site and returned to the original location
- *ex situ:* contaminated soil is excavated and transported for treatment/processing elsewhere

The type, extent and depth of the contamination are thus very important: not all technologies work for all contaminants, nor are they suitable for both high and residual concentrations, or can be used both in the vadose and the saturated zone.

The processes that act on the contaminants might be physical, chemical or biological, or a combination of various processes. They are employed to: (a) degrade or alter the contaminants to less toxic products; (b) stabilize the contaminants (precipitation, complexation, adsorption, absorption); (c) extract/separate the contaminant from environmental media by physical (*e.g.* gravity, sieving, thermal evaporation, steam stripping) or chemical means (*e.g.* solvent extraction); and (d) contain the contaminants by physical barriers (*e.g.* secured setting, slurry walls) or physical processes (*e.g.* solidification, vitrification).[140]

Biological processes offer various advantages over physical and chemical processes. Biological processes are: (a) environmentally sound; (b) minimize soil disruption; (c) are less expensive; and (d) require less energy. One of the main disadvantages is that they may require longer timeframes to meet the remediation goal and that some unwanted by-products may form. Different processes or technologies can be combined to meet remediation goals in what is called a treatment train. In this sense, bioremediation is often employed as a polishing technique to treat residual contamination at sites where another technique has been previously used (*e.g.* thermal treatment of a NAPL source and biological treatment of residual plume).

Before an *in situ* project is attempted in the field, treatability studies may need to be conducted to ensure success. Evaluating the success of *in situ* bioremediation is quite challenging because: (a) there is no one single measure or parameter that can be universally applied to determine the success of bioremediation; (b) residual contamination can be extremely difficult to find; and (c) the definition of success may vary across the various parties involved (regulators, contractors, stakeholders, public, *etc.*). In general, techniques used to evaluate bioremediation are aimed at linking microbial activity to the observed loss of contaminant (Table 4.1).

Table 4.1 Techniques for evaluating biodegradation.

Technique	Notes
Evaluation of microbial activity	
Decrease in the concentration of target compound over time	Ongoing biodegradation
Daughter products or intermediary metabolites	Evidence for biodegradation of parent compound
Ratio of non-degradable to degradable components	Consumption of biodegradable components
Electron acceptor concentration	Decrease in the electron acceptor used during contaminant oxidation
By-products of anaerobic metabolism	e.g. concentrations of sulfide resulting from sulfate reduction or Fe(II) from Fe(III) reduction
Compound Specific Isotope Analysis (CSIA)	Distinguish between contaminant transformation processes (e.g. abiotic vs biotic); estimation of *in situ* activity
^{14}C labelling of target compounds and $^{14}CO_2$ production	Rate of degradation of a specific compound
Cell counting techniques in microcosms/enrichment cultures	
Microscopy counts	Detection of microorganisms growing on substrate
Most Probable Number – Plate/Vial counts	Estimation of abundance of specific viable degraders
Turbidometry	Density of biomass in microbial enrichments via spectrometry
Electrical resistance	Cell density in an electrolyte using a Coulter counter
Flow cytometry	Counting of cells, evaluation of their morphology biochemical composition
Molecular biology techniques	
Real-time PCR (structural e.g. 16S rRNA gene and functional genes)	Presence and abundance of organism of interest/key catabolic genes
Real-time PCR (mRNA)	Information on ctive microorganisms/gene expression (functional genes)
PCR-DGGE/PCR-tRFLP	Microbial community fingerprinting
FISH (Fluroscence *in situ* hybridization)	Visual information on spatial distribution of organism of interest
Microarrays	Presence/absence of functional/structural genes
Pyrotaqsequencing (16S rRNA)	Characterization of microbial communities to the genus level
Metagenomic libraries	Reconstruction of whole genomes and metabolic pathways
Stable isotope probing SIP-DNA, RNA, protein	Following the fate of ^{13}C from labelled substrate into biomolecules

CSIA = Compound Specific Isotope Analysis; DGGE = denaturing gradient gel electrophoresis; FISH = fluorescence *in situ* hybridization; PCR = polymerase chain reaction; SIP = stabe isotope probing; tRFLP = terminal restriction fragment length polymorphism.

4.2.2.4.1 *In Situ* Bioremediation. *In situ* treatment allows soil to be treated without being excavated and transported, resulting in potentially significant cost savings. On the contrary, remediation times are generally longer compared with on-site or *ex situ* approaches. This is partly due to the spatial heterogeneity encountered at field sites and the challenges in achieving sufficient contact between contaminants, microorganisms, nutrients and electron acceptors/donors.

Monitored natural attenuation (MNA) consists of a variety of physical, chemical or biological processes that, under favourable conditions, act without human intervention to reduce the mass, toxicity, mobility, volume or concentration of contaminants in soil or groundwater to acceptable levels. However, it requires modelling to predict degradation at non-sampled locations and long-term monitoring to ensure that degradation or transformation rates are consistent with the clean-up goals.[142] MNA is considered to be well established as a remediation approach for only a few contaminants, primarily BTEX. There are, however, examples of successful stories for other contaminants. MNA was successfully employed as a polishing technique for treatment of PAH contamination following source removal at South Glenn Falls (NY, USA). Both naphthalene (average concentration 52 000 mg kg^{-1} of source) and phenanthrene (average concentration 17 000 mg kg^{-1} of source) were significantly reduced to levels below their drinking water standards in about three years due to biodegradation.[143] At the Dover Air Force Base (DL, USA) natural sequential anaerobic-aerobic biotransformation of the chlorinated solvents PCE and TCE was chosen as the remediation. In the anaerobic portion of the aquifer, concentrations of PCE and TCE were found to decrease in favour of the daughter products cDCE, VC and ethene, whereas near the anaerobic/aerobic interface, concentrations of cDCE and VC decreased to below detection limits likely as a result of aerobic biotransformation processes.[144]

Bioventing involves introducing gas (air, oxygen or methane) into soil above the water table to enhance the aerobic degradation of organic compounds. Bioventing is most useful for degradation of adsorbed fuel residuals, nonvolatile solvents and some pesticides. It works best in soils with relatively good permeability as air or gas needs to be sufficiently distributed to sustain microbial activity. In a recent study, a microbioventing system comprising many small rod air-injection ports was installed to support aerobic biodegradation of diesel fuel spilled at different sites on the sub-Antarctic Macquarie Island.[145]

Biosparging is based on the same principle but the contaminated soil is below the water table, that is, the saturated zone (Figure 4.9(a)). As with bioventing it is important that air flow rates are low to prevent volatile organic compounds (VOCs) passing through the vadose zone without treatment. Biosparging is commonly used to treat groundwater and soil contaminated with jet fuel and BTEX compounds.[146] Biosparging was used to enhance degradation of kerosene and BTEX at a former air force base in the Czech Republic. Degradation of contaminants was attributed to Pseudomonads as they prevailed under high aromatic contamination, whereas greater diversity of Proteobacteria and Gram-positive taxa were observed in remediated soil.[147]

Enhanced *in situ* bioremediation is a process in which the degradation/transformation of organic contaminants is enhanced under aerobic/anaerobic conditions *via* stimulating the activity of indigenous microorganisms (biostimulation), for example, by addition of nutrients, appropriate electron donors/acceptors or other amendments (*e.g.* surfactants for contaminant desorption); or *via* inoculating microorganisms capable of carrying out the reaction of interest

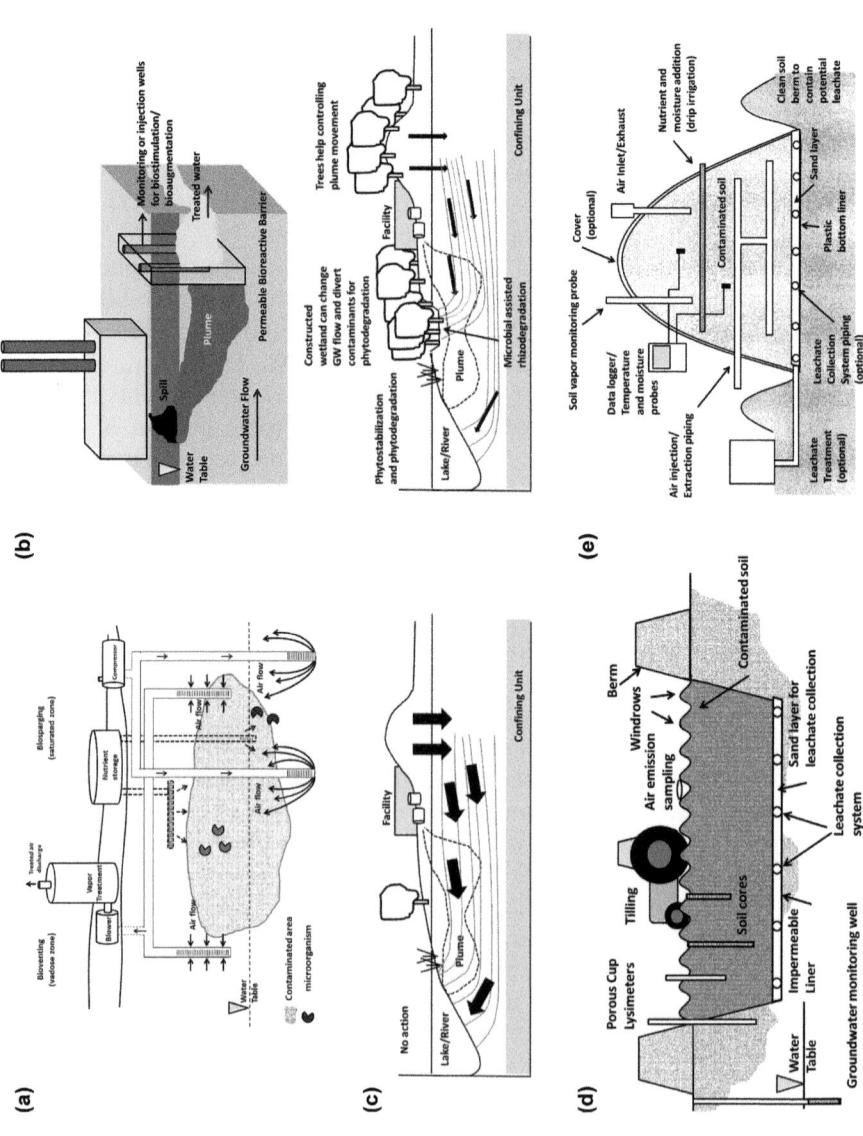

Figure 4.9 (a) biosparging;[169,170] (b) permeable reactive barrier; (c) phytoremediation and microbial assisted rhizoremediation;[168] (d) landfarming;[153] and (e) biopiles.[171]

(bioaugmentation). Enhanced bioremediation of soil typically involves the percolation or injection of uncontaminated water mixed with nutrients and saturated with dissolved oxygen through a gallery or spray irrigation system in the case of shallow contaminated soils. For deeper contaminated soils (saturated zone) injection wells are used. Enhanced *in situ* bioremediation was successfully employed to treat chlorinated hydrocarbons such as PCE, TCE and cDCE in the subsurface at Kelly Air Force Base (TX, USA) using methanol and acetate as electron donors and a microbial inoculum capable of transforming the contaminants all the way to ethene.[62] It should be noted that both biostimulation and bioaugmentation approaches can be used in combination with other remediation technologies such as landfarming,[148] bioreactive barriers and bioreactors.

Bioreactive barriers follow the same principle as permeable reactive barriers (PRBs) but involve the presence of microorganisms in the PBR material which actively transform the target compounds. PBRs are walls placed belowground which allow groundwater to pass through so that contaminants can be trapped and treated (Figure 4.9(b)). In this sense it is not a technology to directly clean up contaminated soil but to stop further spreading of the contamination. PBRs consist of a narrow trench dig in the soil in the path of the groundwater flow which is filled with reactive material (iron, lime, active carbon).[149] The reactive material might be combined with sand to ease groundwater circulation. Microbes can colonize the solid surfaces of the PBR and contribute to the degradation/transformation of the target compounds. To enhance microbial activity and colonization solid, liquid or gaseous amendments such as wood chips, compost, lactate, molasses (C Sources), oxygen, air and H-releasing compounds may be employed. In this case, the system becomes less passive and greater operation and maintenance costs may be involved.[150] Biobarriers have been mostly used for the treatment of chlorinated solvents, BTEX and MTBE. A bioactive PBR with oxygen supply was successfully used in reducing initial concentrations of MBTE and TBA of 10 000 $\mu g\ L^{-1}$ in the centre of the plume to less than 5 $\mu g\ L^{-1}$ down gradient of the PBR after 18 months.[151] In a different scenario, anaerobic conditions were created by supplying acetate and incorporating wood chips soaked in soybean oil/mushroom compost mixture into the PBR. This enabled indigenous bacteria to use perchlorate (ion) as respiratory terminal electron acceptor until nontoxic chloride remained.[152]

4.2.2.4.2 *Ex situ* Bioremediation. On-site or *ex situ* treatments require excavation of the soil and transportation to the treatment facility (*ex situ*). This generally leads to increased costs in terms of energy, equipment, transportation and safety of working personal. The volume of soil that can be processed is also limited. However, better treatment uniformity and better control of conditions can be achieved (moisture, pH, temperature, *etc.*), resulting in shorter remediation times (see examples in Table 4.2.).

Landfarming is a full-scale technology in which contaminated soil is applied into lined beds to control leaching and periodically turned over for aeration (Figure 4.9(d)). Optimal conditions for biodegradation can be achieved by irrigation, frequent aeration, pH adjustment and nutrient addition. Soil is generally disposed into lifts around 40–50 cm wide. Once the desired level of

Table 4.2 PAH removal efficiency in soil contaminated inoculated and non-inoculated with *A. discolor*, in absence and presence of soil indigenous microorganisms, after 60 days. The values are means ± confidence interval (CI).[206]

| Compound | Removal efficiency (%) | | | |
| | Autoclaved soil | | Non-autoclaved soil | |
	Non-inoculated	Inoculated	Non-inoculated	Inoculated
Phenanthrene	39.5 ± 10	61.9 ± 2	98.3 ± 1	95.4 ± 2
Anthracene	27.3 ± 5	72.9 ± 2	83.1 ± 6	61.5 ± 11
Fluoranthene	27.7 ± 8	54.3 ± 3	82.5 ± 12	43.1 ± 6
Pyrene	24.9 ± 10	59.7 ± 2	82.5 ± 12	43.1 ± 6
Benzo[a]pyrene	28.1 ± 7	75.4 ± 1	14.2 ± 6	15.8 ± 6

treatment has been achieved, lifts can be removed. Part of the remediated soil may stay in place to enhance remediation of newly applied non-treated soil. Volatiles, semi-volatiles and dust dispersion might need to be controlled. Landfarming has been mostly employed to treat soils contaminated with petroleum hydrocarbons, particularly in warm regions,[153] for instance, total petroleum hydrocarbons were reduced from 4644 mg kg^{-1} to less than 100 mg kg^{-1} in a one-year trial in Australia.[154] The potential of landfarming has been also shown in cold regions such as Alaska for diesel-range organics, trimethylbenzenes and BTEX compounds,[155] and the Canadian Artic for petroleum hydrocarbons.[156] In both cases the success was attributed to N and K fertilization and intensive tilling.

Biopiles are similar to landfarming and consist of aerated static heaps (2–3 m high) placed in aboveground enclosures in which excavated soil is irrigated, mixed with nutrients and aerated with blowers or vacuum pumps (Figure 4.9(e)).[157] This is the main difference with landfarming, where aeration is provided through tillage. An impermeable liner prevents leaching of contaminants; leachates can be collected and processed in a bioreactor. Biopiles may be covered with plastic to control evaporation and prevent run-off and volatilization of contaminants; coverage also promotes heating. If VOCs are present, gas emissions should be collected and treated before release to the atmosphere. Biopiles are most commonly employed to treat soil contaminated with fuel or petroleum hydrocarbons.[158,159] Aeration, moisture, nutrients, pH and temperature can be adjusted to optimize biodegradation and minimize volatilization. In addition, biopiles can be bioaugmented with suitable microorganisms as shown in a biopile for the treatment of soil contaminated with BTEX, pristane and JP-5 fuel.[160] The biopile was bioaugmented with HC-degrading strains namely *Acinetobacter* sp., *Pseudomonas* sp. and *Rhodococcus* sp., achieving a removal of 90% of total HC in 15 days.

Composting is a process in which organic material (*e.g.* hay, manure, vegetative wastes) is decomposed and recycled in the presence of bulking agents (*e.g.* sawdust) and soil (contains beneficial microorganisms and worms which help aerating) to produce a humus-like end product.[1] Optimal composting requires moisture, temperature and aeration control. During composting organic contaminants are biodegraded or converted into stable innocuous by-products under aerobic or anaerobic conditions (anaerobic microniches also exist). Typically, thermophilic conditions (54–65 °C) must be maintained to properly compost soil contaminated with hazardous organic contaminants. Composting can be applied using different designs: (a) aerated static piles (aerated with blowers or vacuum); (b) mechanically agitated vessel; and (c) windrows (long piles/narrow rows which are periodically mixed with mobile equipment). Windrow composting has been particularly effective in the treatment of soil contaminated with explosives (*e.g.* TNT, HDX, RDX) with removal efficiency close to 100% after 20–40 days at a Superfund site at an army Depot (OR, USA), where potato waste and manure were combined with the contaminated soil.[161]

Bioreactors are engineered devices that support a biologically active environment for a specific process. In bioreactors, contaminated soil is mixed with water to create a slurry (*e.g.* soil : water 1 : 1) improving contact of solids, contaminants and microorganisms. Upon completion of the process, the slurry is dewatered and the treated soil is disposed of. To enhance bioremediation, nutrients can be added (biostimulation) or microorganisms might be inoculated (bioaugmentation). Depending on the type of contaminants bioreactors might be run under aerobic (*e.g.* petroleum hydrocarbons) or anaerobic (*e.g.* TNT) conditions.[125,126,162,163] For treatment of certain contaminants (*e.g.* polychlorinated benzoates) anaerobic–aerobic reactors can be coupled to partially dechlorinate the original compounds which are then amenable to aerobic mineralization.[164] It is also possible to wash the soil to remove the contaminant of interest and to treat the washing solution using slurry-phase bioreactors or other types of bioreactors used for wastewater treatment where biofilms grow on some kind of solid support (*e.g.* trickling filter, fixed-film bioreactor, rotating disk bioreactor).

4.3 Fungal Bioremediation

4.3.1 Introduction

Although chemical and physical treatments are more rapid than biological treatments, they are generally more destructive and intrusive to affected soils, more energy intensive, and often more expensive than bioremediation. Considering the bioremediation options, several bacteria and white-rot fungi (WRF) have been shown to enhance degradation processes in soil, using both pure and mixed cultures.[172–179]

Some fungi have remarkable degradative properties. Lignin-degrading white-rot fungi, like *Phanerochaete chrysosporium*, can degrade several xenobiotics, including PAHs, PCP, PCBs, chlorinated organics, nitrogen-containing aromatics, dyes, heavy metal, xenobiotics and many other pesticides.[180] Some ligninolytic fungi have been used to treat soil contaminated with substances like PCP, PCB, PAHs, DTT or heavy metal for potential bioremediation activities.[183] Treatment generally involves inoculation of the contaminated soil, followed by nutrient addition, irrigation and aeration, and maintenance by general landfarming procedures.[182] In many cases, xenobiotic-transforming fungi need additional utilizable carbon sources because, although capable of degradation, they cannot utilize these substrates as an energy source for growth. Therefore, inexpensive utilizable lignocellulosic wastes such as corncobs, straw and sawdust are used as nutrients to obtain enhanced pollutant degradation.

Wood-rotting and other fungi are also receiving attention for the bleaching of dyes and industrial effluents, and the biotreatment of agricultural wastes such as forestry, pulp and paper by-products, sugar cane bagasse, coffee pulp, sugar beet pulp, apple and tomato pulp, and cyanide.[181,183–185] The purpose of this section is to provide a general description of the different kind of pollutants or contaminants degraded by fungi and an overview of the different remediation process using fungi.

4.3.2   Pollutant Degradation by Fungi and Application for Bioremediation

4.3.2.1   Polycyclic Aromatic Hydrocarbons

4.3.2.1.1 Degradation of PAHs by Fungi. Polycyclic aromatic hydrocarbons (PAHs) are an all-pervading group of hydrophobic organic compounds consisting of two or more combined benzene rings in linear, angular or cluster arrangements. PAHs are the most important components of crude oil, creosote, asphalt and coal tar. They contaminate the soil *via* many routes, including the burning of fossil fuels, the manufacture of gas and coal tar, wood processing, escaped automobile gasoline and the incineration of waste.[186] There are more than 100 diverse PAH compounds and most of them persist in the ecosystem for many years owing to their low aqueous solubility and their absorption to solid particles.[187] The presence of PAHs in contaminated soil and sediment poses a significant risk to the soil, because many PAH compounds are known or suspected to be toxic, mutagenic and, in some cases, carcinogenic.[188]

Diverse fungi capable of utilizing PAHs have been investigated. Some filamentous fungi, basidiomycetes, white-rot fungi and deuteromycetes have been shown to remove PAHs more competently than bacteria. PAHs susceptible to fungal biodegradation include naphthalene, phenanthrene, anthracene, pyrene, benzo[a]pyrene, fluorene, dibenzothiophene, catechol, benzo[a]anthracene, benzo[ghi]-perylene, chrysene, benzo[b]fluoranthene and benzo[k]-fluoranthene.[189,190] As fungi cannot generally use PAHs as their sole carbon and energy source,[191] they must be supplied with nutrients to allow cometabolism. The ligninolytic group is mostly composed of white-rot fungi, *e.g. Pleurotus ostreatus, Phanerochaete chrysosporium* and *Irpex lacteus*. Ligninolytic fungi are capable of cleaving the aromatic rings of PAHs by using extracellular ligninolytic enzymes and this can lead to mineralization.

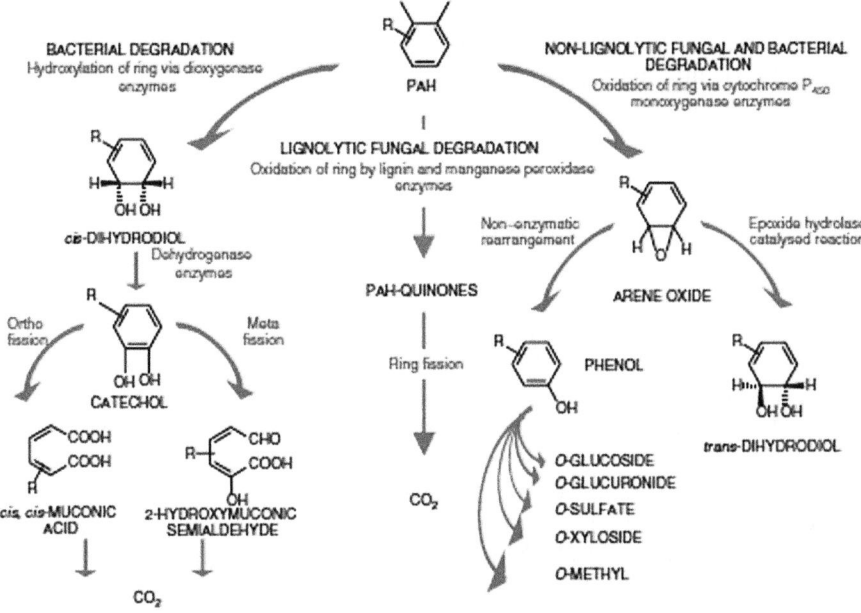

Figure 4.10 The three main pathways for polycyclic aromatic hydrocarbon degradation by fungi and bacteria.[192]

The transformation of PAHs by ligninolytic wood-decaying fungi involves different enzymes including lignin peroxidase, manganese peroxidase, laccase, cytochrome P450 and epoxide hydrolase (Figures 4.10 and 4.11).[192,193]

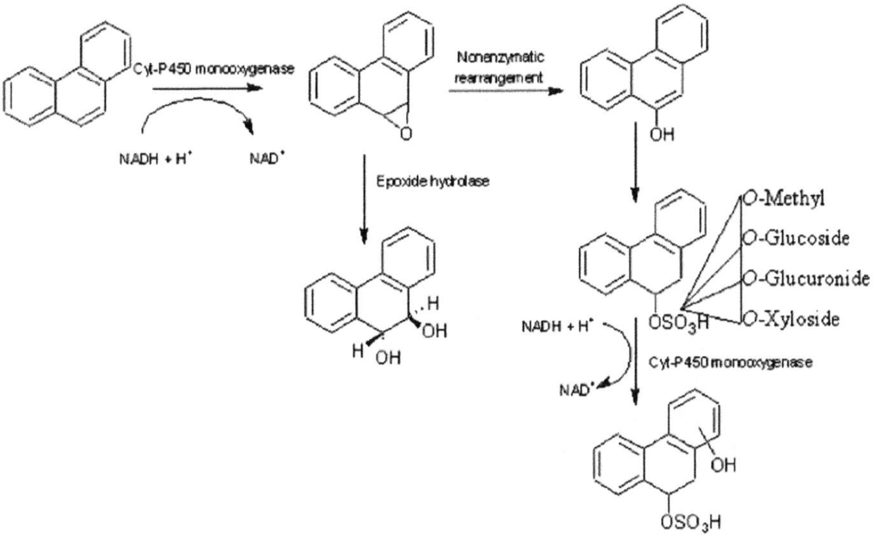

Figure 4.11 Proposed pathway for phenanthrene degradation by fungi (adapted from ref. 193).

Non-ligninolytic fungi such as *Cunninghamella elegans*, *Penicillium janthinellum*, Mucor, Fusarium, *Syncephalastrum* sp. and Penicillium spp. can only oxidize PAHs into hydrophilic or detoxified products. Hence, mineralization is not detected;[194] these non ligninolytic fungi have also been considered for PAHs bioremediation.[195–198]

4.3.2.1.2   Bioremediation of PAHs in Soil by Fungi.   Two white-rot fungi, *Irpex lacteus* and *Pleurotus ostreatus*, were used as inoculum for bioremediation of petroleum hydrocarbon contaminated soil from the vicinity of a manufactured gas plant. After a 10-week *in situ* treatment (using 250 mL Erlenmeyer flaks filled with 10 g of soil), out of 12 different PAHs, the concentration of phenanthrene, anthracene, fluoranthene and pyrene decreased up to 66%. In addition, the ecotoxicity of the soil after bioremediation did not reveal any effect on the survival of *Daphnia magna*, a crustacian.[199]

Bioremediation in soil by fungi composting, a process typically used to degrade solid waste materials, has also been studied as a remediation technology for PAH-contaminated soils. Mushroom compost consisting of wheat straw, chicken manure and gypsum was added to manufactured gas plant (MGP) soil in a thermally insulated composting chamber. Degradation efficiencies of individual PAHs after 54 days of composting in the chamber fell by 20–60%, with further removal of PAHs (37–80%) after an additional 100 days of compost maturation in an open air field.[200] Composting was also found to degrade three- and four-ring PAHs more effectively than five- and six-ring PAHs.[201]

Some authors have shown that the biodegradation percentage of individual PAH in soil infused with the selected PAH compounds and basidiomycetes fungi ranged between 0.88% fluorene by the *Polyporus sulphureus* and 99.52% anthracene by *Phellinus* sp. after 10 days of incubation. *Phellinus* sp. preferentially degraded anthracene (99.52%) and its degradation efficiency declined in the order of anthracene > naphthalene (83.6%) > pyrene (66.2%) > acenaphthene (28.02%) in PAH infused soil after 10 days of incubation. A co-culture with a PAH-degrading bacteria isolated from soil affected by an oil spill, *Bacillus pumilus*, and the selected fungi were also effective in degrading acenaphthene and fluorene in the soil. The co-cultures of fungi and bacteria (*Phellinus* sp. and *Bacillus pumilus,* and *P. sulphureus* and *Bacillus pumilus*) were effective in degrading naphthalene (99.95%) in the soil by the 20th day. In all the treatments, naphthalene (containing two aromatic rings) was efficiently degraded while pyrene (containing four aromatic rings) was poorly degraded. The co-culture technique could not significantly improve PAH biodegradation over the monoculture treatment in liquid state fermentation (LSF) and solid state fermentation (SSF). But in soil experiments, the biodegradation level of naphthalene (9.12%), anthracene (3%) and acenaphthene (5%) increased in co-culture treatment over their corresponding monoculture treatment.[202]

Another study investigated the ability to remove PAHs by fungi and bacteria isolated from petroleum hydrocarbon contaminated soil. The co-culture of *Penicillium* sp. and *Rhodococcus* sp. IC10 was found to be the optimal combination for the removal of different PAHs from contaminated soil and for the reduction of soil toxicity, as evaluated by mustard seed germination and the oxidoreductase activity of *Bacillus cereus*.[203]

The removal efficiency of three-, four- and five-ring PAHs in contaminated soil bioaugmented with *Anthracophyllum discolor*, a white-rot fungus isolated from a forest in southern Chile, was investigated in the absence and presence of indigenous soil microorganisms. *A. discolor* was able to degrade PAHs in Kirk medium with the highest removal occurring in a PAH mixture, suggesting synergistic effects between PAHs or possible cometabolism. A high removal capability for phenanthrene (62%), anthracene (73%), fluoranthene (54%), pyrene (60%) and benzo[a]pyrene (75%) was observed in autoclaved soil inoculated with *A. discolor* in the absence of indigenous microorganisms (Table 4.2).[204] Many PAH remediation assays using fungi have mainly used experimental conditions (microcosms, bioreactor) rather than field experiments (Table 4.3).[201,202,205–212]

4.3.2.2 Chlorinated Aromatic Hydrocarbons

4.3.2.2.1 Pentachlorophenol

4.3.2.2.1.1 PCP Degradation by Fungi. Pentachlorophenol (PCP) is a xenobiotic causing great environmental concern. It has been commonly applied for many years as a bactericide, fungicide, defoliant, herbicide, wood preservative and detergent supplement in soaps.[213,214] PCP can also be formed as a by-product in water chlorination processes and as a metabolite from the degradation of some pesticides.[215,216]

PCP is recognized as highly toxic. It uncouples oxidative phosphorylation and has been described as an endocrine disruptor and a probable carcinogen.[217–220] The white-rot fungi, *P. chrysosporium* and *T. versicolor*, have been the most widely used fungi for PCP degradation because of their ability to secrete enzymes involved in the degradation of lignin, including lignin peroxidase (LiP), manganese peroxidase (MnP) and laccase. Nevertheless, diverse studies have been performed to evaluate new fungal strains with a high ability for the degradation of recalcitrant organic compounds: *Anthracophyllum discolor*, a native Chilean fungus isolated from the temperate forests of southern Chile, with a high ligninolytic activity [mainly manganese peroxidase (MnP], has shown a high capability to degrade PCP in both liquid medium and soil slurry.[221,222] Intermediate metabolites—pentachloroanisole (PCA), tetrachloro-1,4-dimetoxybenzene, 2,5-dichloro-1,4-dimetoxibenzene, 2-chloro-1,4-dimetoxibenzene, 3,4-dimethoxybenzaldehyde (veratraldehyde)—were identified from PCP degradation in soil slurry culture by *A. discolor*.[221] A similar pathway was proposed by Reddy and Gold[223] during the degradation of PCP in liquid media by *P. chrysosporium* (Figure 4.12). Other PCP biodegradation mechanisms used by Mucor species (Figure 4.13) have also been proposed.[214]

4.3.2.2.1.2 Bioremediation of PCP in Soil by Fungi. The most commonly used methods for bioremediation of PCP-polluted soils are incineration, disposal in landfill, soil washing and chemical extraction. The development of

Table 4.3 Bibliographic compilation of PAH bioremediation studies using fungi, composting and microbial consortia (fungi-bacteria) (adapted from ref. 208).

Inoculated microorganism	Soil	Process details	PAHs studied	Reference
White button mushroom (*Agaricus bisporus*) compost consisting wheat straw, chicken manure, gypsum	Manufacture Gas plant soil	170 kg dry soil mixed with 800 kg compost and water to give a soil : compost ratio of 1 : 4 on a wet basis. Composting in a thermally insulated composting chamber with aeration for 54 days. Additional open-air field composting for 100 days	Phenanthrene, anthracene, fluoranthene, pyrene, benz[a]anthracene, chrysene, benzo[b]fluoranthene, benzo[k]fluoranthene, benzo[a]pyrene, dibenz[a,h]anthracene, benzo[ghi]perylene, indeno[1,2,3-cd]pyrene	200
White button mushroom (*Agaricus bisporus*) compost consisting wheat straw, chicken manure, gypsum	Manufacture Gas plant soil	88 kg dry soil mixed with 400 kg compost and water to give a soil : compost ratio of 1 : 4 on a wet basis. Composting in a thermally insulated composting chamber with aeration for 42 days. Additional open-air field composting for 100 days	Phenanthrene, anthracene, fluoranthene, pyrene, benz[a]anthracene, chrysene, benzo[b]fluoranthene, benzo[k]fluoranthene, benzo[a]pyrene, benzo[a]perylene, indeno[1,2,3-c,d]pyrene	201
White rot fungi (*Irpex lacteus, Pleurotus ostreatus*)	Former gasholder soil, sandy clay loam from wood treatment plant	5 g soil treated with 5 g straw substrate in incubation flasks at 24 °C in the dark for 6 weeks. Soil moisture kept constant at 15%	Phenanthrene, anthracene, fluoranthene, pyrene, chrysene, benzo[b]fluoranthene, benzo[k]fluoranthene, benzo[a]pyrene, dibenzo[a]anthracene, benzo[ghi]perylene, indeno[1,2,3-c,d]pyrene	206
Fungus (*Cladosporium sphaerospermum*)	Manufacture Gas plant soil	7 g soil incubated in culture tubes at room temperature in the dark for 30 days. 5 g treated in microcosm at room temperature for 60 days. Soil moisture kept at 70% of water holding capacity	Naphthalene, acenaphtylene, acenaphtene, fluorene, phenanthrene, anthracene, pyrene, benz[a]anthracene, chrysene, benzo[b]fluoranthene, benzo[k]fluoranthene, benzo[a]pyrene, dibenz[a,h]anthracene, benzo[1,2,3-c,d]pyrene, benzo[ghi]perylene	207

Microbial consortia (bacteria, fungi, bacteria–fungi mixture)	Oil-contaminated soil	5 g soil inoculated with 40% abiosalt medium and 2% of the microbial consortia shaken in culture tubes at room temperature in the dark for 30 days	Acenaphtene, acenaphtylene, napthalene, anthracene, fluorene, phenanthrene, chrysene, fluoranthene, pyrene, benz[a]anthracene, benzo[a]pyrene, benzo[b]fluoranthene, benzo[k]fluoranthene, dibenz[a,h]anthracene, benzo[ghi]perylene, indeno[1,2,3-c,d]pyrene	208
Spent mushroom compost (*Pleurotus pulmonarius*)	Spiked garden soil	1 g soil treated with straw spent mushroom compost at 4–80 °C	Naphthalene, phenanthrene, benzo[a]pyrene, benzo[ghi]perylene	209
White rot fungus (*Phanerochaete chrysosporium*)	Manufacture Gas plant soil	Bioreactor treatment at 39 °C for 36 days	Acenaphthene, acenaphthylene, anthracene, benzo[a]anthracene, benzo[a]pyrene, benzo[b]fluoranthene, benzo[ghi]perylene, benzo[k]fluoranthene, chrysene, fluorene, indeno[1,2,3-c,d]pyrene, naphthalene, phenanthrene, pyrene	210
Shredded wheat straw cuttings and white rot fungus (*Pleurotis ostreatus*)	Spiked soil	Bioreactor treatment at 25 °C for 249 days	Fluorene, phenanthrene, anthracene, fluoranthene, pyrene, benz[a]anthracene, chrysene, benzo(b)fluoranthenes, benzo()pyrenes	211
White rot fungus (*Bjerkandera adusta* BOS55)	Spiked marsh soil	Bioreactor treatment at 30 °C for 27 days	Dibenzothiophene, fluoranthene, pyrene, chrysene	212

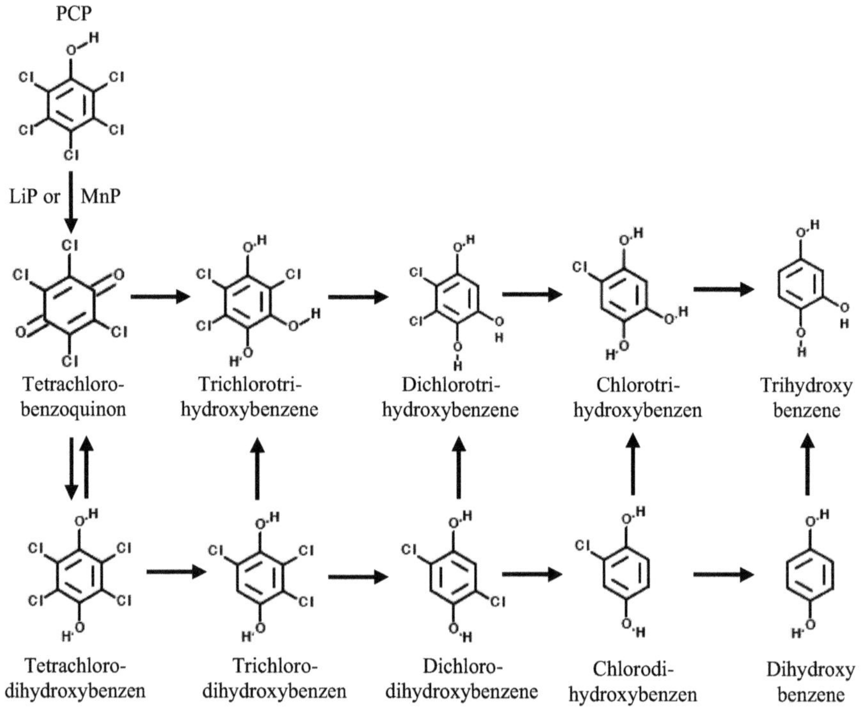

PCP

Figure 4.12   Proposed pathway(s) for the degradation of PCP by *P. chrysosporium*.
Adapted from and reproduced with permission of SOCIETY FOR
GENERAL MICROBIOLOGY[223]

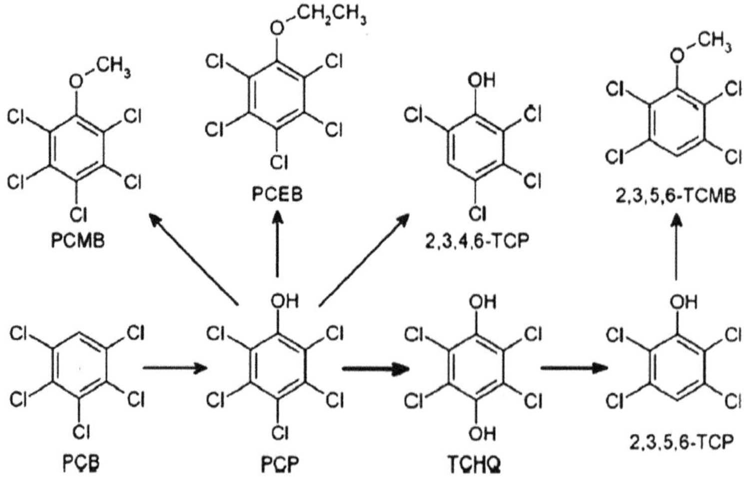

Figure 4.13   Reaction scheme of chlorophenols transformation by *Mucor ramosissimus*
IM.[214]

biological approaches to remediation has generated great interest as an alternative.

Bioremediation of PCP-contaminated soil using the method of inoculating free and immobilized white-rot fungi, *P. chrysosporium,* in PCP-polluted soils has been investigated.[224] The experimental results showed that the degradation rate of PCP by the immobilized fungi exceeded 50% at day 9, while that of the non-immobilized fungi achieved the same rate at day 16, and that the final degradation rates of PCP for both of them were >90% at day 60. Therefore, the method of composting with immobilized *P. chrysosporium* is effective for the bioremediation of PCP-contaminated soil.[224]

P. chrysosporium was cultured in a solid-state system consisting of glass columns packed with crushed fibrous plant residue from sugar cane refining. Once the fungus grew, it was used to inoculate soil spiked with 100 parts per million (ppm) of PCP at a soil to plant residue ratio of 85 : 15. A significant decrease (87%) in PCP levels in contaminated soil was observed four days after inoculation with *P. chrysosporium.* During this period, soil toxicity decreased rapidly, becoming almost negligible after 11 days. These results suggested *P. chrysosporium* application to PCP-contaminated soil is a feasible bioremediation option, considering that this microorganism reduced PCP levels in a short period of time without generation of more toxic compounds. However, more research is needed to evaluate performance of this technology under field conditions.[225]

Application of fungal treatment in beds of contaminated soil (1–4435 mg kg^{-1}) was studied at a site in Oshkosh (WI, USA). The contamination was a wood preservative formulation composed of 5% PCP in mineral spirits. Blended soil, from which the larger stones and rocks were removed, was added to each soil bed. Two fungi *(P. chrysosporium* and *P. sordida)* were added to the contaminated area using spore inoculated/infested wood chips. The PCP concentration was depleted by 82% for *P. chrysosporium* and 85% for *P. sordida,* after 56 days of treatment.[226]

Another treatment effectiveness study for fungal treatment of PCP-contaminated soil was conducted at an abandoned wood treating site at Brookhaven, Mississippi.[227] In this study, *P. sordida* treatment resulted in an overall decrease of 88 to 91% at a PCP concentration of 672 mg kg^{-1} in 6.5 weeks. *P. chrysosporium* treatment reduced PCP by 67–72% in multiple soil beds at PCP concentrations >1000 mg kg^{-1}.[228]

Bioremediation of soil contaminated with PCP (250 and 350 mg kg^{-1} soil) was investigated using *A. discolor* and compared with the reference strain *P. chrysosporium.* Both strains were incorporated as free and immobilized in wheat grains, a lignocellulosic material. For both fungal strains and at the two PCP concentrations, a high PCP removal (70–85%) occurred compared with that measured with the fungus as free mycelium (30–45%). Additionally, the use of wheat grains in soil allowed the proliferation of microorganism PCP decomposers, showing a synergistic effect with *A. discolor* and *P. chrysosporium,* and increasing the PCP removal in the soil.[229]

Chiu *et al.*[230] reported that the spent mushroom substrate (SMS) of *P. pulmonarius* performed better than various mushroom mycelia, namely, *Armillaria gallica, Armillaria mellea, Ganoderma lucidum, L. edodes, P. chrysosporium, P. pulmonarius, Polyporus sp., Coprinus cinereus* and *Volvariella volvacea*, in decontamination of PCP. Similarly, a high biodegradation capacity, *i.e.* 15.5 ± 1.0 mg of PCP by 1 g of SMS from *P. pulmonarius* was reported.[231] The main contribution to the removal process was the immobilized enzymes secreted by the mushroom during production but also the biosorption of PCP by chitin.[230,331]

4.3.2.2.2 Polychlorinated Biphenyls

4.3.2.2.2.1 Fungal Degradation of PCBs. Polychlorinated biphenyls (PCBs) are a family of compounds with a wide range of industrial applications in heat transfer fluids, dielectric fluids, hydraulic fluids, flame retardants, plasticisers, solvent extenders, and organic diluents. In practice, PCB contamination results from commercial preparations composed of complex mixtures of congeners with commercial names such as Aroclor, Fenclor, Kanechlor, Delor and Phenclor. Commercial preparations of Aroclor are specified with a four-digit code. The first two numbers in the code refer to the parent structure (12 indicating biphenyl) and the second two digits refers to the weight percentage of chlorine. For example, Aroclor 1242, 1248, 1254 and 1260 refer to PCB mixtures with an average weight percentage of chlorine of 42, 48, 54 and 60%, respectively.[232] PCBs have entered into soil and sediment environments as a result of improper disposal of industrial PCB wastes and leakage of PCBs from electric transformers. Their chemical inertness, due to a stable molecular structure and hydrophobicity, is believed to be responsible for their low biodegradation in ecosystems, leading to their accumulation in the food chain and persistence in the environment.[233,234] PCBs have been proved to be carcinogenic and to have a number of serious effects on immune, neurological, developmental, respiratory and reproductive functions due to estrogenic activity.[235] The high resistance of these toxic compounds requires drastic conditions for decomposition, either high temperatures or chemical reagents, both of which are very expensive processes.

The degradation of PCBs by white-rot fungi has been known since 1985.[236,237] Many fungi have been tested for their ability to degrade PCBs, including the white-rot fungi *Coriolus versicolor, Funalia gallica, Phlebia brevispora, Poria cinerescens,*[237] *Coriolopsis polysona,*[238] *Hirneola nigricans* and *Lentinus edodes,*[239] *P. chrysosporium,*[234,236,239–242] possibly *Lentinus tigrinus* and *P. ostreatus*[239,241] and *T. versicolor.*[238,241] PCB metabolism have also been studied by ectomycorrhizal fungi[243] and other fungi such as *Aspergillus flavus,*[244] *Cunninghamella elegans,*[245] *Aureobasidium pullulans* and *S. cerevisiae,*[237] *Aspergillus niger,*[246] *Candida boidinii* and *Candida lipolytica,*[239] and fungi from marine sediment.[235]

Although many fungi have been tested for their ability to degrade PCBs, the metabolic mechanisms of PCBs by white-rot fungi were not well understood.[247] Dmochewitz and Ballschmiter[246] detected trace amounts of hydroxylated trichlorobiphenyls as well as dichloro- and trichlorobenzoic acids during Clophen

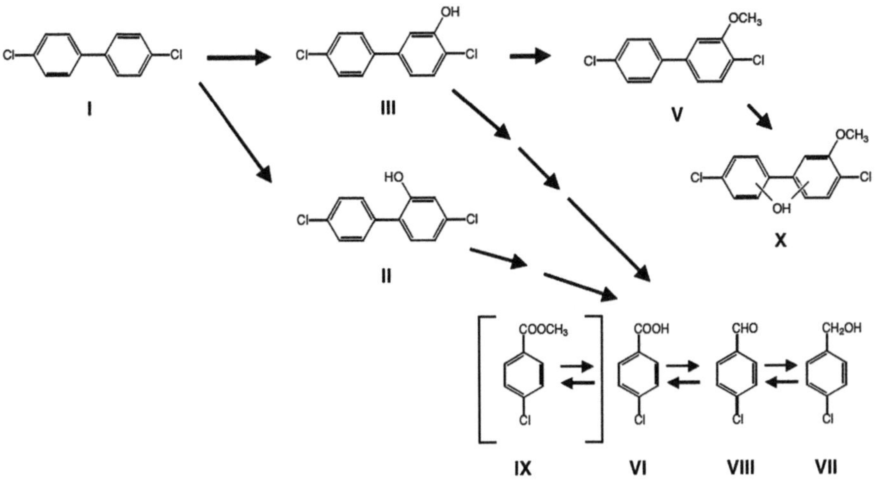

Figure 4.14 Proposed pathway for the degradation of 4,4'-DCB by *Phanerochaete* sp.[250]

A 30 (a technical PCB mixture averaging 42% Cl by weight) degradation by *A. niger*. Generally, the metabolic mechanisms published in the literature suggest possible transformation of PCBs to chlorobenzoic acids (CBAs), or production of two metabolites, 4-chlorobenzoic acid and 4-chlorobenzyl alcohol during the degradation of 4,4'-dichlorobiphenyl by *P. chrysosporium*.[242] Other PCB congeners degradation pathways have been proposed for *P. brevispora*,[249] *Phanerochaete* sp. MZ142[248] (Figure 4.14) and *P. ostreatus*.[250]

4.3.2.2.2.2 Bioremediation of PCBs in Soil by Fungi. It has been shown that PCB degradation depends on the level of biphenyl chlorination (*i.e.* the extent of PCB mineralization appears to decrease as the degree of PCB chlorination increases)[242] on the chlorine position (*ortho, meta, para*)[251] and on concentration; moreover, this degradation could be improve in the presence of surfactant.[252]

In one study, where very low concentrations of 2,4,2',4'-tetrachlorobiphenyl were utilized, 9.6% mineralization could be demonstrated.[240] Similar trends were also observed with technical PCB mixtures where *P. chrysosporium* decreased PCB concentrations of Aroclor 1242, 1254, and 1260 (42, 54 and 60% chlorinated) by 60.9%, 30.5% and 17.6%, respectively. These authors also found that 89%, 54% and 47% of 2,3-DCB, 2,4,5-TCB and 2,2',4,4'-TeCB, respectively, were degraded over a 30-day period by *P. chrysosporium* strains ATCC 24725.[234]

Congeners of lower chlorine number were degraded more extensively compared with those of higher chlorine number during the remediation of the technical mix Delor 103 by *Pleurotus ostreatus*.[251] Similarly, of the three Chlophen mixtures tested (A30, A50 and A60 of 42%, 54% and 60% chlorine content, respectively) by the filamentous fungus, *Aspergillus niger*, only the mixture with the lowest total chlorine content was biodegradable.[246] In a similar study, *A. flavus* could only metabolize Aroclor 1221 but not technical mixtures with greater chlorine content.[244]

Another study showed that, in liquid stationary cultures using Low Nitrogen Mineral Medium (LNMM) for three weeks at 28 °C, *P. Chrysosporium*, *Coriolopsis polyzona* and *T. versicolor* were able to degrade 25%, 40% and 50% of the Delor 106 PCB mixture, respectively. This degradation could be correlated with the levels of extracellular ligninolytic enzymes, manganese-dependent peroxidase, lignin peroxidase and laccase.[253]

In a bioaugmentation experiment on a historically contaminated soil by PCB using a maize stalk-immobilized *Lentinus tigrinus*, PCB depletions (33.6 ± 0.3%) and dechlorination (23.2 ± 1.3%) were found after 60-days' incubation. These results suggest that the bioaugmentation strategy with the maize stalk-immobilized mycelium of this species might be promising in the reclamation of PCB-contaminated soils.[254]

Some studies using composting reported that as much as 75% of a lower chlorinated PCB mixture (Aroclor 1221) added at a concentration of 500 mg kg^{-1} disappeared after 30 days of composting and that there was a 83% disappearance of Aroclor 1232 (11 mg kg^{-1}) accompanied with 1.6% volatilization after 35 days of composting at 50 °C.[255]

4.3.2.3   Other Pesticides

4.3.2.3.1   Fungal Degradation of Pesticides.   The extensive and massive use of pesticides in agriculture activities has serious impacts on the environment, compromising soil and water quality. Several pesticides, chlorinated pesticides, such as dieldrin, endorin, endosulfan, heptachlor, β-γ-HCH and DDT, have been used extensively and simultaneously on most agricultural crops for the protection of crops and prevention of vector-borne disease, which lead to a higher risk and increased pollution.[256–258] Given the public concern about environmental pollution by pesticides, increasing attention is being paid to the development of bioremediation systems for reducing this pollution. Indeed, microbial degradation of pesticides is the most important and effective way to remove these compounds from the environment. Fungi generally biotransform pesticides and other xenobiotics by introducing minor structural changes to the molecule, rendering it non-toxic and releasing the biotransformed pesticide into the soil for further degradation by bacteria.[259]

Many fungi, including the white-rot fungi, have been tested for their ability to degrade pesticides. Studies on pesticide metabolism by fungi such as white-rot fungi degrading lindane[260–262] and others HCH isomers,[263,264] by *F. oxysporum* degrading pendimethalin,[265] by *Fusarium* solani,[266] white-rot fungi[267] and brown-rot fungi[268] degrading DTT, *A. oryzae* degrading organophosphorus pesticide monocrotophos,[269] *A. niger* degrading dimethoate,[270] carbary,[271] pyrethroids[272] and endosulfan,[273] are reported as results for pesticide biodegradation, emphasizing the enormous potential of soil fungi for bioremediation.[274]

4.3.2.3.2 Bioremediation of Pesticides in Soil by Fungi. A spent mushroom substrate (SMS) of *A. bisporus* and its dominating microbes was found to be able to biodegrade agricultural fungicides, namely, carbendazim and mancozeb, both *in vitro* and *in situ*.[275] The indigenous fungi and bacteria associated with the SMS of *A. bisporus* were *Trichoderma* sp. and *Aspergillus* sp; B-1 and B-IV bacterial isolates, respectively. It was the extracellular lignolytic enzymes in SMS that helped in fungicide degradation, although these enzymes were not identified.

In Mexico, enzymes such as laccase, MnP and phenol oxidase were detected in a crude extract of SMS from *P. pulmonarius*. Accordingly, chlorothalonil, a broad-spectrum organochlorine fungicide, at the initial concentration of 2 mg/l, could be 100% degraded by a freshly prepared SMS extract from *P. pulmonarius*.[276]

More recently, the crude enzyme derived from SMS of *Agaricus blazei* was demonstrated to degrade metsulfuron methyl, a sulfonylurea herbicide with no or low phytotoxicity towards oil rape (*Brassica napus* L.).[277]

Collectively, these findings open the possibility of reusing SMS from mushroom industries for degrading herbicides and pesticides without harmful ecotoxicological effects.[278]

4.3.2.3.3 Example of Lindane Degradation and Bioremediation in Soil by Fungi. The organochlorine insecticide, lindane (γ-hexachlorocyclohexane, γ-HCH, Figure 4.15),[279] is extensively used world-ide for the control of agricultural and disease vector pests by farmers and government agencies. It is persistent in the environment, has a tendency to bioaccumulation, and can lead to environmental contamination and health hazards. Lindane residues have been found in soil, air, water, microorganisms, livestock meat, bovine milk, human breast milk and food grains.[236,280]

The capability of white-rot fungi to degrade lindane is based on the production of highly potent, non-specific extracellular enzymes such as peroxidases, phenol oxidases, laccases, manganese and lignin peroxidase.[281] Lindane biodegradation using WRF has been reported using different strains of *Pleurotus* such as *florida*, *sajor-caju*, *eryngii* and *ostreatus*, and by other white-rots such as *Cyathus bulleri*, *Phanerochaete chrysosporium* and *Trametes hirsutus* (Table 4.4).[279,282–287]

A non-white-rot phycomyceteous fungus, *Conidiobolus* 03-1-56, was also able to completely degrade lindane in culture medium by the fifth day of incubation in the presence of lignin-modifying enzymes and lignin peroxidase, which were responsible for lindane degradation. In the degradation of lindane (Figure 4.16), intermediate metabolites such as tetrachlorocyclohexene (TCCH) and tetrachlorocyclohexenol (TCCOL) have been detected for *T. hirsutus* and *P. chrysosporium*,[282,283] *Cyathus bulleri* and *P. sordida*.[262] However, organochloride compounds such as ethanone 1-(3-chloro-4-methoxyphenyl)- and 1-benzenecarbonyl chloride, 2,4-dichloro-3-methoxy were detected from lindane degradation by other white-rot fungi.[264]

Some reports on the degradation of γ-HCH by *P. chrysosporium* show percentage decompositions between 10.6% and 90%,[283,288] 42% with *B. adusta* in

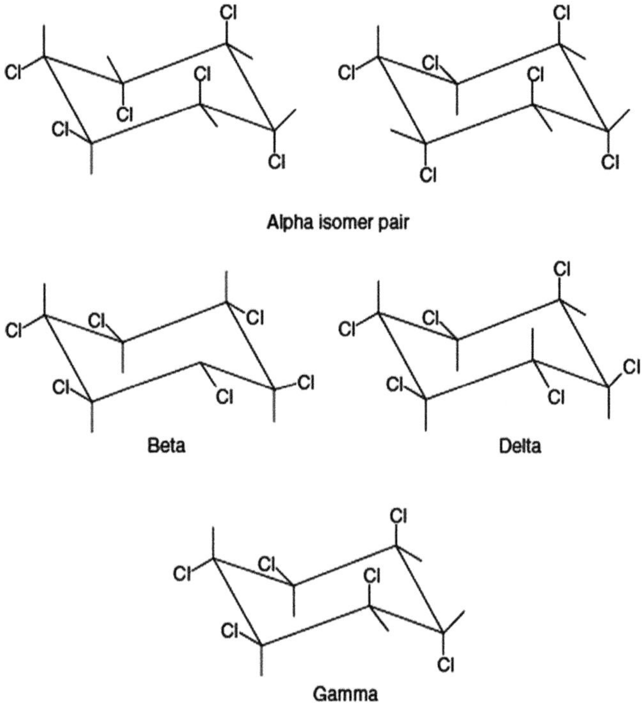

Alpha isomer pair

Beta   Delta

Gamma

Figure 4.15   Structures of HCH isomers, including the two enantiomers of α-HCH.[279]

Table 4.4   Summary of fungi with demonstrated ability to degrade HCHs (adapted from reference 279).

Microorganism	γ-HCH degradation characteristics	Reference
Phanerochaete chrysosporium BKM-F-1767	Aerobic mineralization	236
Phanerochaete chrysosporium	Aerobic mineralization	261
Pleurotus eryngii	Aerobic dechlorination	288
Pleurotus florida	Aerobic dechlorination	288
Pleurotus sajor-caju	Aerobic dechlorination	288
Trametes hirsutus	Aerobic dechlorination	283
Cyathus bulleri	Aerobic dechlorination	262
Phanerochaete sordida	Aerobic dechlorination	262
A sub-tropical white rot fungi (DSPM95)	Aerobic dechlorination	284

liquid medium and 8–17% degradation of the same isomer in a sandy soil system[264] and mineralization values between 3.9%[282] and more than 90%.[236] In soil bioremediation assays, less γ-HCH is eliminated than in liquid medium.[261,264] This is due to the difference in mass transfer rates between soil and liquid. To increase the levels of pesticides degradation in soil, fungal species were immobilized on lignocellulosic material supports such as wood chips, corncobs and wheat straw, and then inoculated into polluted soils.[260,289] These

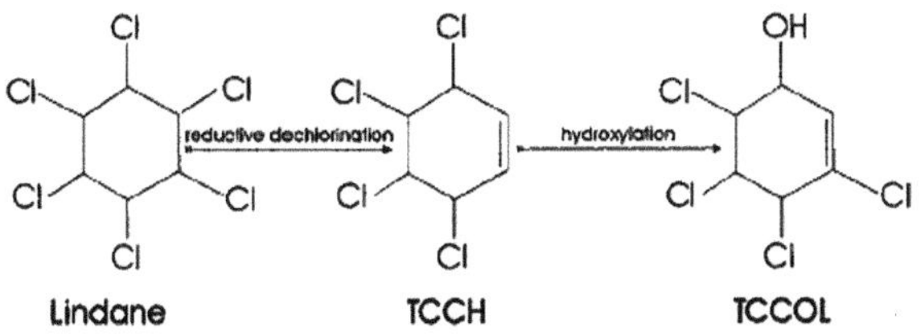

Figure 4.16 Reaction steps of lindane degradation by white-rot fungi.[262]

cellulosic materials are a plentiful source of nutrients that favour fungal growth and soil colonization,[285,290,291] and improve fungal ability to degrade pesticides. Indeed, as an example, the white-rot fungus *P. chrysosporium* immobilized in corncobs achieved 62% degradation of 2,4,5-trichlorophenoxyacetic acid after 30 days of treatment,[286] 10% degradation of DDT after 60 days' treatment,[287] and 23% degradation of chlordane after 60 days' of treatment.[260]

4.2.3.4 Example of DDT Degradation in Historically Contaminated Soil by Spent Mushroom Waste of Pleurotus Ostreatus

Bioremediation of DDT was performed using an environmentally friendly bio-logical approach by means of some new biological sources to degrade DDT, including brown-rot fungi (BRF), cattle manure compost (CMC) and spent mushroom waste (SMW). BRF including *Gloeophyllum trabeum*, *Fomitopsis pinicola* and *Daedalea dickinsii* have shown a good ability to degrade DDT.[292–294] *Mucor circinelloides* and *Galactomyces geotrichumwere* isolated from CMC have been shown to be good at degrading DDT in artificially contaminated soil.[295] In addition, mineralization of DDT by SMW of *Pleurotus ostreatus* in artificially contaminated soil has been reported and proposed as a promising source of DDT bioremediation.[296] Compared with the control, about 70% of DDT was degraded by SMW of *P. ostreatus* in historically contaminated soil. This study provides evidence that SMW of *P. ostreatus* is a potential natural source which can be used for the bioremediation of DDT in 'real world' applications.[294]

4.3.3 Other Examples of Fungi Application for Bioremediation

4.3.3.1 Removal of Heavy Metals

The removal of heavy metals by a living or dead fungal cells from contaminated water or soil solution consists of two main mechanisms. The first, biosorption, is a passive extracellular uptake involving the surface binding of metal ions to the cell wall. The second one, bioaccumulation, is an active intracellular uptake of the metal through the cell membrane and into the cell, depending on the fungal cell metabolism.[297] Basically, metal sorption and accumulation depends on diverse factors including pH, temperature, organic matter content, ionic speciation and the presence of other ions in solutions, which may lead to competition.[298] Generally, in the contaminated soils, the metal-tolerant fungal stains are alive, meaning that both mechanisms (*i.e.* sorption and accumulation) can occur simultaneously. Thus, in this section, we review the research findings based of the targeted heavy metal in order to take into consideration both processes properly to assess the overall myco-remediation process.

Several Cd-tolerant fungi have been isolated from polluted soils, which make them quite useful to remove this toxic metal and to reclaim the degraded soils. Bagot *et al.*[299] tested several microorganisms (four bacteria and a fungus) for their ability to grow and biosorb cadmium in conditions close to that of the soil. The results showed that the fungal biomass *Fomitopsis pinicola* CCBAS 535 was the most efficient with a cadmium uptake close to 50%. The authors related this high Cd removal efficiency to the magnitude of its biomass production.[299]

In Argentina, various strains of filamentous soil fungi were isolated from industrially polluted sediments, including species like *Aspergillus terreus*, *Cladosporium cladosporioides*, *Fusarium oxysporum*, *Gliocladium roseum*, *Penicillium* spp., *Talaromyces helicus* and *Trichoderma koningii*. The most efficient fungi were able to remove 63–70% of the Cd content during the 13-day growth period.[298] The authors also demonstrated that, despite the fact that the biomass developed in static assays represented 5–53% of the yield of stirred cultures (for all fungal species), the cadmium removal capacities were similar in both cases. In Morocco, other filamentous fungi belonging to the genera *Fusarium*,[300] *Alternaria Penicillium*, *Aspergillus* and *Geotrichum* were isolated from heavy metal-contaminated sites in Tangier. They were screened for their resistance to heavy metals and the results revealed that the majority of the isolates were resistant to Pb, Cr, Cu and Zn; whereas for Cd, only the fungus *Penicillium* sp. was able to grow.[301]

In China, a research team managed to isolate and characterize a Cd and Pb tolerant fungal endophyte from rape roots, *Mucor* sp. CBRF59, grown in a heavy metal contaminated soil, and to assess its potential applications to remove Cd and Pb from contaminated solutions and experimental soil. The analyses showed that, under the same operating conditions, the maximum removal capacity of Cd by dead biomass of *Mucor* sp. CBRF59 (*i.e.* biosorption) was 108 mg g^{-1}, while the removal capacity of Cd by active biomass of the strain was 173 mg g^{-1} (*i.e.* bioaccumulation). They also proved that, in a mixed solution, the ratio of Pb to Cd and initial pH values affected the bioaccumulation and biosorption capacities of the metals by CBRF59.[302]

Another study investigated the potential use of *Agaricus macrosporus* for the bioextraction of metals from contaminated substrates. The results revealed that the tested mushroom effectively extracted Cd, Hg and Cu from the contaminated substrates. Although the results suggest that fungi such as *A. macrosporus* may be effective for bioremediation of metal-contaminated substrates, cultivation of mushrooms of this type may be difficult in many contaminated environments.[303]

In Poland, a research study was carried out to analyse the tolerance to, and bioaccumulation of, selected heavy metals by filamentous fungi as a result of natural adaptation and improved resistance of fungal cells.[304] The fungi were selected from soils highly contaminated with Zn, Ba and Fe in the vicinity of a chemical plant. The related results showed that the isolates of *Trichoderma atroviride* and *Mortierella exigua* were frequently present in soil with high contents of metals. It was shown that the tested fungi were able to grow at high Zn^{2+} and Ba^{2+}, but not Fe concentrations. Maximal metal accumulation in *T. atroviride* mycelium were 21.3, 25.0 and 4.5 mg g^{-1} in the case of Zn^{2+}, Ba^{2+} and Fe, respectively. Maximal uptake in *M. exigua* mycelium were equal to 32.9 mg g^{-1} (Zn^{2+}), 15.3 mg/g (Ba^{2+}) and 1.2 mg g^{-1} (Fe^{3+}).

In Japan, two chromate-resistant filamentous fungi *Aspergillus* sp. N2 and *Penicillium* sp. N3, were isolated from Cr(VI) contaminated soil and tested for their ability to reduce Cr(VI) levels in the growth medium. The researchers found that, after 120 h of growth in a medium containing 50 ppm Cr(VI) at near neutral pH, *Aspergillus* sp. N2 reduced the Cr(VI) concentration by about 75%. *Penicillium* sp. N3 was able to reduce the Cr(VI) concentration by 35%. The results also showed that the presence of Cu(II) enhanced the Cr(VI) reducing ability of both fungal strains.[305]

Last but not least, a study by an Indian team proved that *Aspergillus nidulans* isolated from arsenic-contaminated soil has the potential to remove arsenic from soil. The analysis revealed that the biomass production (growth profile) of *A. nidulans* in arsenic-contaminated soil was found to be 0.71 g after 11 days. The arsenic adsorption potential of the fungal biomass from the contaminated soil was found to be 84.35% after 11 days at pH 4 and 35 °C.[306]

4.3.3.2 Dye Removal

The textile industry is known to release vast amount of dyes to the water and soil environments during and after the dyeing process, causing serious problems. Several techniques have been applied to remediate such complication, among them the use of fungi with the interesting ability to biodegrade and/or biosorb synthetic dyes.

In Pakistan, a research team tried to remove azo dye pollutants from the soil using fungal strains.[307] First, the fungi were isolated by using Remazol Black-B azo dye as the sole source of carbon and nitrogen. Then, 10 isolates (unidentified) were selected to test their decolourization potential. It was found that the S4 was the most effective strain in removing the tested azo dye from soil suspension as well as in liquid medium (95% decolourization by the strain S4 in soil under optimal incubation conditions). Overall, the authors found that the dye decolourization was maximal at 100 mg of dye per kg of soil, at pH 7–8 under static conditions.

Another research study was conducted in Korea to evaluate the degradation of aromatic dyes by 10 white-rot fungi. The growth and degradation efficiency of the tested fungi was determined based on their decolourization ability in solid media. The results showed that *Pycnoporus cinnabarinus*, *Pleurotus pulmonarius*, *Ganoderma lucidum*, *Trametes suaveolens*, *Stereum ostrea* and *Fomes fomentarius* have the ability to efficiently degrade Congo Red on solid media. In the case of Malachite Green, the dye inhibited the mycelial growth of many of these organisms, therefore affecting their overall decolourization potential. However, *P. cinnabarinus* and *P. pulmonarius* were able to effectively decolourize Malachite Green on solid media. However, *T. suaveolens* and

F. rosea decolourised methylene blue more effectively than any of the other fungi evaluated in this study in solid media.[308]

Four other fungi (*Aspergillus niger, Curvularia lunata, Fusarium oxysporum, Mucor mucedo*) were reported in the literature for their abilities to degrade dyestuffs. In a study, the researchers started by isolating the fungi from textile and dye contaminated soils, which were then tried for their efficiency in colour removal. The physical, chemical and biological characteristics were analysed and related to the process of decolourization. *A. niger* and *M. mucedo* were resistant in the soils and efficient in decolourization (92% removal). The scientist also revealed that *C. lunata and F. oxysporum* although occurring abundantly, were not as successful in the process of colour removal or in enzyme secretions.[309]

In Italy, 25 basidiomycetes belonging to 17 species and ascribable to different eco-physiological groups were screened for their ability to decolourize, on solid substrate, nine industrial dyes comprising a variety of anthraquinonic, azoic and phtalocyanin chromophores. The related results showed that three strains of *Bjerkandera adusta* were able to degrade all dyes on all media and achieved the highest decolourization yields (67–100%). Then, using simulated effluents composed of single and mixed dyes at high concentration (1000 mg L^{-1}), it was shown that *B. adusta* MUT 3060 was the most effective fungal strain with a decolourization percentage over 90%.[310]

In Germany, a published study assessed the potential of fungal strains newly isolated from diverse freshwater environments for the decolourization of several azo and anthraquinone type textile dyes. The authors also tested different white-rot basidiomycetes for comparison. The results revealed that the majority of dyes were decolourized by several fungal aquatic isolates at rates essentially comparable with those observed with the most efficient white-rot fungus. Under specific conditions, particular aquatic strains decolourized dyes even more efficiently than the best performing white-rot basidiomycete. This is the case for *Cylindrocarpon didymium* P10-10-3, which decolourized Disperse Yellow 3 dye significantly more efficiently than the best performing white-rot fungus (*Stropharia rugosoannulata* DSM 11372).[311]

4.4   Biosurfactants for Soil Bioremediation

One of the major factors limiting the degradation of pollutants in soils such as *n*-alkanes is their low availability to the microbial cells. Microorganisms use several strategies to enhance the availability of hydrophobic pollutants, including biofilm formation and biosurfactant production.[312,313] Growth of microorganisms on oil hydrocarbons has often been related to their capacity to produce polymers with surfactant activity referred to as biosurfactants. Microbial biosurfactants are amphipathic molecules having typical molecular weights of 500–1500 Da. They are composed of peptides, saccharides or lipids, or their combinations. Some of them are biologically produced by yeast or bacteria from various substrates including sugars, oils, alkanes and wastes. The biosurfactant may be located on the cell surface or be secreted into the extracellular medium and facilitate uptake of hydrophobic molecules. The uptake mechanism may involve direct cellular contact with hydrophobic solids, droplets or micellarization.

Biosurfactants increase the apparent water solubility of organic compounds and bioavailability by micelle formation.[314–317] Biosurfactants concentrate at the interfacial boundary existing between interfaces of two immiscible phases (solid–liquid, liquid–liquid or vapour–liquid). The hydrophobic portion concentrates at the surface while the hydrophilic portion is oriented towards the solution. The surfactant lowers the surface tension, which is a measure of the surface free energy per unit area required to bring a molecule from the bulk phase to the surface.[318] The surface tension is correlated to the concentration of the surface-active compound until the critical micelle concentration (CMC) is reached, which is defined as the minimum concentration necessary to initiate micelle formation.[319] The most active biosurfactants can lower the surface tension of water from 72 to 30 mN m^{-1} and the interfacial tension between water and *n*-hexadecane from 40 to 1 mN m^{-1}.[320] The CMC is also the maximum concentration of surfactant monomers and efficient surfactants have a low CMC. Below the CMC, bioavailability may increase by decreasing surface tension and above the CMC, micelles or other aggregates are formed (Figure 4.17). This leads to hydrophobic substrates partitioning and biodegradation being enhanced by higher cell–substrate interactions and direct substrate delivery.[321]

Biosurfactants can be readily used in soil bioremediation since they are biodegradable, have a low toxicity, and are effective in solubilizing and enhancing degradation of insoluble compounds in soil. They accelerate the removal of various soil contaminants such as hydrocarbon contaminants, pesticides and heavy metals.[313,322,323] They also attract great interest because their production can be realized from cheap and renewable resources, including bacteria, yeast, fungi and plants.[324]

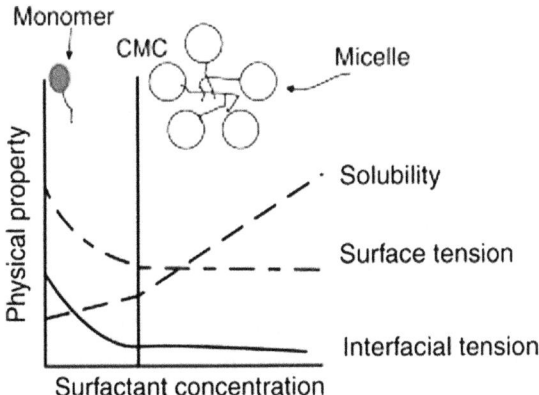

Figure 4.17 Relationship between biosurfactant concentration, surface tension and formation of micelles.[355]

4.4.1   Classes and Origin of Biosurfactants

Biosurfactant activity is highly related to its chemical nature and structure. However, the chemical composition and activity of the biosurfactant depends not only on the producer strain, but also on the growth and nutrient conditions suggesting that the chemical composition of biopolymers produced depends of the different hydrophobic compounds added to culture media. The high diversity of biosurfactant produced by numerous microorganisms is noteworthy.[325] Biosurfactants are classified according to their chemical structure, into the following main groups:[326]

(i) Glycolipids, containing carbohydrates such as sophorose, trehalose or rhamnose groups attached to a long-chain aliphatic;
(ii) Rhamnolipids consisting of one or two sugar moieties joined to one or two caprilic acid moieties *via* a glycosidic linkage;[325,327]
(iii) Surfactin composed of seven amino-acid ring structure coupled to one molecule of 3-hydroxy-13-methyl tetradecanoic acid;
(iv) Polysaccharide–lipid complexes;[328]
(v) Protein-like substances such as liposan composed of protein and carbohydrates.

According to Mulligan,[322] using substrates including C11 and C12 alkanes, succinate, pyruvate, citrate, fructose, glycerol, olive oil, glucose and mannitol,[329] *Pseudomonas aeruginosa* is able to produce rhamnolipids (Figure 4.18). Rhamnolipids produced from *P. aeruginosa* has been studied extensively[330,331] and seven homologues have now been identified.[332] They have a surface tension of 29 mN m^{-1}. Two types of rhamnolipids contain either two rhamnoses attached to β-hydroxydecanoic acid or one rhamnose linked to the identical fatty acid.[322] Rhamnolipids were also produced by a *Pseudoxanthomonas* sp. PNK-04 strain; a mixture of mono- and di-rhamnose units was identified.[333]

Sophorose lipids (Figure 4.19) can lower the surface tension to 33 mN m^{-1}. The interfacial tensions of *n*-hexadecane and water were lowered from 40 mN m^{-1} to 5 mN m^{-1} with 10 mg L^{-1} of pure sophorolipid.[334] Sophorolipid are produced by *Candida bombicola* (formerly *Torulopsis bombicola*) yeast.[322] Only few types of yeasts are known to produce biosurfactants.[334]

A lipopeptide (called surfactin) (Figure 4.20) containing seven amino acids bonded to the carboxyl and hydroxyl groups of a 14-carbon acid is produced by *Bacillus subtilis*.[24] Surfactin concentrations as low as 0.005% reduce the surface tension to 27 mN m^{-1}, making surfactin one of the most powerful biosurfactants. The structure of surfactin consists of a heptapeptide with a β-hydroxy fatty acid within a lactone ring structure.[335] Its three-dimensional structure, as determined by ^{1}H nuclear magnetic resonance (NMR) techniques,[336] folds into a β-sheet structure having a horse saddle shape.[337] The solubility and surface-active properties of the surfactin are dependent on the orientation of the residues.

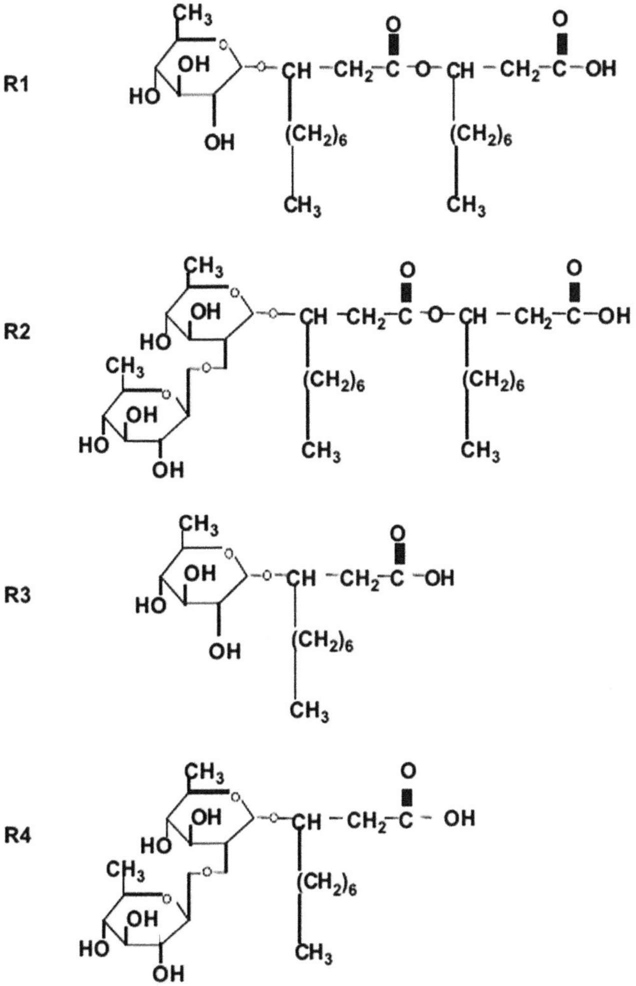

Figure 4.18 Structure of four different rhamnolipids produced by *P. aeruginosa*.[322]

New biosurfactants are also produced from other organisms.[322] A biosurfactant (JE-1058) was produced by *Gordonia* sp. and used as oil spill dispersant in sea water. Somayeh *et al.*[338] isolated biosurfactant-producing bacteria, *P. aeruginosa*, from oil-contaminated soil and water using paraffin and glucose as substrate. Its surface tension was less than 40 mN m^{-1}.

Warranaphon *et al.*[339] isolated *Burkholderia cenocepacia* BSP3, a biosurfactant-producing bacteria, from a fuel oil contaminated soil. The strain produced a glycolipid with a very low surface tension (25 mN m^{-1}) and good emulsification ability. The fungus *Aspergillus fumigatus* grown on rice bran and husk studied was isolated by Pirollo *et al.*[340] as a biosurfactant producer from hydrocarbon-contaminated soil. Plant-based biosurfactants, such as soya

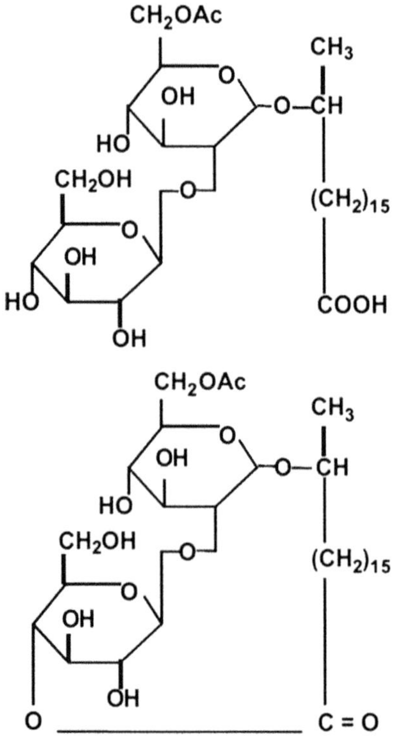

Figure 4.19   Structure of sophorolipid from *C. bombicola*.[322]

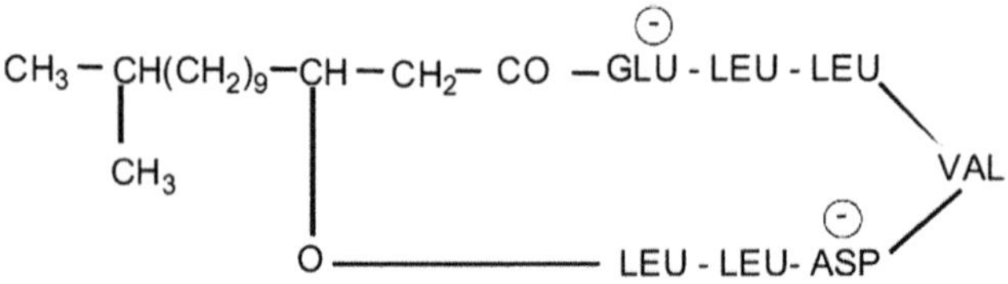

Figure 4.20   Structure of surfactin.[322]

lecithin 30,[341] a saponin,[342] or a biosurfactant extracted from cactus named BOD Balance[343] have also been reported.

4.4.2   Mechanism of Action of Biosurfactants

Hydrophobic contaminants in soil have very low water solubility and are adsorbed on soil particles, and therefore have poor bioavailability for bioremediation. Surfactants can increase their bioavailability and be transferred from soil particles to the soil aqueous phase. Thus they can be easily biodegraded by microorganisms or removed by soil washing. Soil washing is known to remove hydrocarbon contaminants with a low impact on soil constitution.

Mobilization and solubilization are two mechanisms by which hydrocarbons are removed by biosurfactants during soil washing. Mobilization of the pollutant occurs below the CMC, which lowers the interfacial tension and causes pollutant flushing. Above the CMC, the hydrocarbon is solubilized by association to the micelles of the surfactants; it can be further washed.

Determining how microorganisms interact with hydrophobic organic chemicals, metals and biosurfactants is challenging. The mechanism by which microorganisms can enhance access to poorly soluble substrates prior to either passive, or possibly active uptake, into the cell has long been discussed. Microorganisms can access a substrate *via* direct contact with soluble molecules, direct contact with solid crystals or liquid droplets, or by contact with pseudo-solubilized substrate in surfactant micelles.[344–346] During direct contact the hydrophobicity of both the cell surface and the substrate surface are key factors and biosurfactants play a role in mediating these interactions.

The structure of a biosurfactant is also a key factor for soil bioremediation, *e.g.* the lactonic form of sophorolipids was shown to inhibit hexadecane biodegradation while the acidic form stimulated it.[346,347] The methyl ester form of rhamnolipids was a more effective biodegrader of hexadecane and octadecane than the acid form.[348] The methyl ester form of rhamnolipid lowers interfacial tension to <0.1 dyne cm^{-1} while the acid form lowers it only up to 5 dyne cm^{-1}.[348] The carboxyl group of the acid form of rhamnolipid confers on it a negative charge which interacts better with water than with alkanes; thus it is less effective in reducing the interfacial tension. However, the methyl ester form of rhamnolipid lowers the interfacial tension more and acts as a better alkane dispersant. Nevertheless, because of its low water solubility, the methyl ester form of rhamnolipids may not be suitable for environmental applications. Structural differences may impart different physicochemical properties to biosurfactants that may guide the efficiency of biodegradation and bioremediation.

Due to their ability to form complexes with metals, biosurfactants can be used for the bioremediation of heavy metal contaminated soil. The anionic biosurfactants form ionic bonds complexes with metals. The cationic biosurfactants compete with metals, replacing the charged metal ions on negatively charged surfaces by an ion exchange mechanism. The polar groups of micelles can also bind metals, thus removing them from soil surfaces; the metals are then mobilized in water.[349–351]

4.4.3 Removal of Hydrocarbon and Oil by Biosurfactants

Addition of biosurfactants can enhance hydrocarbon biodegradation in soil[352] by stimulation, dissolution or desorption, or solubilization of hydrocarbons. Rhamnolipids stimulate different processes involved in the degradation of organic substrates; rhamnolipid was shown to stimulate the degradation of hexadecane to a greater extent when it was entrapped in matrices with pore sizes larger than 300 nm.[353] Addition of nutrients to the soil such as nitrogen and phosphorous, allowed the native microbial population to develop and expand. Optimum activity occurred when the soil moisture was 50–80% of saturation.

Temperature can also play a key role. Indeed, a thermophilic *Bacillus* strain NG80-2 was shown to produce emulsifying agents increasing the degradation of long-chain *n*-alkanes.[354] When oil-contaminated soils were subjected to very low temperatures, biosurfactants limited the viscosity and water solubility of the hydrocarbons at lower temperatures.

Rhamnolipid surfactants were used to release oil from beaches in Alaska after the Exxon Valdez tanker spill.[355] Removal efficiency varied according to contact time and biosurfactant concentration.[356] Scheibenbogen *et al.*[357] found that the rhamnolipids from *P. aeruginosa* UG2 were able to effectively remove a hydrocarbon mixture from a sandy loam soil. The percentage of removal varied from 23–59% depending on the type of hydrocarbon removed and the concentration of the surfactant used.

Lafrance and Lapointe[358] also showed that injection of low concentrations of *P. aeruginosa* UG2 rhamnolipid (0.25%) enhanced transport of pyrene more than sodium dodecyl sulfate (SDS) with less impact on the soil. At pH 6, 60% of hexadecane could be removed by a solution of rhamnolipid at 500 mg L^{-1}.[359] Various biological surfactants (aescin, lecithin, rhamnolipid, saponin, tannin) were compared at various temperature from 5 to 50 °C by Urum *et al.*[360] for their ability to wash a crude oil contaminated soil at concentrations ranging from 0.004 to 0.5 M. The optimal temperature was 50 °C and 79% of the oil was removed by rhamnolipid and saponin; similar results were found when sodium dodecyl sulfate was used.

Shulga *et al.*[361] examined the use of the biosurfactants and a biopolymer produced by strain *Pseudomonas* PS-17 in cleaning oil from coastal sand, and from the feathers and furs of marine birds and animals. The biosurfactant and biopolymer reduced the surface tension of water to 29.0 mN m^{-1} and the interfacial tension against heptane to between 0.01 and 0.07 mN m^{-1}. The results indicated that this biosurfactant/biopolymer could be used to remove oil from marine birds and animals contaminated by oil.

In 1984, sophorolipids were used to release bitumen from tar sands.[334] Later, several works reported its potential to enhance the remediation of hydrocarbon-contaminated soils. A study by Oberbremer *et al.*[362] show that 90% of a hydrocarbon mixture was degraded in the presence of sophorolipids in 79 h compared with only 81% in 114 h without the biosurfactant.

Schippers *et al.*[363] showed that sophorolipids enhanced the degradation of phenanthrene by *Sphingomonas yanoikuyae*. The CMC of the sophorolipid in water was 4 mg L^{-1}, but this increased to 10 mg L^{-1} in the presence of 10% soil suspensions, indicating adsorption of the surfactant onto the soil. Solubilization tests showed a 10-fold higher removal than with a synthetic surfactant such as SDS. Solubilization was involved in increased biodegradation of the phenanthrene.

Kang *et al.*[364] tested sophorolipids for the remediation of soil polluted by aliphatic and aromatic hydrocarbons and Iranian light crude oil. Biodegradation increased the rate of degradation; 85–97% of the total amount of hydrocarbons was biodegraded.

Surfactin produced by *Bacillus subtilis* O9[365] was added to ship bilge waste; Aliphatic and aromatic compounds were degraded more quickly in the presence of the biosurfactant than without biosurfactant and, after 10 days, only 6.8 and 7.2% of aliphatic and aromatic compounds, respectively, remained.

A strain of *Bacillus subtilis*, producing biosurfactant, was able to remove 62% of the oil in a sand pack saturated with kerosene, allowing *in situ* oil removal and cleaning sludge from sludge tanks.[366,367] Eliseev *et al.*[368] also showed that a *Bacillus* species could release oil in oil-saturated columns.

4.4.4 Removal of PAHs by Biosurfactants

PAHs have the general form $C_{4n}C_2H_{2n}C_4$, n being the number of rings. They are produced during petroleum refining, coke production and wood preservation. As the number of rings increases, the compounds become more difficult to degrade due to decreasing volatility and solubility and increased sorption. The rhamnolipids from the *Pseudomonas* UG2 strain was up to five times more effective than SDS, as they could enhance the solubilization of four-ring PAHs in a bioslurry. However, high molecular weight PAHs were not biodegraded.[369]

Biosurfactants produced by *Pseudomonas marginalis* solubilized PAHs such as phenanthrene and enhanced biodegradation.[370] A comparison of the solubilization of another PAH, naphthalene, by a rhamnolipid and synthetic surfactants including SDS, an anionic surfactant, and Triton X-100, a non-ionic surfactant, found that the biosurfactant increased the solubility of naphthalene by 30 times.[371] However, biodegradation of naphthalene (30 mg L^{-1}) took 40 days in the presence of biosurfactant (10 g L^{-1}) compared with 100 h for Triton X-100 (10 g L^{-1}); the biosurfactant was used as a carbon source instead of the naphthalene, which did not occur in the case of Triton X-100.

Other research by Garcia-Junco *et al.*[372] indicated that addition of rhamnolipids led to attachment to phenanthrene, which enhanced its bioavailability and consequently increased degradation of the contaminant by *P. aeruginosa*. Addition of rhamnolipid produced by *Pseudomonas* sp. DS10-129 showed that it enhanced *ex situ* bioremediation of a gasoline-contaminated soil along with poultry litter and coir pith.[373]

Barkay *et al.*[374] studied the effect of a bioemulsifier alasan produced by *Acinetobacter radioresistens* KA53 on the solubilization of PAHs, phenanthrene (PHE) and fluoranthene (FLA). They also studied the influence of alasan on mineralization of PHE and FLA by *Sphingomonas paucimobilis* EPA505. They showed a linear increase of phenanthrene and fluoranthene aqueous solubility with increasing alasan concentration. Mineralization of PAHs by *S. paucimobilis* EPA505 was stimulated by alasan addition; the degradation rate of fluoranthene doubled and increased the degradation rate of phenanthrene.

4.4.5 Removal of Chlorinated Compounds and Pesticides by Biosurfactants

When rhamnolipid R1 from *P. aeruginosa* was added to an acclimated culture of *Alcaligenes eutrophus*, 4,4-chlorobiphenyl was mineralized by 213 times more than the control.[375] Fiebig *et al.*[376] showed that glycolipids (GL-K12) from *Pseudomonas cepacia* enhanced the degradation of Arochlor 1242 by mixed cultures.

Mata-Sandoval *et al.*[377] studied the biodegradation of trifluralin, coumaphos and atrazine in the presence of rhamnolipid and with Triton X-100. Trifluralin biodegradation was enhanced in the presence of both surfactants. Coumaphos biodegradation increased at rhamnolipid concentrations above 3 mM, but declined when Triton concentrations were above that of the CMC. Awashti *et al.*[378] evaluated the ability of surfactin produced by *Bacillus subtilis* to enhance biodegradation of the pesticide endosulfan.

4.4.6 Removal of Metals

Many contaminated sites are also co-contaminated with metals and so metal chelation by biosurfactants has also been studied. Surfactin, rhamnolipids and sophorolipids were shown to remove heavy metals from contaminated soil.[379] Due to their anionic character, rhamnolipids are able to remove metals and ions such as cadmium, copper, lanthanum, lead and zinc from soil by complexation.[380,381] Rhamnolipid from *P. aeruginosa* was shown to form stable complexes with metals in the order:[382]

$$Al^{3+} > Cu^{2+} > Pb^{2+} > Cd^{2+} > Zn^{2+} > Fe^{3+} > Hg^{2+} > Ca^{2+} > Co^{2+}$$

$$> Ni^{2+} > Mn^{2+} > Mg^{2+} > K^+$$

Stability constants were established by an ion exchange resin technique. The affinities values calculated for rhamnolipid were similar or higher than those obtained for organic acids like acetic, citric, fulvic and oxalic acids have for metals. Molar ratios values of the rhamnolipid to metal were 2.31, 2.37, 1.91, 1.58, 0.93 for copper, lead, cadmium, zinc and respectively. Molar ratios were 0.84 and 0.57, respectively for common soil cations, magnesium and potassium. When rhamnolipids were applied to a soil in the presence of oil,[383] 80–100% of cadmium and lead were removed from artificially contaminated soil. In field samples, however, only 20–80% were removed,[384] as clay and iron oxide affect the efficiency of the biosurfactants. Biosurfactant can also be added to excavated soil during washing.

Neilson *et al.*[385] studied lead removal by rhamnolipids. They show that 15% of the lead could be removed after 10 washes by a rhamnolipid solution. These very low removal levels could be improved by biosurfactants recycling. Similar values were found for zinc.[379,383]

Rhamnolipid was able to reduce cadmium toxicity for *Burkholderia* sp. when grown on naphthalene as sole carbon source due to rhamnolipid complexation of cadmium and rhamnolipid-induced lipopolysaccharide removal from the cell surface.[386] The removal of lipopolysaccharides and cadmium complexation may decrease the overall negative charge, reducing cadmium uptake by of the cell.

Rhamnolipid type I and type II, with surface tensions of 29 mN m^{-1}, were found to be suitable for soil washing and heavy metal removal.[387] The best removal rates were 73.2% for cadmium and 68.1% for nickel, obtained at pH 10. Ascy *et al.*[388] also used rhamnolipids for Cd(II) and Zn(II) extraction from quartz; 91.6% and 87.2% of sorbed Cd(II) and Zn(II), respectively, were recovered.

Surfactin can be used to bind metals such as magnesium, manganese, calcium, barium, lithium and rubidium. It has been demonstrated that binding is due to the two negative charges on glutamate and aspartate of surfactin.[389] Hong *et al.*[390] used saponin, a plant-derived biosurfactant, for the removal of cadmium and zinc from three types of soils (Andosol, Cambisol and Regosol); 90–100% removal for cadmium and 85–98% for zinc were obtained. Hong *et al.*[391] also showed that a derivative obtained from dehydration of spiculisporic acid produced by *Penicillium spiculisporum* led to sodium salt of 2-(2-carboxyethyl)-3-decyl maleic anhydride that could remove cadmium, copper, zinc and nickel. Highest rejection rates were achieved when the molar ratio of the biosurfactant to metal ions was equal.

4.4.7 Field Studies

Straube *et al.*[392] evaluated the addition of *Pseudomonas aeruginosa* strain 64, a biosurfactant, with a bulking agent (ground rice hulls) to enhance the bioremediation of PAH-contaminated soil during landfarming. This biostimulation/bioaugmentation approach led to an 87% decrease in total PAHs; larger scale pan experiments showed a reduction of 86% in the PAHs. Overall, biosurfactants were produced in the soil by the bacterial strain *Pseudomonas aeruginosa* 64, enabling PAH biodegradation to occur.

Lin *et al.*[393] used an integrated field landfarming system to assess the relative effectiveness of five biological approaches on diesel degradation. The consortium consisted of four diesel-degrading bacteria strains. Rhamnolipid was used as the biosurfactant. The diesel concentration, bacterial population, evolution of CO_2 and bacterial community in the soil were measured periodically. In the early stage, total petroleum hydrocarbon was degraded 10 times faster than the degradation rates measured during the period from day 30 to 100.

Kildisas *et al.*[394] described a technology based on bioremediation principles for cleaning up soil contaminated by oil pollutants in a large scale, using physico-chemical treatment by washing the contaminated soil. The pilot plant for washing the contaminated soil consisted of an area of 340 m^2 in which 1000 m^3 of contaminated soil was cleaned up. The migration fraction was washed by application of biosurfactants, separation of water, oil and soil. Biodegradation of the residual non-migrating oil fraction was achieved by use of specific bacteria degrading crude oil and oil products, and phytoremediation. Before treatment, oil concentrations were between 180 and 270 g kg^{-1} of soil; after washing, the concentrations had reduced to 34–59 g kg^{-1} of soil.

After degradation, the pollutant concentrations had decreased to 3.2–7.3 g g^{-1} of soil.

4.4.8 Conclusions

Biosurfactants such as rhamnolipids, sophorolipids and surfactin and saponin have shown potential for the remediation of soil contaminated by either organic compounds or heavy metals. Biosurfactants increase the contaminants' solubilization and emulsification. Due to their biodegradability, their low toxicity and their potential for *in situ* production by microorganisms, biosurfactants are promising for eco-friendly remediation technology. Further research regarding the prediction of their efficiency will focus on soil constitution, pollutant chemical structure, as well as the fate and transport of contaminants. Methods to optimize the cost of biosurfactant utilization should be developed regarding the scaling-up and cost reduction for *ex situ* production.[322] Optimized growth conditions using cheap renewable substrates such as agro-industrial wastes and efficient methods for isolation and purification of biosurfactants need to be developed for their production to be more economically feasible.

4.5 Perspectives

Microorganisms offer multiple possibilities for the remediation of contaminated soils. The mechanisms and catabolic activities behind many bacterial processes have been thoroughly investigated. However, there are still a significant number of pathways which remain poorly characterized, particularly those of anaerobic organisms (*e.g.* for TNT) and those related to emerging contaminants. At the same time, the potential of fungi to deal with both organic and inorganic contaminants under circumstances where bacteria are at a disadvantage (*e.g.* low substrate availability, high heavy metal concentrations) needs to be exploited; this includes the utilization of both fungal mycelia and enzymes in bioremediation, as well as mycorrhizal associations which enhance survival of plants in phytoremediation strategies. With the advent of OMICs technologies (genomics, transcriptomics, proteomics, metabolomics) and systems biology approaches, the discovery of new catabolic genes and biomarkers can be accelerated and new insights gained from microbial physiology and metabolism.

Nonetheless, the isolation of microorganisms remains a major bottleneck to fully exploit the information gained from OMIC technologies and to improve the control of biological processes in the field. Engineered microorganisms containing and expressing catabolic genes from other strains or species have shown potential in lab-based or controlled studies. However, their use in full-scale systems is still very limited as most of the engineered strains are unable to survive or out-compete the native organisms. Enrichment techniques allow mining of mixed microbial communities from natural environments which are robust enough for commercial applications. A better understanding of the inherent syntrophic interactions within such communities is important to improve their performance and to develop cultures with new capabilities. The utilization of molecular biology tools such as real-time polymerase chain reaction (PCR), microarrays and next-generation sequencing might significantly improve the characterization, monitoring and modelling of microbial activity at contaminated sites. Such information will hopefully improve predictions on the outcome of bioremediation and facilitate decision-making at contaminated sites.

Acknowledgements

Dr Pérez-de-Mora thanks the European Commission for financial support through an OIF Marie Curie Fellowship (contract 334 234974/ AnDeMiC) within the Seventh Framework for Research and Technological Development.

References

1. N. C. Brady and R. R. Weil, *The Nature and Properties of Soils*, Prentice Hall, New Jersey, NJ, 2002.
2. European Commission, *Towards a Thematic Strategy for Soil Protection*, COM(2002) 179 final, European Commission, Brussels, 2002.
3. D. C. Adriano, *Trace Elements in Terrestrial Environments*, Springer, New York, 2001.
4. European Commission, *Proposal for a Directive of the European Parliament and of the Council Establishing a Framework for the Protection of Soil and Amending Directive 2004/35/EC*, COM(2006) 232 final, European Commission, Brussels, 2006.
5. USEPA, *Cleaning Up the Nation's Waste Sites: Markets and Technology Trends, 2004 Edition*, EPA 542-R-04-015, US Environmental Protection Agency, Office of Solid Waste and Emergency Response, Washington DC, 2004.
6. J. I. Prosser, in *Modern Soil Microbiology*, ed. J. D. van Elsas, J. K. Jansson and J. T. Trevors, CRC Press, Boca Raton, FL, 2007, pp. 237–261.
7. G. M. Gadd, *Geoderma*, 2004, **122**, 109.
8. S. M. Ross, in *Toxic Metals in Soil-Plant Systems*, ed. S. M. Ross, Wiley & Sons, Chichester, 1994, pp. 3–25.
9. R. K. Poole and G. M. Gadd, *Metals: Microbe Interactions*, IRL Press, Oxford, UK, 1989.
10. P. Koedrith and Y. R. Seo, *Int. J. Mol. Sci.*, 2011, **12**, 9576.
11. R. Clay, *Environ. Health Perspect.*, 2001, **109**, 162.

12. J. R. Lloyd, J. C. Renshaw, I. May, F. R. Livens, I. T. Burke, R. J. G. Mortimer and K. Morris, *Nuc. Radiochem. Sci.*, 2005, **6**, 17.

13. USEPA, *Commonly Encountered Radionuclides* [online], US Environmental Protection Agency, Washington DC, 2012, available at: www.epa.gov/rpdweb00/radionuclides/.

14. A. Kabata-Pendias, *Trace Elements in Soils and Plants*. CRC Press, Boca Raton, FL, 2001.

15. A. Kabata-Pendias, *Geoderma*, 2004, **122**, 143.

16. H. H. Tabak, P. Lens, E. D. van Hullebush and W. Dejonghe, *Rev. Environ. Sci. BioTechnol.*, 2005, **4**, 115.

17. G. M. Gadd, *Geoderma*, 2004, **122**, 109.

18. G. M. Gadd and J. A. Sayer, in *Environmental Microbe–Metal Interactions*, ed. D. R. Lovley, American Society of Microbiology, Washington DC, 2000, pp. 237–256.

19. L. Birch and R. Bachofen, *Experientia*, 1990, **46**, 827.

20. K. Bosecker, *FEMS Rev.*, 1997, **591**, 20.

21. D. R. Lovley, in *Environmental Microbe–Metal Interactions*, ed. D. R. Lovley, American Society of Microbiology, Washington DC, 2000, pp. 3–30.

22. W. L. Smith and G. M. Gadd, *J. Appl. Microbiol.*, 2000, **88**, 983.

23. E. D. van Hullebusch, P. N. L. Lens and H. H. Tabak, *Rev. Environ. Sci BioTechnol.*, 2005, **4**, 185.

24. J. A. Sayer, J. D. Cotter-Howells, C. Watson, S. Hillier and G. M. Gadd, *Curr. Biol.*, 1999, **9**, 691.

25. G. M. Gadd, *Microbiology*, 2010, **156**, 609.

26. H. Brandl and M. A. Faramarzi, *China Particuol.*, 2006, **4**, 93.

27. C. White, J. S. Dennis and G. M. Gadd, *Biodegradation*, 2003, **14**, 139.

28. US Geological Survey, *Bioremediation: Arsenic* [online], US Geological Survey, Reston, VA, available at: http://microbiology.usgs.gov/bioremediation_arsenic.html.

29. B. R. Mohapatra, O. Dinardo, W. D. Gould and D. W. Koren, *Miner. Eng.*, 2010, **23**, 591.

30. US Department of Energy, *Genomics: GTL Roadmap*, US Department of Energy, Office of Science, Washington DC, 2005, Appendix B.

31. R. T. Anderson, H. A. Vrionis, I. Ortiz-Bernad, C. T. Resch, P. E. Long, R. Dayvault, K. Karp, S. Marutzky, D. R. Metzler, A. Peacock, D. C. White, M. Lowe and D. R. Lovley, *Appl. Environ. Microbiol.*, 2003, **69**, 5884.

32. L. N. Moyes, M. J. Jones, W. A. Reed, F. R. Livens, J. M. Charnock and J. F. W. Mosselmans, *Environ. Sci. Technol.*, 2002, **36**, 179.

33. J. D. Istok, J. M. Senko, L. R. Krumholz, D. Watson, M. A. Bogle, A. Peacock, Y.-J. Chang and D. C. White, *Environ. Sci. Technol.*, 2004, **38**, 468.

34. A. Pérez-de-Mora, P. Madejón, P. Burgos, F. Cabrera, N. W. Lepp and E. Madejón, *Environ. Pollut.*, 2011, **159**, 3018.

35. B. E. Rittman and P. L. McCarty, *Environmental Biotechnology: Principles and Applications*, McGraw-Hill, New York, 2001.
36. R. P. Schwarzenbach, P. M. Gchwend and D. M. Imboden, *Environmental Organic Chemistry*, John Wiley & Sons, New York, 1993.
37. M. Alexander, *Biodegradation and Bioremediation*, Academic Press, New York, 1999.
38. L. Adrian, U. Szewzyk and H. Görisch, *Biodegradation*, 2000, **11**, 73.
39. J. Vangronsveld and S. D. Cunningham, in *Metal-contaminated Soils: In-situ Inactivation and Phytorestoration*, ed. J. Vangronsveld, Springer, Berlin, 1998, pp. 1–15.
40. B. Mahro, R. Müller and V. Kasche, in *Treatment of Contaminated Soil: Fundamentals, Analysis and Applications*, ed. R. Stegman, G. Brunner, W. Calmano and G. Matz, Springer, Berlin, 2001, pp. 181–196.
41. M. Carmona, M. A. Prieto, B. Galán, J. L. García and E. Díaz, in *Microbial Biodegradation Genomics and Molecular Biology*, ed. E. Díaz, Caister Academic Press, Norfolk, UK, 2008, pp. 97–144.
42. P. Grathwohl, in *Mikrobieller Schadstoffabbau: Ein Interdisziplinärer Ansatz*, ed. C. Knorr and T. von Schell, Springer-Verlag, Berlin and Heidelberg, 1997.
43. H. Steinhart, T. Käcker, S. Meyer and G. Biernoth, in *Treatment of Contaminated Soil: Fundamentals, Analysis and Applications*, ed. R. Stegman, G. Brunner, W. Calmano and G. Matz, Springer, Berlin, 2001, pp. 37–48.
44. J. A. Labinger and J. E. Bercaw, *Nature*, 2002, **417**, 507.
45. J. B. van Beilen and E. G. Funhoff, *Appl. Microbiol. Biotechnol.*, 2007, **74**, 13.
46. T. Kotani, H. Yurimoto, N. Kato and Y. Sakai, *J. Bacteriol.*, 2007, **189**, 886.
47. J. B. van Beilen, S. Panke, S. Lucchini, A. G. Franchini, M. Rothlisberger and B. Witholt, *Microbiology*, 2001, **147**, 1621.
48. F. Widdel and R. Rabus, *Curr. Opin. Biotechnol.*, 2011, **12**, 259.
49. F. Rojo, *Environ. Microbiol.*, 2009, **11**, 2477.
50. V. Grossi, C. Cravo-Laureau, J.-F. Rontani, M. Cros and A. Hirschler-Réa, *Res. Microbiol.*, 2011, **162**, 915.
51. W. H. Lee, J. B. Park, K. Park, M. D. Kim and J. H. Seo, *Appl. Microbiol. Biotechnol.*, 2007, **76**, 329.
52. N. A. Donoghue and P. W. Trudgill, *Eur. J. Biochem.*, 1975, **60**, 1.
53. D. Koma, Y. Sakashita, K. Kubota, Y. Fujii and F. Hasumi, *Appl. Microbiol. Biotechnol.*, 2005, **66**, 92.
54. T. Yi, E. H. Lee, Y. G. Ahn, G.-S. Hwang and K.-S. Choa, *J. Hazard. Mater.*, 2011, **191**, 393.
55. F. Musat, H. Wilkes, A. Behrends, D. Woebken and F. Widdel, *ISME J.*, 2010, **4**, 1290.
56. R. E. Doherty, *J. Environ. Forensics*, 2000, **1**, 69.
57. P. L. McCarty, in *In Situ Remediation of Chlorinated Plumes*, ed. H. F. Stroo and C. H. Ward, Springer, New York, 2010, pp. 1–28.
58. J. T. Wilson and B. H. Wilson, *Appl. Environ. Microbiol.*, 1985, **49**, 242.

59. P. M. Bradley and F. H. Chapelle, *Environ. Sci. Technol.*, 2000, **34**, 221.
60. R. J. Oldenhuis, J. Y. Oedzes, J. J. Vanderwaarde and D. B. Janssen, *Appl. Environ. Microbiol.*, 1991, **57**, 7.
61. D. L. Freedman and J. M. Gossett, *Appl. Environ. Microbiol.*, 1989, **55**, 2144.
62. D. W. Major, M. M. McMaster, E. E. Cox, E. A. Edwards, S. M. Dworatzek, E. R. Hendrickson, M. G. Starr, J. A. Payne and L. W. Buonamici, *Environ. Sci. Technol.*, 2002, **36**, 5106.
63. A. Pérez-de-Mora, A. Zila, M. McMaster and E. A Edwards, presented at the American Society for Microbiology General Meeting, **21–24** May 2011, New Orleans, USA.
64. H. J. Badham and L. M. Winn, *Toxicology*, 2007, **229**, 177.
65. M. D. Mikesell, J. J. Kurkor and R. H. Olsen, *Biodegradation*, 1993, **4**, 249.
66. M. P. McLeod and L. D. Eltis, in *Microbial Biodegradation Genomics and Molecular Biology*, ed. E. Diaz, Caister Academic Press, Norfolk, UK, 2008, pp. 1–24.
67. C. S. Harwood, G. Burchhardt, H. Herrmann and G. Fuchs, *FEMS Microbiol. Rev.*, 1999, **22**, 439.
68. S. A. B. Weelink, M. H. A. van Eekert and A. J. M. Stams, *Rev. Environ. Sci. Biotechnol.*, 2010, **9**, 359.
69. J. B. Coates, R. Chakraborty, J. G. Lack, S. M. O'Connor, K. A. Cole, K. S. Bender and L. A. Achenbach, *Nature*, 2011, **411**, 1039.
70. A. C. Ulrich and E. A. Edwards, *Environ. Microbiol.*, 2003, **5**, 92.
71. A. K. Haritash and C. P. Kaushik, *J. Hazard. Mater.*, 2009, **169**, 1.
72. B. W. Bogan and W. R. Sullivan, *Chemosphere*, 2003, **52**, 1717.
73. C. E. Cerniglia, *Biodegradation*, 1992, **3**, 351.
74. D. T. Gibson and R. E. Parales, *Curr. Opin. Biotechnol.*, 2000, **11**, 236.
75. R. H. Peng, A. S. Xiong, Y. Xue, X. Y. Fu, F. Gao, W. Zhao, Y. S. Tian and Q. H. Yao, *FEMS Microbiol. Rev.*, 2008, **32**, 927.
76. R. U. Meckenstock, M. Safinowski and C. Griebler, *FEMS Microbiol. Ecol.*, 2004, **49**, 27.
77. R. U. Meckenstock and H. Mouttaki, *Curr. Opin. Biotechnol.*, 2011, **22**, 406.
78. J. A. Field and R. Sierra-Alvarez, *Biodegradation*, 2008, **19**, 463.
79. J. L. Barber, A. J. Sweetman, D. van Wijk and K. C. Jones, *Sci. Total Environ.*, 2005, **349**, 1.
80. J. C. Spain and S. F. Nishino, *Appl. Environ. Microbiol.*, 1987, **53**, 1010.
81. M. Jechorek, K. D. Wendlandt and M. Beck, *J. Biotechnol.*, 2003, **102**, 93.
82. J. M. Fung, B. P. Weisenstein, E. E. Mack, J. E. Dumsky, A. E. Tom and S. H. Zinder, *Environ. Sci. Technol.*, 2009, **43**, 2302.
83. T. J. Gentry, G. Wang, C. Rensing and I. L. Pepper, *Microbial. Ecol.*, 2004, **48**, 90.
84. J. A. Field and R. Sierra-Alvarez, *Rev. Environ. Sci. Biotechnol.*, 2008, **7**, 191.
85. L. C. M. Commandeur, R. J. May, H. Mokross, D. L. Bedard, W. Reineke, H. A. J. Govers and J. R. Parsons, *Biodegradation*, 1997, **7**, 435.

86. J. Dolfing and J. M. Tiedje, *FEMS Microbiol. Ecol.*, 1986, **38**, 293.
87. USEPA, *Consumer Factsheet on: Pentachlorophenol*, US Environmental Protection Agency, Washington DC, 2006, pp. 11–28.
88. I. P. Solyanikova and L. A. Golovleva, *J. Environ. Sci. Health B*, 2004, **39**, 333.
89. J. H. A. Apajalahti and M. S. Salkinoja-Salonen, *J. Bacteriol.*, 1997, **169**, 5125.
90. V. Matus, M. A. Sanchez, M. Martinez and B. Gonzalez, *Appl. Environ. Microbiol.*, 2003, **69**, 7108.
91. B. L. Sun, J. R. Cole, R. A. Sanford and J. M. Tiedje, *Appl. Environ. Microbiol.*, 2000, **66**, 2408.
92. J. A. Field and R. Sierra-Alvarez, *Chemosphere*, 2008, **71**, 1005.
93. H. A. Arfmann, K. N. Timmis and R. M. Wittich, *Appl. Environ. Microbiol.*, 1997, **63**, 3458.
94. H. Nojiri and T. Omori, *Biosci. Biotechnol. Biochem.*, 2002, **66**, 2001.
95. D. E. Fennell, I. Nijenhuis, S. F. Wilson, S. H. Zinder and M. M. Haggblom, *Environ. Sci. Technol.*, 2004, **38**, 2075.
96. J. A. Field and R. Sierra-Alvarez, *Biodegradability of Cholorinated Aromatic Compounds*, Science Dossier 12, Euro Chlor, Brussels, 2007.
97. USPA, *Health Effects of PCBs* [online], US Environmental Protection Agency, Washington DC, 2012, available at: www.epa.gov/osw/hazard/tsd/pcbs/pubs/effects.htm.
98. J. J. Arensdorf and D. D. Focht, *Appl. Environ. Microbiol.*, 1994, **60**, 2884.
99. D. L. Bedard, in *Dehalogenation: Microbial Processes and Environmental Applications*, ed. M. M. Haggblom and I. D. Bossert, Kluwer Academic Publishers, Boston, MA, 2003, pp. 443–465.
100. A. Kubatova, P. Erbanova, I. Eichlerova, L. Homolka, F. Nerud and V. Sasek, *Chemosphere*, 2001, **43**, 207.
101. O. Segev, A. Kushmaro and A. Brenner, *Int. J. Environ. Res. Public Health*, 2009, **6**, 478.
102. L. S. Birnbaum and D. F. Staskal, *Environ. Health Perspect.*, 2004, **112**, 9.
103. T. An, L. Zu, G. Li, S. Wan, B. Mai and P. K. Wong, *Bioresour. Technol.*, 2011, **102**, 9148.
104. L. Zu, G. Li, T. An and P. K. Wong, *Bioresour. Technol.*, 2012, **110**, 153.
105. L. K. Lee, C. Ding, K. L. Yang and J. He, *Environ. Sci. Technol.*, 2011, **45**, 8475.
106. Y. M. Kim, K. Murugesan, Y. Y. Chang, E. J. Kim and Y. S. Chang, *J. Chem. Technol. Biotechnol.*, 2012, **87**, 216.
107. N. Saito and K. Yamaguchi, in *Chromatographic Analysis of Environmental and Food Toxicants*, ed. T. Shibamoto, CRC Press, Boca Raton, FL, 1998, pp. 169–212.
108. R. Winkler and C. Hertweck, *ChemBioChem*, 2007, **8**, 973.
109. K. S. Ju and R. E. Parales, *Microbiol. Mol. Biol. Rev.*, 2010, **74**, 250.
110. A. Estevez-Nuñez, A. Caballero and J. L. Ramos, *Mol. Biol. Rev.*, 2001, **65**, 335.

111. R. Boopathy and C. F. Kulpa, *Curr. Microbiol.*, 1992, **25**, 235.
112. B. Z. Li, X. Y. Xu and L. J. Zhu, *J. Zhejiang Univ., Sci. B*, 2010, **11**, 177.
113. J. F. Wu, C. Y. Jiang, B. J. Wang, Y. F. Ma, Z. P. Liu and S. J. Liu, *Appl. Environ. Microbiol.*, 2006, **72**, 1759.
114. D. Zhen, H. Liu, S. J. Wang, J. J. Zhang, F. Zhao and N. Y. Zhou, *Appl. Microbiol. Biotechnol.*, 2006, **72**, 797.
115. H. Z. Wu, C. H. Wei, Y. Q. Wang, Q. C. He and S. Z. Liang, *J. Environ. Sci. China*, 2009, **21**, 89.
116. P. K. Arora, C. Sasikala and C. V. Ramana, *Appl. Microbiol. Biotechnol.*, 2012, **93**(6), 2265.
117. K. Takagi, A. Iwasaki, I. Kamei, K. Satsuma, Y. Yoshioka and N. Harada, *Appl. Environ. Microbiol.*, 2009, **75**, 4452.
118. USEPA, *Perchlorate* [online], US Environmental Protection Agency, Washington DC, 2012, available at: http://water.epa.gov/drink/contaminants/unregulated/perchlorate.cfm.
119. E. Cox, J. Allan, S. L. Neville, presented at the Division of Environmental Chemistry American Chemical Society, New Orleans 22–26 August, 1996. Extended abstract Vol. 39, No. 2, 102–104.
120. J. D. Coates, U. Michaelidou, R. A. Bruce, S. M. O'Connor, J. N. Crespi and L. A. Achenbach, *Appl. Environ. Microbiol.*, 1999, **65**, 5234.
121. M. Balk, T. van Gelder, S. A. Weelink and A. J. M. Stams, *Appl. Environ. Microbiol.*, 2008, **74**, 403.
122. M. Balk, F. Mehboob, A. H. van Gelder, W. I. C. Rijpstra, J. S. Sinninghe-Damsté and A. J. M. Stams, *Appl Microbiol. Biotechnol.*, 2010, **88**, 595.
123. E. E. Cox, E. A. Edwards and S. Neville, in *Perchlorate in the Environment* ed. E. T. Urbansky, Kluwer Academic, New York, 2000, pp. 231–240.
124. National Science and Technology Council, Committee on Environment and Natural Resources, *Interagency Assessment of Oxygenated Fuels*, Office of Science & Technology Policy, Executive Office of the President, Washington, DC, 1997, ch. 2, pp. 231–240.
125. L. C. Davis and L. E. Erickson, *Environ. Prog.*, 2004, **23**, 243.
126. J. R. Hanson, C. E. Ackerman and K. M. Scow, *Appl. Environ. Microbiol.*, 1999, **65**, 4788.
127. P. B. Hatzinger, K. McClay, S. Vainberg, M. Tugusheva, C. W. Condee and R. J. Steffan, *Appl. Environ. Microbiol.*, 2001, **67**, 5601.
128. A. François, H. Mathis, D. Godefroy, P. Piveteau, F. Fayolle and F. Monot, *Appl. Environ. Microbiol.*, 2002, **68**(2), 2754.
129. M. M. Häggblom, L. K. G. Youngster, P. Somsamak and H. H. Richnow, *Adv. Appl. Microbiol.*, 2007, **62**, 1.
130. USEPA, *MtBE Treatment Profiles* [online], US Environmental Protection Agency, Technology Innovation and Field Services Division, Washington DC, available at: www.clu-in.org/products/mtbe/usersearch/mtbe_search.cfm.
131. US Geological Society, http://toxics.usgs.gov/regional/emc/ [last accessed 22 August 2013].

132. K. Fischer, E. Fries, W. Körner, C. Schmalz and C. Zwiener, *Appl. Microbiol. Biotechnol.*, 2011, **94**, 11.
133. T. Yoshimoto, *Appl. Environ. Microbiol.*, 2004, **70**, 5283.
134. Z. Yu and W. Huang, *Environ. Sci. Technol.*, 2005, **39**, 4878.
135. Hughes, B. R. Clark and C. D. Murphy, *Biodegradation*, 2011, **22**, 741.
136. F. E. Löffler, R. Sohn and R. Song, *Final Report: Microbial Transformation of Fluorinated Environmental Pollutants*, EPA 5535, US Environmental Protection Agency, Washington DC, 2004.
137. F. E. Löffler, R. Sohn and R. Song, *2003 Progress Report: Microbial Transformation of Fluorinated Environmental Pollutants*, EPA 5535, US Environmental Protection Agency, Washington DC 2003.
138. F. A. Caliman, B. M. Robu, C. Smaranda, V. L. Pavel and M. Gavrilescu, *Clean Technol. Environ. Policy*, 2011, **13**, 241.
139. C. N. Mulligan, R. N. Yong and B. F. Gibbs, *Eng. Geol.*, 2001, **60**, 193.
140. Member Agencies of the Federal Remediation Technologies Roundtable, *Abstracts of Remediation Case Studies*, EPA 542-R-00-006, US Environmental Protection Agency, Washington DC, 2000, vol. 4.
141. ESTCP 2005, Bioaugmentation for remediation of chlorinated solvents: technology development, status and research needs clu-in.org/download/remed/Bioaug2005.pdf? [last accessed 30 September 2013].
142. USEPA, *Use of Monitored Natural Attenuation at Superfund, RCRA Corrective Action, and Underground Storage Tank Sites*, US Environmental Protection Agency, Washington DC, 1999.
143. E. F. Neuhauser, J. A. Ripp, N. A. Azzolina, E. L. Madsen, D. M. Mauro and T. Taylor, *Ground Water Monit. Remed.*, 2009, **29**(3), 66.
144. M. E. Witt, G. M. Klecka, E. J. Lutz, T. A. Ei, N. R. Grosso and F. H. Chapelle, *J. Contam. Hydrol.*, 2002, **57**, 61.
145. J. L. Rayner, I. Snape, J. L. Walworth, P. M. A. Harvey and S. H. Ferguson, *Cold Reg. Sci. Technol.*, 2007, **48**, 139.
146. J. M. Lambert, T. Yang, N. R. Thomson and J. F. Barker, *Int. J. Soil Sediment Water*, 2009, **2**(3), Article 6.
147. N. Kabelitz, J. MacHackova, G. Imfeld, M. Brennerova, D. H. Pieper, H. J. Heipieper and H. Junca, *Appl. Microbiol. Biotechnol.*, 2009, **82**, 565.
148. W. L. Straube, C. C. Nestler, L. D. Hansen, D. Ringleberg, P. H. Pritchard and J. Jones-Meehan, *Acta Biotechnol.*, 2003, **23**(2–3), 179.
149. USEPA, *A Citizen's Guide to Permeable Reactive Barriers*, EPA 542-F-01-005, US Environmental Protection Agency, Washington DC, 2001.
150. ITRC Permeable Reactive Barriers Team, *Technical/Regulatory Guidelines. Permeable Reactive Barriers: Lessons Learned/New Directions*, Interstate Technology & Regulatory Council, Washington DC, 2005.
151. P. C. Johnson, K. D. Miller and C. L. Bruce, *A Practical Approach to the Design, Monitoring, and Optimization of In Situ MTBE Aerobic Biobarriers*, Technical Report TR-2257-ENV, Naval Facilities Engineering Council, Engineering Service Center, Port Hueneme, CA, 2004.

152. M. Craig, *Technology News and Trends*, 2004, US EPA Newsletter February Issue www.clu-in.org/products/newsletters/tnandt/#archives [last accessed 30 September 2013].

153. USEPA, *How To Evaluate Alternative Cleanup Technologies For Underground Storage Tank Sites: A Guide For Corrective Action Plan Reviewers*, US Environmental Protection Agency, Washington DC, 1994, ch. 5 [Available at: http://www.epa.gov/oust/pubs/tum_ch5.pdf last accessed 25 September 2013].

154. M. A. Line, C. D. Garland and M. Crowley, *Waste Manage.*, 1996, **16**, 567.

155. K. McCarthy, L. Walker, L. Vigoren and J. Bart., *Cold Region Sci. Technol.*, 2004, **40**, 31.

156. K. Paudyn, A. Rutter, K. Rowe and J. S. Poland, *Cold Region Sci. Technol.*, 2008, **53**, 102.

157. USEPA, *How To Evaluate Alternative Cleanup Technologies For Underground Storage Tank Sites: A Guide For Corrective Action Plan Reviewers*, US Environmental Protection Agency, Washington DC, 1994, ch. 4.

158. J. W. Quinn and D. R. Reinhart, *Pract. Period. Hazard. Toxic, Radioact. Waste Manage.*, 1997, **1**(1), 18.

159. D. Scanscartier, B. Zeeb, I. Koch and K. Reimer, *Cold Region Sci. Technol.*, 2009, **55**, 167.

160. M. Genovese, R. Denaro, S. Cappello, G. Di Marco, G. La Spada, L. Giuliano, L. Genovese and M. M. Yakimov, *J. Appl. Microbiol.*, 2008, **105**, 1694.

161. H. Craig and W. Sisk, *The Composting Alternative to Incineration of Explosives Contaminated Soils*, EPA Tech Trends, November 5, 1998.

162. J. G. Stefanoff and M. B. García, Jr., *Environ. Prog.*, 1995, **14**, 104.

163. S. B. Funk, *Appl. Biochem. Biotechnol.*, 1995, **51**, 625.

164. J. Gerritse, V. Renard, J. Visser and J. C. Gottschal., *Appl. Microbiol. Biotechnol.*, 1995, **43**, 920.

165. S. A. B. Weelink, M. H. A. van Eekert and A. J. M. Stams, Degradation of BTEX by anaerobic bacteria: physiology and application, *Rev. Environ. Sci. Biotechnol.*, 2010, **9**, 359.

166. M. D. Mikesell, J. J. Kukor and R. H. Olsen, *Biodegradation*, 1993, **4**, 249–259.

167. A. K. Haritash and C. P. Kaushik, *J. Hazard. Mater.*, 2009, **169**, 1.

168. W. L. O'Niell and V. A. Nzengung, *Environ. Res. Eng. Manage.*, 2004, **4**(30), 49.

169. USEPA, Biosparging, 1994. http://www.epa.gov/oust/pubs/tum_ch8.pdf (last accessed 23 Aug 2013).

170. USEPA 1994. http://www.epa.gov/oust/pubs/tum_ch3.pdf (last accessed 26 September 2013).

171. E. Cox, E. A. Edwards, S. Neville, M. Girard. Cost-Effective Bioremediation of perchlorate in soil and groundwater, Geosyntec Consultants. http://www.denix.osd.mil/edqw/upload/GEOSYNTEC_BIO_P2.PDF

[Available from: DoD Environment, Safety and Occupational Health Network and Information Exchange, last accessed 25 September 2013].

172. M. Megharaj, B. Ramakrishnan, K. Venkateswarlu, N. Sethunathan and R. Naidu, *Environ. Int.*, 2011, **8**, 1362.

173. H. Harms, D. Schlosser and L. Y. Wick, *Nat. Rev. Microbiol.*, 2011, **3**, 177.

174. X. Y. Lu, T. Zhang and H. H. Fang, *Appl. Microbiol. Biotechnol.*, 2011, **5**, 1357.

175. A. Srinivasan and T. Viraraghavan, *J. Environ. Manage.*, 2010, **10**, 1915.

176. F. Fernández-Luqueño, C. Valenzuela-Encinas, R. Marsch, C. Martínez-Suárez, E. Vázquez-Núñez and L. Dendooven, *Environ. Sci. Pollut. Res. Int.*, 2011, **1**, 12.

177. R. Lal, G. Pandey, P. Sharma, K. Kumari, S. Malhotra, R. Pandey, V. Raina, H. P. Kohler, C. Holliger, C. Jackson and J. G. Oakeshott, *Microbiol. Mol. Biol. Rev.*, 2010, **1**, 58.

178. C. Novotný, T. Cajthaml, K. Svobodová, M. Susla and V. Sasek, *Folia Microbiol. (Praha)*, 2009, **5**, 375.

179. A. L. Leitão, *Int. J. Environ. Res. Public. Health*, 2009, **4**, 1393.

180. S. Pointing, *Appl. Microbiol. Biotechnol.*, 2001, **57**, 20.

181. G. M. Gadd, *Mycol. Res.*, 2007, **111**, 3.

182. I. Singleton, in *Fungi in Bioremediation*, ed. G. M. Gadd, Cambridge University Press, Cambridge, 2001, pp. 79–96.

183. J. S. Knapp, E. J. Vantoch-Wood and F. Zhang, in *Fungi in Bioremediation*, ed. G. M. Gadd, Cambridge University Press, Cambridge, 2001, pp. 242–304.

184. M. Barclay and C. J. Knowles, in *Fungi in Bioremediation*, ed. G. M. Gadd, Cambridge University Press, Cambridge, 2001, pp. 335–358.

185. R. Cohen and Y. Hadar, in *Fungi in Bioremediation*, ed. G. M. Gadd, Cambridge University Press, Cambridge, 2001, pp. 305–334.

186. R. G. Harvey, *Polycyclic Aromatic Hydrocarbons: Chemistry and Carcinogenicity*, Cambridge University Press, Cambridge, 1991.

187. T. N. P. Bosma, P. J. M. Middeldorp, G. Schraa and A. J. B. Zehnder, *Environ. Sci. Technol.*, 1997, **31**, 248.

188. P. Patnaik, *A Comprehensive Guide to the Hazardous Properties of Chemical Substances*, Von Nostrand Reinhold, London and New York, 1992, p. 425.

189. C. E. Cerniglia, *J. Ind. Microbiol. Biotechnol.*, 1997, **19**, 324.

190. K. T. Steffen, S. Schubert, M. Tuomela, A. Hatakka and M. Hofrichter, *Biodegradation*, 2007, **18**, 359.

191. C. E. Cerniglia and J. B Sutherland, in *Fungi in Bioremediation*, ed. G. M. Gadd, Cambridge University Press, Cambridge, 2001, pp. 136–187.

192. S. M. Bamforth and I. Singleton, *J. Chem. Technol. Biotechnol.*, 2005, **80**, 723.

193. R. H. Peng, A. S. Xiong, Y. Xue, X. Y. Fu, F. Gao, W. Zhao, Y. S. Tian and Q. H. Yao, *FEMS Microbiol. Rev.*, 2008, **6**, 927.

194. X. Wang, Z. Q. Gong, P. J. Li, L. H. Zhang and X. M. Hu, *Environ. Eng. Sci.*, 2008, **25**, 677.

195. J. C. Colombo, M. Cabello and A. M. Arambarri, *Environ. Pollut.*, 1996, **94**, 355.

196. L. J. Pinto and M. M. Moore, *Environ. Toxicol. Chem.*, 2000, **19**, 1741.

197. A. Saraswathy and R. Hallberg, *FEMS Microbiol. Lett.*, 2002, **2**, 227.

198. S. Chulalaksananukul, G. M. Gadd, P. Sangvanich, P. Sihanonth, J. Piapukiew and A. S. Vangnai, *FEMS Microbiol. Lett.*, 2006, **1**, 99.

199. V. Sasek, T. Cajthaml and M. Bhatt, *Water Air Soil Pollut. Focus*, 2002, **3**, 5.

200. V. Sasek, M. Bhatt, T. Cajthaml, K. Malachová and D. Lednická, *Arch. Environ. Contam. Toxicol.*, 2003, **3**, 336.

201. T. Cajthaml, M. Moder, P. Kacer, V. Sasek and P. Popp, *J. Chromatogr. A*, 2002, **974**, 213.

202. A. Arun and M. Eyini, *Bioresour. Technol.*, 2011, **102**, 8063.

203. J. D. Kim and C. G. Lee, *Biotechnol. Bioprocess Eng.*, 2007, **12**, 410.

204. F. Acevedo, L. Pizzul, M. D. P. Castillo, R. Cuevas and M. C. Diez, *J. Hazard. Mater.*, 2011, **185**, 212.

205. S. Gan, E. V. Lau and H. K. Ng, *J. Hazard. Mater.*, 2009, **172**, 532.

206. V. Leonardi, V. Sasek, M. Petruccioli, A. D'Annibale, P. Erbanova and T. Cajthaml, *Int. Biodeterior. Biodegrad.*, 2007, **60**, 165.

207. O. Potin, E. Veignie and C. Rafin, *FEMS Microbiol. Ecol.*, 2004, **51**, 71.

208. X. Li, P. Li, X. Lin, C. Zhang, Q. Li and Z. Gong, *J. Hazard. Mater.*, 2008, **150**, 21.

209. K. L. Lau, Y. Y. Tsang and S. W. Chiu, *Chemosphere*, 2003, **52**, 1539.

210. S. Harayama, *Curr. Opin. Biotechnol.*, 1997, **8**, 268.

211. H. H. Richnow, R. Seifert, M. Kastner, B. Mahro, B. Horsfield, U. Tiedgen, S. Bohm and W. Michaelis, *Chemosphere*, 1995, **8**, 3991.

212. L. Valentin, T. A. Lu-Chau, C. Lopez, G. Feijoo, M. T. Moreira and J. M. Lema, *Process Biochem.*, 2007, **42**, 641.

213. N. R. Nascimento, S. M. C. Nicola, M. O. O. Rezende, T. A. Oliveira and G. Öberg, *Geoderma*, 2004, **121**, 221.

214. R. Szewczyk and J. Dlugonski, *Int. Biodeterior. Biodegrad.*, 2009, **63**, 123.

215. D. G. Crosby, *Pure Appl. Chem.*, 1981, **53**, 1051.

216. M. Czaplicka, *Sci. Total Environ.*, 2004, **322**, 21.

217. R. S. Chhabra, R. M. Maronpot, J. R. Bucher, J. K. Haseman, J. D. Toft and M. R. Hejtmancik, *Toxicol. Sci.*, 1999, **48**, 14.

218. A. T. Proudfoot, *Toxicol. Rev.*, 2003, **22**, 3.

219. J. Michałowicz, M. Posmyk and W. Duda, *J. Plant Physiol.*, 2009, **166**, 559.

220. J. Michałowicz, H. Urbanek, B. Bukowska and W. Duda, *Biol. Plant.*, 2010, **54**, 597.

221. O. Rubilar, G. Feijoo, C. Diez, T. A. Lu-Chau, M. T. Moreira and J. M. Lema, *Ind. Eng. Chem. Res.*, 2007, **46**, 6744.
222. G. R. Tortella, O. Rubilar, L. Gianfreda, E. Valenzuela and M. C. Diez, *World J. Microbiol. Biotechnol.*, 2008, **24**, 2805.
223. G. V. B. Reddy and M. H. Gold, *Microbiology*, 2000, **146**, 405.
224. X. Y. Jiang, G. M. Zeng, D. L. Huang, Y. Chen, F. Liu, G. H. Huang and H. L. Liu, *World J. Microbiol. Biotechnol.*, 2006, **22**, 909.
225. A. Mendoza-Cantú, A. Albores, L. Fernández-Linares and R. Rodriguez-Vázquez, *Environ. Toxicol.*, 2000, **15**, 107.
226. R. T. Lamar, *Soil Biol. Biochem.*, 1994, **26**, 603.
227. R. T. Lamar, J. W. Evans and J. A. Glaser, *Environ. Sci. Technol.*, 1993, **27**, 566.
228. J. A. Glaser and R. T. Lamar, in *Bioremediation: Science and Applications*, SSSA Special Publication 43, Soil Science Society of America, Madison, WI, 1995, pp. 117–133.
229. O. Rubilar, G. Tortella, M. Cea, F. Acevedo, M. Bustamante, L. Gianfreda and M. C. Diez, *Biodegradation*, 2011, **22**, 31.
230. S. W. Chui, M. L. Ching, K. L. Fong and D. Moore, *Mycol. Res.*, 1998, **102**, 1553.
231. W. Law, W. Lau, K. Lo, L. Wai and S. Chiu, *Chemosphere*, 2003, **52**, 1531.
232. J. A. Field and R. Sierra-Alvarez, *Environ. Pollut.*, 2008, **155**, 1.
233. P. De-voogt and J. C. Klamer, *Bull. Environ. Contam. Toxicol.*, 1984, **1**, 45.
234. J. S. Yadav, J. F. Quensen, J. M. Tiedje and C. A. Reddy, *Appl. Environ. Microbiol.*, 1995, **7**, 2560.
235. Y. Yin, J. Guo, Z. Li, T. Li and X. Wang, *World J. Microbiol. Biotechnol.*, 2011, **27**, 2567.
236. J. A. Bumpus, M. Tien, D. Wright and S. D. Aust, *Science*, 1985, **228**, 1434.
237. D. Eaton, *Enzyme Microbiol. Technol.*, 1985, **7**, 194.
238. B. R. M. Vyas, V. Sasek, M. Matucha and M. Bubner, *Chemosphere*, 1994, **28**, 1127.
239. V. Sasek, O. Volfova, P. Erbanova, B. R. M. Vyas and M. Matucha, *Biotechnol. Lett.*, 1993, **15**, 521.
240. D. R. Thomas, K. Carlswell and G. Georgiou, *Biotechnol. Bioeng.*, 1992, **40**, 1395.
241. A. Zeddel, A. Majcherczyk and A. Huttermann, *Toxicol. Environ. Chem.*, 1993, **40**, 255.
242. D. Dietrich, W. J. Hickey and R. Lamar, *Appl. Environ. Microbiol.*, 1995, **61**, 3904.
243. P. K. Donnelly and J. S. Fletcher, *Bull. Environ. Contam. Toxicol.*, 1995, **54**, 507.
244. M. A. Murado, M. C. Tejedor and G. Baluja, *Bull. Environ. Contam. Toxicol.*, 1976, **15**, 768.
245. R. H. Dodge, C. E. Cerniglia and D. T. Gibson, *Biochem. J.*, 1979, **178**, 223.
246. S. Dmochewitz and K. Ballschmiter, *Chemosphere*, 1988, **17**, 111.

247. L. A. Beaudette, S. Davies, P. M. Fedorak, O. P. Ward and M. A. Pickard, *Appl. Environ. Microbiol.*, 1998, **64**, 2020.
248. I. Kamei, R. Kogura and R. Kondo, *Appl. Microbiol. Biotechnol.*, 2006, **72**, 566.
249. I. Kamei, S. Sonoki, K. Haraguchi and R. Kondo, *Appl. Microbiol. Biotechnol.*, 2006, **73**, 932.
250. M. Cvančarová, Z. Křesinová, A. Filipová, S. Covino and T. Cajthaml, *Chemosphere*, 2012, **11**, 1317.
251. A. Kubatova, P. Erbanova, I. Eichlerova, L. Homolka, F. Nerud and V. Sasek, *Chemosphere*, 2001, **43**, 207.
252. G. M. Ruiz-Aguilar, J. M. Fernández-Sánchez, R. Rodríguez-Vázquez and H. Poggi-Varaldo, *Adv. Environ. Res.*, 2002, **6**, 559.
253. C. Novotny, K. Svobodova, P. Erbanova, T. Cajthamla, A. Kasinatha, E. Lang and V. Sasek, *Soil Biol. Biochem.*, 2004, **36**, 1545.
254. E. Federici, M. Giubilei, G. Santi, G. Zanaroli, A. Negroni, F. Fava and A. D'Annibale, *Microb. Cell Fact.*, 2012, **11**, 35.
255. F. C. Michel, J. Quensen and C. A. Reddy, *Compost Sci. Util.*, 2001, **4**, 274.
256. E. Roca, E. D'Errico, A. Izzo, S. Strumia, A. Esposito and A. Fiorentino, *Int. Biodeterior. Biodegrad.*, 2009, **63**, 182.
257. N. Pino and G. Penuela, *Int. Biodeterior. Biodegrad.*, 2011, **65**, 827.
258. M. Köck-Schulmeyer, A. Ginebreda, S. González, J. L. Cortina, M. López de Alda and D. Barceló, *Chemosphere*, 2012, **86**, 8.
259. M. C. Diez, *J. Soil. Sci. Plant. Nutr.*, 2010, **10**, 244.
260. D. W. Kennedy, S. D. Aust and J. A. Bumpus, *Appl. Environ. Microbiol.*, 1990, **56**, 2347.
261. C. Mougin, C. Pericaud, J. Dubroca and M. Asther, *Soil Biol.Biochem.*, 1997, **29**, 1321.
262. B. Singh and R. Kuhad, *Pest Manag. Sci.*, 2000, **56**, 142.
263. J. C. Quintero, T. Lu-Chau, M. T. Moreira, G. Feijoo and J. M. Lema, *Int. Biodeterior. Biodegrad.*, 2007, **60**, 319.
264. J. C. Quintero, M. T. Moreira, G. Feijoo and J. M. Lema, *Cien. Inv. Agr.*, 2008, **35**, 159.
265. G. Kulshrestha, S. B. Singh, S. Lal and N. T. Yaduraju, *Pest Manage. Sci.*, 2000, **56**, 202.
266. J. Mitra, P. K. Mukherjee, S. P. Kale and N. B. K. Murthy, *Biodegradation*, 2001, **12**, 235.
267. Y. Zhao, X. Yi, M. Li, L. Liu and W. Ma, *Chin. J. Chem. Eng.*, 2010, **18**, 486.
268. A. S. Purnomo, T. Mori, K. Takagi and R. Kondo, *Int. Biodeterior. Biodegrad.*, 2011, **65**, 691.
269. T. S. Bhalerao and P. R. Puranik, *Int. Biodeterior. Biodegrad.*, 2009, **63**, 503.
270. Y. H. Liu, Y. C. Chung and Y. Xiong, *Appl. Environ. Microbiol.*, 2001, **67**, 3746.
271. Q. Zhang, Y. Liu and Y. H. Liu, *FEMS Microbiol. Lett.*, 2003, **228**, 39.

272. W. Q. Liang, Z. Y. Wang, H. Li, P. C. Wu, J. M. Hu and N. Luo, *J. Agric. Food Chem.*, 2005, **53**, 7415.

273. T. S. Bhalerao and P. R. Puranik, *Int. Biodeterior. Biodegrad.*, 2007, **59**, 315.

274. A. P. Pinto, C. Serrano, T. Pires, E. Mestrinho, L. Dias, D. M. Teixeira and A. T. Caldeira, *Sci. Total Environ.*, 2012, **435**, 402.

275. O. P. Ahlawat, P. Gupta, S. Kumar, D. K. Sharma and K. Ahlawat, *Ind. J. Microbiol.*, 2010, **50**, 390.

276. R. A. Córdova Juarez, L. L. Gordillo Dorry, R. Bello-Mendoza and J. E. Sánchez, *J. Environ. Manage.*, 2011, **92**, 948.

277. R. Gonzalez Matute, D. Figlas, G. Mockel and N. Curvetto, *Bioremed. J*, 2012, **16**, 31.

278. C. W. Phan and V. Sabaratnam, *Appl. Microbiol. Biotechnol.*, 2012, **96**, 863.

279. T. M. Phillips, A. G. Seech, H. Lee and J. T. Trevors, *Biodegradation*, 2005, **16**, 363.

280. A. K. Johri, M. Dua, R. Tuteja, D. M. Saxena and R. Lal, *FEMS Microbiol. Rev.*, 1996, **537**, 1.

281. G. M. Gadd, *Fungi in Bioremediation*, Cambridge University Press, Cambridge, 2001.

282. C. Mougin, C. Pericaud, C. Mollosse, C. Laugero and M. Asther, *Pestic. Sci.*, 1996, **47**, 51.

283. B. K. Singh and R. C. Kuhad, *Lett. Appl. Microbiol.*, 1999, **28**, 238.

284. M. Tekere, I. Ncube, J. Read and R. Zvauya, *Environ. Technol.*, 2002, **23**, 99.

285. X. Fujian, C. Hongzhang and L. Zuohu, *Bioresour. Technol.*, 2001, **80**, 149.

286. T. P. Ryan and J. A. Bumpus, *Appl. Microbiol. Biotechnol.*, 1989, **31**, 302.

287. J. A. Bumpus, F. Tudor, M. A. Jurek and S. D. Aust, in *Biotechnology Applications in Hazardous Waste Treatment*, Engineering Foundation, New York, 1988, pp. 363–381.

288. M. Arisoy, *Bull. Environ. Contam. Toxicol.*, 1998, **60**, 872.

289. B. E. Andersson and T. Henrysson, *Appl. Microbiol. Biotechnol.*, 1996, **46**, 647.

290. J. C. Quintero, G. Feijoo and J. M. Lema, *Vitae*, 2006, **13**, 61.

291. M. Castillo, A. Andersson, P. Ander, J. Stenström and L. Torstensson, *World J. Microbiol. Biotechnol.*, 2001, **17**, 627.

292. A. S. Purnomo, I. Kamei and R. Kondo, *J. Biosci. Bioeng.*, 2008, **105**, 614.

293. A. S. Purnomo, F. Koyama, T. Mori and R. Kondo, *Chemosphere*, 2010, **80**, 619.

294. A. S. Purnomo, T. Mori, I. Kamei and R. Kondo, *Int. Biodeterior. Biodegrad.*, 2011, **65**, 921.

295. A. S. Purnomo, T. Mori, I. Kamei, T. Nishii and R. Kondo, *Int. Biodeterior. Biodegrad.*, 2010, **64**, 397.

296. A. S. Purnomo, T. Mori and R. Kondo, *Int. Biodeterior. Biodegrad.*, 2010, **64**, 560.

297. B. Volesky, in *Biosorption of Heavy Metals*, CRC Press, Boca Raton, FL, 1990, pp. 3–43.
298. E. Fourest and J. Roux, *Microbiol. Biotechnol.*, 1992, **37**, 399.
299. D. Bagot, T. Lebeau and K. Jezequel, *Environ. Chem. Lett.*, 2006, **4**, 207.
300. G. Massaccesi, M. C. Romero, M. C. Cazau and A. M. Bucsinszky, *World J. Microbiol. Biotechnol.*, 2002, **18**, 817.
301. L. Ezzouhri, E. Castro, M. Moya, F. Espinola and K. Lairini, *African J. Microbiol. Res.*, 2009, **3**, 35.
302. Z. Deng, L. Cao, H. Huang, X. Jiang, W. Wang, Y. Shi and R. Zhang, *J. Hazard. Mater.*, 2011, **185**, 717.
303. M. A. Garcia, J. Alonso and M. J. Melgar, *J. Chem. Technol. Biotechnol.*, 2005, **80**, 325.
304. M. Kacprzak and G. Malina, *Can. J. Soil Sci.*, 2005, **85**, 283.
305. T. Fukuda, Y. Ishino, A. Ogawa, K. Tsutsumi and H. Morita, *J. Gen. Appl. Microbiol.*, 2008, **54**, 295.
306. S. Maheswari and A. G. Murugesan, *Environ. Technol.*, 2009, **30**, 921.
307. A. Khalid, S. Batool, M. T. Siddique, Z. H. Nazli, R. Bibi, S. Mahmood and M. Arshad, *Soil Environ.*, 2011, **30**, 1.
308. C. Jayasinghe, A. Imtiaj, G. W. Lee, K. H. Im, H. Hur, M. W. Lee, H. S. Yang and T. S. Lee, *Mycobiology*, 2008, **36**, 114.
309. N. Kousar and M. A. Charya, *Ind. J. Environ. Health*, 2002, **44**, 65.
310. A. Anastasi, V. Prigione and G. C. Varese, *J. Hazard. Mater.*, 2010, **177**, 260.
311. C. Junghanns, G. Krauss and D. Schlosser, *Bioresour. Technol.*, 2008, **99**, 1225.
312. G. Bognolo, *Colloids Surf.*, 1999, **152**, 41.
313. N. Christofi and I. B. Ivshina, *J. Appl. Microbiol.*, 2002, **93**, 915.
314. S. Boonchan, M. L. Britz and G. A., *Biotechnol. Bioeng.*, 1998, **59**, 482.
315. A. R. Johnsen and U. Karlson, *Appl. Microbiol. Biotechnol.*, 2004, **63**, 452.
316. K. H. Shin, K. W. Kim and E. A. Seagren, *Appl. Microbiol. Biotechnol.*, 2004, **65**, 336.
317. B. Zhao, L. Zhu, W. Li and B. Chen, *Chemosphere*, 2005, **58**, 33.
318. M. J. Rosen, *Surfactants and Interfacial Phenomena*, John Wiley & Sons, New York, 1978.
319. P. Becher, *Emulsions, Theory and Practice*, Reinhold Publishing, New York, 2nd edn, 1965.
320. M. P. Plociniczak, G. A. Plaza, Z. P. Seget and S. S. Cameotra, *Inter. J. Mol. Sci.*, 2011, **12**, 633.
321. R. M. Miller and R. Bartha, *Appl. Environ. Microbiol.*, 1989, **55**, 269.
322. C. N. Mulligan, *Environ. Pollut.*, 2005, **133**, 183.
323. A. Singh, J. D. Van Hamme and O. P. Ward, *Biotechnol. Adv.*, 2007, **25**, 99.
324. C. N. Mulligan, *Colloid Interface Sci.*, 2009, **14**, 372.

325. E. Rosenberg and E. Z. Ron, *Appl. Microbiol. Biotechnol.*, 1999, **52**, 154.
326. C. Calvo, M. Manzanera, G. A. Silva-Castro, I. Uad and J. Gonzalez-Lopez, *Sci. Total Environ.*, 2009, **407**, 3634.
327. S. Lang and D. Wullbrandt, *Appl. Microbiol. Biotechnol.*, 1999, **51**, 22.
328. E. Rosenberg, A. Zuckerberg, C. Rubinovitz and D. L. Gutnick, *Appl. Environ. Microbiol.*, 1979, **37**, 402.
329. M. Robert, M. E. Mercade, M. P. Bosch, J. L. Parra, M. J. Espiny, M. A. Manresa and J. Guinea, *Biotechnol. Lett.*, 1989, **11**, 871.
330. K. Hitsatsuka, T. Nakahara, N. Sano and K. Yamada, *Agri. Biol. Chem.*, 1971, **35**, 686.
331. L. H. Guerra-Santos, O. Kappeli and A. Fiechter, *Appl. Environ. Microbiol.*, 1984, **48**, 301.
332. A. Abalos, A. Pinaso, M. R. Infante, M. Casals, F. Garcia and A. Manresa, *Langmuir*, 2001, **17**, 1367.
333. A. S. Nayak, M. H. Vijaykumar and T. B. Karegoudar, *Int. Biodeterior. Biodegrad.*, 2009, **63**, 73.
334. D. G. Cooper and D. A. Paddock, *Appl. Environ. Microbiol.*, 1984, **47**, 173.
335. A. Kakinuma, A. Oachida, T. Shima, H. Sugino, M. Isano, G. Tamura and K. Arima, *Agri. Biol. Chem.*, 1969, **33**, 1669.
336. J. M. Bonmatin, M. Genest, H. Labbe, I. Grangemard, F. Peypoux, R. Maget-Dana, M. Ptak and G. Michel, *Lett. Peptide Sci.*, 1995, **2**, 41.
337. Y. Ishigami, M. Osman, H. Nakahara, Y. Sano, R. Ishiguro and M. Matusumoto, *Colloids Surf. B*, 1995, **4**, 341.
338. V. Somayeh, A. S. Abbas and A. S. Nouhi, *J. Bacteriol.*, 2008, **1365**, 678.
339. H. T. Wattanaphon, A. Kerdsin, C. Thammacharaoen, P. Sangvanich and A. S. Vangnai, *J. Appl. Microbiol.*, 2008, **105**, 416.
340. M. P. Pirollo, A. P. Mariano, R. B. Lovaglio, S. Costa, V. Walter, R. Hausmann and J. Contiero, *J. Appl. Microbiol.*, 2008, **105**, 1484.
341. F. Occulti, G. C. Roda, S. Berselli and F. Fava, *Biotech. Bioeng.*, 2008, **99**, 1525.
342. J. J. Hong, S. M. Yang, C. H. Lee, Y. K. Choi and T. Kajiuchi, *J. Colloid Interface Sci.*, 1998, **202**, 63.
343. G. Nakhla, M. Al-Sabawi, A. Bassi and V. Liu, *J. Hazard. Mater. B*, 2003, **102**, 243.
344. F. Volkering, A. M. Breure and W. H. Rulkens, *Biodegradation*, 1998, **8**, 401.
345. J. D. Van Hamme, A. Singh and O. P. Ward, *Microbiol. Mol. Biol. Rev.*, 2003, **67**, 503.
346. J. D. Van Hamme, A. Singh and O. P. Ward, *Biotechnol. Adv.*, 2006, **24**, 604.
347. S. Ito and S. Inoue, *Appl. Environ. Microbiol.*, 1982, **43**, 1278.
348. S. Ito and S. Inoue, *Agri. Biol. Chem.*, 1980, **44**, 2221.
349. Y. M. Zhang and R. M. Miller, *Appl. Environ. Microbiol.*, 1995, **61**, 2247.
350. C. N. Mulligan and B. F. Gibbs, *Proc. Ind. Natl. Sci. Acad.*, 2004, **1**, 31.

351. P. Singh and S. S. Cameotra, *Biochem. Biophy. Res. Commun.*, 2004, **319**, 291.

352. Y. Asci, M. Nurbas and Y. S. Acikel, *J. Hazard. Mater.*, 2008, **154**, 663.

353. N. Kosaric, *Food Technol. Biotechnol.*, 2001, **39**, 295.

354. R. S. Norman, R. Frontera-Suau and P. J. Morris, *Appl. Environ. Microbiol.*, 2002, **68**, 5096.

355. L. Wang, Y. Tang, S. Wang, R. L. Liu, M. Z. Liu, Y. Zhang, L.-F. Liang and L. Feng, *Extremophiles*, 2006, **10**, 347.

356. S. Harvey, I. Elashi, J. J. Valdes, D. Kamely and A. M. Chakrabarty, *Biotechnol.*, 1990, **8**, 228.

357. K. Scheibenbogen, R. G. Zytner, H. Lee and J. T. Trevors, *J. Chem. Technol. Biotechnol.*, 1994, **59**, 53.

358. P. Lafrance and M. Lapointe, *Ground Water Monitor. Remed.*, 1998, **18**, 139.

359. G. Bai, M. L. Brusseau and R. M. Miller, *J. Contam. Hydrol.*, 1998, **30**, 265.

360. K. Urum, T. Pekdemir and M. Gopur, *Transact. Instit. Chem. Eng.*, 2003, **81**, 203.

361. A. Shulga, E. Karpenko, R. Vildanova-Martishin, A. Turovsky and M. Soltys, *Adsorp. Sci. Technol.*, 2000, **18**, 171.

362. A. Oberbremer, R. Muller-Hurtig and F. Wagner, *Appl. Microbiol. Biotechnol.*, 1990, **32**, 485.

363. C. Schippers, K. Gessner, T. Muller and T. Scheper, *J. Biotechnol*, 2000, **83**, 189.

364. S. W. Kang, Y. B. Kim, J. D. Shin and E. K. Kim, *Appl. Biochem. Biotechnol.*, 2010, **160**, 780.

365. N. L. Olivera, M. G. Commendatore, A. C. Moran and J. L. Esteves, *J. Ind. Microbiol. Biotechnol.*, 2000, **25**, 70.

366. R. S. Makkar and S. S. Cameotra, *J. Am. Oil Chem. Soc.*, 1997, **74**, 887.

367. R. S. Makkar and S. S. Cameotra, *J. Ind. Microbiol. Biotechnol.*, 1997, **18**, 37.

368. S. A. Eliseev, R. Vildanova-Martisishin, A. Shulga, A. Shabo and A. Turovsky, *Microbiol. J.*, 1991, **53**, 61.

369. L. Deschenes, P. Lafrance, J. P. Villeneuve and R. Samson, presented at the Fourth Annual Symposium on Groundwater and Soil Remediation, 21–23 September, Calgary, Alberta, 1994.

370. G. Burd and O. P. Ward, *Biotechnol. Tech.*, 1996, **10**, 371.

371. C. Vipulanandan and X. Ren, *J. Environ. Eng.*, 2000, **126**, 629.

372. M. Garcia-Junco, E. De Olmedo and J. J. Ortego-Calvo, *Environ. Microbiol.*, 2001, **3**, 561.

373. K. S. Rahman, I. M. Banat, T. J. Rahman, T. Thayumanavan and P. Lakshmanaperumalsamy, *Bioresour. Technol.*, 2002, **81**, 25.

374. T. Barkay, S. Navon-Venezia, E. Z. Ron and E. Rosenberg, *Appl. Environ. Microbiol.*, 1999, **65**, 697.

375. K. G. Robinson, M. M. Ghosh and Z. Shi, *Water Sci. Technol.*, 1996, **34**, 303.

376. R. Fiebig, D. Schulze, J. C. Chung and S. T. Lee, *Biodegrad.*, 1997, **8**, 67.

377. J. C. Mata-Sandoval, J. Karns and A. Torrents, *J. Agri. Food Chem.*, 2001, **49**, 3296.

378. N. Awashti, A. Kumar, R. Makkar and S. Cameotra, *J. Environ. Sci. Health*, 1999, **34**, 793.

379. C. N. Mulligan, R. N. Yong and B. F. Gibbs, *Environ. Progress*, 1999, **18**, 50.

380. D. C. Herman, J. F. Artiola and R. M. Miller, *Environ. Sci. Technol.*, 1995, **29**, 2280.

381. H. Tan, J. T. Champion, J. F. Artiola, M. L. Brusseau and R. M. Miller, *Environ. Sci. Technol.*, 1994, **28**, 2402.

382. F. J. Ochoa-Loza, J. F. Artiola and R. M. Maier, *J. Environ. Qual.*, 2001, **30**, 479.

383. C. N. Mulligan, R. N. Yong and B. F. Gibbs, *J. Hazard. Mater.*, 2001, **85**, 111.

384. L. Fraser, *Environ. Health Perspect.*, 2000, **108**, 320.

385. J. W. Neilson, J. F. Artiola and R. M. Maier, *J. Environ. Qual.*, 2003, **32**, 899.

386. T. R. Sandrin, A. M. Chech and R. M. Maier, *Appl. Environ. Microbiol.*, 2000, **66**, 4585.

387. C. N. Mulligan and S. Wang, *Eng. Geol.*, 2006, **85**, 75.

388. Y. Asci, M. Nurbas and Y. S. Acikel, *J. Environ. Manage.*, 2010, **91**, 724.

389. L. Thimon, F. Peypoux and G. Michel, *Biotechnol. Lett.*, 1992, **14**, 713.

390. J. J. Hong, S. M. Yang, C. H. Lee, Y. K. Choi and T. Kajiuchi, *J. Colloid Interface Sci.*, 1998, **202**, 63.

391. K. J. Hong, S. Tokunaga and T. Kajiuchi, *Chemosphere*, 2002, **49**, 379.

392. W. L. Straube, C. C. Nestler, L. D. Hansen, D. Ringleberg, P. J. Pritchard and J. Jones-Meehan, *Acta Biotechnol.*, 2003, **2**, 179.

393. T. C. Lin, P. T. Pan, C. C. Young, J. S. Chang, T. C. Chang and S. S. Cheng, *Environ. Sci. Pollut. Res.*, 2011, **18**, 1487.

394. V. Kildisas, D. Levisauskas and S. Grigiskis, *Environ. Res. Eng. Manage.*, 2003, **25**, 87.

CHAPTER 5

Biological Waste Gas Treatments

PIERRE LE CLOIREC,*[a,b] ABDELTIF AMRANE,[a,b]
BENOIT ANET[a] AND CATHERINE COURIOL[a,b]

[a] Ecole Nationale Supérieure de Chimie de Rennes, CNRS, UMR 6226, 11
Allée de Beaulieu, CS 50837, 35708 Rennes Cedex 7, France; [b] Université
européenne de Bretagne, Boulevard Laennec, 35000 Rennes, France
*Email: Pierre.le-Cloirec@ensc-rennes.fr

5.1 Introduction

Human health and welfare (the population in general and plant operators) and
environmental protection (domestic and wild animals, plants, paintwork or
damage to buildings) are strong arguments for the development and use of new
and original processes to control waste gas emissions from agricultural, industrial or domestic activities. International treaties for environmental protection (Rio, Kyoto) have been transcribed and applied in many countries.
From these ratifications of international agreements, local legislation has been
written, particularly for solid waste management, water and wastewater
treatment, and air quality. Air pollution control regulations reflect the concern
of governments for the protection of people and the environment. Bravo-
Alvarez[1] mentioned two fundamental reasons for cleaning up the waste gas
stream: profit and protection. This is the case, for example, for the upgrading of
biogas, the cleaning of waste incinerator flue gas[2] or the treatment of industrial
process emissions.

RSC Green Chemistry No. 25
Biomass for Sustainable Applications: Pollution Remediation and Energy
Edited by Sarra Gaspard and Mohamed Chaker Ncibi
© The Royal Society of Chemistry 2014
Published by the Royal Society of Chemistry, www.rsc.org

According to the nature of the contaminants and/or the complex mixture of pollutants in the gaseous phase, their concentrations and the flow to be cleaned, removing non-particulate pollutants from a gas stream is achieved by different processes involving different mechanisms.[3–6] These processes can be classified into three categories:

- Thermal and/or catalytic oxidation, biological transformation
- Transfer into a liquid phase (absorption) or onto a solid phase (adsorption) with or without chemical reactions such as acid–base interaction, oxidation, complexation, physisorption or chemisorption
- Phase change (condensation).

Depending on the emission characteristics in terms of concentrations and flow, one of these technologies will be chosen with the aim of achieving the required performance for the lowest investment and operating costs.

These processes are widely used in industrial applications to remove single toxins or a mixture of contaminants. Many activities are concerned such as chemistry, petrochemistry, pharmacy, cosmetics, surface cleaning, polymer production, printing, painting, mechanical and car manufacture, waste and wastewater treatments.

Biological treatments of gas streams are relatively recent technologies compared with thermal destruction or mass transfer systems. However, researchers have been paying attention to these promising and interesting processes for several years and indeed bioprocesses appear to be a very competitive way to treat the waste gas stream before its discharge into the atmosphere. The removal of a large number of soluble and biodegradable volatile organic compounds (VOCs) or odorous molecules has been the subject of many previous studies and industrial applications.[7,8] The optimal range of pollutant concentration goes from a very diluted pollutant present in the gas stream (from some $\mu g\ m^{-3}$ to $mg\ m^{-3}$) to above $1\ g\ m^{-3}$. The installation designs cater for an air flow from a few $m^3\ h^{-1}$ to $100\ 000\ m^3\ h^{-1}$, or even more in some systems.

This chapter presents general approaches to the bioreactors used in waste gas stream treatments and describes more specifically the different biosystems such as biofilters, biological trickling beds and bioscrubbers. The general presentation, operating conditions, yields and industrial applications of these bioprocesses are discussed.

5.2 General Approaches to Biological Treatment of Waste Gases

Clearly, the final objective of a biological treatment is to transform the contaminants present in the gaseous phase and used as substrates by microorganisms (bacteria, fungi and yeast) into innocuous compounds. A very simplified reaction pathway expressing the principle of biodegradation is given in Table 5.1. According to the oxygen level provided by the air to be treated, the

Table 5.1 Simplified reaction system of the degradation of contaminants by enzyme (E) of aerobic bacteria (X).

Contaminants (substrates) + O_2, N, P, trace elements H_2O	E, X →	$CO_2 + H_2O + X$ Metabolites Internal energy

Table 5.2 Examples of particular bacterial families degrading specific contaminants present in air.

Contaminants	*Microorganisms*	*References*
Ethanol	*Pseudomonas* sp. *Pseudomonas putida*	68
Phenol, benzene, toluene, xylene, ethylbenzene, isopropylbenzene	*Pseudomonas putida*	69, 70
Toluene	*Pseudomonas putida* To11A	71, 72
Styrene	*Exophilia jeanselmei, Tsukamurella, Pseudomonas, Sphigomonas, Xanthomonas*	73, 74, 75
Styrene and Xylene	*Nocardia*	76
1-Chlorobutane	*Rhodococcus*	77
Butanal	*Pseudomonas fluorescens*	78
Hydrogen Sulfide	*Pseudomonas putida* CH11	79, 80
Methyl sulfide	*Hydromicrobium* sp.	81
Dimethyl sulfur	*Pseudomonas acidovorans*	82, 83
Dichloromethane	*Hydromicrobium* sp.	84, 85
Dichloroethane	*Xanthobacter*	86
Trichloroethylene	*Methylosinus*	87
Ammonia	*Actinomadura nitrigens* sp.	88, 89

biomass used is mainly composed of aerobic species. Nevertheless, some anaerobic microorganisms are suspected to be present in potentially anoxic zones caused by a non-ideal gas or liquid flow through the reactor, which leads to the occurrence of plug flow with a non-active zone, favouring the creation of channels and by-passing others.[9] To maintain biological activity, the addition of water is required. The equilibrium of substrates is necessary to stimulate biomass activity and thus nitrogen compounds, phosphate and trace elements are often added through incorporation or solution injection into the biosystem. The degradation of organic contaminants gives mostly carbon dioxide (CO_2) and water (H_2O). Anionic metabolites, such as nitrates, sulfates and chlorides, accompanied by oxonium cations (H_3O^+), are frequently produced. Microorganism growth (X) is controlled by the substrate load applied to the bioreactor and by operating parameters such as pH, temperature and moisture content. Internal energy is also produced by the bacteria.

A large number of previous publications have mentioned the degradation of specific contaminants by a particular microorganism family. Table 5.2 presents examples of pollutant-degrading biomass for some organic or inorganic volatile compounds. Although some researchers advise the utilization of such pure strains, it is difficult to avoid a possible external bio-contamination. In other words, the reactor is not under sterile conditions and can be naturally colonized by a large number of bacteria. As a consequence, a consortium of microorganisms generally grows in the biological reactor depending on the nature of the available biodegradable substrates. These microorganisms, responsible for this bio-contamination, can be initially present on the packing materials, induced by polluted gas or provided by water or the initial inoculation step. For industrial applications, diluted activated sludge coming from a wastewater treatment plant is inoculated in the early days of bioreactor implementation or when restart is required. During the acclimatization period, when the performance increases gradually, a microbial selection occurs. The most adapted microorganisms grow to the detriment of other species.

In terms of the process, an important initial question is how to put microorganisms in contact with water and substrates present in the gas phase. Taking into account the mobility of the aqueous phase and the microorganisms, three processes are generated as shown in a conventional matrix: biofilters, trickling beds and bioscrubbers (Table 5.3).

As degradation of the compound is carried out by microorganisms present in the aqueous phase or structured as a biofilm rich in water, the water solubility of the pollutants appears to be a major factor in choosing a technology. A diagram of decision support, based on the pollutant concentration and its air–water partition coefficient, has been proposed.[10] As a result, biofilters are recommended for the treatment of pollutants having an air–water partition coefficient <1. Bioscrubbing is useful for gaseous pollutants with a Henry's constant or partition coefficient of <0.01, while trickling filters can be used efficiently for the treatment of compounds characterized by an air/water partition coefficient <0.1.[11]

More recently, other types of more complex bioreactor have been studied in the laboratory or at pilot unit scale: membrane bioreactors or multiphase bioscrubbers. The schematic presentations of the main different biological processes are given in Figure 5.1. Biomass is attached to the packing material of

Table 5.3 Biological treatment processes: classification according to the mobility of the liquid phase and biomass.

		Liquid phase		
		Mobile	Immobile	
Biomass	Free	Bioscrubber		Scrubber + treatment tank
	Fixed	Biotrickling	Biofilter	Packed column
		Nutrient solution	Nutrient solution, packing assimilation	**Reactor**
		Nutrient supply		

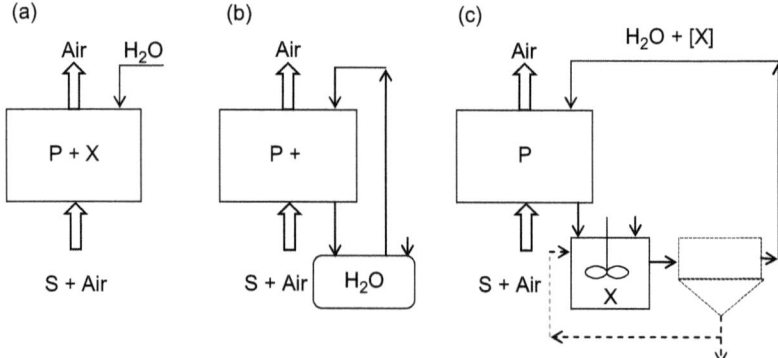

Figure 5.1   Schematic presentations of the different biological processes used in gas
stream treatment: (a) biofilter; (b) trickling bed; and (c) bioscrubber
(S = substrate; X = biomass; P = packing materials).

the bioreactor or in the form of an activated sludge. Water moistens the media
in a biofilter and is recycled for the other two processes. In these cases, absorption of the pollutant into water or an organic solution is required before
degradation by microorganisms.

5.3   Biofilters

In recent decades, biofiltration has shown interesting development prospects in
the field of VOC and odour treatment.[12] It is particularly suitable for the
treatment of high air-flow rates at low concentrations of a large number of
pollutants. This, along with its low implementation and maintenance costs, has
made biofiltration one of the most used methods for the treatment of industrial
gaseous emissions. Many successful industrial applications include the treatment of gaseous emissions from water treatment plants, composting platforms,
rendering plants and printing factories.[13,14]

5.3.1   Process Description and Mechanism

5.3.1.1   *Process Description*

In its implementation, a biofilter consists of a porous organic or inorganic bed
through which a moist polluted gaseous stream passes (Figure 5.2). A complex
microbial consortium of pollutant-degrading microorganisms is immobilized at
the material surface and carries out the degradation of VOCs and odours under
given operational parameters (moisture content, pH, nutrient availability,
temperature).[11,15]

To maintain the packing material humidity and avoid biofilter drying, the air
flow is generally moistened to bring the relative humidity above 98%.

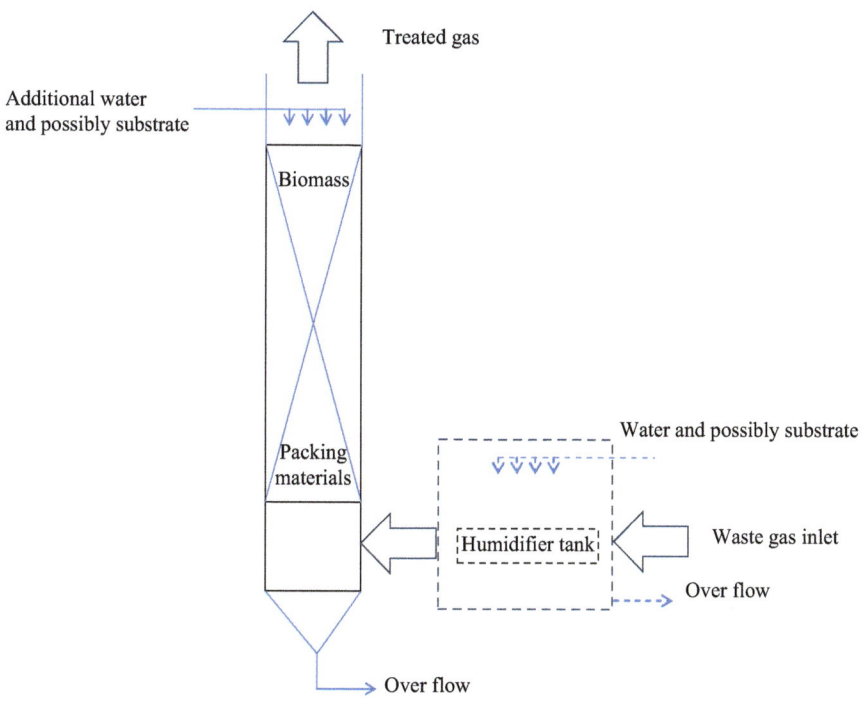

Figure 5.2 Diagram of a biofilter for waste gas stream treatment.

Humidification has various positive impacts since it reduces the temperature by energy consumption through water evaporation, promoting the development of mesophilic microorganisms at the packing material surface. It also removes particles (dust, fat vesicles), thus limiting the clogging of the reactor, the packing material and the air distribution system.[16] This step can be carried out by a cyclone, a wet electrofilter or a venturi scrubber.[17] In addition to the air stream humidification, a superficial watering is generally performed to control the moisture content.

As far as the microorganisms are concerned, they are mainly mesophilic, with an optimum development temperature close to 37 °C, even though some applications have been carried out in psychrophilic and thermophilic conditions. Considering their carbon sources, autotrophic and heterotrophic species related to the pollutants to be removed are observed. Their development depends on the nutrient availability. For instance, odorous sulfur compounds are treated by autotrophic microorganisms on inert inorganic material which require two main nutrients for microbial growth such as carbon (HCO_3^- or the carbon dioxide contained in the air) and nitrogen (use of NH_4^+).

An example of a biofilter treating an odorous gas is presented in Figure 5.3. The flow is 50 000 m³ h⁻¹. The filter surface is 2×50 m² and the depth is close to 2 m. The packing material was initially an inorganic solid waste reacting with

Figure 5.3   Example of a biofilter used to treat odorous molecules: treated air is emitted into the atmosphere through a stack.

acidic compounds such as H_2S and mercaptans, inducing clogging. Because of this problem, the biofilter was refilled with peat. The purified air is evacuated into the atmosphere by a chimney.

5.3.1.2   Specific Mechanisms

The elimination of pollutants in a biofilter involves several successive steps, which are briefly schematized in Figure 5.4:

- Pollutant transfer from the gaseous phase to the gas-liquid interface
- Diffusion of pollutants within the liquid phase and the biofilm
- Biodegradation of the pollutants by the microbial population of the biofilm; they are used as carbon and energy sources by the microorganisms (Figure 5.4)
- Diffusion of the metabolites produced into the liquid phase and then into the gas phase.

The diffusion of the pollutants, as well as their microbiological degradation, is a complex phenomenon. The sorption process of contaminants at the material surface is frequently observed at the startup of a biofilter, namely when biofilm formation is still weak. During this acclimation phase of the microorganisms, the packing material acts as a conventional adsorbent and a significant decrease in the pollution load can be observed. However, biofilm development induces a decrease in the available adsorption sites, leading to a decrease in the adsorption capacity.

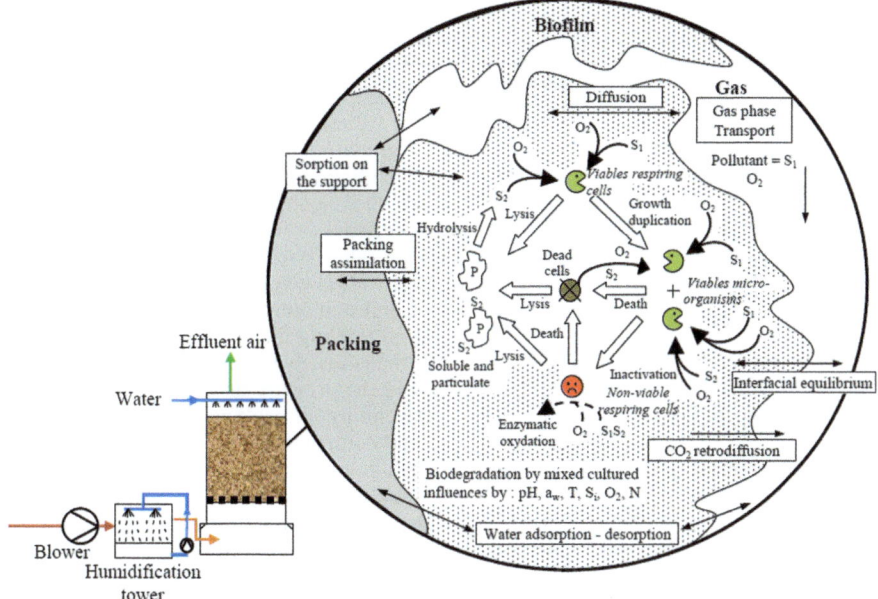

Figure 5.4 Representation of the biological and physical mechanisms involved in the biofiltration process (adapted from ref. 37).

5.3.2 Operating Conditions and Performance

5.3.2.1 Operating Conditions

General operating conditions are given in Table 5.4. The performances of a biofilter are dependent not only on its design (gas residence time as a function of bed surface, depth and void fraction) but also on the operating conditions such as watering, pH and especially waste gas velocity, which is low ($100-500$ m h^{-1}) due to the slow biodegradation kinetics.

5.3.2.2 Factors Affecting Biofilter Performance

The performance of a biofilter depends on many parameters such as pH, humidity, nutrient supply and concentrations, and the structural properties of the packing material.[18]

5.3.2.2.1 Packing Material. Although the effectiveness of a biofilter is associated with the microbial activity, it should be emphasized that the material is at the heart of the process because it is where the purifying biomass develops.[16,19] The intrinsic properties of the packing material induce the establishment of a more or less conducive environment for the development of an effective microbial consortium and a homogeneous gas distribution

Table 5.4   General operating conditions for waste gas stream treatment in biofilters (adapted and updated from ref. 4).

Parameter	Value	Comments
Gas phase velocity (U_G)/m. h^{-1}	50–500	Low values due to low kinetics of degradation
Air residence time/s	15–90	Depending on the molecule degradation kinetics alcohols > ketones > n-alkanes > aromatics
Bed porosity/-	0.4–0.95	High values avoid clogging
Specific surface area (S)/ m^2 m^{-3}	100–400	High values give a better mass transfer and a higher biomass concentration (biotrickling filter)
Filter depth (H)/m	0.5–2.5	Compromise between the residence time and the pressure drop
Pressure drop/m H_2O	0.1–0.5	Depending on packing material, clogging and compaction
Air humidity/%	60–100	High values are advised to maintain biofilter humidity
Water pH	5–9	Depending on the pollutant solubility
Temperature (T)/°C	10–40	Mesophilic microorganisms Thermophilic microorganisms at temperature ranging from 50 to 70 °C
Acclimation time/day	8–30	Function of the biodegradability of pollutants Utilization of an inoculum (diluted activated sludge)
Pollutant concentration/ mg m^{-3}	1–1000	Possible inhibition at higher concentration by molecules or degradation by-products
Efficiencies/%	90–99	Depending on molecules
Lifetime/year	2–5	Structured and inorganic materials increase lifetime

throughout the bed. For these reasons, numerous laboratory and industrial studies have focused on the influence of the packing material on biofilter performances[20–23] or on the improvement of hydrodynamic properties.[24,25] Several authors have highlighted the following characteristics as suitable for biofilter material:[17,19,26,27]

- A large porosity and a high void fraction for promoting the development of microorganisms and gas flow distribution
- A large surface area to improve the transfer of pollutants, nutrients and oxygen, and to promote water retention—the accumulation of microorganisms is optimal in the presence of pores with sizes ranging between one and five times that of the microorganisms (1–10 μm)[28]
- A dense and diverse microbial population and chemical characteristics favouring the development of microorganisms (pH, buffering capacity and nutrients)
- A good mechanical stability to avoid bed compaction
- Low investment costs and a long life.

The most commonly used organic packing materials are peat, compost, bark and wood chips. Peat and compost are good supports for bacterial development in terms of nutritive supplement, but their mechanical resistance is insufficient and their hydrodynamic properties vary with time. This dysfunction can be avoided by the use of more structured packing, such as bark and wood chips which remain, like peat and compost, biodegradable but which have to be replaced more often than inorganic ones. Inorganic supports such as activated carbon, which is expensive for this application, or pozzolan have better mechanical properties enabling bed depths above 2 m and providing a longer life time. Nevertheless, their implementation necessarily requires a nutrient supply.[29]

There is no universal material that meets all these criteria, whether organic, inorganic or synthetic. It is often necessary to achieve a compromise between these different properties, or to consider a mixture of different materials, which may be in the form of layers or mixtures, to benefit from the advantages of each.

5.3.2.2.2 Biofilter Design. Filter depth is in the range of 0.5 to 2.5 m with a usual value of 1 m for organic packing materials, giving a significant residence time to treat pollutants while minimizing footprint and pressure drop. Greater filter depths can result in support compaction, thus increasing pressure drop and energy consumption as well as favouring the formation of preferential paths. These can induce a rise in local flow rates, leading to a lower residence time and consequently lower treatment efficiency.

In terms of design, and especially for the direction of gas and liquid flow, two modes are observed. The open counter-current configuration allows easy access for maintenance but the effectiveness of these units is affected by heavy rainfall, sometimes justifying coverage. Although it enables better management of moisture and nutrient dispersion, the co-current mode is less employed due to the seal required at the top of the biofilter. Configurations in parallel or in series are also possible in order to isolate units during maintenance operations or to remove compounds in specific biofilters combining special operating conditions (pH, material and humidity).

5.3.2.2.3 Water Requirements and Air Moisture. Among the major operating parameters, biofilter moisture is considered the most important for biofilter management.[30] The presence of water is essential for microorganisms. Humidity evaporation by the unsaturated air flow through the packing material can result in localized bed drying, leading to a negative impact on the microbial community and density, and also on gas distribution. With high evaporation rates, cracks can appear leading to the creation of preferential paths resulting in a lower removal capacity.[31] It should be noted that water reclamation can increase for the treatment of high loads of components because of the exothermicity of the biological reactions involved.

Conversely, excessive wetting of packing material can induce oxygen transfer limitations and hydrophobic volatile compounds at the packing material and biofilm surface. It can also promote the appearance of anaerobic zones in the biofilter, decreasing the reaction kinetics and transfer rates, and hence the removal efficiency.[32] Furthermore, over-watering can lead to compaction of the bed and opposition to the flow of gas, increasing the pressure drop.[33]

The optimal moisture of a biofilter varies with the type of packing material implemented, and especially with characteristics like porosity and specific surface area; the amount of water in packing materials is generally in the range of 30% to 60%. To optimize the material moisture, air to be treated can be humidified before the biofilter to maintain a relative humidity in the range of 95% to 99%. A periodic spraying of the material can also be carried out; the spraying solution can also contain nutrients to supply microorganism growth or can be alkaline to increase its buffer capacity.

5.3.2.2.4 Oxygen and Nutrient Supply. Microorganisms found on the material are mainly aerobic and hence require oxygen for their metabolism. For instance, heterotrophic aerobic bacteria found on biofilter material require between 5 and 15% oxygen.[34] In the case of flow gases heavily loaded with organic compounds, biofiltration performance can be negatively impacted by the amount of oxygen in the gas to be treated.[35] However, in most biofiltration processes, the oxygen content does not constitute a real problem owing to its abundance in the gaseous effluent and the low biofilm thickness, which avoids diffusional limitations.

Microorganisms require various resources to cover their energy and nutritional needs. In situations of nutrient deficiency, microorganisms reduce their metabolic activity[36] while an excessive dose, from a continuous supply, induces uncontrolled development of the biofilm leading to clogging.[37,38] Some authors suggest that the metabolic activity of microorganisms is stimulated in the presence of key elements in certain proportions, *i.e.* in a C : N : P ratio of 100 : 15 : 3 or 100 : 5 : 1.[39–41] In general, whatever the material used, a nutritional supplement is often required to maintain a satisfactory treatment efficiency. It has been observed that an extended use of compost leads to a gradual depletion of nutrient resources,[42,43] which can become a limiting factor in the long term.[44]

5.3.2.2.5 Temperature. Accurate control of gas temperature is required to optimize pollutant transfer from the gaseous phase to the biofilm and to promote microbial growth. Mass transfer is favoured by low temperatures while degradation kinetics are favoured by high temperatures.[32] Based on growth temperature, three groups of microorganisms can be found in biofiltration: psychrophilic, mesophilic and thermophilic, growing below 20 °C, between 20 and 40 °C and above 45 °C, respectively. The temperatures observed in biofiltration are generally in the range of 20 to 40 °C,[16,32] showing that the microbial population is mainly mesophilic. Even though the biological activity within the packing material can induce an increase in temperature, in the range of 2 to 10 °C,[16] the temperature is generally set by the gas to be treated[32] and hence upstream gases may need to be cooled or heated to avoid temperature shocks which can be harmful to cells. Consequently, in the case of cold or hot air treatment, the costs associated with temperature control can be a major drawback to the use of biofiltration.

5.3.2.2.6 pH. Not only does pH have a major impact on microbial metabolism it also seems to play a role in microbial attachment to the packing material.[32] Generally, biofiltration is implemented in a pH range of 5 to 9, and microorganisms do not easily support more than 2–3 units of pH variation.[16] Drastic changes in pH can damage the plasma membrane and also inhibit enzymatic activity and transmembrane proteins. They can also limit the bioavailability of certain nutrients. For example, the bioavailability of ammonia, which is an excellent source of nitrogen, is increased in its protonated form (NH_4^+; pKa $NH_4^+/NH_3 = 9.2$). As a consequence, mass transfer and hence pollutant absorption can lead to improved performance, for example, due to biofilter acidification during ammonia treatment.[45]

The pH of the filter medium depends on several factors, primarily the nature of the pollutants and their ability to generate acidic by-products.[16] Microbial degradation of pollutants can affect the pH by an acidification due to CO_2 production or the formation of acidic metabolites.[16,17] For example, the treatment of H_2S concentrations above 15 ppm leads to a rapid acidification by sulfuric acid formation.[46]

To maintain appropriate values, the pH in the packing material can be adjusted by the addition of an alkaline solution sprayed above the material, or the material itself can have a buffer capacity, for example, through addition of lime[47] or crushed oyster shell.[48]

5.3.2.3 Advantages, Drawbacks and Limitations of the Process

Biofilters show clear advantages for the treatment of gaseous effluents due to their simplicity of implementation, their rusticity and their efficiency for low pollutant concentrations. In addition, they are characterized by low operating costs and, unlike biotrickling filters, do not generate large amounts of effluent, liquid or sludge. However, the residence times needed for efficient degradation require large-size facilities, leading to high civil engineering costs. Moreover, the packing material needs to be replaced periodically (after 2–5 years of use). In addition, biofilters show decreasing efficiency in the case of high pollutant concentrations or recalcitrant compounds. Finally, and despite biofilter rusticity, special attention should be paid to process optimization.[16]

5.3.2.4 Fate of By-products Generated during Biofiltration

Biofilter use leads to the production of liquid and solid wastes, which should be treated or exploited. The liquid effluents are mainly leachates and water run-off due to rainfall for non-covered biofilters. These effluents, which contain large amounts of organic matter, salts from microbial metabolism (sulfates, chlorides, carbonates), acids (HCl, H_2SO_4) and microorganisms are sent to the wastewater treatment plant of the industrial site.[16] The solid wastes are mainly components from the packing material. Since these materials can be organic, mineral or inert, various waste treatments can be implemented. Energy recovery from organic material (peat, pine bark) can be considered through incineration, pyrolysis or methanogenesis/biogas, or can be sent to compost units. Non-biodegradable inorganic materials (*e.g.* pozzolan) are generally recovered and sent to landfill sites.

5.3.3 Modelling a Biofilter

The model generally applied to simulate the different steps of diffusion and biodegradation was proposed by Ottengraf.[49] Presented succinctly, this model is based on the following assumptions:

- Biodegradation occurs in the biofilm liquid phase.
- The liquid phase in the biofilm is assimilated to water.
- The biofilm thickness is small compared with the packing material dimension, so that the surface of the biofilm is considered flat.
- The biomass concentration is assumed to be homogeneous in the reactor volume and constant as a function of time.
- The gas flow is a plug flow.
- The gas phase is considered ideal and there are no reactions between the chemical species.
- The mass transfer resistance in the gas phase is negligible.
- The regime is steady-state, thus there is no variation in operating conditions or in the biofilm (concentration of microorganisms, depth).
- Equilibrium occurs at the gas–biofilm interface.

Three cases can be considered for the mechanisms of biodegradation (Figure 5.5):

- First-order kinetics: the substrate concentration is a limiting factor. The reaction rate is controlled by the pollutant diffusion inside the biofilm.
- Zero-order kinetics—biological regime: the substrate concentration in the biofilm is high ($S_L \gg K_s$) but the diffusion into the biofilm is not a limiting factor.

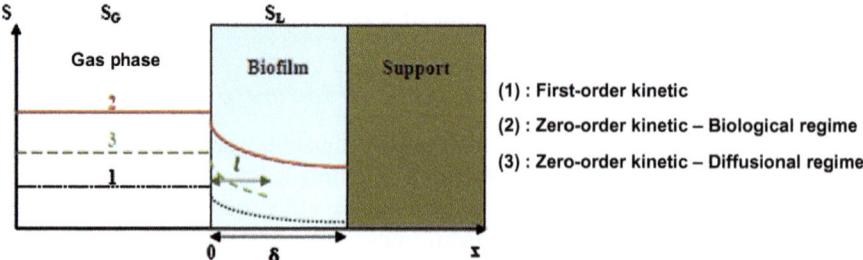

Figure 5.5 Substrate concentration profile in the biofilm as a function of the kinetic order.[38]

Table 5.5 Overview of kinetic constants and biodegradation yield as a function of the kinetic order.[38]

	Kinetic constant (K_i)	Biodegradation yield (BY)	Total biodegradation condition
First-order kinetics	$K_1 = \dfrac{a.D}{\delta} \phi_1 \tanh \phi_1$ (s^{-1})	$BY = 1 - \exp\left(\dfrac{-K_1 Z}{m.U_G}\right)$	$-\dfrac{K_1 Z}{m.U_G} \to \infty$
Zero-order kinetics: biological regime $(\phi_{cr} < \sqrt{2})$	$K_0 = R_0.a.\delta$ $(g\ m^{-3}\ s^{-1})$	$BY = \dfrac{K_0 Z}{U_G.S_G^e}$	$\dfrac{K_0 Z}{U_G.S_G^e} \geq 1$
Zero-order kinetics: diffusional regime $(\phi_{cr} > \sqrt{2})$	$K_0 = R_0.a.\delta$ $(g\ m^{-3}\ s^{-1})$	$BY = 1 - \left(1 - \dfrac{Z}{U_G}\sqrt{\dfrac{K_0.D.a}{2S_G^e.m.\delta}}\right)^2$	$\dfrac{K_0 Z}{U_G.S_G^e} \geq 2$

a specific surface area ($m^2\ m^{-3}$)
BY biodegradation yield
D diffusivity in the liquid phase ($m^2\ s^{-1}$)
K_i kinetic constant
m partition coefficient (Henry's law) (dimensionless; $g\ m^{-3}$ gas and $g\ m^{-3}$ water)
R_0 maximal biodegradation rate ($g\ m^{-3}\ s^{-1}$)
SG substrate concentration in the gas phase ($g\ m^{-3}$)
U_0 empty bed velocity ($m\ s^{-1}$)
U_G gas velocity ($m\ s^{-1}$)
Z biofilter length (m)
δ biofilm thickness (m)
ϕ_i Thiele number for an i-th order reaction (dimensionless)

– Zero-order kinetics—diffusional regime: the substrate concentration in the biofilm is high ($S_L \gg K_s$) and the diffusion into the biofilm is a limiting factor.

The substrate concentration evolution in the biofilm is $S_L(g\ m^{-3})$. K_s ($g\ m^{-3}$) is the Monod affinity constant. For carbon substrates, K_s is low and ranges from 10^{-3} to $10^{-2}\ g\ m^{-3}$. Table 5.5 gives an overview of the biofiltration yields for the three cases.

5.4 Biotrickling Filters

5.4.1 Process Description and Mechanism

As shown in Figures 5.6 and 5.7, a biotrickling filter is a column packed with solid materials (inorganic grains, Rashig rings, Berl saddles) which is covered by a biofilm formed by predominantly aerobic microorganisms. The waste gas stream is introduced at the bottom of the system and flows through the packed bed; treated gas is vented to the atmosphere. Counter-current recycled water trickles over the packing and is recovered in a buffer tank. Fresh water and possibly

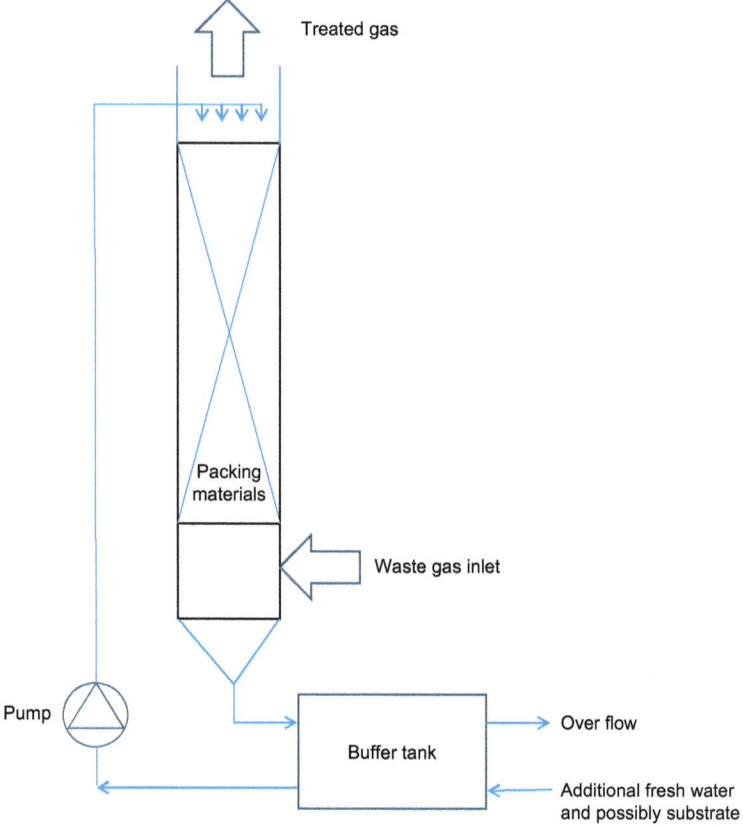

Figure 5.6   Diagram of a biotrickling filter for waste gas stream treatment.

substrates are added to this tank. These operations are required because of water evaporation by the gas stream and an increase in inorganic ions and/or metabolites in the aqueous solution. The water pH is also controlled by the addition of base or acid to the fresh solution to obtain appropriate bacterial growth.

The contaminants present in the gaseous phase are first transferred into the aqueous phase and then degraded in the biofilm present at the surface of the packing. Metabolites are removed by the solution flowing in the column. Some bacteria are also detached and are carried away by water. Washing the packing materials is sometimes necessary, due to clogging of the column by un-controlled biofilm growth.

5.4.2   Operating Conditions and Performance

An adapted and updated summary of previously published information[4,50] on the operating conditions and performance of biotrickling filters is given in Table 5.6. To obtain optimal performance, a column design taking into account

Figure 5.7   Example of a biotrickling filter used to remove ammonia present in air extracted from a waste storage room.

the residence time and the liquid to gas (L/G) mass ratio has to be optimized according to the properties of the contaminants, especially their solubility and biodegradability. L/G is a value adjusted to minimize the pressure drop and to optimize the mass transfer and thus the removal efficiency. Temperature in the column ranges from 10 to 40 °C and is similar to the temperature of the inlet gas. pH has to be close to neutral for good microorganism growth. Acid or base is added to the solution to control the mass transfer (slightly basic for acid pollutants such as H_2S and mercaptans or slightly acidic for basic molecules like ammonia and amines). A pH value ranging from 5 to 9 is also required for biofilm stability.

Table 5.7 presents data on the performance of a biotrickling filter in removing various pollutants. The majority of compounds are carbon and energy sources for heterotrophic microorganisms in aerobic conditions. The removal rates are correlated to the properties of the molecules such as solubility, given by the Henry's law coefficient (m), and biodegradability. For soluble compounds (weak m), mass transfer is easier and for biodegradable molecules, the removal rate is significant. This step of gas–liquid transfer is essential to obtain good performance from a biotrickling filter.

Table 5.6 Ranges of operating conditions applied in biotrickling filters (adapted and updated from ref. 4 and 33).

Parameter	Value	Remarks
Liquid velocity (U_L)/m h^{-1}	0.05–20	Compromise between the solubility of contaminants in water, the flooding point and the biomass loss due to significant flow of the solution
Liquid hold-up/%	<5	Value to be optimized for a better mass transfer
Air velocity (U_G)/ m h^{-1}	100–1000	Limited by the flooding point (*cf.* liquid velocity) and the clogging of the column due to microorganism growth by a supply of substrate
Air residence time/s	<60	Residence time lower than for biofilters depending on the biodegradation kinetics of pollutants
Bed porosity/ dimensionless	0.5–0.95	High value decreases clogging and head loss
Pressure drop/m H$_2$O or bar	0.005–0.5 0.0005–0.05	Depending on mass flow ratio of air and solution, bed porosity and concentration of biofilm
Specific surface area (S)/m^2 m^{-3}	100–400	Lower value compared with biofilters Better mass transfer due to an increase in external surface area.
Filter depth (H)/m	2–15	Light packing materials allow taller columns.
Solution pH	5–9	Relatively easy to control by addition of acid or base to the liquid phase
Temperature/°C	10–40	Mesophilic microorganisms. Treatments with temperatures ranging from 50 to 70°C could be efficient with thermophilic populations
Lag time/days	5–210	Function of the biodegradability of the molecule
Optimal performances/% removal	90–99	Depending on solubility and biodegradability of contaminants

5.4.3 Modelling a Biotrickling Filter

A multiscale approach integrating different mechanisms is required to model biological treatments and especially biotrickling filters. The hydrodynamic approaches such as holdup, head loss and residence time distribution are performed using classical equations used in multiphase reactors.[51–53] In terms of mass transfer and biodegradation, the assumptions are as follows:

- The column is working in a counter-current gas–liquid.
- For the gas phase, the column is considered as a plug flow reactor.
- The system is steady-state in terms of pollution loads, operating conditions (L/G), biofilm depth and bacteria concentrations.
- The double-film theory is applied in the contaminant transfer from the gas to liquid. In this case the transfer resistance in the gas phase is negligible.

Table 5.7 Removal rate of different contaminants in a biotrickling filter (adapted from ref. 7, 90 and 91).

Pollutant	Biodegradability	Henry's law coefficient (m)	Removal rate/g m^{-3} h^{-1})	Ref.
Alkanes				
n-Hexane	++	74.0	7.5	52, 92
n-Heptane	++	83.2	24	7
Aromatics				
Styrene	++	0.11	32	93
Toluene	+++	0.28	80	7, 94
Oxygenated compounds				
Methanol	+++	0.00019	100	95
n-Butanol	+++	0.00035	100	96
Propionaldehyde	+++	0.0025-4	300	97
Acetone	+++	0.0016	500	57
Methyl ethyl ketone	+++	0.0024	40	98
Diethylether	+	0.028	60	99
Methylterbutylether	+	0.023	45	100
Chlorinated compounds				
Dichloromethane	+++	0.093	150–200	101, 102
Chlorobenzene	++	0.18	60–300	103–105
Nitrogenous compounds				
Ammonia	+++	0.0007	4.9–22.6	106, 107
Trimethylamine	+++	6.6	13.3	108
Nitrobenzene	+	0.00098	50	61
Nitrogen oxides (NOx)	+		25	109, 110
Sulfur compounds				
Hydrogen sulfide	+++	0.94	100	111, 112
Carbon disulfide	++	0.39	220	113
Dimethyldisulfide	+	0.054	22.1	114

- The transfer resistance at the liquid-biofilm interface is not taken into account. Biodegradation in the biofilm is processed *via* a Monod kinetic equation.

Three phases are present: (i) the biofilm coating an inert packing material; (ii) the aqueous solution absorbing the contaminants; and (iii) the gas stream to be treated. The multiphase mechanisms of transfer and biodegradation and the concentration evolution profile are presented in Figure 5.8.

The equation system of the engineering model is as follows:[54,55]

- Global transfer coefficient for the absorption of contaminants into the aqueous solution:

$$\frac{1}{K_L} = \frac{1}{mk_G} + \frac{1}{k_L} \qquad (5.1)$$

where K_L (m s^{-1}) is the global transfer coefficient, m (dimensionless g m^{-3} gas/g m^{-3} water) is the partition coefficient between the gas and the liquid

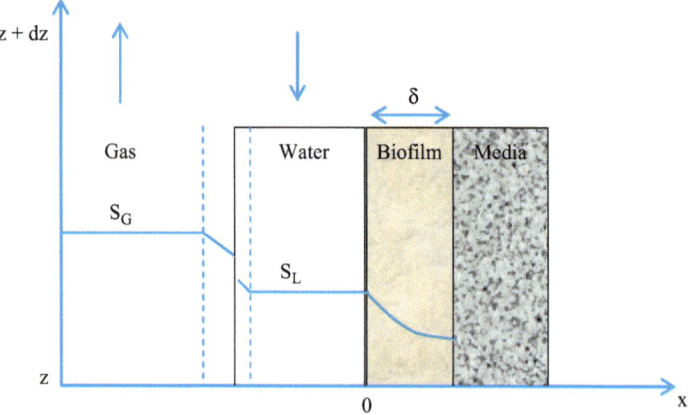

Figure 5.8 Schematic representation of the gas-water-biofilm-media system: evolution of substrate concentrations.

(Henry's law), k_L and k_G are the transfer coefficients (m s^{-1}) in the liquid and gas phase, respectively.

– Substrate mass flow (J) into the biofilm with Monod kinetics (first-order):

$$J = \sqrt{2DK_sR_0\left[\frac{S_L}{K_s} - \ln\left(1 + \frac{S_L}{K_s}\right)\right]} \tag{5.2}$$

In the previous equation, D (m^2 s^{-1}) is the diffusion coefficient of the substrate into the biofilm. S_L is the substrate concentration at the liquid-biofilm interface (g m^{-3}), K_s is the Monod constant (g m^{-3}) and R_0 (g substrate s^{-1}) is the maximum substrate consumption rate.

In cases where all the depth of the biofilm is used and the concentration is high, compared with the Monod constant K_S, the biodegradation kinetics are zero-order. The maximal transfer flux (J_{max}) is then:

$$J_{max} = R_0\delta \tag{5.3}$$

where δ (m) is the biofilm thickness.

– Mass balance in the gas phase

$$U_0\frac{dS_G}{dz} = \pm K_La\left(\frac{S_G}{m} - S_L\right) \tag{5.4}$$

where a is the packing material specific surface area, S_G (g m^{-3}) is the substrate concentration in the gaseous phase and z (m) the height of the packing.

– Mass balance in the liquid phase (transfer and biodegradation) Ja : mass flow of substrate transferred in the biofilm (g/s)

$$U_L\frac{dS_L}{dz} = -K_La\left(\frac{S_G}{m} - S_L\right) - Ja \tag{5.5}$$

The initial conditions are the inlet pollutant concentration in the gas phase and the recirculation of the liquid phase, which has the same substrate concentration in the inlet and outlet of the column. These conditions are written:

$$S_G^{z=0} = S_{Ge} \quad S_L^{z=0} = S_L^{z=Z} \tag{5.6}$$

where Z (m) is the total height of the packing. The differential equation system is solved using a numerical method (the Runge–Kutta method, for example).

The model has been used for parametric studies. It was shown that performance is limited not by the gas–liquid transfer but by the compound's Henry's law coefficient and the substrate biodegradation step. However, this model has some limitations. The substrate consumption in the biofilm is overestimated due to a negligible transfer resistance between the liquid and the biofilm, and thus performance is overestimated. A variation in the biodegradation kinetics, due to inhibition by some substrate or metabolite concentrations, is not integrated into the model's equations.[56]

5.5 Bioscrubbers

5.5.1 Process Description and Mechanism

A schematic presentation of a bioscrubber is given in Figure 5.9. The removal of contaminants present in the gas stream is performed by an association of two steps:

- Mass transfer in a gas–liquid column. An empty column with a spray injector, a packed column or a venturi scrubber are the principal types of gas–liquid contactor.[14] Water is generally used but an organic solution can be used for hydrophobic molecules.[57–59]
- An activated sludge basin, possibly with a settling tank. The pollutant, transferred previously into the solution, is degraded in the bioreactor. The microbial suspension may settle in a sedimentation basin. The water is recycled at the top of the gas–liquid column.

Consequently, the mechanisms are similar to absorption and to biodegradation in an activated sludge reactor. To avoid clogging of the gas–liquid contactor, a settling tank separates the solid and liquid phases.

This system is particularly useful for hydrophilic molecules. In the case of hydrophobic compounds, an organic solution (silicone oil, for example) is used as the absorption solution in the transfer column.[60,61] The bioreactor is a multiphase system comprising organic and aqueous solutions, air injected into the solutions and a bacterial suspension.[62,63]

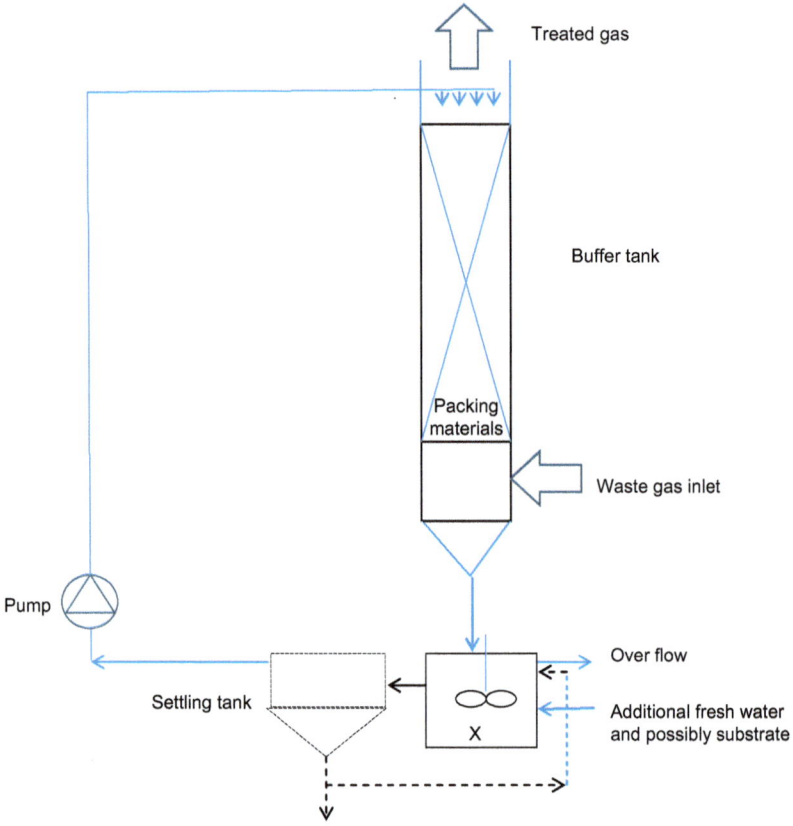

Figure 5.9   Diagram of a bioscrubber for waste gas stream treatment. X : activated sludge – biomass.

5.5.2   Operating Conditions and Performance

The operating conditions in a bioscrubber are similar to these in a biotrickling filter.[64] Table 5.6 gives some data for the absorber column. The activated sludge basins are managed as classical bioreactors found in wastewater treatment plants. The residence time ranges from 30 min to 2 h depending on the mass load of pollutants in the solution coming from the absorber.[65] Generally, the mass load is small and the activated sludge concentration is also at a low level (from 0.1 to 1 g L^{-1}).[66]

An example is given in Figure 5.10. Air extracted from a unit used to prepare food has to be treated to remove odorous emissions. A column (2.5 m in diameter and 4 m in length) is packed with Berl saddles. Waste gas flow is 7500 m^3 h^{-1} and is loaded with odorous molecules (aldehydes, ketones, organic acids) at low concentrations (between 1 and 10 mg m^{-3}) and fat aerosols. To remove fat particles, a cyclone is positioned just before the absorber. This system, which has been working for three years, gives good results with a removal percentage close to 90–95% in terms of pollutant concentrations and standard odour units.

Figure 5.10   Example of a bioscrubber used to remove odorous molecules present in air extracted from a semi-prepared food factory.

5.5.3   Modelling a Bioscrubber

To describe the bioscrubber behaviour and especially the mass transfer column numerically, a set of transfer equations with hydrodynamic parameters is required.[67] An extensive description has already been published.[50] A succinct presentation is given here. The assumptions retained are as follows:

- The regime is assumed to be steady-state.
- Biodegradation of transferred pollutant is negligible in the gas–liquid column.
- There is no change in concentration with time in the liquid phase.
- A plug flow model, with and without axial dispersion, is used to represent the gas flow in the column.
- The tanks-in-series with a mass exchange model is used to represent the washing solution flow.

The model is based on the partial mass balance in the gas phase and liquid phase and is written as:

$$-Q_G \cdot \frac{dC_G^i}{dz} + D_a \cdot S \cdot \frac{d^2C_G^i}{d^2z} - K_G \cdot \left(C_{Gz}^i - C_G^{iE}\right) \cdot a \cdot S \cdot dz = 0 \qquad (5.7)$$

$$Q_L\left(C_L^{i+1} - C_L^i\right) + \int_0^{\Delta h} N_A^i \cdot a \cdot S \cdot dz = 0 \qquad (5.8)$$

with dz the height of a compartment i, Q_G is the gas flow rate ($m^3\ s^{-1}$), C_G is the volatile compound concentration in the gas phase ($mol\ m^{-3}$), D_a is the axial dispersion of the gas flow ($m^2 \, s^{-1}$), K_G is the overall transfer coefficient in the gas phase ($m \, s^{-1}$), C_G^E is the gas concentration at equilibrium with the liquid phase concentration ($mol \, m^{-3}$), a is the specific surface area (m^{-1}) and S is the surface area of the column (m^2), Q_L is the liquid flow rate ($m^3 \, s^{-1}$), C_L is the liquid concentration of the compound ($mol.m^{-3}$) and N_A is the density of transfer flux ($mol.m^{-2}.s^{-1}$).

An analytical solution is proposed for an elementary compartment i:

$$\text{Gas}: \; C_G^i - C_G^{i-1} = \left(H' \cdot C_L^i - C_G^i\right) \cdot \left[1 - \exp\left(\alpha \cdot \frac{Z}{J}\right)\right] \qquad (5.9)$$

$$\text{Liquid}: \; C_L^{i+1} - C_L^i = \gamma \cdot \left(C_G^i - C_G^{i-1}\right) \qquad (5.10)$$

where H' is the Henry's law coefficient (dimensionless), Z the height of the transfer column (m), and J the number of tanks-in-series. α and γ are coefficients whose expressions vary according to the type of transfer and reaction:

– For a plug flow of the gas phase:

$$\alpha = \frac{-K_G.a.S}{Q_G} \qquad (5.11)$$

– For a plug flow with axial dispersion of the gas phase:

$$\alpha = \frac{U_G}{2 \cdot D_a} \cdot \left(-1 + \sqrt{1 + 4 \cdot \frac{K_G \cdot a \cdot D_a}{U_G^2}}\right) \qquad (5.12)$$

– For a transfer without reaction in the liquid phase:

$$\gamma = \frac{-K_G \cdot a \cdot S}{\alpha \cdot Q_L} \qquad (5.13)$$

– For a transfer with reaction in the liquid phase:

$$\gamma = \frac{K_G \cdot a \cdot S}{\alpha \cdot Q_L} \cdot \left(1 + \frac{K_2}{[H_3O^+]} + \frac{K_1 \cdot K_2}{[H_3O^+]^2}\right)^{-1} \qquad (5.14)$$

This model has been applied to a bioscrubber used to remove ethanol (a very biodegradable molecule) and hydrogen sulfide. A statistical approach shows that the prediction of the transfer efficiency has a 95% confidence interval.

5.6 Conclusions and Trends

Different bioprocesses used to remove contaminants found in waste gas have been presented: biofilters, biotrickling filters and bioscrubbers. The technologies have been described according to their principles, operating conditions and performances. Although numerous publications have appeared in recent decades, more research into these complex systems is still required to:

- determine the mechanisms of mass transfer into aqueous solution for complex mixtures of contaminants in the gas phase;
- study the biodegradation in the biofilters or in activated sludge basins;
- examine the engineering and design of these bioprocesses;
- model and simulate the multireactor systems.

Some laboratory studies have been published on multiphase reactors using organic/aqueous solutions to capture hydrophilic and hydrophobic contaminants before biodegradation. However, more work is still necessary to scale up the pilot unit to an industrial process.

Strong collaborations and research programmes between researchers in microbiology, biochemistry, chemistry and chemical or environmental engineering are required to obtain realistic mechanisms and performances. A good design of processes and simulation of these systems will then be achieved.

References

1. H. Bravo Alvarez, in *Air Pollution Control Equipment Calculations*, L. Theodore, Wiley, Hoboken, NJ, USA, 2008, Ch. 1.
2. J. Chevalier, P. Rousseaux, P. V. Benoit and B. Benadda, *Chem. Eng. Sci.*, 2003, **58**, 2053.
3. P. L. Lens, C. Kennes, P. Le Cloirec and M. Deshusses, *Waste Gas Treatment for Resource Recovery*, IWA Publishing, London, 2006.
4. P. Le Cloirec, Y. Andrès, C. Gérente and P. Oré, in *Biotechnology for Odour and Air Pollution Control*, ed. Z. Shareefdeen and A. Singh, Springer-Verlag, Heidelberg, Germany, 2005, p. 302.
5. P. Le Cloirec, *Les Composés Organiques Volatils (COV) dans l'Environnement*, Tech. & Doc, Paris, 1998.
6. L. Theodore, *Air Pollution Control Equipment Calculations*, Wiley, Hoboken, NJ, USA, 2008.
7. C. Kennes and M. C. Veiga, *Bioreactors for Waste Gas Treatment*, Kluwer Academic Dordrecht, The Netherlands, 2001.
8. H. H. J. Cox and M. A. Deshusses, *Biotrickling Filters, Bioreactors for Waste Gas Treatment*, Kluwer Academic, Dordrecht, The Netherlands, 2001.
9. O. Levenspiel, *Chemical Reaction Engineering*, John Wiley and Sons, New York, 1972.
10. H. J. G. Kok, A. J. Dragt and J. van Ham, *Studies Environ. Sci.*, 1992, **51**, 77.
11. C. Kennes and F. Thalasso, *J. Chem. Technol. Biotechnol.*, 1998, **72**, 303.
12. F. Gaudin, Y. Andres and P. Le Cloirec, *Chemosphere*, 2008, **70**, 966.

13. J. S. Devinny, M. A. Deshusses and T. S. Webster, *Biofiltration for Air Pollution Control*, Lewis Publishers, Boca Raton, FL, 1999.

14. R. Iranpour, H. H J. Cox, M. A. Deshusses and E. D. Schroeder, *Environ. Progress*, 2005, **24**, 254.

15. M. A. Deshusses, *Curr. Opin. Biotechnol.*, 1997, **8**, 335.

16. P. Pré, Y. Andrès, C. Gérente and P. Le Cloirec, *Les techniques de l'ingénieur Traité Environnement*, 2004, G1780.

17. Z. Shareefdeen and A. Singh, *Biotechnology for Odour and Air Pollution Control*, Springer-Verlag, Heidelberg, 2005.

18. S. Mudliar, B. Giri, K. Padoley, D. Satpute, R. Dixit, P. Bhatt, R. Pandey, A. Juwarkar and A. Vaidya, *J. Environ. Manage.*, 2010, **91**, 1054.

19. A. Elias, A. Barona, A. Arreguy, J. Rios, I. Aranguiz and J. Peñas, *Process Biochem.*, 2002, **37**, 813.

20. J. Luo and S. Lindsey, *Bioresour. Technol.*, 2006, **97**, 1461.

21. J. P. Maestre, X. Gamisans, D. Gabriel and J. Lafuente, *Chemosphere*, 2007, **67**, 684.

22. J. L. Filho, L. T. Sader, M. H. Damianovic, E. Foresti and E. L. Silva, *Chem. Eng. J*, 2010, **158**, 441.

23. N. Akdeniz, K. A. Janni and I. A. Salnikov, *Bioresour. Technol.*, 2011, **102**, 4974.

24. X. Jin-Ying, H. Hong-Ying, Z. Hong-Bo and Q. Yi, *Biochem. Eng. J.*, 2005, **23**, 123.

25. J. Hernández, Ó. J. Prado, M. Almarcha, J. Lafuente and D. Gabriel, *J. Hazard. Mater.*, 2010, **178**, 665.

26. H. Jorio, R. Brzezinski and M. Heitz, *J. Chem. Technol. Biotechnol.*, 2005, **80**, 796.

27. D. Gabriel, J. P. Maestre, L. Martín, X. Gamisans and J. Lafuente, *Biosyst. Eng.*, 2007, **97**, 481.

28. J. Tampion and M. D. Tampion, *Immobilized Cells: Principles and Applications*, Cambridge University Press, Cambridge, 1987.

29. C. Corre, C. Couriol, A. Amrane, E. Dumont, Y. Andrès and P. Le Cloirec, *Environ. Technol.*, 2012, **33**, 1671.

30. T. O. Williams and F. C. Miller, *BioCycle*, 1992, **33**, 75.

31. S. P. P. Ottengraf and A. H. C. Van der Oever, *Biotechnol. Bioeng.*, 1983, **25**, 3089.

32. D. McNevin and J. Barford, *Biochem. Eng. J.*, 2000, **5**, 231.

33. A. Wani and R. M. R. Branion, *J. Hazard. Mater.*, 1998, **60**, 287.

34. S. Dharmavaram, presented at 84th Annual Meeting Exhibition Air and Waste Management Association, Pittsburgh, PA, 1991.

35. Y. Yang, B. Minuth and E. R. Allen, *J. Air Waste Manage. Assoc.*, 2002, **52**, 279.

36. R. M. Atlas, *Microbiol. Mol. Biol. Rev.*, 1981, **45**, 180.

37. M. A. Deshusses, *Curr. Opin. Biotechnol.*, 1997, **8**, 335.

38. C. Alonso, X. Zhu, M. T. Suidan, B. R. Kim and B. J. Kim., *J. Environ. Eng.*, 2001, **127**, 655.

39. S. Curtis, H. J. McDowell, T. G. Bourgeois and T. Zitride, Polybac Corporation, Allentown, Penna. 1980.

40. E. Rosenberg, in *The Procaryotes*, ed. A. Balows, H. G. Trüper, M. Dworkin, W. Harder and K.-H. Schleifer, Springer, New York, 1992, pp. 446–459.

41. M. Heitz, M. Rothenbühler, M. Beerli and B. Marcos, *Water Air Soil Pollut.*, 1995, **83**, 37.

42. E. Morgenroth, E. D. Schroeder, D. P. Y. Chang and K. M. Scow, *J. Air Waste Manage. Assoc.*, 1996, **46**, 300.

43. M. J. Gribbins and R. C. Loehr., *J. Air Waste Manage. Assoc.*, 1998, **48**, 216.

44. M. C. Delhoménie, L. Bibeau, S. Roy, R. Brzezinski and M. Heitz, *J. Chem. Technol. Biotechnol.*, 2001, **76**, 997.

45. Y. Jun and X. Wenfeng, *Bioresour. Technol.*, 2009, **100**, 3869.

46. A. Vincent and J. Hobson, *Odour Control, CIWEM Monographs on Best Practice No. 2*, Terence Dalton Publishing, Suffolk, 1998.

47. Y. Zhang, S. N. Liss and D. G. Allen, *Chem. Eng. Sci.*, 2007, **62**, 2474.

48. W. F. Wright, E. D. Schroeder, D. P. Y. Chang and K. Romstad, *J. Environ. Eng.*, 1997, **123**, 547.

49. S. P. Ottengraf, in *Biotechnology: A Comprehensive Treatise*, ed. H.-J. Rhem and G. Reed, Verlag Chemie, Weinheim, 1986, Ch. 8.

50. P. Le Cloirec, in *Waste Gas Treatment for Resource Recovery*, ed. P. L. Lens, C. Kennes and P. Le Cloirec, IWA Publishing, London, 2006, Ch. 5.

51. M. Roustan, *Transferts gaz: liquide dans les procédés de traitement des eaux et des effluents gazeux*, Tec & Doc, Paris, 2003.

52. M. Coulson and J. F. Richardson, *Chemical Engineering*, Butterworth Heinemann, London, 4th edn, 1997.

53. R. E. Treybal, *Mass Transfer Operations*, McGraw-Hill, Singapore, international edn, 1980.

54. B. E. Rittmann and P. L. McCarty, *J. Environ. Eng.*, 1978, **104**, 889.

55. H. F. Ockeloen, T. J. Overcamp and C. P. L. Grady, *J. Environ. Eng.*, 1996, **122**, 191.

56. C. J. Gantzer, *J. Environ. Eng.*, 1989, **115**, 302.

57. G. Darracq, A. Couvert, C. Couriol, A. Amrane and P. Le Cloirec, *Can. J. Chem. Eng.*, 2010, **88**, 655.

58. E. Dumont, G. Darracq, A. Couvert, C. Couriol, A. Amrane, D. Thomas, Y. Andres and P. Le Cloirec, *Chem. Eng. J.*, 2011, **168**, 241.

59. E. Dumont, G. Darracq, A. Couvert, C. Couriol, A. Amrane, D. Thomas, Y. Andres and P. Le Cloirec, *Chem. Eng. Sci.*, 2012, **71**, 146.

60. G. Quijano, A. Couvert, A. Amrane, G. Darracq, C. Couriol, P. Le Cloirec, L. Paquin and D. Carrie, *Chem. Eng. Sci.*, 2011, **66**, 2707.

61. G. Quijano, A. Couvert, A. Amrane, G. Darracq, C. Couriol, P. Le Cloirec, L. Paquin and D. Carrie, *Chem. Eng. J.*, 2011, **174**, 27.

62. G. Darracq, A. Couvert, C. Couriol, A. Amrane, D. Thomas, E. Dumont, Y. Andres and P. Le Cloirec, *J. Chem. Technol. Biotechnol.*, 2011, **86**, 324.

63. G. Darracq, A. Couvert, C. Couriol, A. Amrane and P. Le Cloirec, *J. Chem. Technol. Biotechnol.*, 2011, **85**, 1156.

64. P. Le Cloirec, P. Humeau and J. Bourcier, in *Environmental Odour Management: Odour Emission – Odour Nuisance – Olfactometry – Electronic Sensors – Odour Abatement*, Proceedings of the European Conference on Environmental Odour Management, 17–19 November, Cologne, Germany, VDI Report No. 1850, 2004, pp. 569–572.

65. P. Humeau, J. N. Baleo, F. Raynaud, J. Bourcier and P. Le Cloirec, *Water Sci. Technol.*, 2000, **41**, 191.

66. P. Le Cloirec, P. Humeau and E. M. Ramirez-Lopez, *Water Sci. Technol.*, 2001, **44**, 219.

67. P. Humeau, P. Pre and P. Le Cloirec, *J. Environ. Eng.*, 2004, **130**, 314.

68. S. Roy, J. Gendron, M. C. Delhomenie, L. Bibeau, M. Heitz and R. Brzezinski, *Applied Microbiology and Biotechnology*, 2003, **61**, 366.

69. M. Converti, DelBorghi and M. Zilli, *Bioprocess Engineering*, 1997, **16**, 105.

70. Y. S. Oh and R. Bartha, *World Journal of Microbiology & Biotechnology*, 1997, **13**, 627.

71. S. J. Ergas, K. Kinney, M. E. Fuller and K. M. Scow, *Biotechnology and Bioengineering*, 1994, **44**, 1048.

72. S. Roy, J. Gendron, M. C. Delhomenie, L. Bibeau, M. Heitz and R. Brzezinski, *Applied Microbiology and Biotechnology*, 2003, **61**, 366.

73. H. H. J. Cox, R. E. Moerman, S. vanBaalen, W. N. M. vanHeiningen, H. J. Doddema and W. Harder, *Biotechnology and Bioengineering*, 1997, **53**, 259.

74. M. Arnold, A. Reittu, A. von Wright, P. J. Martikainen and M. L. Suihko, *Applied Microbiology and Biotechnology*, 1997, **48**, 738.

75. J. H. Jang, M. Hirai and M. Shoda, *J. Haz. Mats*, 2006, **129**, 223.

76. M. Zilli, A. Converti, A. Lodi, M. Delborghi and G. Ferraiolo, *Biotechnology and Bioengineering*, 1993, **41**, 693.

77. M. C. Delhomenie and M. Heitz, *Critical Reviews in Biotechnology*, 2005, **25**, 53.

78. K. Kirchner, G. Hauk and H. J. Rehm, *Applied Microbiology and Biotechnology*, 1987, **26**, 579.

79. Y. C. Chung, C. P. Huang and C. P. Tseng, *Biotechnology Progress*, 1996, **12**, 773.

80. C. P. Huang, Y. C. Chung and B. M. Hsu, *Biotechnology Techniques*, 1996, **10**, 595.

81. E. Smet, H. Van Langenhove and W. Verstraete, *Biodegradation*, 1997, **8**, 53.

82. L. Zhang, M. Hirai and M. Shoda, *J. Ferm. Bioeng*, 1991, **72**, 392.

83. L. Zhang, I. Kuniyoshi, M. Hirai and M. Shoda, *Biotechnology Letters*, 1991, **13**, 223.

84. S. P. P. Ottengraf and J. H. G. Konings, *Bioprocess Engineering*, 1991, **7**, 89.

85. W. J. H. Okkerse, S. P. P. Ottengraf, R. M. M. Diks, B. Osinga-Kuipers and P. Jacobs, *Bioprocess Engineering*, 1999, **20**, 49.

86. R. M. M. Diks and S. P. P. Ottengraf, *Bioprocess Engineering*, 1991, **6**, 93.
87. K. Sun and T. K. Wood, *Biotechnology and Bioengineering*, 1997, **55**, 674.
88. Lipski and K. Altendorf, *Int. J. Syst. Bacteriol.*, 1995, **45**, 717–723.
89. E. S. Lee, J. Y. Park, S. H. Yeom and Y. J. Yoo, *Korean Journal of Chemical Engineering*, 2008, **25**, 139.
90. P. Pré, Y. Andrès, C. Gérente and P. Le Cloirec, *Techniques de l'Ingénieurs*, 2004, **G1**, 780.
91. Kennes, M. Montes, M. Estefania Lopez and M. C. Veiga, *Canadian Journal of Civil Engineering*, 2009, **36**, 1887.
92. T. Plaggemeier, O. Lammerzahl and K. H. Engesser, *International Symposium on Biological Waste Gas Cleaning*, Düsseldorf, Germany, 1997.
93. E. R. Rene, M. Montes, M. C. Veiga and C. Kennes, *Bioresource Technology*, 2011, **102**, 6791; R. Lebrero, J. M. Estrada, R. Munoz and G. Quijano, *Biochem. Eng. J.*, 2012, **60**, 44.
94. R. Lebrero, J. M. Estrada, R. Munoz and G. Quijano, *Biochem. Eng. J.*, **60**, 44.
95. G. Allen, Z. Kong, R. R. Fulthorpe and L. Farhana, in *93rd Annual Meeting & Exhibition of the Air & Waste Manage. Assoc.*, Pittsburg, USA, 2000.
96. U. Heinze and C. G. Freindrich, *Appl. Microbiol. Biotechnol*, 1997, **48**, 411.
97. K. Kirchner, S. Wagner and H. J. Rehm, *Appl. Microbiol. Biotechnol.*, 1992, **37**, 277.
98. M. S. Chou and J. J. Huang, *J. Environ. Eng.*, 1997, **123**, 579.
99. X. Zhu, M. J. Rihn, M. T. Suidan, B. J. Kim and B. R. Kim, *Water Sci. Technol.*, 1996, **34**, 573.
100. N. Y. Fortin and M. A. Deshusses, *Environ. Sci. Technol.*, 1999, **33**, 2980.
101. S. Hartmans and J. Tramper, *Bioproc. Eng.*, 1991, **6**, 83.
102. R. M. M. Diks and S. P. P. Ottengraf, *Bioproc. Eng.*, 1991, **6**, 131.
103. Y. S. Oh and R. Bartha, *J. Appl. Environ. Microbiol.*, 1994, **60**, 2717.
104. J. Mpaniasc and B. C. Baltzis, *Biotechnol. Bioeng.*, 1998, **59**, 328.
105. L. L. Zhang, S. Q. Leng, R. Y. Zhu and J. M. Chen, *Applied Microbiology and Biotechnology*, 2011, **91**, 407.
106. F. Wu, Q. H. Wang, X. H. Sun, N. T. Xue, S. Liu and W. M. Xie, *Waste Manage.*, 2011, **31**, 1702.
107. G. Moussavi, A. Khavanin and A. Sharifi, *Bioresource Technology*, 2011, **102**, 2517.
108. S. G. Wan, G. Y. Li, L. Zu and T. C. An, *Bioresource Technology*, 2011, **102**, 6757.
109. M. S. Chou and J. J. Huang, *J. Environ. Eng.*, 1997, **123**, 569.
110. Y. L. Yang, S. B. Huang, W. Liang, Y. Q. Zhang, H. X. Huang and F. Q. Xu, *J. Haz. Mats*, 2012, **203**, 326.
111. N. J. R. Kraakman, R. W. Melse, B. Koers and J. Van Dijk, USC-TRG Conference on Biofiltration, Tustin, USA, 1998.

112. M. Fortuny, X. Gamisans, M. A. Deshusses, J. Lafuente, C. Casas and D. Gabriel, *Water Research*, 2011, **45**, 5665.
113. W. C. Hugler, J. Cantu De La Garza and M. Villa Garcia, in *89th Annual Meeting & Exhibition of the Air & Waste Manage. Assoc.*, Pittsburg, USA, 1996.
114. M. Fortuny, X. Gamisans, M. A. Deshusses, J. Lafuente, C. Casas and D. Gabriel, *Water Research*, 2011, **45**, 5665.

CHAPTER 6

Bacteria for Bioenergy: A Sustainable Approach Towards Renewability

S. VENKATA MOHAN,*[a,b] M. VENKATESWAR REDDY,[b]
RASHMI CHANDRA,[a,b] G. VENKATA SUBHASH,[a,b]
M. PRATHIMA DEVI[a,b] AND S. SRIKANTH[b]

[a] Academy of Scientific and Innovative Research; [b] Bioengineering and
Environmental Centre (BEEC), CSIR-Indian Institute of Chemical
Technology (CSIR-IICT), Hyderabad 500 007, India
*Email: vmohan_s@yahoo.com

6.1 Introduction

Bioenergy is signified as a sustainable alternative to fossil fuels and is deemed to be the futuristic energy source which is going to change the visage of energy scenario in the near future. Renewable energy is viewed as one of the ways to compensate for the escalating future fuel demands and increasing global warming scenario. Bioenergy derived from microorganisms is of great interest in the present world's energy scenario due to its renewability. Bacteria have flexible and diverse metabolic capability to convert/synthesise variety of organics to various forms of bioenergy. Establishing a practicable link between a terminal electron acceptor limited microorganism and an electron sink is the basis for bioenergy generation. This chapter describes various metabolic options of bacteria as biocatalyst towards harnessing different forms of bioenergy, *viz.* biohydrogen, biomethane, bioelectricity (through microbial fuel cells),

RSC Green Chemistry No. 25
Biomass for Sustainable Applications: Pollution Remediation and Energy
Edited by Sarra Gaspard and Mohamed Chaker Ncibi
© The Royal Society of Chemistry 2014
Published by the Royal Society of Chemistry, www.rsc.org

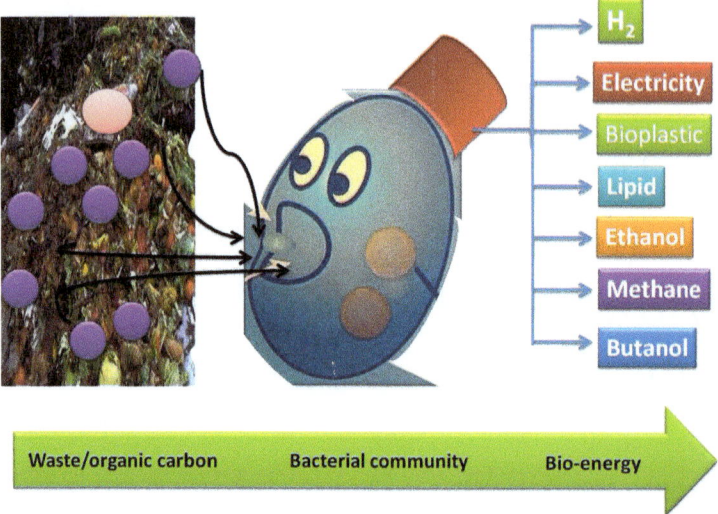

Figure 6.1   Bioenergy generation from bacteria using wastewater.

bioplastics, lipids, bioethanol and biobutanol (Figure 6.1). Fundamental aspects of biochemical reactions towards bioenergy generation are discussed in detail.

6.2   Bioenergy Generation through Microbial Fermentation

Fermentation is the biochemical process that helps in the generation of energy-rich reducing powers—nicotinamide adenine dinucleotide (NADH), flavin adenine dinucleotide (FADH), *etc.*—simultaneously with the conversion of different substrates to their end products through microbial metabolic pathways by involving several enzymes. The reducing powers generated during fermentation process will be oxidised during respiration with simultaneous generation of biological energy molecules (adenosine triphosphate, ATP) in the presence of terminal electron acceptor. However, the respiration is varied based on the electron acceptor available.

During aerobic respiration, the electrons pass through a redox cascade of respiratory/electron transport chain where their energy is gradually transformed to biological energy (ATP) through phosphorylation and are reduced in the presence of an externally available terminal electron acceptor (TEA, oxygen).[1] Oxygen does not act as a TEA in the anaerobic microenvironment and hence the electrons flow through a series of interconversion reactions and finally lead to the formation of energy-rich reduced end products.

During anaerobic respiration, the bacteria have the ability to utilise a wide range of compounds (*viz.* NO_3^-, SO_4^{2-}, organic, inorganic, wastewater, *etc.*) as electron acceptors for ATP generation with their simultaneous reduction. However, if multiple components are present at the same time in the system, their function as electron acceptor depends on the electrochemical and thermodynamic hierarchy. For example, NO_3^- is a more favourable electron acceptor than SO_4^{2-} when both are present due to the thermodynamic favourability. The anaerobic fermentation process helps in generating energy-rich metabolic intermediates and maintains the carbon flow in the metabolic intermediates for longer periods without releasing the end products, unlike aerobic fermentation. Due to the less positive redox potentials of oxidants, the energy gain for the anaerobic metabolism is considerably lower than aerobic respiration.[1] The electrons available under anaerobic microenvironment can be harnessed in various forms of bioenergy.

Both aerobic and anaerobic pathways share the initial glycolysis but aerobic metabolism proceeds with the tricarboxylic acid (TCA) cycle and oxidative phosphorylation, whereas anaerobic process continues with the interconversion (dehydrogenation), decarboxylation and methanogenesis. Glycolysis is the primary metabolic process for both the processes where the glucose molecule is converted to pyruvate, which is the central molecule of the microbial fermentation. The function of the microbial fuel cell (MFC) is based on harnessing these available electrons by artificially introduced electrodes as intermediary electron acceptors. Under aerobic fermentation, the pyruvate enters the TCA cycle and the carbon is removed as CO_2 along with the generation of reducing powers; and it further re-oxidises in the presence of oxygen-generating ATP. Pyruvate has a diverse fate under anaerobic fermentation based on the operational conditions. However, the generation and consumption of reducing powers is common irrespective of the metabolic pathway it enters. In anaerobic fermentation, NADH transfers its electrons to a product of the catabolism, whereas in respiration, NADH transfers electrons to molecules that are abundant in the organism. Pyruvate may enter the acidogenic pathway for the production of biohydrogen (H_2) associated with volatile fatty acids (VFA) such as acetic acid, propionic acid, butyric acid, valeric acid, *etc.*, which are finally converted to biomethane (CH_4) and CO_2 through methanogenesis by acetoclastic methanogens.[2,3] H_2 generated by proton-reducing reactions is a common fermentation by-product generated during electron acceptor limited microbial processes.[4] The protons (H^+) and electrons associated with the reducing powers will be harnessed as bioelectricity when artificial respiration conditions are created by introducing electrodes separated by ion selective membrane. Pyruvate gets converted to ethanol through sequential decarboxylation and dehydrogenation, while pyruvate is converted to butanol through hydration followed by reduction. Pyruvate can also be converted to lactic acid individually (homolactic fermentation) or in combination with other products (heterolactic fermentation). Organic acids generated during fermentation are activated to the corresponding acyl-CoA which further converted to polyhydroxyalkanoates (PHAs, bioplastics). The metabolic intermediates formed during anaerobic fermentation will act as precursor molecules for generating different forms of bioenergy, *viz.* biohydrogen (H_2), CH_4, bioelectricity (electron discharge from the bacterial metabolism to the externally available TEA through electrically connected solid electrodes in MFC), bioplastics, lipids,

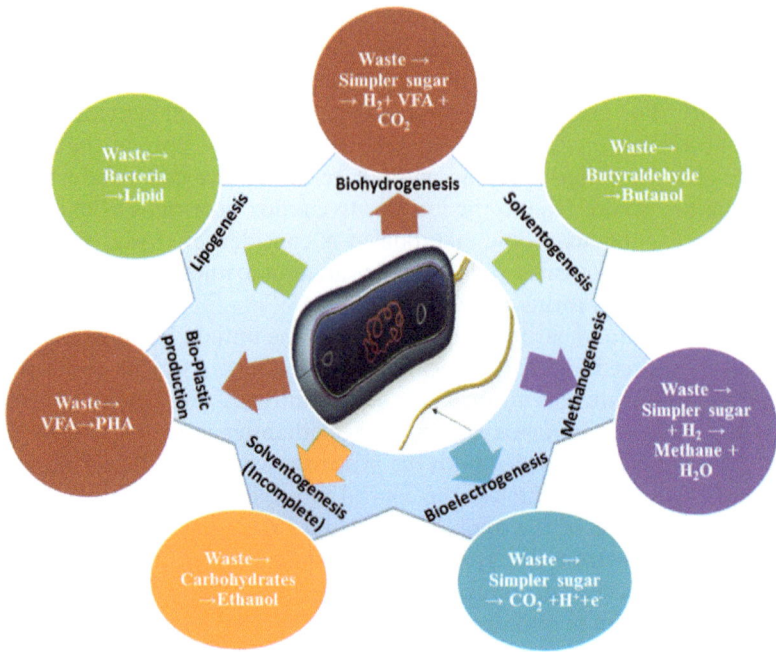

Figure 6.2 Diverse biochemical reactions associated with the bacterial metabolism towards various forms of bioenergy generation using wastewater.

bioethanol and biobutanol (Figure 6.2). Irrespective of the pathway, the reducing powers will be re-oxidised with the help of different electron acceptors forming respective reduced end products, *viz.* CH_4, H_2, NH_3, *etc.*

6.2.1 Nutritional Mode

Nutrition is the biological process by which an organism obtains energy from the surrounding environment and utilises at the cellular level for its metabolism.[5] Nutritional behaviour can be differentiated based on the carbon, energy and electron sources (Table 6.1). Based on the carbon source, the nutritional behaviour of microbes is classified as autotrophic (inorganic) and heterotrophic (organic). Energy requirement is generally available from two sources to the microbes, either from light (phototrophs) or from the oxidation of chemical (organic or inorganic) compounds (chemotrophs). Microorganisms also require electrons which can be obtained from either inorganic substances (lithotrophs) or organic compounds (organotrophs), which are considered as a source of energy. Based on unifying these three parameters, microbes are classified as: photolithotrophic autotrophy (light as energy, inorganic hydrogen as electron donor); chemolithotrophic autotrophy (chemical as energy source, inorganics as electron donor); photoorganotrophic heterotrophy (light as energy source, organic as electron donor); and chemoorganotrophic heterotrophy (chemical energy source, organic as electron donor). Photolithotrophic

Table 6.1 Various nutritional modes with their functional requirements.

| | Autotrophic | | Heterotrophic | |
	Photolithotrophic	Chemolithotrophic	Photoorganotrophic	Chemoorganotrophic
Energy	Light	Chemical energy (inorganic)	Light	Chemical energy (organic)
Carbon	CO_2	CO_2	Organic compound	Organic compound
Electron	H_2O or inorganic	Inorganic	Organic compound	Organic
Representative group	Algae, PSB and cyanobacteria	Sulfur-oxidising bacteria	PSB	Most of the bacteria which includes methanogens and acidogenic bacteria

autotrophy is attributed to the algae, purple and green sulfur bacteria and cyanobacteria, while sulfur and iron oxidising bacteria along with the nitrifying bacteria represent the chemolithotrophic autotrophy (Table 6.1). Photo-organotrophic heterotrophy can be implied to purple and green non-sulfur bacteria, while chemoorganotrophic heterotrophy represents all the other groups of bacteria comprising major fraction of existing microbes (almost all non-photosynthetic bacteria).

Depending on substrate availability, the nutrition mode varies and likewise the dominance of the genera varies, which finally reflects the bioenergy generation. In the case of phototrophy including litho- and organotrophs, the nutrition mode is governed by the photosynthetic mechanism. However, based on the source of electrons and carbon, the metabolic pathway will be diversified into different products. These microbes possess the photosynthetic pigment, chlorophyll (oxygenic photosynthesis) or bacteriochlorophyll (anoxygenic photosynthesis) which helps to trap the sunlight. The electron from the photolysis of water (H_2O) acts as a source of electron for algae and cyanobacteria, while in the case of some photosynthetic bacteria, the electron comes from hydrogen sulfide (H_2S) releasing oxygen and sulphur, respectively. The electron source for these two categories comes from inorganic source and hence they are considered as litho-trophs. During these processes, bioenergy can be generated in the form of H_2 and bioelectricity, apart from the accumulation of carbohydrates which may further converted into lipids depending on the conditions provided. However, some photosynthetic bacteria utilise organic compounds (*viz.* glucose, acetate, butyrate, *etc.*) as electron donor and come under organotrophs of phototrophs. Most of the existing photosynthetic bacteria follow this mode of nutrition and obtain energy from light and carbon from the organic matter. This group is considered as potent H_2 producers through photofermentation (see the section for mechanism of photofermentation) and equally efficient in bioelectricity generation. The photosynthetic mechanism is anoxygenic, which does not release any O_2 and provides a suitable condition for harvesting the reducing equivalents [H^+ and electron] in the form of bioelectricity.[6]

The chemotrophic microbes utilise chemical energy from inorganic or organic sources and hence are classified as lithotrophs and organotrophs, respectively (Table 6.2). In the case of lithotrophs, the electron will generally

Table 6.2 Various metabolic reactions pertaining to different nutritional types and their respective energy products.

Nutritional mode	Expected reaction	Energy produced
Photolithotrophic	1. $6CO_2 + 6H_2O + Light \rightarrow C_6H_{12}O_6 + 6O_2$ 2. $H_2S + Light \rightarrow H_2 + S°$	Biohydrogen, lipid, bioelectricity
Photoorganotrophic	1. $C_6H_{12}O_6 + 6H_2O + Light \rightarrow 6CO_2 + 24H^+ + 24e^-$ 2. $CH_3COOH + 2H_2O + Light \rightarrow 2CO_2 + 8H^+ + 8e^-$ 3. $CH_3CH_2COOH + 3H_2O + Light \rightarrow 3CO_2 + 14H^+ + 14e^-$ 4. $CH_3CH_2COOH + 3H_2O + Light \rightarrow HCO_3^- + CH_3COOH + 7H^+ + 7e^-$ 5. $CH_3CH_2\ CH_2COOH + 6H_2O + Light \rightarrow 4CO_2 + 20H^+ + 20e^-$ 6. $HCOOH + Light \rightarrow CO_2 + 2H^+ + 2e^-$	Biohydrogen, bioelectricity
Chemolithotrophs	1. $2H_2 + O_2 \rightarrow 2H_2O$ 2. $2NO_2^- + O_2 \rightarrow 2NO_3^-$ 3. $2NH_4^+ + 3O_2 \rightarrow 2NO_2^- + 2H_2O + 4H^+$ 4. $2S° + 3O_2 + 2H_2O \rightarrow 2H_2SO_4$ 5. $S_2O_3^{2-} + 2O_2 + H_2O \rightarrow 2SO_4^{2-} + 2H^+$ 6. $4Fe^{2+} + 4H^+ + O_2 \rightarrow 4Fe^{3+} + 2H_2O$	Bioelectricity
Chemoorganotrophs	1. $CH_3COOH + 2H_2O \rightarrow 2CO_2 + 8H^+ + 8e^-$ 2. $CH_3CH_2COOH + 4H_2O \rightarrow 3CO_2 + 14H^+ + 14e^-$ 3. $CH_3CH_2COOH + 3H_2O \rightarrow HCO_3^- + CH_3COOH + 7H^+ + 7e^-$ 4. $CH_3CH_2\ CH_2COOH + 6H_2O \rightarrow 4CO_2 + 20H^+ + 20e^-$ 5. $CH_3CH_2\ CH_2COOH + 2H_2O \rightarrow 2CH_3COOH + 4H^+ + 4e^-$ 6. $HCOOH \rightarrow CO_2 + 2H^+ + 2e^-$	Biohydrogen, bioelectricity, biomethane, butanol, ethanol

come from the sulfur (oxidation of H_2S), H_2 (H_2 to water), iron (oxidation of ferrous compounds into ferric forms), methane (methane into water and CO_2), nitrogenous compounds (ammonia and nitrogen compounds into nitrates), *etc.* These organisms are majorly found in wastewater treatment and during treatment there is a possibility of harnessing bioenergy in the form of bioelectricity. Several reports are available on the simultaneous bioelectricity generation along with treatment of pollutants by MFC and a bioelectrochemical system (BES).[7,8]

Chemolithoautotrophs can be further divided into different groups (*viz.* halophiles, sulfur oxidisers and reducers, nitrifiers, anammox bacteria and thermoacidophiles, *etc.*). Chemoorganotrophs are unable to fix inorganic carbon and they obtain energy by the oxidation of organic compounds in their residing environments. The majority of the light-independent microbes (aerobic and anaerobic) fall in this category, which mainly depend on the degradation of carbon rich organic energy sources such as carbohydrates, lipids and proteins.

Aerobic bacteria involved in the oxidation of the carbon source with simultaneous utilisation of reducing equivalents do not yield any chance to harness bioenergy. On the contrary, anaerobic bacteria (*viz.* methanogenic bacteria, acidogenic bacteria, *etc.*) are not involved in the oxidation reactions that help in conserving the reducing equivalents towards the bioenergy generation.

Similar to the above mentioned nutritional modes, bacteria also undergo a phenomenon called mixotrophy where bacteria rely on both inorganic energy sources and organic carbon sources for growth and survival. Few microbial species show great metabolic flexibility and alter their metabolic patterns in response to the environmental changes. For example, many purple non-sulfur bacteria act as photoorganotrophs in the presence of light, while in the absence of light, they function as chemoorganotrophs. Similarly, *Beggiatoa* rely on both organic and inorganic energy sources. These microbes are called as mixotrophic organisms (mixotrophs) as they combine both autotrophic and heterotrophic metabolic processes. This sort of flexibility seems beneficial for the breakdown of complex chemicals and provides a definite advantage towards bioenergy generation. The syntrophic association of microbes present in mixed culture can also be considered as mixotrophy as the mutual interactions among their metabolic activities will have the added advantage of converting multiple components into products, especially when considering bioenergy from wastewater. The beauty of this process relies on the fact that the metabolic products of one organism will be the precursor substrate molecules for the other organisms. The flux/flow of electron through a series of metabolic reactions results in harnessing of the bioenergy.

6.2.2 Carbon Flow

Initially glucose containing six carbon atoms (6C) is converted to two molecules of pyruvate with three carbons (3C) through glycolysis in both the aerobic and anaerobic metabolism (Figure 6.3). In extension to that, during aerobic metabolism, the pyruvate enters the TCA cycle, where it passes through

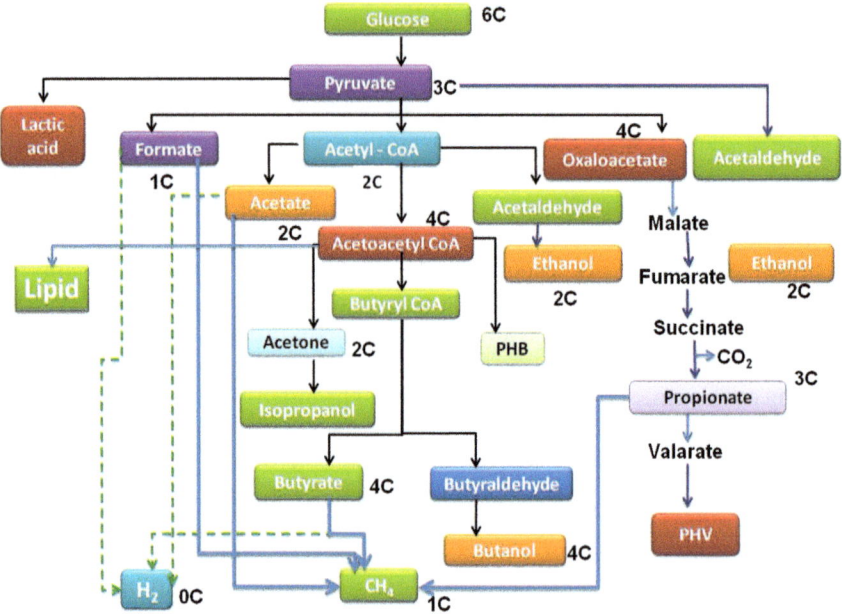

Figure 6.3   Carbon flow through various metabolic routes of microbial fermentation.

several dehydrogenation and decarboxylation reactions generating the energy rich reducing powers (NAD^+/FAD^+). These reducing powers will be delivered to an electron transport chain driven by oxygen as the terminal electron acceptor generating a proton motive force which helps in the addition of inorganic phosphate to adenosine diphosphate (ADP) generating biological energy (ATP) molecules. As long as there is an extracellular terminal electron acceptor available to drive these reducing powers through the electron transport chain, ATP production is assured. This energy is essential for cell growth and metabolic activities, but cannot be conserved. On the contrary, the 3C molecule, pyruvate, has a diverse fate under anaerobic metabolism, based on the operational conditions provided. It may pass through acidogenic process where the volatile acids (2C to 4C) will be generated along with the reducing powers or else it may pass through the solventogenesis (aldehydes and alcohols, 2C to 3C) where the reducing powers will be consumed. The reducing powers generated during acidogenesis have a higher energy value where they can be conserved in the form of biohydrogen or bioelectricity.

Furthermore these acids may pass through diverse energy-accumulating pathways such as methane (1C), acetone (2C), butyraldehyde (4C), butanol (4C) and polyhydroxyalkanoates which have higher energy. Microbes will have some fine-tuned enzymatic pathways towards oxidation and reduction of the inorganic (chemolithotrophs) as well as organic (chemoorganotrophs) compounds delivering the electron to the electron transport systems where the energy can be generated either in the biological form (ATP) or in the bioenergy form (biohydrogen, bioelectricity, *etc.*). Anaerobic metabolism provides the feasibility to harness bioenergy over aerobic metabolism. If oxygen is not available as electron acceptor, the bacteria become enabled to use other terminal electron acceptors such as oxidised organic compounds (resulting in reduced energy-rich compounds) and pollutants (resulting in treatment). However, based on the thermodynamic hierarchy of the reactions, the microbes will switch over to other biochemical pathways. Many anaerobic processes are not catalysed by a single microbe but need a syntrophic association of physiologically distinctive organisms that leads to organic carbon driven reduction of iron, manganese, sulfate, *etc.*, along with the acetogenesis and methanogenesis.

6.3   Bioenergy

6.3.1   Biohydrogen (H_2)

H_2 has been recognised as a promising, green and an ideal energy carrier for the future due to its cleaner efficiency, high energy yield (122 kJ g^{-1}), renewable and sustainable nature.[9] Currently, global H_2 production is mainly from fossil sources and electrolysis of water. In recent times, biological routes of H_2 production gained promising attention among the research community around the globe.[10–22]

Broadly speaking, biological process of H_2 production can be classified into biophotolysis (direct and indirect) and fermentation (dark and photo) processes, or combination of these processes. Irrespective of the mechanism, H_2 production occurs only in the absence of O_2 during fermentation in wide group of microbes. Cyanobacteria and microalgae undergo direct and indirect biophotolysis and produce H_2 by utilising inorganic CO_2 in the presence of sunlight and water, while photosynthetic bacteria (PSB) manifest H_2 production through photofermentation consuming a wide variety of substrates ranging from inorganic to organic acids in the presence of light. However, PSB are considered to be more promising H_2 producers over microalgae due to the variation in their metabolisms. However, dark fermentation by anaerobic (acidogenic) bacteria is the most widely understood process for H_2 production using a wide variety of organic carbon. Both obligate and facultative bacteria are capable of H_2 production in an oxygen-free environment.

6.3.1.1   Photobiological H_2 Production

Photobiological H_2 production is feasible through two diverse photosynthetic mechanisms (*viz.* oxygenic and anoxygenic) catalysed by three physiologically distinct groups of microbes. Microalgae and cyanobacteria facilitate oxygenic photosynthesis, while PSB catalyse anoxygenic photosynthesis. In the oxygenic photosynthesis, O_2 as a by-product is released by biophotolysis of water, where the H_2 is generated either of the two mechanisms, *viz.* direct and indirect photolysis (Figures 6.4 and 6.5).

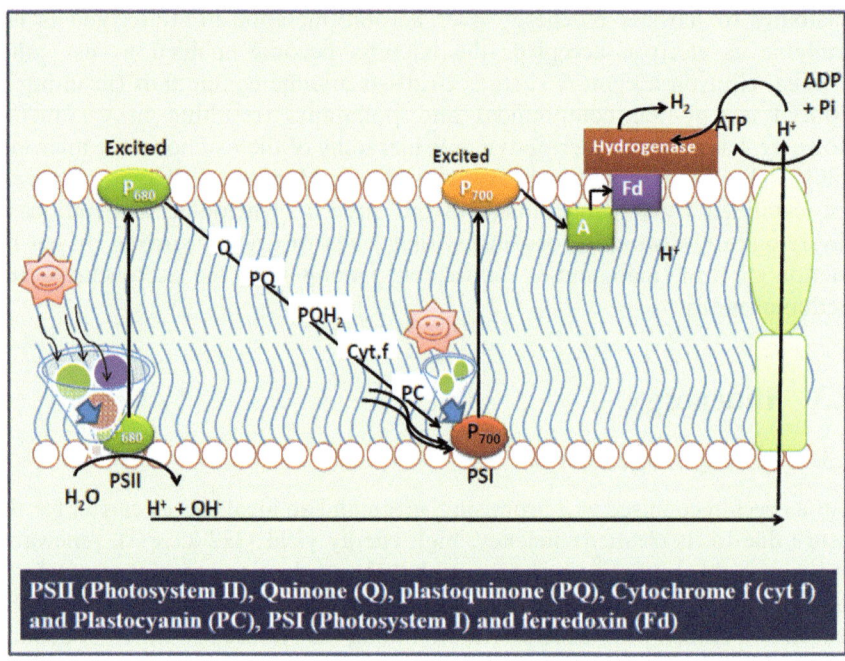

Figure 6.4 Mechanism of direct and indirect biophotolysis involved in oxygenic photosynthesis.

In both direct and indirect photolysis, photosystem II (PSII) drives the first stage of the process by splitting water molecule into H^+, electrons and O_2. The H^+ and electrons are transferred through a cascade of membrane proteins and finally are reduced to H_2 by the Fe-Fe hydrogenase.[23–25] However, the components involved in the electron transfer from the excitation (PSII) to the sink (Fe-Fe hydrogenase) are varied from a direct to an indirect biophotolysis mechanism. In the case of direct photolysis, photons will be captured by the light-harvesting complex (LHC) and transferred to the P_{680} (PSII) which becomes exited followed by splitting one water molecule to H^+, electron and O_2, discharging the electrons into quinine pool. These electrons will flow through a series of proteins (by redox reactions), plastoquinone, cytochrome f (Cyt f) and plastocyanin, finally reaching P_{700} (photosystem I, PSI) and making it exited. Electrons from the exited P_{700} will be accepted by ferredoxin (an iron-containing electron carrier) with the help of unknown mediators and finally reduced to H_2 by the Fe-Fe hydrogenase in the presence of H^+ (*cf.* Figure 6.4). Whereas, in the case of indirect photolysis, the H^+ and electrons generated from the photolysis will pass through the direct photolysis mechanism, but these electrons govern the formation of carbohydrates instead of being accepted by the Fe-Fe hydrogenase. The H^+ and electrons generated from the oxidation of the carbohydrates will also be accepted by the plastoquinone, which is reduced to dihydroplastoquinone (PQH_2) and then transferred to the interior of the thylakoid. These H^+ and electrons will then be accepted

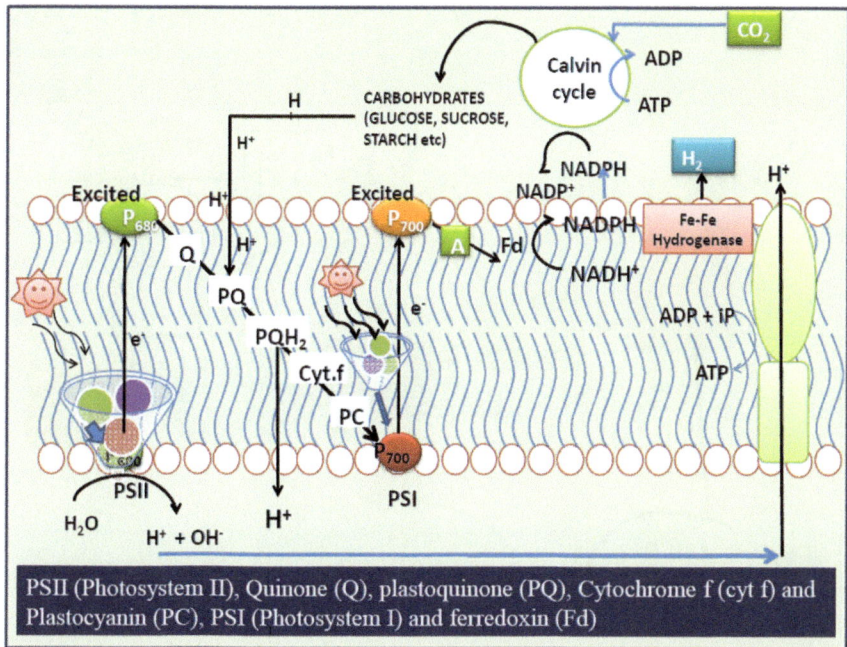

PSII (Photosystem II), Quinone (Q), plastoquinone (PQ), Cytochrome f (cyt f) and Plastocyanin (PC), PSI (Photosystem I) and ferredoxin (Fd)

Figure 6.5   Mechanism of indirect biophotolysis involved in oxygenic photosynthesis.

by the Fe-Fe hydrogenase through the ATPase complex and reduced to H_2 (*cf.* Figure 6.5).

The hydrogenase function is based on the O_2 availability in the system irrespective of the photolysis mechanism. Hydrogenase activity also depends on the supply of electrons and H^+ derived either directly by photosynthetic water splitting (driven by PSII) or indirectly from the degradation of organic molecules (starch). Thus, hydrogenase essentially acts as a H^+/electron release valve by recombining H^+ (from the medium) and electrons (from reduced ferredoxin) to produce H_2. Alternatively, the H_2 production can also be catalysed by nitrogenase during nitrogen fixation and the process is extremely O_2 sensitive. Cyanobacteria develop mechanisms for protecting nitrogenase from O_2 through localisation of nitrogenase in the heterocysts of filamentous cyanobacteria.[26] Investigations into prolongation and optimisation of H_2 production have revealed that the H_2-producing activity of cyanobacteria is stimulated by nitrogen starvation. The scientific challenge associated with H_2 production at cellular level is the consumption of H_2 by the enzyme 'reversible hydrogenase' or 'bidirectional hydrogenase'.[27] Bidirectional hydrogenases are mainly involved in the disposal of excess reducing power derived from fermentation and photosynthesis resulting in H_2 evolution.

On the contrary, anoxygenic photosynthesis by PSB will not yield any oxygen and hence it is advantageous for H_2 production. PSB are potent H_2 producers compared with cyanobacteria in the presence of sunlight through the

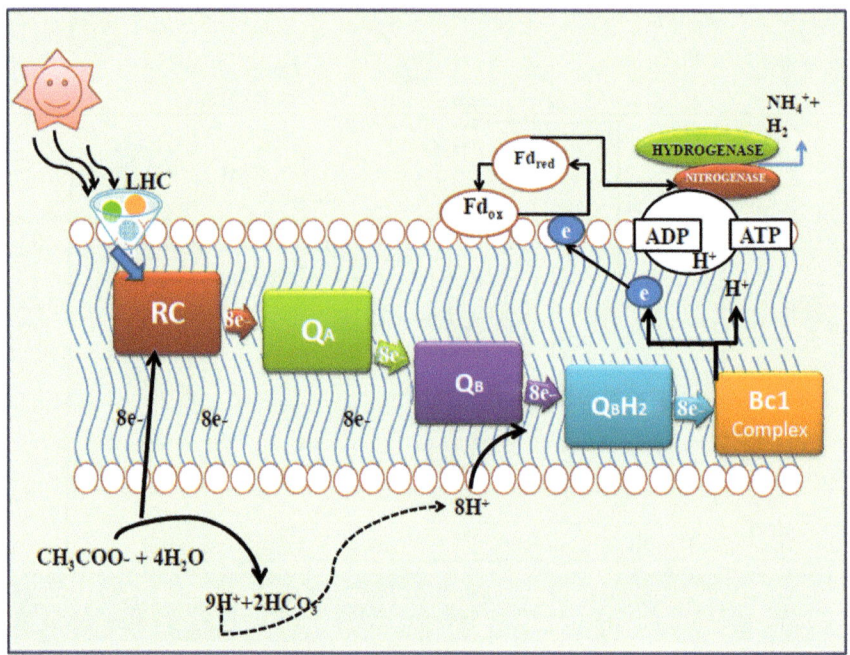

Figure 6.6   Mechanism of anoxygenic photosynthesis.

photofermentation process by utilising variety of carbon sources (carbo-hydrates and VFA). PSB neither use water as an electron source nor produce O_2 photosynthetically, and thus they are strictly anoxygenic in nature.[28] H_2 production in PSB is mediated by hydrogenase and nitrogenase activities which do not become inhibited due to the non-production of O_2. Light absorption by a dimer of bacteriochlorophyll (BChl) molecules instigates the reaction forming bacteriopheophytin (BPh)[1,29] (Figure 6.6). The electron transfer proceeds from BPh to a quinine pool and then to the cytochrome subunit of the reaction centre generating a H^+ gradient which drives the ATP generation and is finally reduced to H_2.[28,30] The efficiency of light energy conversion to H_2 by PSB is much higher than that by cyanobacteria because of the lower quantum of light energy requirement than water photolysis.[26,31,32] Among the photosynthetic processes of H_2 production, photofermentation route seems to be favourable due to the availability of relatively higher substrate to H_2 yields, its ability to trap energy over a wide range of the light spectrum, and its versatility in terms of sources of metabolic substrates.[33]

6.3.1.2   Dark Fermentation

Both obligate and facultative acidogenic bacteria catalyse H_2 production in association with the production of VFA and CO_2 by a dark fermentation process utilising organic carbon as primary substrate.[16,17,34–39] Fermentative conversion requires a series of interrelated steps (hydrolysis, acidogenesis and acetogenesis) with five physiologically distinct groups of microorganisms to convert organic compounds from complex to simple molecules (Figure 6.7).

Hydrolytic microorganisms convert complex organic compounds to monomers through hydrolysis. Obligatory H_2 producing acidogenic bacteria ferment the monomers to a mixture of low molecular weight organic acids and H_2 through acidogenesis (eqns (6.1) to (6.6)). This also includes acetate production from H_2 and CO_2 by acetogens and homoacetogens, and finally acetoclastic methanogens convert the organic acids to CH_4 and CO_2 through methanogenesis.[2,3] Acidogenic bacteria grow by syntrophic association with the hydrogenotrophic methanogens (H_2 consuming, methanogenic bacteria) and keep the H_2 partial pressure low enough to allow acidogenesis to become thermodynamically favourable by interspecies H_2 transfer. Methanogenic activity needs to be restricted to make H_2 as the metabolic end product. Considering glucose as substrate, fermentative H_2 production starts with the conversion of glucose to pyruvate through glycolysis by both obligate and facultative anaerobic bacteria (Figure 6.7).

$$CH_3COOH + 2H_2O + Light \rightarrow 2CO_2 + 8H^+ + 8\,electron \qquad (6.1)$$

$$CH_3CH_2COOH + 3H_2O \rightarrow HCO_3^- + CH_3COOH + 7H^+ + 7\,electron \quad (6.2)$$

$$C_6H_{12}O_6 + 2H_2O \rightarrow 2CH_3COOH + 2CO_2 + 4H_2 \qquad (6.3)$$

$$C_6H_{12}O_6 \rightarrow CH_3CH_2CH_2COOH + 2CO_2 + 2H_2 \qquad (6.4)$$

$$C_6H_{12}O_6 + 2H_2 \rightarrow 2CH_3CH_2COOH + 2H_2O \qquad (6.5)$$

$$C_6H_{12}O_6 \rightarrow 2CH_3CH_2OH + 2CO_2 \qquad (6.6)$$

Interconversion of substrates takes place through degradation that increases the availability of H^+ and electron in the cell under anaerobic conditions. The H^+ and electrons associated with redox mediators are the main sources of fermentative H_2 production. In bacterial system the electron transport chain has membrane bound protein complexes (NADH dehydrogenase and Cyt bc1) and a mobile carrier (quinine and cytochrome C) to facilitate electron transport (Figure 6.7(b)). Proton shuffling and electron transport are mainly governed by membrane-bound proteins and redox mediators. Initially the H^+ from redox mediators (NADH/FADH) are detached in the presence of NADH-dehydrogenase enzyme. This protein complex (NADH-dehydrogenase) transfers the H^+ to the intermembrane space and electron to the quinine pool. In the quinine pool there is a continuous inter conversion of quinine and H^+ to QH_2 and *vice versa*. During interconversion the electrons are transferred to the cytochrome bc1 (Cyt bc1) complex and again the H^+ are transferred to the intermembrane space. The electron from the Cyt bc1 complex is again excited and transferred to the cytochrome, then to the iron-containing protein ferredoxin and is reduced to H_2 in presence of hydrogenase.[35] In facultative anaerobes, pyruvate gets converts to acetyl-CoA and formate by the catalytic function of pyruvate formate lyase (PFL).[35] H_2 is produced from formate by

(a)

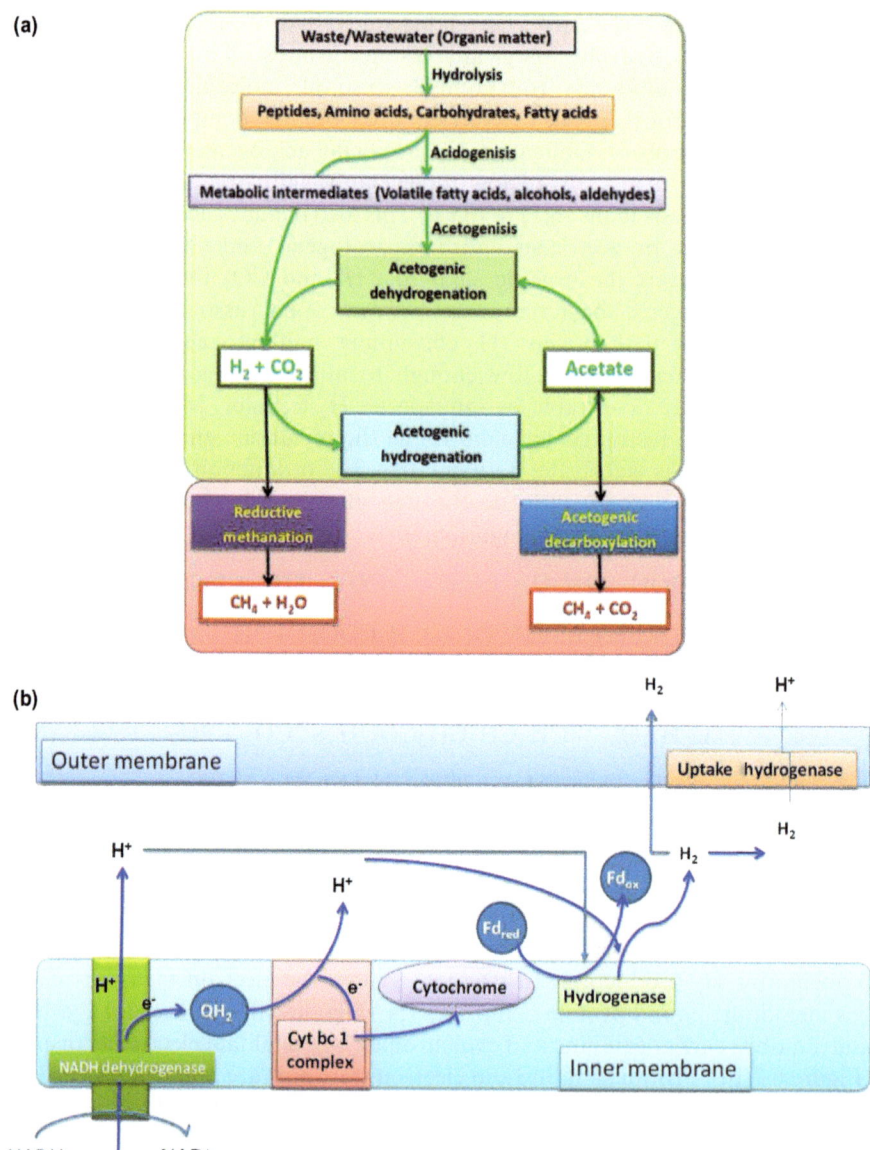

Figure 6.7 Schematic illustration of: (a) substrate conversion to its end product during anaerobic fermentation; and (b) proton transfer through membrane-bound components during acidogenic fermentation associated with H_2 production mechanism.

the formate hydrogen lyase (FHL) complex. In obligate anaerobes, pyruvate is converted to acetyl-CoA and CO_2 through pyruvate ferredoxin oxidoreductase (PFOR) and this oxidation process requires reduction of ferredoxin.[35–40] Generally, in the anaerobic microenvironment, interconversion of substrates takes place through degradation that increases the availability of H^+ in the cell. The H^+ and electrons associated with redox mediators are the main sources of fermentative H_2 production. The H^+ from redox mediators (NAD^+/FAD^+) are detached in the presence of NADH-dehydrogenase and are reduced to H_2 in presence of the hydrogenase enzyme with the help of electrons donated by oxidized ferredoxin (co-factor).[35,41] Proton shuttling takes place between metabolic intermediates with the help of various redox mediators. Redox mediators are capable of carrying H^+ and electrons, otherwise known as energy carriers, as they are involved in ATP generation.[42,43]

Hydrogenase and nitrogenase are two important enzymes involved in H_2 production as they catalyse the reversible reduction of H^+ to H_2.[34] Both enzymes contain complex metal clusters at their active site with diverse subunits. In hydrogenase, mainly two types of enzymes are present, *i.e.* Fe-hydrogenase and Ni-Fe hydrogenase.[34,44] Fe-hydrogenase involves the removal of excess reducing equivalents throughout the metabolism. Ni-Fe hydrogenase is also called an uptake hydrogenase and is mainly involved in the quinine pool, where electrons are used directly or indirectly for reduction of nicotinamide adenine dinucleotide (NAD)/nicotinamide adenine dinucleotide phosphate ($NADP^+$) reduction.[34] The nitrogenase enzyme contains two component protein systems (Mo-Fe and Fe protein) which uses Mg-ATP and electrons to reduce various substrates with simultaneous H_2 production.[44] Dehydrogenase is another important enzyme carrying out interconversion of metabolites and the transfer of H^+ between metabolic intermediates through redox reactions using several mediators (NAD^+, FAD^+, *etc.*).[41,42] Both dehydrogenase and hydrogenase functions are important to maintain H^+ equilibrium in the cell and to reduce H^+ to H_2.

The soluble acid metabolites or VFA formed during fermentation helps in understanding the pathway used by the microorganism and the corresponding H_2 yields.

Using mixed consortia as a biocatalyst is a promising and practical option for scaling up the biohydrogen technology, especially when wastewater is used as substrate.[45,46] From an engineering point of view, producing H_2 by mixed culture offers lower operational costs and ease of control in concurrence to the possibility of using waste/wastewater as feedstock.[39] The feasibility of H_2 production with typical anaerobic consortia is limited as it is rapidly consumed by methanogens.[39] Application of pretreatment to the parent inoculum facilitates the selective enrichment of acidogenic bacteria capable of producing H_2 as the end-product with the simultaneous prevention of hydrogenotrophic methanogens.[47–49]

Physiological differences between H_2 producing bacteria (acidogenic bacteria) and H_2 uptake bacteria (methanogenic bacteria) form the fundamental basis for the methods used for the preparation of H_2 producing inoculum.[42,47–50] Under extreme conditions/environment (*viz.* high temperature, extreme acidity and alkalinity, *etc.*), H_2 producing bacteria usually form protective spores; as methanogens do not have such capability, this forms the basis for the pretreatment methods developed.[42,47–49] Pretreatment also prevents competitive growth and co-existence of other H_2 consuming bacteria. Different pretreatment methods (*viz.* heat-shock, chemical, acid-shock, alkaline–shock, oxygen-shock, load-shock, infrared irradiation, microwave irradiation, freezing and thawing, *etc.*) are reported for the selective enrichment of H_2 producing inoculums.[42,47,49,51] The fate of pyruvate depends on the operating pH. Under acidic conditions, pyruvate is converted to VFA along with H_2 by acidogenic bacteria. Neutral pH operation leads to the formation of CH_4 and CO_2 by methanogenic bacteria (Figure 6.7(a)). At basic pH, anaerobic digestion leads to solventogenesis. Hydrogenase activity is higher at acidic pH but with an increase in pH, the metabolic pathway proceeds to the next step of anaerobic digestion where H^+ are reduced to CH_4 (methanogenesis) or ethanol (solventogenesis).

Biohydrogen is presently at the developmental stage. The passing decade produced significant work on biohydrogen in both basic and applied fields. During the initial phase of biohydrogen research, much attention was paid to the photobiological routes using specific strains and defined medium. Low rates of H_2 production due to the presence of a complex reaction system, the inhibitory effect of O_2 on hydrogenase and nitrogenase,[52,53] and lower utilisation of waste[18,54] are some of the inherent disadvantages linked with the photobiological process. The high activation energy to drive hydrogenase and the low solar conversion efficiencies are also considered major limitations for this process.

The past decade has witnessed significant progress in the dark fermentation process due to its feasibility of utilising a broad range of wastewater as substrate with mixed cultures as biocatalysts. This process does not rely on the light, requires less energy, is technically much simpler, requires low operating costs and is more stable.[17,19,34,36,40,55,56] Exploitation of wastewater as substrate for H_2 production with simultaneous wastewater treatment is leading to open a new avenue for the utilisation of renewable and inexhaustible energy sources.[11–15,18–20,57–59] In conjunction with the wastewater treatment, this process is capable of solving two issues, *viz.* reduction of pollutants in waste and the generation of a clean alternative fuel.[60,61]

The process simplicity, efficiency and smaller footprint are some of the striking features of the dark fermentation process which makes it practically more feasible for the mass production of H_2.[21] Much of the literature on biohydrogen production is concerned with wastewater utility by the dark fermentation process. However, optimization of various process parameters is essential for the up-scaling of the technology.[39,62–65] On the contrary, photosynthetic bacteria can readily utilize the organic acids generated form dark fermentation process to produce additional H_2.[30,33]

6.3.2   Biomethane (CH_4)

Biomethanisation (fermentative methane production or anaerobic digestion) is complimentary to the H_2 production process (Figure 6.8). Methanogens are both obligate and facultative anaerobes, functioning on a heterotrophic or an autotrophic mechanism based on the carbon source utilised. Autotrophs grow

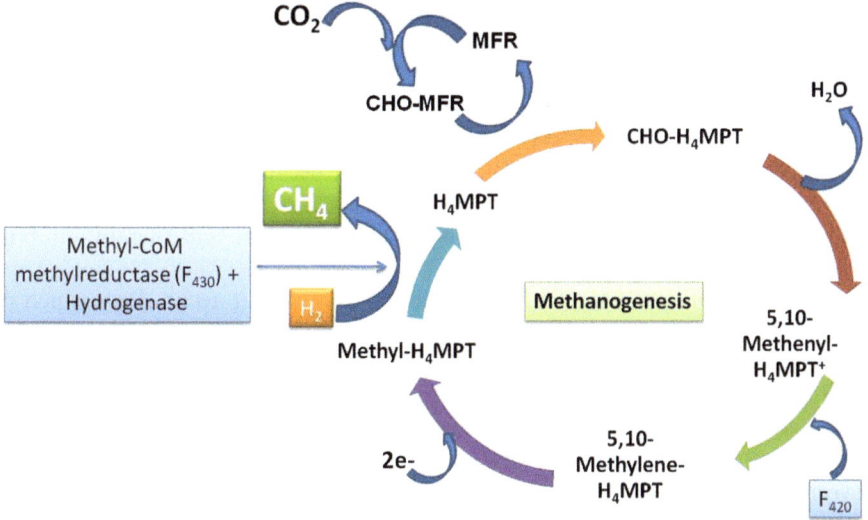

Figure 6.8   Mechanism of methanogenesis during methane production.

on H_2 and CO_2.[5] They obtain energy by converting CO_2, H_2, formate, methanol, acetate and other compounds to either CH_4 or CH_4 and CO_2. Methanogenic bacteria produce CH_4 as the major end product of methanogenic metabolism. The primary step is the reduction where CO_2 is reduced to a methyl group with electrons derived from the oxidation of electron donors (primarily H_2 or formate) (Figure 6.8). The methanogenesis starts when the CO_2 combines with the coenzyme methanofuran (MFR) and forms CHO-MFR. This CHO-MFR binds with the co-factors tetrahydromethanopterin (H_4MPT) and form CHO-H_4MPT associated with the release of MFR. MFR again binds to the other CO_2 molecule. CHO-H_4MPT releases H_2O and is converted to 5,10-methylene-H_4MPT$^+$. F420 is a complex metalloorganic molecule which helps in the conversion of 5,10-methylene-H_4MPT$^+$ to methyl-H_4MPT by gaining two electrons. The co-factor F420 plays a major role in this step as carrier for electrons and H^+. Methyl-H_4MPT reacts with the H_2 and releases CH_4 by allocating its methyl group. In this reaction the co-factor F430 (nickel tetrapyrrole) for the enzyme methyl-CoM methylreductase works collaboratively with bidirectional hydrogenase (uptake hydrogenase) and liberates the CH_4. H_4MPT again binds to another CHO-MFR molecule and makes itself cyclic. H_4MPT and MFR play an important role in CO_2 reduction to CH_4. The electrons coming from a membrane-bound electron transport chain coupled to proton extrusion provide a chemical gradient that drives ATP synthesis by a membrane-bound ATPase.[66] The redox reactions, catalysed by the membrane-bound electron transport chains, are coupled to H^+ translocation across the cytoplasmic membrane.[4,67] The CH_4 production process is energy expensive in terms of ATP consumption. CH_4 formation creates an electrochemical gradient across cell membrane, used to generate ATP through chemiosmosis.[2]

Methanogens are the largest group of archaea having five orders (*Methanobacteriales, Methanococcales, Methanomicrobiales, Methanosarcinales* and *Methanopyrales*) and 26 genera. Anaerobic digestion allows the co-existence of acidogenic bacteria and methanogenic bacteria which differ widely in terms of physiology, nutritional needs, growth kinetics and environmental sensitivity.[68] Methanogens are sensitive to temperature and pH, where the optimum range is between 15 °C to 100 °C and 6.5 to 7.5, respectively. Shifting either side of the optimum range affects metabolic rates and thereby inhibits CH_4 production.

The anaerobic process of CH_4 production is a mature technology which is extensively employed for the treatment of sewage and high strength industrial wastewater.[3,69,70] The process offers significant advantages, *viz.* low amounts of sludge production, less energy requirements, and maximum energy recovery.[71] The disadvantages of anaerobic process are slow growth of methanogenic organisms and the instability of anaerobic processes upset by toxic substrates or overloading. Most of the advances in anaerobic technology have been achieved during the last four decades with the development of high-rate reactors and a better knowledge of the microbiology of methanogenic ecosystems.

6.3.3   Bioelectricity Production in a Microbial Fuel Cell

More recently harnessing of bioenergy in the form of bioelectricity using microbial fuel cells has garnered significant interest in both basic and applied research due to its sustainable and renewable nature. Electrons in reduced substrates (*e.g.* carbohydrates, organic acids, H_2S, CH_4) begin an 'electric circuit' that ends when the electrons reach the electron sink furnished by a final electron acceptor.[4] Establishing an electrical circuit between the electron source (from the substrate metabolism) and the electron sink (molecular oxygen) by placing non-catalysed electrodes (as intermediary electron acceptor) along the membrane facilitates potential development which can be harnessed as bioelectricity.[72–75] The oxidation reaction (generating the reducing equivalents), separated by a selectively permeable ionic membrane (generally a proton exchange membrane), allows capture of the electron by a hybrid bio-electrochemical system called a microbial fuel cell (MFC).[43,76–79] The MFC facilitates direct transformation of chemical energy stored in the bioconvertible substrate to electrical energy *via* microbial catalysed redox reactions (fermentative) under ambient temperature/pressure in the absence of oxygen. The MFC functions on the basis of anodic oxidation and cathodic reduction reactions. The anode chamber is a biofactory which facilitates the generation of reducing equivalents [H^+ and electrons] through a series of bio-electrochemical redox reactions during substrate degradation under anaerobic microenvironment (eqn (6.7)). Protons reach the cathode through a proton exchange membrane (PEM), resulting in a potential difference against which the electrons flow through the circuit (current) towards the counter electrode (cathode) and are reduced in presence of oxygen forming water (eqns (6.8) and (6.9)).

$$C_6H_{12}O_6 + 6H_2O \rightarrow 6CO_2 + 24H^+ + 24e^-(\text{anode}) \tag{6.7}$$

$$4e^- + 4H^+ + O_2 \rightarrow 2H_2O(\text{cathode}) \tag{6.8}$$

$$C_6H_{12}O_6 + 6H_2O + 6O_2 \rightarrow 6CO_2 + 12H_2O(\text{overall}) \tag{6.9}$$

Most of the research on MFCs has been confined to the operation of the anodic chamber with anaerobic metabolic function. Very few reports are available in the literature pertaining to the application of aerobic metabolic function.[72,80,81] In the presence of oxygen (under prevailing oxygenated microenvironment), the H^+ are reduced to form water. Here, higher energy gains are possible with the positive redox potential of a terminal electron acceptor. But in the presence of low oxygen (dissolved) concentration, it may not be possible to reduce all the protons released. The remaining protons cross the PEM and form a gradient against which the electrons will flow through the circuit in the MFC.

Electrons derived from substrate oxidation are donated to either organic electron acceptors/carriers (E^+/E–H) (fermentation) or to O_2 by way of an electron transport chain (aerobic respiration). Under an anoxic micro-environment the facultative aerobic/anaerobic bacteria which survive in either oxygenated or deoxygenated microenvironments can switch between aerobic oxidation and fermentation processes.[72] In anoxic operation, the protons will not be completely neutralised due to the lower levels of O_2. In the absence of O_2, reduced coenzymes are not oxidized by the electron transport chain due to the unavailability of a terminal electron acceptor. This condition brought forward the necessity of searching for alternative electron acceptors and the electrodes (anode and cathode) help the biocatalyst to re-oxidize the reducing powers thereby serving as electron acceptors. Recently, utilisation of waste-water[8,76,77,82–85] as well as solid waste[86] as anodic fuel was established which also facilitates their simultaneous treatment.

Apart from power generation, MFC application has been extended to multiple functions—including waste remediation, removal of specific pollutants and the recovery of value added products based on the electron donor and acceptor conditions—and renamed a bio-electrochemical treatment (BET) system. When a BET system is operated with wastewater composed of diverse components, some of these act as an electron acceptors either in the anode (between the biocatalyst and electrode), or in the cathode chamber (between electrode and TEA), and are reduced to their respective end products.[6,7,77,85] For example, nitrate-rich wastewater can be treated by providing conditions so that the nitrate will act as a TEA in the cathode chamber. Similarly, sulfates and other organic and inorganic pollutants can be treated by providing the requisite electron acceptor conditions.

The principle of BET relies on the fact that electrochemically active micro-organisms can transfer electrons from a reduced electron donor to an electrode and finally to an oxidised electron acceptor, generating power. The anode chamber of a BET system resembles the conventional anaerobic bioreactor and mimics the function of a conventional electrochemical cell used for wastewater treatment where the gradient created between anodic oxidation and cathodic reduction reactions helps in the degradation of complex organic matter and toxic/xenobiotic pollutants.[8,85] The synergistic interaction between the MFC components and the biocatalyst needs to be understood and optimised to fully exploit the capacities of these systems to maximise product recovery and energy generation.[87]

Similar to dark fermentation, photosynthetic fuel cells can also be operated using algae/cyanobacteria or PSB based on their photosynthetic mechanism (oxygenic or anoxygenic). In anoxygenic photosynthesis, the organisms use Bchl molecules to exploit sunlight as an energy source and organic waste/CO_2 as the electron source to direct the reaction centre of the photosystem.[6] Light energy captured by pigment molecules channels to bacterial chlorophyll *a* (Bchl *a*) and is excited from the ground state, triggering a series of photochemical reactions and thus separating the positive and negative charges across the membrane without O_2 generation (anoxygenic). Charge separation initiates a series of electron transfer reactions that are coupled to the translocation of protons across the membrane, generating an electrochemical proton gradient called proton motive force (PMF) which results in the development of bio-potential (biological mediated voltage), facilitating electrogenesis.[27,88] However, oxygenic photosynthesis requires three membrane-bound protein complexes (PSI, PSII and cytochrome bf complex) operating in series for the transport of electrons from water to $NADP^+$ generating O_2 (oxygenic). Electrons are transferred between these large protein complexes by small mobile molecules (plastoquinone and plastocyanin). These small molecules carry electrons over relatively long distances and play a unique role in photosynthetic energy conversion.

Unlike the dark fermentative condition, very few and specific reports are available on the use of the photosynthetic mechanism for fuel cell operation. Most of these studies relate to single strains of green algae, cyanobacteria and PSB acting as an anodic biocatalyst by adapting either the photoautotrophic or photoheterotrophic mode of nutrition. The metabolic activity of the biocatalyst used and nutrition mode adopted generally govern the efficiency of the photosynthetic fuel cell (PhFC). The photoautotrophic mode integrated with the oxygenic photosynthesis condition was studied with *Chlamydomonas reinhardtii, Phormidium, Nostoc, Spirulina, Anabaena, Synechocystis* PCC-6803, *etc.* as the anodic biocatalyst.[89–91] The photoheterotrophic mode of nutrition has been mostly evaluated with the anoxygenic photosynthesis condition using photosynthetic bacteria like *Rhodopseudomonas palustris, Rhodobacter sphaeroides, Rhodobacter, Rhodopseudomonas, etc.* as biocatalyst.[90–94] Anoxygenic photosynthesis is more favourable for electron recovery than oxygenic photosynthesis due to O_2 generation during oxygenic photosynthesis.

6.3.4 Biodegradable Plastics

Synthetic plastics make up a large fraction of buried waste due to their non-biodegradable nature.[95] Bioplastics in the form of polyhydroxyalkanoates

(PHAs) are an alternative to these plastics that could lower their contribution to municipal landfills by about 20% of the total waste by volume and 10% by weight.[96] PHAs are high molecular weight (50–100 kDa) carbonaceous, cellular reserve storage products that guard against starvation in bacteria. They occur in many types of Gram-positive and Gram-negative bacteria under excess carbon and nutrient-deprived conditions.[4] They possess many traits similar to modern synthetic plastics: a high degree of polymerisation, highly crystalline, optically active and insoluble in water.[4,97] PHAs are non-toxic polymers that are biocompatible (non-allergenic, non-toxic) and biodegradable. Degradation of organic compounds produces different precursors which involved in PHA production. On hydrolysis these complex organic molecules are degraded and are converted into simple molecules like glucose. This glucose is converted to VFA, which are the key precursors for PHA production. VFA generated during fermentation are transported across the cell membrane and then activated to the corresponding acyl-CoA (Figure 6.9). Two molecules of acetyl-CoA condense to form acetoacetyl-CoA, which is then reduced to 3-hydroxybutyryl-CoA by utilising NADPH (the reduced form of NADP$^+$). Propionate is converted to acetyl-CoA *via* propionyl-CoA; both molecules (acetyl-CoA and propionyl-CoA) can produce poly-β-hydroxy butyrate (PHB) and poly-β-hydroxy valerate (PHV), respectively (Figures 6.8 and 6.9). Whereas butyrate is activated by coenzyme A (CoA) and forms butyryl-CoA, this is oxidised to acetyl-CoA through the β-oxidation pathway and continues into the PHA production pathway. ATP is required for the activation of VFA while reducing equivalents (NADPH) are required for the formation of hydroxyacyl-CoA, which is the precursor for PHA production. ATP and reducing equivalents are acquired from respiration under aerobic operation and are generated in the glycogen metabolism under anaerobic operation.[98]

Three main enzymes are involved in the PHA biosynthetic pathway, *i.e.* β-ketothiolase, acetoacetyl-CoA reductase and PHB synthase. PHB is synthesised in a three-step reaction starting with acetyl-CoA, when bacteria are cultivated on carbohydrates, pyruvate or acetate (Figure 6.10).[98] Two acetyl-CoA molecules are coupled to form acetoacetyl-CoA in a condensation reaction catalysed by β-ketothiolase. The product is subsequently stereoselectively reduced to (*R*)-3-hydroxybutyryl-CoA in a reaction catalysed by NADPH-dependent acetoacetyl-CoA reductase. Finally, PHB is synthesised from (*R*)-3-hydroxybutyryl-CoA by the PHB synthase and free coenzyme-A is released. PHB

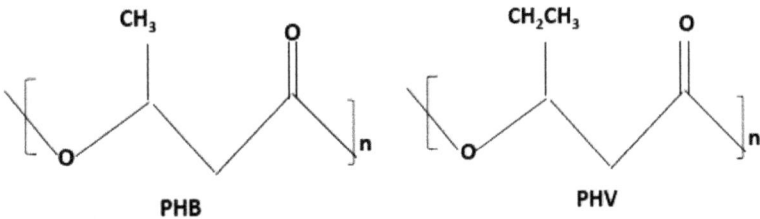

Figure 6.9 Structure of PHB and PHV.

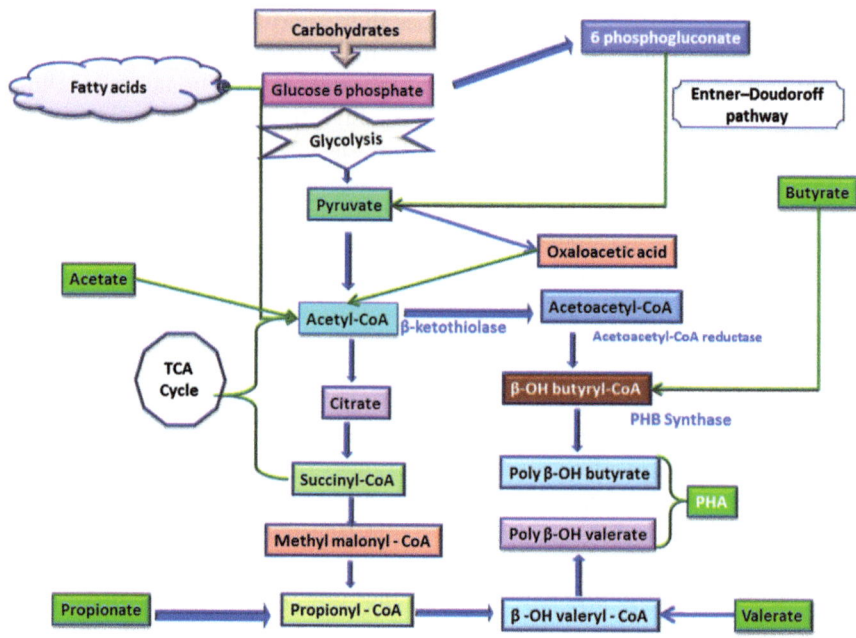

Figure 6.10 Possible routes of bacterial metabolic pathways for PHA production.

synthesis is regulated at the enzymatic level[99] and the intracellular concentration of acetyl-CoA and free coenzyme-A play a central role in the regulation of polymer synthesis.[100] The enzyme β-ketothiolase catalyses the conversion of two molecules of acetyl-CoA to acetoactyl Co-A and controls PHB biosynthesis with CoA as the key effector metabolite.[101] β-Ketothiolase is a homotetrameric enzyme; its active site contains a cysteine-containing nonapeptide and the enzyme is susceptible to inactivation by iodoacetamide. It has the ability to catalyse H⁺ abstraction from the C2 methyl group of the acetyl portion of substrate in a transition state separated from C–C bond formation.[98] The enzyme has a highly active cysteine residue at the 89th position. Studies suggest that there are three cysteine residues, and possibly a histidine residue in or near the active site of the enzyme, the rate-limiting step for which in the thiolytic direction is the enzymatic deacylation half-reaction and, in the condensation direction, the acetoacetyl-CoA carbon–carbon bond-forming step.[98] β-Ketothiolase has been purified from various PHB synthesising bacteria including *Azotobacter beijerinckii*.[99]

The enzyme acetoacetyl-CoA reductase involved in PHB synthesis functions in the conversion of acetoacetyl-CoA to 3-hydroxybutyryl-CoA. Acetoacetyl-CoA is reduced to D-3-hydroxyacyl-CoA by the action of NADPH reductase during PHB synthesis.[100–103] NADPH reductase is active with C4 to C6 compounds, D-3-hydroxyacyl-CoAs, and has no activity with L-substrates and the reduction of acetoacetyl-CoA yields only D-3-hydroxybutyryl-CoA. On the contrary, the NADH reductase produces only L-3-hydroxybutyryl-CoA from acetoacetyl-CoA and is active with both D- and L- substrates in the reverse reaction. Only the NADPH reductase was found to participate in PHB synthesis from acetyl-CoA in reconstituted systems containing purified β-keto-thiolase, acetoacyl-CoA reductase and PHB synthase; this is because the PHB synthase is specific for D-substrates.[100] It is conceivable that L-3-hydroxyacyl-CoAs produced during the β-oxidation of fatty acids could be converted to D-3-hydroxyacyl-CoAs (*via* acetoacetyl-CoA) by the sequential reactions catalysed by the NADH and NADPH reductases, thus enabling their use in PHB biosynthesis.

PHB synthase activity has been observed in two forms (soluble or granule bound), each attributed to the same protein, whose location was dependent on prevailing growth conditions.[104,105] Organisms with high polymer contents possess granule-bound activity, whereas those with low polymer contents possess a soluble form of the enzyme. During growth PHB synthase exists mainly in the soluble form under nitrogen limitation conditions where PHB accumulation occurs followed by granule-associated PHB synthase formation. Disappearance of soluble synthase under these conditions is consistent with its association with PHB granules. The PHB synthase of *Alcaligenes eutrophus* is specific for D-enantiomers, and when tested with 3-hydroxyacyl-CoAs, it was active only with C4 and C5 substrates.[100] The incorporation of 4-hydroxybuterate (4HB) and 5-hydroxyvalerate (5HV) into PHA by *A. eutrophus* suggests that the synthase is also active with 4-hydroxybutyryl-CoA and 5-hydroxyvaleryl-CoA respectively.[106–108] The mechanism of PHB synthase action remains ambiguous. A two-stage polymerisation reaction involving an acyl-*S*-enzyme intermediate has been proposed.[109] A subsequent model suggested the involvement of two thiol groups, one locating the incoming 3-hydroxy butyrate monomer and the other locating the growing polymer chain.[110] Condensation occurs through a four-membered transition state, leaving one of the thiol groups vacant for the next monomer. It is presumed that the chain transfer role performed by the synthase controls the molecular weight of the polymer produced, which is a characteristic of a given organism.[98]

The chemical composition of PHA polymers can be manipulated by varying the organic substrates fed to the PHA-producing bacteria. Existing processes use pure culture, single substrate and sterile conditions for PHA production. Some bacteria have the capacity to accumulate PHA up to 60–90% of their dry weight. PHA is four to nine times more expensive than conventional plastics due to its high production cost, which includes substrate cost, culture procurement and sterilisation.[111] The substrate contributes majorly to the overall cost in PHA production. A number of substrates including carbohydrates, oils, alcohols, acids and hydrocarbons can be used as substrate by various bacteria.

An interesting alternative to pure cultures and high-cost substrates is the use of mixed cultures in conjunction with wastewater (low-cost substrates), which allows both less investment and low operating costs.[112,113] Along with mixed cultures different wastewaters such as municipal wastewater,[114] sugar cane molasses,[112] paper mill wastewater,[113] tomato cannery wastewater,[115] olive oil mill effluent,[116] biohydrogen reactor effluent,[46] food waste[117] and pea shells[118] have been used for PHA production. The use of mixed cultures and wastewater for PHA production has the advantage of operating under non-axenic conditions, thus saving energy and equipment costs. PHA production in mixed cultures has been studied in the past along with wastewater treatment, as it allows for a greater variety of substrates to be used due to the presence of a number of PHA producing organisms. PHA in activated sludge is mainly composed of homopolymer PHB and/or co-polymer [poly-3(hydroxy butyrate-*co*-hydroxy valerate), P3(HB-*co*-HV)]. The co-polymer had high strength and elasticity, and is less brittle than homopolymer PHB.

PHA accumulation was found to be directly proportional to the organic load up to a certain point, where a higher carbon concentration indicates higher accumulation and a lower carbon concentration indicates lower PHA accumulation.[119,120] A lower concentration of nitrogen shows higher PHA accumulation, while an increase in its concentration increases biocatalyst growth but not PHA accumulation.[112] A minimal level of internal phosphate is essential for PHA accumulation. A lower phosphorous concentration shows good PHA accumulation, rather than complete phosphorous deficiency.[121] Phosphorous and nitrogen limitations appear to be suitable stimulants for PHA accumulation.[121] High nitrogen and phosphorous concentrations lead to protein synthesis, while phosphate deprivation leads to a reduction of the protein synthesis rate and diverts the metabolic pathway towards PHA accumulation. Under balanced growth conditions CoASH levels are high and PHA synthesis is inhibited. When growth is limited by an essential nutrient other than the carbon and energy source, this reduces the complexity of the metabolism that occurs in the cell and the flow of carbon is channelled to a unidirectional path such as PHA synthesis and the NADH concentration increases, resulting in inhibition of the early enzymes of the TCA cycle such as citrate synthase and isocitrate dehydrogenase.[122] This leads to the accumulation of acetyl-CoA, which relieves the inhibition exerted by CoASH, leading to PHB formation.

Microaerophilic conditions facilitate a higher fraction of substrate accumulation as PHA.[123] Activated sludge showed 20% (cell dry weight) of PHA accumulation under anaerobic conditions.[124–126] When the microenvironment was shifted to a microaerophilic condition, PHA content increased to 62%. About 77% of the reducing equivalents available from acetate were converted to PHB under oxygen limitation, but in oxygen excess conditions, only 54% of the reducing equivalents were available when a higher fraction of acetate was used for biomass growth.[127] At low dissolved oxygen (DO) concentrations, the limited availability of ATP prevents biomass growth and most of the ATP is used for acetate transport inside the cell.[122] In contrast, high DO provided additional ATP, leading to the higher growth rates and resulting in decreased PHB accumulation. Oxygen management is crucial for conserving the reducing power and an excessive aeration rates decrease PHB in spite of higher biomass growth.[124,127] Pyruvate formed through the glycolysis is converted to acetyl-CoA resulting in PHB production, while the propionyl-coA formed through the propionate succinate pathway forms PHV.[124,128] ATP is required for VFA activation and reducing equivalents (NADPH) are required for the formation of hydroxyacyl-CoA, which is a precursor of PHA production.[98]

Iron is the major trace element for all bacteria and normally represents about 0.02% of bacterial dry weight.[129] PHB production is also dependent on iron concentration. Under a low iron concentration, the PHB content in cells was low compared with that under a higher iron concentration.[130] A reduction in aeration level under iron-replete conditions resulted in highest PHB production (70–72.5% of the dry cell weight). Sufficient iron is required for enzymatic protection against toxic oxygen products.[131] Citrate synthase activity is significantly inhibited by both high intracellular concentrations of NADH and NADPH, indicating that PHB accumulation is enhanced by facilitating the metabolic flux of acetyl-CoA to the PHB synthetic pathway.[132] This suggests that citrate synthase is a potentially important control point in the whole PHB synthesis process based on its ability to control the availability of CoA,[133] which regulates the activity of β-ketothiolase. An isocitrate dehydrogenase leaky mutant of *Ralstonia eutropha*, which exhibits low TCA cycle activity, produced PHB at a faster rate than the wildtype organism.[134] Bacteria grown on fatty acids metabolise substrate *via* the β-oxidation pathway and produce PHA polymer.[135] The key enzyme of the PHB polymerisation process is PHA synthase, often called PHA polymerase. Based on *in vitro* studies, the ultimate substrate for PHA polymerase is 3-hydroxyacyl-CoA.[136] With addition of acrylic acid, β-oxidation slows down and shows the negative effect on PHA production.[137,138]

6.3.5 Microbial Lipids

Some bacteria, microalgae and fungi are reported to be oleaginous (Table 6.3), *i.e.* they have the capability to accumulate significant amounts of lipids under specific cultivated conditions.[139–141] Microorganisms that can produce a high content of lipid (>20%) may be termed as 'oleaginous'.[142–144] Bacteria use these lipids for the essential functioning of membranes and membranous structure formation. The process of lipid accumulation can be seen as a two-phase batch system where the first phase consists of balanced growth with all the nutrients being available; the subsequent 'fattening' or 'lipogenic' stage occurs after the exhaustion of a key nutrient other than the carbon.[145]

Lipid accumulation begins when nutrient is exhausted (usually nitrogen), but an excess of carbon is still assimilated by the cells and is converted into triacylglycerols (TAGs). Oils and fats are primarily composed of TAGs (Figure 6.11), which serve as a primary storage form of carbon and energy in the microorganisms.[146] There are three possible mechanisms for the initiation of fatty acid biosynthesis in bacteria.[147] Condensation of acetyl-CoA with malonyl-acetyl carrier protein (malonyl-ACP) to yield acetoacetyl-ACP catalysed by the β-ketoacyl-ACP synthase III is one route. Alternatively, the acetate moiety is first transferred from acetyl-CoA to acyl-ACP by either

Table 6.3   Different oleaginous microorganisms and their oil storage efficiency reported in literature.

	Microorganism	Oil content/% dry weight	Reference
Bacteria	Arthrobacter sp.	>40	148
	Acinetobacter calcoaceticus	27–38	148
	Rhodococcus opacus	24–25	149
	Bacillusalcalophilus	18–24	149
Yeast	Candida curvata	58	148
	Cryptococcus albidus	65	148
	Lipomyces starkeyi	64	148
	Rhodotorula glutinis	72	148
Molds	Aspergillus sp.	23	150
	Aspergillus oryzae	57	148
	Mortierella isabellina	86	151
	Humicola lanuginose	75	140
	Mortierella vinacea	66	152
Microalgae	Mixed microalgae	28	63
	Botryococcus braunii	25–75	148
	Cylindrotheca sp.	16–37	141
	Chlorella sp.	28–32	153
	Crypthecodinium cohnii	20	153
	Dunaliella primolecta	23	153
	Isochrysis sp.	25–33	153
	Monallanthus salina	>20	153
	Nanochloris sp.	20–35	153
	Nanochloropsis sp.	31–68	148
	Neochloris oleoabundans	35–54	148
	Nitzschia sp.	45–47	141
	Phaeodactylum tricornutum	20–30	141
	Schizochytrium sp.	50–77	148
	Tetraselmis sueica	15–23	148

acetyl-CoA-ACP transacylase and then condensed with malonyl-ACP. Similarly, acetyl-ACP is formed from the decarboxylation of malonyl-ACP followed by its subsequent condensation with malonyl-ACP. Irrespective of the mechanism, the first step is the condensation of malonyl-ACP with a growing acyl chain by β-ketoacyl-ACP synthase. This is the only irreversible step in the elongation cycle, and therefore it is not surprising that the 3-ketoacyl-ACP synthases play a key role in regulating the product distribution of the pathway.

The final step in the fatty acid biosynthetic pathway is the transfer of the acyl chains of the acyl-ACP end products into membrane phospholipids by the glycerolphosphate acyltransferase system. Two critical regulated enzymes, including malate enzyme and ATP: citrate lyase (ACL), have effect on the lipid accumulation.[144] Fatty acids are highly reduced materials which use reductant (NADPH) for its synthesis.[144] Acetyl-CoA carboxylase (ACC) is a biotin-dependent enzyme that catalyses the irreversible carboxylation of acetyl-CoA to produce malonyl-CoA through its two catalytic activities, biotin carboxylase (BC) and carboxyltransferase (CT).[147] The most important function of ACC is to provide the malonyl-CoA substrate for the biosynthesis of fatty acids.

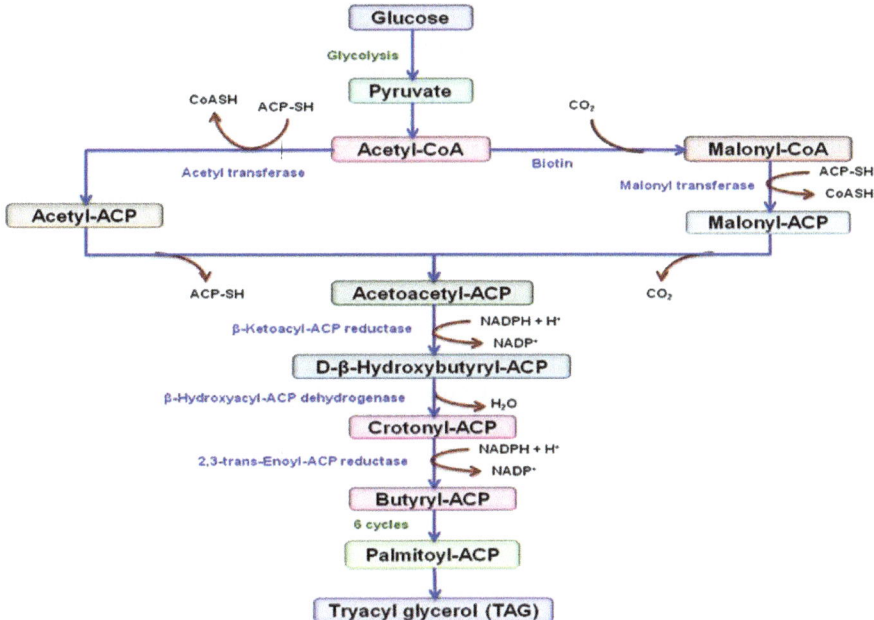

Figure 6.11 Mechanism of triacylglycerol (TAG) synthesis in oleaginous microorganism.

In oleaginous microorganisms, the lipid accumulation starts when carbon source is present in excess and there is a limiting element in the growth medium. The limitation of many elements can induce lipid accumulation, but usually it is nitrogen which is used for this purpose because it is the most efficient type of limitation for inducing lipid accumulation.[154,155] In nitrogen limited conditions, the organisms continue to assimilate the carbon source but the cell proliferation stops as nitrogen is required for the protein and nucleic synthesis. Under these conditions the carbon flux is diverted towards lipid synthesis, leading to an accumulation of TAG within discrete lipid bodies in the cells. The lipid that is formed has to be stored within the existing cells which can no longer be divided.[155]

Bacterial fatty acid composition is also superior to that of the other cellular lipids (phospholipids and glycolipids).[156] The *Actinomycete* group are capable of synthesising remarkably high amounts of fatty acids (up to 70% of the cellular dry weight) from simple carbon sources like glucose under growth-restricted conditions and accumulate them intracellularly as TAGs.[157] Bacterial species such as *Mycobacterium, Streptomyces, Rhodococcus,* and *Nocardia* can also accumulate TAGs at high concentrations. The composition and structure of bacterial TAG vary considerably depending on the microorganism utilising the carbon source.[157] *Rhodococcus* and *Nocardia corallina* accumulate primarily TAGs with minor amounts of diacylglycerols (DAGs) and wax esters under nitrogen-limiting conditions.[158] The microbial lipids contain high fractions of polyunsaturated fatty acids (PUFAs).

Microbial lipid production has many advantages such as short lifecycles, lower labour requirement, less demand on space, season and climate, and ease of scaleup.[159] Almost all studies on microbial lipid accumulation have been conducted using batch cultivation.[160] Various low-cost feedstocks have been explored[161] including soluble starch[162,163] and industrial glycerol.[164] Microbial lipids offer a potential source for sufficient production of renewable fuels to impact on the consumption of fossil fuels. To succeed in this endeavour, a suit of autotrophic, heterotrophic and mixotrophic microbial systems utilising diverse substrates are available. To be cost-effective, it is necessary to use innovative combinations of cultivation systems involving all types of carbonaceous materials including wastes. Several recent efforts in this area are utilising domestic and industrial wastes. Economic analyses have indicated the need to minimise the costs of media components and for further research on the microbial systems capable of producing lipids at relatively high productivities in minimal media.

6.3.6 Bioethanol

Ethanol has high commercial value due to its wide spectrum of applications and usage in diverse fields. Hydration of ethylene ($CH_2{=}CH_2$) for ethanol production is the oldest synthetic process dating back more than one hundred years. Fermentative production of ethanol gained importance in the early 1990s due to the low economics of the process and high yields from simple substrates. Yeast is a well-known ethanol producer from complex carbohydrate molecules. Some of the bacteria also have the capability of producing ethanol through fermentation. Ethanol fermentation is usually carried by three steps, *viz.* formation of fermentable sugars through hydrolysis, fermentation of these sugars to ethanol, and finally their separation and purification.[165] Microbial metabolism of ethanol production along with the other metabolites is depicted in Figure 6.12. Similar to yeast metabolism, alcohol dehydrogenase (ADH) is also the critical enzyme for ethanol production in the bacterial metabolism:

$$C_6H_{12}O_6 + 2ADP + 2Pi \rightarrow 2C_2H_5OH + 2CO_2 + 2ATP \qquad (6.10)$$

Bacteria are able to ferment a wide range of substrates to generate pentoses and hexoses that may be further converted to pyruvate. Pyruvate produced from the bacterial metabolism is converted to acetyl Co-A and then to acetaldehyde through decarboxylation. Acetaldehyde is further reduced to ethanol by the enzyme ADH along with the regeneration of NAD^+. Accumulation of reducing powers ($NADH$, $FADH_2$, *etc.*) results in the necessity for their regeneration process, but the absence of an efficient electron acceptor and unfavourable conditions for the reduction of H^+ creates a favourable environment for the onset of solventogenesis where the carbon as well as electron flux is diverted towards solvent production.

The generation of ATP in the fermentation necessitates the utilization of the organic compounds as terminal electron acceptors and reduces the energy-rich organic acids to the respective solvents as end products of metabolism.[166] Some

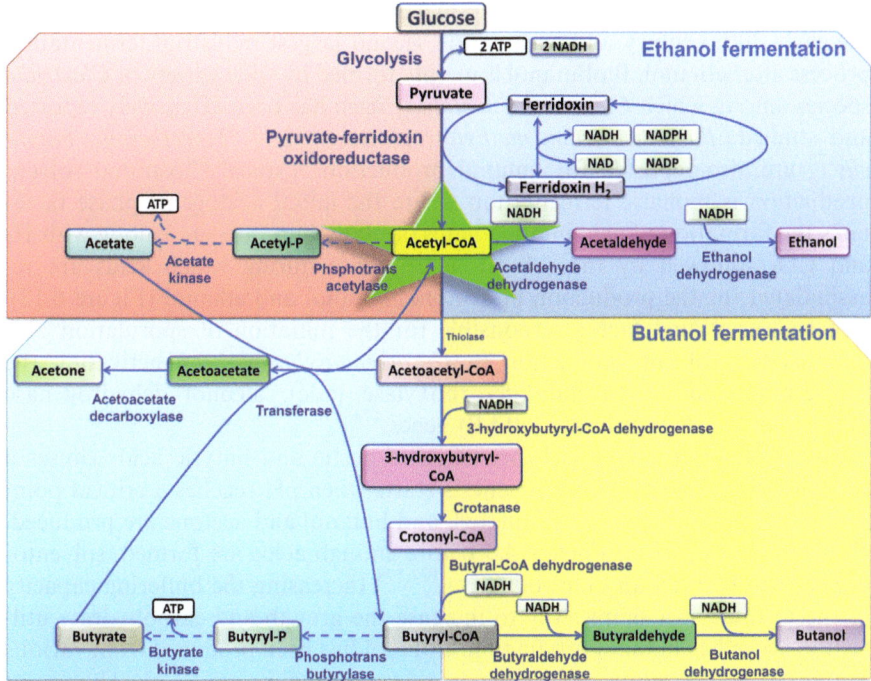

Figure 6.12 Biphasic acetone–butanol–ethanol (ABE) fermentation mechanism.

of the bacterial population such as *Zymononas mobilis* and *Pichia stipites* converts pyruvate to ethanol by pyruvate decarboxylase/alcohol dehydrogenase (PDC/ADH) enzyme complexes, while some species such as *Escherichia coli* produces ethanol through the mixed fermentation involving the production of acetyl-CoA as intermediate of ethanol production.[166] Operational parameters such as temperature, pH, alcohol tolerance, growth rate, productivity, osmotic tolerance, specificity, yield, genetic stability and inhibitor tolerance influence the process performance of ethanol production.[167] Although bacteria grow faster than fungi and yeast, they are not considered the correct choice for ethanol production at large scale due to their substrate specificity.[168] Polysaccharides such as starch from grains, potatoes and root crops, and cellulose from wood, agricultural residue, municipal solid wastes, paper waste and crop residues are also used as substrate for ethanol production along with simple sugars. Bioethanol production from the fermentable sugars produced during the pretreatment of cellulosic substrates has gained tremendous importance over the last decade due to its projection as a feasible alternative to petroleum based on the transportation of fuels.[169,170]

6.3.7 Biobutanol

Butanol is also considered an efficient alternative to existing fuels due to its high combustion value and other fuel properties. After Pasteur discovered bacterial butanol production in 1861, fermentative butanol production prospered during the early 20th century and became the second largest industrial fermentation process after ethanol. Biobutanol is mainly formed by wide variety of *Clostridia* species among which *Clostridium acetobutylicum* has been extensively reported and studied. *Butyribacterium methylotrophicum*[171] and *Hyperthermus butylicus*[172] are also reported for butanol production. Typical Clostridial solvent production is biphasic fermentation where the initial acidogenic phase facilitates the formation of acid intermediates (acetate, butyrate, *etc.*) along with H_2 and CO_2 followed by the solventogenic phase during which acids are re-assimilated for the production of acetone, butanol and ethanol (Figure 6.12). The transcriptional factor responsible for the initiation of sporulation also initiates the solvent production in *C. acetobutylicum*[173] by activating the transcription of acetoacetate decarboxylase (adc), alcohol dehydrogenase (adhE), and CoA transferase (ctfAB) genes.[174–176]

During acidogenesis, rapid formation of acetic and butyric acids causes a sudden fall in the pH. Solventogenesis starts when pH reaches a critical point beyond which acids are re-assimilated, and butanol and acetone are produced. However, if the pH falls below 4.5 before enough acids are formed, solventogenesis will be brief and unproductive.[177,178] Increasing the buffering capacity of the medium is a simple way to increase the growth and carbohydrate utilisation as well as butanol production. Butanol has been produced from various raw materials including molasses, whey permeate and corn.[179–182] The use of excess carbon under nitrogen limitation is required to achieve high levels of solvent production.[183] Iron is an important mineral supplement since the conversion of pyruvate to acetyl-CoA involves a ferredoxin oxidoreductase iron–sulfur protein.[178] Similarly, pH is also very important to the biphasic acetone–butanol fermentation.

6.4 Conclusions

Bioenergy generation from bacteria has been established as a prospective alternative to conventional energy and an integral component of green sustainable energy. The transition from a fossil fuel-based economy to a biofuel-based economy depends on the improvement of biofuel yields *via* the successful development of suitable microorganisms capable of efficiently fermenting a variety of sugars while simultaneously displaying tolerance to high end-product concentrations.[184]

There exist some scientific and technical barriers to implement many of these processes which need to be resolved. Multidisciplinary research approach is vital to establish sustainable technologies. Fundamental understanding of the microbiology will help in further development of the processes. Low yield of end products is one of the critical issues where much research attention needs to be focused. The selection and enrichment of suitable biocatalyst are also considered to be crucial. Process engineering and optimisation of operational factors govern the performance of any biological system and also have considerable influence on bioenergy generation. The design and development of a bioreactor is one of the vital domains where considerable focus is required for enhancement of these processes. Scaling up the process to pilot- or large-scale to generate baseline engineering data will impart relevance from the view of commercialisation. The separation of soluble products from the fermentation broth and the stability of pure/co-culture processes require significant attention.[185] The persistence of an acidic microenvironment due to the production of (volatile fatty acid generation) soluble acid metabolites and their presence as a non-utilised organic fraction influences the process performance with low substrate conversion and creates a problem in disposing of these effluents to the environment.[15]

More recently waste/wastewater is being considered as a potential/ideal feedstock/substrate for bioenergy generation due to its intrinsic advantage of possessing huge quantities of degradable organic material that can result in a net positive energy gain apart from its remediation. Reducing the treatment cost of waste and finding ways to produce useful or value-added products from treatment is a current research perspective. Integrating anaerobic waste treatment process with bioenergy generation seems to be a potential and viable platform which has attracted considerable interest due to its renewable nature. Bioenergy generation keeping waste remediation as an integral element enables a reduction in the overall treatment cost associated with addition revenue generation.

Acknowledgements

The authors acknowledge the Director, CSIR-IICT, for support and encouragement. This work was supported by the funding from CSIR-XII five year plan project-SETCA. M.V.R., G.V.S., R.C. and M.P.D. thank the Council of Scientific and Industrial Research (CSIR) for providing research fellowship.

References

1. D. Nelson and M. Cox, *Lehninger Principles of Biochemistry*, W. H. Freeman, New York, 5th edn, 2008.
2. L. T. Angenent, K. Karim, A. M. H. Dahhan, B. A. Wrenn and E. R. Domiguez, *Trends. Biotechnol.*, 2004, **22**, 477.
3. S. Venkata Mohan, N. C. Rao, K. K. Prasad, P. Muralikrishna, R. S. Rao and P. N. Sarma, *Biotechnol. Bioeng.*, 2005, **90**, 732.
4. E. L. Madsen, *Environmental Microbiology*, Blackwell Publishing, Oxford, 2008.
5. J. W. Prescott, L. Sherwood and C. Woolverton, *Prescott's Principles of Microbiology*, McGraw-Hill Higher Education, New York, 2008.
6. R. Chandra, G. Venkata Subhash and S. Venkata Mohan, *Bioresour. Technol.*, 2012, **109**, 46.
7. S. Venkata Mohan and S. Srikanth, *Bioresour. Technol.*, 2011, **102**, 10210.
8. G. Velvizhi and S. Venkata Mohan, *Bioresour. Technol.*, 2011, **102**, 10784.

9. Y. Mu and H. Q. Yu, *Int. J. Hydrogen Energy*, 2007, **32**, 3308.
10. C. Y. Lin and C. H. Lay, *Int. J. Hydrogen Energy*, 2004, **29**, 275.
11. S. V. Ginkel and B. Logan, *Water Res.*, 2005, **39**, 3819.
12. Y. T. Fan, Y. H. Zhang, S. F. Zhang, H. W. Hou and B. Z. Ren, *Bioresource Technol.*, 2006, **7**, 500.
13. H. N. Gavala, I. V. Skiadas and B. K. Ahring, *Int. J. Hydrogen Energy*, 2006, **31**, 1164.
14. K. Vijayaraghavan, D. Ahmad and M. K. B. Ibrahim, *Int. J. Hydrogen Energy*, 2006, **31**, 569.
15. H. Yang, P. Shao, T. Lub, J. Shena, D. Wang and Z. Xub, *Int. J. Hydrogen Energy*, 2006, **1**, 1306.
16. S. Venkata Mohan, *Int. J. Hydrogen Energy*, 2009, **34**, 7460.
17. S. Venkata Mohan, in *Sustainable Biotechnology: Sources of Renewable Energy*, O. V. Singh and S. P. Harvey, Springer, Dordrecht, The Netherlands, 2010, pp. 129–164.
18. S. Venkata Mohan, G. Mohanakrishna, S. V. Raghuvulu and P. N. Sarma, *Int. J. Hydrogen Energy*, 2007, **32**, 3284.
19. S. Venkata Mohan, Y. V. Bhaskar and P. N. Sarma, *Water Res.*, 2007, **41**, 2652.
20. S. Venkata Mohan, G. Mohankrishna and P. N. Sarma, *Int. J. Hydrogen Energy*, 2008, **33**, 2156.
21. S. Venkata Mohan, P. Chiranjeevi and G. Mohanakrishna, *Int. J. Hydrogen Energy*, 2012, **37**, 3130.
22. G. Mohanakrishna, G. V. Subhash and S. Venkata Mohan, *Int. J. Hydrogen Energy*, 2011, **36**, 8943.
23. O. Kruse, J. Rupprecht, K. P. Bader, S. T. Hall, P. M. Schenk, G. Finazzi and B. Hankamer, *J. Biol. Chem.*, 2005, **280**, 34170.
24. L. L. Beer, E. S. Boyd, J. W. Peters and M. C. Posewitz, *Curr. Opin. Biotechnol.*, 2009, **20**, 264.
25. O. Kruse and B. Hankamer, *Curr. Opin. Biotechnol.*, 2010, **21**, 238.
26. D. Vyas and H. D. Kumar, *Int. J. Hydrogen Energy*, 1995, **20**, 163.
27. M. Horch, L. Lauterbach, O. Lenz, P. Hildebrandt and I. Zebger, *FEBS Letters*, 2012, **586**, 545.
28. R. E. Blankenship, M. T. Medigan and C. E. Bauer, *Anoxygenic Photosynthetic Bacteria*, Kluwer Academic, Sordrecht, The Netherlands, 1995.
29. J. M. Berg, J. L. Tymoczko and L. Stryer, *Biochemistry*, W. H. Freeman, New York, 2002.
30. R. Chandra and S. Venkata Mohan, *Int. J. Hydrogen Energy*, 2011, **36**, 12211.
31. K. A. Batyrova, A. A. Tsygankov and S. N. Kosourov, *Int. J. Hydrogen Energy*, 2012, **37**, 8834.
32. A. Melis, *Energy. Environ. Sci.*, 2012, **5**, 5531.
33. S. Srikanth, S. Venkata Mohan, M. P. Devi, M. Lenin Babu and P. N. Sarma, *Int. J. Hydrogen Energy*, 2009, **34**, 1771.
34. P. C. Hallenbeck and J. R. Benemann, *Int. J. Hydrogen Energy*, 2002, **27**, 1185.

35. G. Vardar-Schara, T. Maeda and T. K. Wood, *Microbiol. Biotechnol.*, 2008, **1**, 107.
36. G. Dinopoulou, R. M. Sterritt and J. N. Lester, *Biotechnol. Bioeng.*, 1988, **31**, 969.
37. D. W. Klein, L. M. Prescott and J. Harley, *Microbiology*, McGraw-Hill, New York, 2005.
38. V. Lalit Babu, S. Venkata Mohan and P. N. Sarma, *Int. J. Hydrogen Energy*, 2009, **34**, 3305.
39. S. Venkata Mohan, L. Agarwal, G. Mohanakrishna, S. Srikanth, A. Kapley, H. J. Purohit and P. N. Sarma., *Int. J. Hydrogen Energy*, 2011, **36**, 8234.
40. J. T. Kraemer and D. M. Bagley, *Biotechnol. Lett.*, 2007, **29**, 685.
41. S. Venkata Mohan, S. Srikanth, M. Lenin Babu and P. N. Sarma, *Bioresour. Technol.*, 2010, **101**, 1826.
42. S. Srikanth, S. Venkata Mohan, V. Lalit Babu and P. N. Sarma, *Int. J. Hydrogen Energy*, 2010, **35**, 10693.
43. S. Srikanth, S. Venkata Mohan and P. N. Sarma, *Bioresour. Technol.*, 2010, **101**, 5337.
44. J. W. Chang, S. Chen, S. Shih, J. Yu, F. Lay, J. Wen and C. Huang, *App. Microbiol. Biotechnol.*, 2006, **5**, 598.
45. S. Venkata Mohan, M. Lenin Babu, G. Mohanakrishna and P. N. Sarma, *Int. J. Hydrogen Energy*, 2009, **34**, 6149.
46. S. Venkata Mohan, M. Venkateswar Reddy, G. Venkata Subhash and P. N. Sarma, *Bioresour. Tecnol*, 2010, **23**, 9382.
47. S. Venkata Mohan and R. K. Goud, in *Biogas Production: Pretreatment Methods in Anaerobic Digestion*, A. Mudhoo, Wiley and Scrivener Publishing, Hoboken, NJ, and Salem, MA, 2012, pp. 291–311.
48. R. K. Goud and S. Venkata Mohan, RSC Adv., 2012, DOI: 10.1039/C2RA20526B.
49. S. Venkata Mohan, V. Lalit Babu and P. N. Sarma, *Bioresour. Technol.*, 2008, **99**, 59.
50. H. Zhu. and M. Beland, *Int. J. Hydrogen Energy*, 2006, **31**, 1980.
51. J. Wang and W. Wan, *Int. J. Hydrogen Energy*, 2008, **33**, 2934.
52. E. R. Orskav, W. P. Flatt and P. W. Moe, *J. Dair. Sci*, 1968, **51**, 1429.
53. B. E. Rittmann, *Biotechnol. Bioeng.*, 2008, **100**, 203.
54. Energy Information Administration, Annual Energy Outlook: With Projection to 2030, US Department of Energy, Washington DC, 2005.
55. V. V. Idania, R. Sparling, D. Risbey, R. S. Noemi, M. Hec and H. M. P. Varaldo, *Bioresour. Technol.*, 2005, **96**, 1907.
56. D. V. Gustavo, A. M. Felipe, L. R. Antonio de and R. F. Elias, *Int. J. Hydrogen Energy*, 2008, **33**, 4989.
57. Y. V. Bhaskar, S. Venkata Mohan and P. N. Sarma, *Bioresour. Technol.*, 2008, **99**, 6941.
58. A. Y. Atif, A. F. Razia, M. A. Ngan, M. Morimoto, S. E. Iyukec and N. T. Veziroglu, *Water Sci. Technol.*, 2005, **53**, 271.

59. G. Mohanakrishna, R. Kannaiah Goud, S. Venkata Mohan and P. N. Sarma, *Int. J. Hydrogen Energy*, 2010, **35**, 533.
60. M. J. Cuetos, X. Gomez, A. Escapa and A. Moran, *Power Sources*, 2007, **169**, 131.
61. X. Gomez, A. Moran, M. J. Cuetos and M. E. Sanchez, *J. Power Sources*, 2006, **157**, 727.
62. M. P. Devi, S. Venkata Mohan, G. Mohanakrishna and P. N. Sarma, *Int. J. Hydrogen Energy*, 2010, **35**, 10701.
63. S. Venkata Mohan, M. P. Devi, G. M. Krishna, N. Amarnath, M. L. Babu and P. N. Sarma, *Bioresour. Technol.*, 2011, **102**, 1109.
64. S. Srikanth and S. Venkata Mohan, RSC. Adv., 2012, DOI:10.1039/C2RA20383A.
65. S. Venkata Mohan and M. Lenin Babu, *Bioresour. Technol.*, 2011, **102**, 8457.
66. P. E. Jablonski and J. G. Ferry, *FEMS Microbiol. Lett.*, 1992, **1**, 55.
67. A. D. Goldman, J. A. Leigh and R. Samudrala, *BMC Evol. Biol.*, 2009, **9**, 199.
68. F. G. Pohland and S. Ghosh, *Environ. Lett.*, 1971, **1**, 255.
69. S. Venkata Mohan, K. K. Prasad, N. Chandrasekhar Rao, Y. V. Bhaskar, V. Lalit Babu and P. N. Sarma, *J. Sci. Ind. Res.*, 2005, **64**, 771.
70. S. Venkata Mohan, N. Chandrasekhara Rao, K. K. Prasad and P. N. Sarma, *Process. Biochem*, 2005, **40**, 2849.
71. S. Ghosh, J. P. Ombregt and P. Pipyn, *Water Res.*, 1985, **19**, 1083.
72. S. Venkata Mohan, G. Mohanakrishna and P. N. Sarma, *Environ. Sci. Technol.*, 2008, **42**, 8088.
73. S. Venkata Mohan, S. Veer Raghuvulu and P. N. Sarma., *Biosens. Bioelectron.*, 2008, **23**, 1326.
74. S. Venkata Mohan, R. Sarvanan, S. Veer Raghuvulu, G. Mohankrishna and P. N. Sarma., *Bioresour. Technol.*, 2008, **99**, 596.
75. S. Venkata Mohan, G. Mohanakrishna, B. P. Reddy, R. Sarvanan and P. N. Sarma, *Biochem. Eng. J.*, 2008, **39**, 121.
76. G. Mohanakrishna, S. Venkata Mohan and P. N. Sarma, *Int. J. Hydrogen Energy*, 2010, **35**, 3440.
77. G. Mohanakrishna, S. Venkata Mohan and P. N. Sarma, *J. Hazard. Mater.*, 2010, **177**, 487.
78. M. Lenin Babu and S. Venkata Mohan, *Bioresour. Technol.*, 2012, **110**, 206.
79. S. Venkata Mohan, G. Mohanakrishna, S. Srikanth, in *Biofuels: Alternative Feedstocks and Conversion Processes*, A. Pandey, C. Larroche, S. C. Ricke, C.-G. Dussap and E. Gnansounou, Academic Press, Oxford, 2011, ch. 22, pp. 499–524.
80. B. R. Ringeisen, R. Ray and B. Little, *J. Power Sources*, 2007, **165**, 591.
81. M. A. Rodrigo, P. Canizares, J. Lobato, R. Paz, C. Saez and J. J. Linares, *J. Power Sources*, 2007, **169**, 198.
82. S. Veer Raghavulu, S. Venkata Mohan, R. K. Goud and P. N. Sarma, *Electrochem Comm.*, 2009, **11**, 371.

83. S. Veer Raghavulu, P. S. Babu, R. K. Goud, G. V. Subhash, S. Srikanth and S. Venkata Mohan, *RSC. Advances.*, 2012, **2**, 677.
84. R. K. Goud, P. Suresh Babu and S. Venkata Mohan, *Int. J. Hydrogen Energy*, 2011, **36**, 6210.
85. S. Venkata Mohan, S. Srikanth, G. Velvizhi and M. Lenin Babu, in *Biofuel Technologies: Recent Developments*, V. K. Gupta and M. G. Tuohy, Spinger-Verlag, Berlin and Heidelberg, 2013, ch. 14, pp. 335–368.
86. S. Venkata Mohan and K. Chandrasekhar, *Bioresour. Technol.*, 2011, **102**, 7077.
87. S. Srikanth, T. Pavani, P. N. Sarma and S. Venkata Mohan, *Int. J. Hydrogen Energy*, 2011, **36**, 2271.
88. N. Ivashin, B. Kallebring, S. Larsson and O. Hansson, *J. Phys. Chem.*, 1998, **102**, 5017.
89. J. M. Pisciotta, Y. Zou and I. V. Baskakov, *PLoS ONE*, 2010, 5.
90. M. Rosenbaum, U. Schröder and F. Scholz, *Environ. Sci. Technol.*, 2005, **39**, 6328.
91. Y. Zou, J. Pisciotta, R. B. Billmyre and I. V. Baskakov, *Biotechnol. Bioeng.*, 2009, **104**, 939.
92. D. Xing, Y. Zuo, S. Cheng, J. M. Regan and B. E. Logan, *Environ. Sci. Technol.*, 2008, **42**, 4146.
93. S. Scheuring, R. P. Goncalves, V. Prima and J. N. Sturgis, *J. Mol. Biol.*, 2006, **358**, 83.
94. A. Yeliseev, J. M. Eraso and S. Kaplan, *J. Bacteriol.*, 1996, **178**, 5877.
95. G. Swift, *Acc. Chem. Res.*, 1993, **26**, 105.
96. USEPA, Municipal Solid Waste Generation, Recycling and Disposal in the United States: Facts and Figures for 1998, EPA 530-F-00-024, US Environmental Protection Agency, Office of Solid Waste and Emergency Response, Washington DC, 2000.
97. H. Satoh, T. Mino and T. Matsuo, *Int. J. Biolog. Macromol*, 1999, **25**, 105.
98. A. J. Anderson and E. A. Dawes, *Microbiol. Rev.*, 1990, **54**, 450.
99. P. J. Senior and E. A. Dawes, *Biochem. J.*, 1971, **125**, 55.
100. G. W. Haywood, A. J. Anderson and E. A. Dawes, *FEMS Microbiol. Lett.*, 1989, 571.
101. V. Oeding and H. G. Schlegel, *Biochem. J.*, 1973, **134**, 239.
102. T. Saito, T. Fukui, F. Ikeda, Y. Tanaka and K. Tomita., *Arch. Microbiol.*, 1977, **114**, 211.
103. H. Shuto, T. Fukui, T. Saito, Y. Shirakura and K. Tomita, *Eur. J. Biochem.*, 1981, **118**, 53.
104. T. Fukui, A. Yoshimoto, M. Matsumoto, S. Hosokawa, T. Saito, H. Nishikawa and K. Tomita, *Arch. Microbiol.*, 1976, **110**, 149.
105. I. Tomita, T. Saito and T. Fukui, in *The Biochemistry of Metabolic Processes*, D. L. F. Lennon, F. W. Stratman and R. N. Zahlten, Elsevier Science, New York, 1983, pp. 353–366.
106. Y. Doi, A. Segawa and M. Kunioka, *Int. J. Biol. Macromol.*, 1990, **12**, 106.

107. Y. Doi, A. Tamaki, M. Kunioka and K. Soga, *Makromol. Chem. Rapid Commun.*, 1987, **8**, 631.
108. Y. Doi, M. Kunioka, Y. Nakamura and K. Soga, *Macromol*, 1988, **21**, 2722.
109. R. Griebel and J. M. Merrick, *J. Bacteriol.*, 1971, **108**, 782.
110. D. G. H. Ballard, P. A. Holmes and P. J. Senior, in Recent Advances in Mechanistic and Synthetic Aspects of Polymerization, ed. M. Fontanille and A. Guyot, Reidel (Kluwer) Publishing, Lancaster, UK, 1987, vol. 215, pp. 293–314.
111. C. Kasemsap and C. Wantawin, *Bioresour. Technol.*, 2007, **98**, 1020.
112. M. G. E. Albuquerque, M. Eiroa, C. Torres, B. R. Nunes and M. A. M. Reis, *J. Biotechnol.*, 2007, **130**, 411.
113. S. Bengtsson, A. Werker, M. Christensson and T. Welander, *Bioresour. Technol.*, 2008, **99**, 509.
114. A. S. M. Chua, H. Takabatake, H. Satoh and T. Mino, *Water Res.*, 2003, **37**, 3602.
115. H. Y. Liu, P. V. Hall, J. L. Darby, E. R. Coats, P. G. Green, D. E. Thompson and F. J. Loge, *Water Environ. Res.*, 2008, **80**, 367.
116. U. K. M. Beccari, L. Bertin, D. Dionisi, F. Fava, S. Lampis, M. Majone, F. Valentino, G. Vallini and M. Villano, *J. Chem. Technol. Biotechnol.*, 2009, **84**, 901.
117. M. Venkateswar Reddy and S. Venkata Mohan, *Bioresour. Technol.*, 2012, **103**, 313.
118. S. K. S. Patel, M. Singh, P. Kumar, H. J. Purohit and V. C. Kalia, *Biomass. Bioener*, 2012, **36**, 218.
119. M. Venkateswar Reddy and S. Venkata Mohan, *Bioresour. Technol.*, 2012, **114**, 573.
120. S. Venkata Mohan and M. Venkateswar Reddy, *Bioresour. Technol.*, 2013, **128**, 409.
121. B. Panda, P. Jain, L. Sharma and N. Mallick, *Bioresour. Technol.*, 2006, **97**, 1296.
122. E. A. Dawes, in *Microbial Energetics*, E. A. Dawes, Blackie & Son, Glasgow, 1986, pp. 145–165.
123. S. Pratt, A. Werker, S. F. Morgan and P. Lant, *Water Sci. Technol.*, 2012, **65**, 243.
124. H. Salehizadeh and M. C. M. Van Loosdrecht, *Biotechnol. Adv.*, 2004, **22**, 261.
125. H. Satoh, Y. Iwamoto, T. Mino and T. Matsuo, *Water Sci. Technol.*, 1998, **38**, 103.
126. H. Takabatake, H. Satoh, T. Mino and T. Matsuo, *Water Sci. Technol.*, 2002, **45**, 119.
127. A. Third, M. Newland and R. C. Ruwisch, *Biotechnol. Bioeng.*, 2003, **82**, 238.
128. S. Bengtsson, *Biotech. Bioeng.*, 2009, **104**, 698–708.
129. S. C. Andrews, *Adv. Microbiol. Physiol.*, 1998, **40**, 281.
130. S. Krallish, S. Gonta and L. Savenkova, *Proc. Biochem*, 2009, **44**, 369.

131. B. A. Qurollo, P. E. Bishop and H. M. Hassan, *Can. J. Microbiol.*, 2001, **47**, 63.
132. Y. Lee, M. K. Kim, H. N. Chang and Y. H. Park, *FEMS Microbiol Lett.*, 1995, **131**, 35.
133. R. A. Henderson and C. W. Jones, *Arch. Microbiol.*, 1997, **168**, 486.
134. S. Park and Y. H. Lee, *J. Ferment. Bioeng.*, 1996, **81**, 197.
135. G. N. M. Huijberts, T. C. D. Rijk, P. D. Waard, and G. Eggink, *J. Bacteriol.*, 1995, 176, 1661.
136. M. N. Kraak, B. Kessler and B. Witholt, *Eur. J. Biochem.*, 1997, **250**, 432.
137. Q. S. Qi, A. Steinbuchel and B. H. A. Rehm, *FEMS Microbiol. Lett.*, 1998, **167**, 89.
138. B. Kessler and B. Witholt, *J. Biotech.*, 2001, **86**, 97.
139. C. Huang, M. Zong, H. Wu and Q. Liu, *Bioresour. Technol.*, 2009, **100**, 4535.
140. S. Papanikolaou, M. Komaitis and G. Aggelis, *Bioresour. Technol.*, 2004, **95**, 287.
141. Y. Chisti, *Biotechnol. Adv.*, 2007, **25**, 294.
142. G. Knothe, J. Krahl and J. V. Gerpen, *The Biodiesel Handbook*, AOCS Press, Champaign, IL, 2005.
143. M. Mittelbach and C. Remschmidt, *Biodiesel: The Comprehensive Handbook*, Martin Mittelbach, Graz, Austria, 2004.
144. C. Ratledge, *Biochem. Soc. Trans.*, 2002, **32**, 1047.
145. S. Ramalingam, D. Stephen, Z. Mark and B. Rakesh, *J. Ind. Microbiol. Biotechnol.*, 2010, **37**, 1271.
146. Q. Hu, M. Sommerfeld, E. Jarvis, M. Ghirardi, M. Posewitz, M. Seibert and A. Darzins, *Plant. J.*, 2008, **54**, 621.
147. M. Kelly, J. Suzanne, R. Charles and C. John, *Microbiol. Rev.*, 1993, 522.
148. X. Meng, J. Yang, X. Xu, L. Zhang, Q. Nie and M. Xian, *Renew. Energy*, 2009, **34**, 1.
149. M. K. Gouda, S. H. Omar and L. M. Aouad, *World J. Microbiol. Biotechnol.*, 24, 1703, **2008**.
150. G. Venkata Subhash and S. Venkata Mohan, *Bioresour. Technol.*, 2011, **102**, 9286.
151. S. Fakas, A. Makri, M. Mavromati, M. Tselepi and G. Aggelis, *Bioresour. Technol.*, 2009, **100**, 6118.
152. G. Vicente, L. F. Bautista, R. Rodrıguez, F. J. Gutierrez, I. Sadaba, R. M. Ruiz-Vazquez, T. S. Martınez and V. Garre, *Biochem. Eng. J.*, 2009, **48**, 22.
153. A. M. Illman, A. H. Scragg and S. W. Shales, *Enzyme Microb. Technol.*, 2000, **27**, 631–635.
154. A. P. Nigam, in *Encyclopedia of Food Microbiology*, R. K. Robinson, C. A. Batt and P. D. Patel, Academic Press, London and San Diego, 2000, pp. 718–729.
155. A. Beopoulos, J. Cescut, R. Haddouche, J. L. Uribelarrea, C. M. Jouve and J. M. Nicaud, *Progress Lipid Res.*, 2009, **48**, 375.

156. J. Pruvost, G. V. Vooren, G. Cogne and J. Legrand, *Bioresour. Technol.*, 2009, **100**, 5988.
157. H. M. Alvarez and A. Steinbuchel, *Appl. Microbiol. Biotechnol.*, 2002, **60**, 367.
158. H. M. Alvarez, R. Kalscheuer and A. Steinbuchel, *Fett/Lipid J*, 1997, **99**, 239.
159. M. Xin, Y. Jianming, X. Xin, Z. Lei, N. Qingjuan and X. Mo, *Renew. Energy*, 2009, **34**, 1.
160. O. Hiruta, K. Yamamura, H. Takebe, T. Futamura, K. Iinuma and H. Tanaka, *J. Ferment. Bioeng.*, 1997, **83**, 79.
161. C. Gwendoline, V. Kumar, R. Nouaille, G. Gaudet1, P. Fontanille1, A. Pandey, C. R. Soccol and C. Larroche, *Braz. Arch. Biol. Technol.*, 2012, **55**(29).
162. H. C. Chen and T. M. Liu, *Enzyme Microb. Technol.*, 1997, **21**, 137.
163. S. Patil, Thesis, UL Lafayette, Chemical Engineering Department, Lafayette, LA, 2010.
164. S. Papanikolaou and G. Aggelis, *Bioresour. Technol.*, 2002, **82**, 43.
165. F. M. Girio, C. Fonseca, F. Carvalheiro, L. C. Duarte, S. Marques and R. B. Łukasik, *Bioresour. Technol.*, 2010, **101**, 4775.
166. M. Foster and S. McLaughlin., *J. Membr. Biol.*, 1974, **17**, 155.
167. K. Ohta, D. S. Beall and L. O. Ingram, *Biotechnol. Bioeng.*, 1991, **38**, 296.
168. V. Senthilkumar and P. Gunasekaran, *J. Sci. Ind. Res.*, 2005, **64**, 845.
169. U. S. Aswathy, K. S. Rajeev, G. Lalitha Devi, K. P. Rajasree, R. S. Reeta and A. Pandey, *Bioresour. Technol.*, 2010, **101**, 925.
170. P. Binod, R. Sindhu, R. R. Singhania, S. Vikram, L. Devi, S. Nagalakshmi, N. Kurien, R. K. Sukumaran and A. Pandey, *Bioresour. Technol.*, 2010, **101**, 4767.
171. A. J. Grethlein, *J. Ferment. Bioeng.*, 1991, **72**, 58.
172. W. Zillig, I. Holz and S. Wunderl, *Int. J. Syst. Bacteriol.*, 1991, **41**, 169.
173. H. Bahl, H. Muller, S. Behrens, H. Joseph and F. Narberhaus, *FEMS Microbiol. Rev.*, 1995, **17**, 341.
174. L. Sullivan and G. N. Bennett., *J. Ind. Microbiol. Biotechnol.*, 2006, **33**, 298.
175. L. M. Harris, N. E. Welker and E. T. Papoutsakis, *J. Bacteriol.*, 2002, **184**, 3586.
176. K. Thormann, L. Feustel, K. Lorenz, S. Nakotte and P. Durre, *J. Bacteriol.*, 2002. 184, 1966.
177. P. D. Boyer, H. Lardy and K. Myrback, in *The Enzymes*, Academic Press, New York, 2nd edn, 1963, vol. 7, pp. 447–466.
178. J. Kim, R. Bajpai and E. L. Iannotti, *App. Biochem. Biotechnol.*, 1988, **18**, 175.
179. T. C. Ezeji, N. Qureshi and H. P. Blaschek, *Curr. Opin. Biotech*, 2007, **18**, 220.
180. T. C. Ezeji, N. Qureshi and H. P. Blaschek, *J. Ind. Microbiol. Biotechnol.*, 2007, **34**, 771.
181. D. T. Jones and D. R. Woods., *Microbiol. Rev.*, 1986, **50**, 484.

182. N. Qureshi, T. C. Ezeji, J. Ebener, B. S. Dien, M. A. Cotta and H. P. Blaschek, *Bioresour. Technol.*, 2008, **99**, 5915.
183. M. S. Madihah, A. B. Ariff, K. M. Sahaid, A. A. Suraini and M. I. A. Karim., *World J. Microbiol. Biotechnol.*, 2001, **17**, 567.
184. K. B. Cantrell, T. Ducey, K. S. Ro and P. G. Hunt, *Bioresour. Technol.*, 2008, **99**, 7941.
185. W. W. Li and H. Q. Yu, *Biotechnol. Adv.*, 2011, **29**, 972.

CHAPTER 7

Plantae and Marine Biomass for Biofuels

MOHAMED CHAKER NCIBI,*[a]
AICHA MENYAR BEN HAMISSA[b] AND
SARRA GASPARD[a]

[a] Department of Chemistry, COVACHIM-M2E Laboratory, University of Antilles and Guyane, UFR Exact and Natural Sciences, Pointe à Pitre 97159, Guadeloupe; [b] Department of Chemistry, University of Sousse, High Institute of Agronomy, Chott Meriam 4042, Tunisia
*Email: ncibi_mc@yahoo.com

7.1 Introduction

Many plantae and marine biomasses are cheap and available resources for industrial utilisation. These biomaterials have the composition, properties and structure that make them suitable to be used in many fields among with liquid and gaseous biofuels (*viz.* bioethanol, biodiesel and biomethane) for transportation.[1,2] Research endeavours on this crucial energetic field are of great importance in the development of an efficient and sustainable way of producing alternative automotive fuels away from the highly controversial fuels derived from food crops such as oil seeds, sugarcane and corn.[3–6] Recourse to these first generation biofuels is falling drastically mainly due to the severe worldwide food crisis.[7]

The production of biofuels from non-food bioresources is gradually becoming a vital issue due to the rarefaction of the fossil fuels and the urgent need to reduce CO_2 emissions,[8,9] along with the technological problems of

RSC Green Chemistry No. 25
Biomass for Sustainable Applications: Pollution Remediation and Energy
Edited by Sarra Gaspard and Mohamed Chaker Ncibi
© The Royal Society of Chemistry 2014
Published by the Royal Society of Chemistry, www.rsc.org

converting the lignocellulosic biomasses into biofuel due to the need to remove resistant lignin and break rigid polysaccharides polymers into free sugar molecules.[10,11] Energy concerns, growing environmental awareness and economic considerations are the major driving forces behind the worldwide direction towards utilising renewable bioresources and agro-wastes for various and fast-growing industrial applications.

This chapter uses recent research works to review the possibilities of using renewable, low-cost and highly available bioresources and/or wastes to produce secure and eco-friendly liquid biofuels. The 'centrepiece' of this chapter is the biomass itself. Thus, a number of types of exploitable biomasses and wastes are investigated for bioethanol, biodiesel and biomethane production, along with technological aspects of converting the biomass into fuel. Given the dynamic nature of the research endeavours related to sustainable biofuel production, several perspectives regarding this important issue are presented at the end of the chapter.

7.2 Bioethanol

7.2.1 Bioethanol from Lignocellulosic Biomass

Basically, lignocellulosic biomass is composed of three major components: cellulose, hemicellulose and lignin. Cellulose and hemicellulose (*i.e.* holocelluose) can be chemically and/or enzymatically hydrolysed into monomeric sugars (hexose and pentose), which can subsequently be converted biologically into bioethanol.

One of the advantages of using lignocellulosic resources for biofuel production is their availability, which has been a subject of debate in the related literature. Indeed, all or nearly all of the quantity of woody biomass was reported as 'potentially available'.[12] Other researchers used the term 'technically available'[13] to point out the difference between the amount of biomass that is expected to be recoverable using current or expected technology and the potentially available quantity of woody biomass. Other studies have related the amount of woody biomass that could be available to a given market price.[14]

There are many methods of grouping lignocellulosic biomass into categories. In this chapter, bioresources are divided into: (i) woody biomass; (ii) herbaceous biomass; and (iii) agro-industrial wastes. The biochemical composition of several lignocellulosic bioresources is reviewed throughout the text.

7.2.1.1 Woody Biomass

7.2.1.1.1 Poplar Trees (*Populus* sp.). Poplar trees are timber species ubiquitous to the Northern hemisphere. They belong to the Salicaceae family and *Populus* genus, with 30–40 species.[15] *Populus* sp. has attracted substantial attention and it can be intensively grown. Its productivity can be almost 9 Mg ha^{-1} yr^{-1}; selection of genotypes adapted to climatic conditions has resulted in nearly 2.5 times as much growth.[16] These trees also provide associated ecological services such as carbon sequestration, contaminant remediation and soil stabilisation.[16] In the United States, *Populus* species are the most abundant fast-growing species suitable for bioethanol production. *Populus deltoids* species cover most of North America from the east to the mid-west US, while *Populus trichocarpa* covers primarily the western US.[17]

To enhance the woody biomass conversion into bioethanol, all steps of the production process could be targeted (*i.e.* pretreatment, saccharification and fermentation). Thus, Negro *et al.*[18] studied two pretreatment methods of poplar (*Populus nigra*) chipped biomass: steam explosion and liquid hot water. The best results were obtained in steam explosion pretreatment at 483 K for 4 min, taking into account a cellulose recovery above 95%, an enzymatic hydrolysis yield of about 60%, and 41% xylose recovery in the liquid fraction. Other studies investigated the effect of pretreating hybrid poplar by aqueous ammonia using both ammonia recycle percolation (ARP) and aqueous ammonia (SAA). The enzymatic hydrolysis of ammonia-treated hybrid was studied applying cellulase enzyme supplemented with additional xylanase or pectinase. Conversion of ARP-treated hybrid poplar to ethanol was carried out by simultaneous saccharification and fermentation (SSF) and saccharification and co-fermentation (SSCF). The maximum ethanol yield observed from the SSCF experiment was 78% of theoretical maximum based on the total carbohydrate (glucan + xylan).[19]

Furthermore, dilute acid and sulfite pretreatment to overcome recalcitrance of lignocelluloses (SPORL) pretreatments were tested to enhance to bioconversion of four poplar wood samples to ethanol. Results showed that the highest bioconversion efficiency with total monomeric sugar yields of 47% and 55% theoretical and ethanol yields of 0.17 and 0.20 L kg^{-1} was attainted for aspen wood under five dilute acids and SPORL pretreatments, respectively.[20]

Another key factor in converting lignocellulosic materials into bioethanol is the enzyme source. For the case of poplar wood, several enzyme source have been investigated among them a combination of Celluclast 1.5 L (cellulase from *Trichoderma reesei* ATCC 26921) and Novozyme 188 (cellobiase from *Aspergillus niger*). Studies showed that for this enzymatic combination, *Populus tormentosa* Carr.[21] and *Populus nigra*[18] had different saccharification yields with 49and 60%, respectively. Other fungal enzyme sources can be found in the literature such as *Agaricus arvensis* and *Pholiota adiposa*. The first helped to obtain a saccharification yield of 37.6% from *Populus balsamifera*.[22] For the second source, the yield was 83.4% using *Populus nigra* as substate.[23]

7.2.1.1.2 Willow Trees (*Salix* sp.). Willow trees belong to the botanical family of Salicaceae, the *Salix* genus, containing 400 species. Based on their proven environmental benefits, willows are intensively cultivated plantations with the aim of reducing the CO_2 content of the neighbouring atmosphere. Willows are also useful in wastewater management and even in remediating high metal contents in soils.[24]

In the UK, lifecycle assessment (LCA) has been used to investigate the environmental and economic sustainability of a potential operation in which bioethanol is produced from the hydrolysis and subsequent fermentation of coppice willow. If the willow were grown on idle arable land in the UK or indeed in Eastern Europe and imported as wood chips into the UK, it was found that savings of greenhouse gas emissions of 70–90% would be possible compared with fossil-derived gasoline on an energy basis.[25]

As a fast-growing energy crop, *Salix* or willow trees have been, like poplar, the subject of numerous research studies attempting to optimise the conversion of their holocellulosic content into valuable bioethanol. In Sweden, wood chips of the *Salix* hybrid (*Salix schwerinii* × *Salix viminalis*) were steam pretreated with sulfuric acid.[26] High sugar recoveries were obtained after pretreatment, and the highest yields of glucose and xylose after the subsequent enzymatic hydrolysis step were 92 and 86% of the theoretical, respectively, based on the glucan and xylan contents of the raw material. The most favourable pretreatment conditions regarding the overall sugar yield were 200 °C for 6 min on average using 0.5% sulfuric acid. The result was a total of 55.6 g glucose and xylose per 100 g of dry raw biomass. Simultaneous saccharification and fermentation were performed and the overall theoretical ethanol yield was about 79%.

In Norway, different steam explosion pretreatments were applied to *Salix* chips and the effect was evaluated by running an enzymatic hydrolysis. After 24 h, the total enzymatic release of glucose and xylose increased with pretreatment harshness, with maximum values being obtained after pretreatment at 210 °C for 10 min for glucose (17.5 g L^{-1}) and at 210 °C for 5 min for xylose (4.5% g L^{-1}). Harsher pretreatment conditions did not increase glucose release and led to the degradation of xylose and the formation of furfurals.[27]

7.2.1.1.3 Eucalyptus Trees (*Eucalyptus* sp.). Eucalyptus trees belong to the Myrtaceae family. They are native to Australia. One of most important valorisation ways of this woody biomass is its use as a feedstock for bioethanol production. Indeed, in addition to their fast growth, Eucalyptus woods possess high cellulose content that can reach at about 45%.[28] The cellulose yield of the Eucalyptus bark is approximately 15% of the total cellulose yield of the entire tree.[29]

In order to optimise the efficiency of conversion of this woody biomass to bioethanol, dilute sulfuric acid was used to pretreat eucalyptus chips prior to enzymatic hydrolysis. The cellulose and hemicellulose contents were 42.6 and 19.6%, respectively. The maximum total sugars yield (combined xylose and glucose) was 47.7 g per 100 g of raw material, representing 82% of total sugars in the eucalyptus biomass. This yield was obtained at 160 °C, 0.75% acid concentration and 10 min residence time.[30]

In Australia, a research study proposed using eucalyptus forest thinnings as a potential feedstock option for the local emerging biofuel industry.[31] They found that the best operating conditions yielded 74% of the theoretical conversion. *Saccharomyces cerevisiae* efficiently fermented crude *Eucalyptus dunnii* hydrolysate, within 30 h, yielding 18 g L^{-1} of ethanol, which represents a sugar to ethanol conversion rate of 0.475 g g^{-1} (92%).

In Spain, *Eucalyptus globulus* wood samples were subjected to non-isothermal autohydrolysis in order to solubilise hemicelluloses, leading to treated solids of increased cellulose content and enzyme digestibility. The autohydrolysate was used as a substrate for bioethanol production by SSF. Wood processing resulted in the generation of soluble products from hemicelluloses, with up to 291 L of bioethanol per tonne of oven-dry wood.[32]

In Japan, saccharification of the inner bark of Eucalyptus was carried out by enzymatic hydrolysis to produce bioethanol. To enhance the accessibility of the enzyme to the polysaccharides such as cellulose and holocellulose in the cell wall of the bark, the biomass was subjected to hydrothermal pretreatment with carbon dioxide. Results showed that this pretreatment considerably influenced enzymatic hydrolysis. The main component of the generated monosaccharide was glucose (>90%), and the yield of glucose on the basis of α-cellulose reached about 80%.[29]

7.2.1.1.4   Pine Trees (*Pinus* sp.).   Pines are native to most of the Northern hemisphere. They belong the Pinaceae family and *Pinus* genus. Different species of pine trees have been tested as potential feedstock for bioethanol production.

An interesting research study tried to link the pulp and paper industry to the production of second generation ethanol.[33] The idea was to minimise the cost related to the pretreatment stage since the separation of the wood components already takes place in the chemical pulp mill using the long proven technology in pulp production known as soda cooking. Aspen (*Populus tremula*) and pine (*Pinus sylvestris*) wood from Nordic mills were studied. Ethanol yields between 81.6% and 87.8% on theoretical maximum were obtained.

In the United States, the technical and economic feasibility of producing bioethanol from loblolly pine (*Pinus taeda*) was evaluated.[34] The production process starts with prehydrolysis using dilute sulfuric acid. The aim was to hydrolyse the hemicellulose fraction and make the cellulose more accessible to the enzymatic hydrolysis. After fermentation, results showed ethanol yields of 80 and 102 US gal ton^{-1} of dry biomass, respectively, for the case where 75% and 95% of the carbohydrates in loblolly pine were converted to fermentable sugars. An economic assessment was indicated that ethanol could be produced at a cost of $1.53 per gallon, based on a 75% conversion of the carbohydrates in wood to sugars for ethanol production. Improving the conversion ratio of wood carbohydrates to sugars to 95% would reduce the production cost to $1.29 per gallon.[34]

Another species of pine trees, lodgepole pine (*Pinus contorta*), has also been experimented with to convert its biopolymers into fermentable sugars. One research study pretreated lodgepole wood chips by sulfite pretreatment to overcome recalcitrance of lignocelluloses (SPORL) and the solid fraction was separated from the liquor stream.[35] Quasi-simultaneous enzymatic saccharification of the cellulosic solids was conducted, along with combined fermentation of the concentrated liquor, at up to 20% total solids loading. A maximum ethanol yield of 47.4 g L^{-1} was achieved, resulting in a calculated yield of 285 L t^{-1} of wood using *Saccharomyces cerevisiae* YRH400 at 35 °C and pH 5.5.

7.2.1.1.5 Cedar Trees (*Cedrus* sp.). Cedars are coniferous trees belonging to the Pinaceae family. Several studies have considered cedar trees as an interesting cellulosic material for bioethanol production. In Japan, *Crypto-meria japonica* (Japanese cedar) makes up around 60% of plantation forests in Japan and there is a growing demand for forest thinnings. However, the softwood is one of the most recalcitrant wood species for hydrothermal and thermochemical pretreatments for enzymatic saccharification. Thus, Baba *et al.*[36] applied combined pretreatments by solvolysis and cultivation with white-rot fungi to develop an environmentally benign pretreatment system applicable to recalcitrant softwood. The analyses showed that the enzymatic saccharification yield from ethanolysis pulp was 10.2%, based on the weight of holocellulose. When the softwood was treated for eight weeks with select-ive white-rot fungi (*Ceriporiopsis subvermispora* FP-90031 and a new fungal isolate *Phellinus* sp. SKM2102) prior to the ethanolysis, the sugar yield in-creased to 35.7 and 40.8%, respectively.

Another Japanese research team worked on ethanol production from Japanese cedar using various pretreatment methods.[37] The maximum values of glucose and reducing sugars were produced using consecutive pretreatments with a 2.5 MPa steam explosion and an ionic liquid were 408 and 462 mg g^{-1} initial dry sample, respectively. The most positive effects on the enzymatic saccharification kinetics were observed when the consecutive pretreatment methods were used. However, the process of using the organosolv treatment of cedar wood chips without the steam explosion seems to be a more cost-effective pretreatment method for the enzymatic saccharification of Japanese cedar; it results in 386 and 426 mg g^{-1} initial dry sample of glucose and reducing sugars, respectively.

7.2.1.1.6 Other Woody Bioresources

7.2.1.1.6.1 Black Locust (Robinia pseudoacacia *L.*). Black locust is a tree in the Fabaceae family of *Robinia*. It is native to the southeastern United States and has been naturalised and acclimated elsewhere in Europe, Africa and Asia. It is considered an invasive species in some areas. As a potential feedstock for bioethanol production, a well-to-wheel analysis was conducted for ethanol obtained from black locust trees by means of an LCA method-ology. According to the results, fuel ethanol derived from black locust bio-mass may help to reduce the contributions to global warming, acidification, eutrophication and fossil fuel use due to the low input production regime of the agricultural stage.[38] In 2012, the same Spanish team published another LCA for three fast-growing wood crops: black locust, eucalyptus and poplar. The use of bioethanol derived from black locust was found to be the option with the lowest impact in most environmental categories, with reductions of 97% for global warming potential over 100 years (GWP100), 42% for acidification potential and 76% for fossil fuel use in comparison with conventional gasoline.[39]

7.2.1.1.6.2   Sweetgum Trees (Liquidambar styraciflua L.).   The sweetgum tree, also called Liquidambar, belongs to the Hamamelidaceae family. The tree is native to eastern North America and is distributed over a wide geographical range extending from North America to East Asia.[40] As a feedstock for biofuel production, a study tested the conversion of sweetgum, along with various other hardwood trees, into bioethanol. After kraft pretreatment and enzymatic hydrolysis of wood samples, results showed that for a pretreatment time of 20 min, sweetgum presented the highest sugar recovery with 60.7%. The effect of pretreatment time played a significant role in enhancing the conversion into sugar; from 20 to 60 min, the saccharification efficiency went from 75.1 to 96.3%, corresponding to an increase in the sugar recovery from 33.1 to 40.6 g 100 g^{-1} of sweetgum wood. [41]

7.2.1.1.6.3   Oak Trees (Quercus serrata L.).   Oak trees belong the Fagaceae family and the *Quercus* genus. In South Korea, attempts have been made to convert oak wood into bioethanol. A study tried to enhance the bioethanol production yield from oak chips by reducing the impact of the inhibitory factors during the bioconversion process.[42] The first observation was that the lignin-degradation products were more inhibiting than the sugar-derived products such as furfural and 5-hydroxymethylfurfural (HMF). As a solution, adaptation of yeast cells to the wood hydrolysate and detoxification methods, such as using charcoal and lime, were applied. After treatment with charcoal and low-temperature sterilisation, the yeast cells could utilise the concentrated wood hydrolysate with 170 as well as 140 g L^{-1} glucose, and produce 69.9 and 74.2 g L^{-1} ethanol, respectively, with a yield of 0.46–0.48 g ethanol g^{-1} glucose.

In Japan, Kamei *et al.*[43] proposed a new process of unified aerobic delignification and anaerobic saccharification and fermentation of oak wood by a single microorganism, the white-rot fungus *Phlebia* sp. MG-60. The results of this work confirmed that this fungus was able to selectively degrade lignin under aerobic solid state fermentation conditions and to produce ethanol directly from delignified oak wood under semi-aerobic liquid culture conditions. Thus, after 20 days of anaerobic incubation, 43.9% of the theoretical maximum ethanol was produced, and after 56 days, 40.7% of initial lignin was degraded.

7.2.1.2   Herbaceous Biomass

7.2.1.2.1   Miscanthus (*Miscanthus* sp.).   Miscanthus is a high-yielding perennial grass, belonging to the Poaceae family native to Africa and South Asia. Miscanthus species were paid considerable attention by scientists due to their good high biomass yields (*i.e.* high productivity), low inputs and high polysaccharide content.[44] In Europe, 15 different genotypes of Miscanthus were planted with biomass yields oscillating from 38 t ha^{-1} to a maximum of 41 t ha^{-1} for *Miscanthus × giganteus* and Miscanthus hybrids, respectively.[45] Thus, this biomass represents a potential source for bioethanol production.

In South Korea, 12 Miscanthus genotypes (eight lines of *Miscanthus sinensis*, one line of *Miscanthus × giganteus* and three lines of *Miscanthus sacchariflorus*) were cultivated to assess their suitability for biofuel production.[46] The results suggested that the two genotypes *Miscanthus sacchariflorus* seem to be most suitable for biofuel production, based on high polysaccharide content (close to 65%) and low lignin content (<20%).

In Denmark, a study used dilute sulfuric acid presoaking, wet explosion and enzymatic hydrolysis to evaluate the effect of these different pretreatment methods for bioethanol production using Miscanthus as a substrate.[47] The analysis showed that the combination of presoaking, wet explosion and enzymatic hydrolysis was found to give the highest sugar yields. In the wet explosion, the use of atmospheric air gave the highest xylose yield (94.9% xylose and 61.3% glucose), while hydrogen peroxide gave the highest glucose yield (82.4% xylose and 63.7% glucose).

In the United States, investigations to use Miscanthus as a potential feedstock for lignocellulosic ethanol have been performed. Among them, a comparison was made between Miscanthus and the already used maize crop. The analysis revealed that using maize to produce enough ethanol to offset 20% of US gasoline consumption would divert 25% of US cropland currently in production, while getting the same amount of ethanol from Miscanthus would divert only 9.3%.[48] If the projections of Heaton *et al.*[49] can be proven in the market place, Miscanthus could help the United States to reach its target of replacing 30% of the gasoline it uses with biofuels by 2030.

7.2.1.2.2 Switchgrass (*Panicum virgatum*). Switchgrass is a perennial grass native to America. It is a herbaceous perennial grass that belongs to the Poaceae family. Switchgrass can be used as a forage plant or as an ornamental, and also in protecting the soil from erosion. It is also considered a potential source of biofuel biomass in the United States and Europe.[45,50] Like in the other lignocellulosic materials, the lignin fraction in the switchgrass biomass is a major barrier limiting the accessibility of carbohydrates to hydrolytic enzymes. To overcome this problem, several pretreatment procedures have been investigated experimentally. One study of fungal pretreatment of switchgrass involved SSF to improve saccharification and simultaneously produce enzymes as co-products. The results revealed that the fungus *Pycnoporus* sp. SYBC-L3 can significantly degrade lignin and enhance enzymatic hydrolysis efficiency. After a 36-day cultivation period, a nearly 30% reduction in lignin content was obtained without significant loss of cellulose and hemicellulose.[51]

In the United States, lime pretreatment of switchgrass was explored to improve its enzymatic digestibility at mild temperatures,. The effects of residence time, lime loading and biomass washing on the sugar production efficiency were investigated. Pretreatments were evaluated based on the yields of biomass-derived sugars in the subsequent enzymatic hydrolysis. Under the best pretreatment conditions [50 °C, 24 h, 0.10 g Ca(OH)$_2$ g^{-1} raw biomass and wash intensity of 100 mL water g^{-1} raw biomass], the yields of glucose, xylose and total reducing sugars reached 239, 127, and 433 mg g^{-1} raw biomass, which were respectively 3.1, 5.8, and 3.6 times those of untreated biomass.[52]

Based on the research breakthroughs regarding the optimisation of the saccharification step, several projections have been presented on the bioethanol production yield from switchgrass (*Panicum virgatum*, L.), with conversion efficiency oscillating between 1711 and 2000 L of ethanol ha^{-1}.[53,54] Schmer *et al.*[55] made their projection based on a more productive farms and the proposed yield was 3000 L of ethanol ha^{-1}.

7.2.1.2.3 Reed Canary Grass (*Phalaris arundinacea* L.).   Reed canary grass is a rhizomatous, perennial robust coarse grass about 2 m high,[56] belonging to the Poaceae family. It is widely distributed across temperate regions and grows naturally in Europe, Asia, Africa and North America.[57] It can give high biomass yields of about 8–10 t of dry matter ha^{-1} when harvested in summer.[56]

Kallionien *et al.*[58] studied the enzymatic hydrolysis and fermentation of reed canary grass harvested in spring or autumn. After pretreatments, the washed solid fraction was hydrolysed using commercial cellulases at a dosage of 10 filter paper unit (FPU) g^{-1} dry matter and β-glucosidase (Novozym 188) 100 nkat g^{-1} dry matter. A high hydrolysis yield (95%) was obtained with both raw materials at 10% consistency. After fermentation the highest ethanol yield, 82% of theoretical based on glucose, was obtained on reed canary grass.

In the United States, a study investigated the effect of two kinds of pretreatments on reed canary grass: sulfuric acid and calcium hydroxide (lime).[59] After fermentation using *Saccharomyces cerevisiae* D5A, the results showed that the conversion of the sugar content to ethanol for sulfuric acid-treated reed canary grass ranged from 22 to 83%. However, for the lime pretreatment, the conversion ranged 21 to 55%.

Another team tried to optimise bioethanol production from reed canary grass by applying liquid hot water and dilute ammonia pretreatments prior to fermentation.[60] Dilute ammonia gave higher yield efficiencies than liquid hot water. The optimal condition for dilute ammonia (4% w/v) pretreatment was 170 °C for 20 min. Hydrolysates were converted to ethanol using *Saccharomyces cerevisiae*, along with a blend of commercial cellulases. The final ethanol conversion efficiency was 84% based upon total hexosans, with 72% of the xylan converted to soluble xylan oligomers.

7.2.1.2.4 Giant Reed Grass (*Arundo donax* L.).   Giant reed grass is a perennial rhizomatous plant of the Gramineae family, native to southern Europe and widely distributed in all subtropical and warm temperate areas of the world.[61] It is one of the largest of the herbaceous grasses characterised by a tall erect stem. It can be produced in all types of soils, tolerates difficult environmental conditions and has very rapid growth.[62]

Pretreated *Arundo donax* was investigated for its suitability as a feedstock for bioethanol production by applying either separate hydrolysis and fermentation (SHF) or simultaneous saccharification and fermentation (SSF). For the fermentation, a xylose-fermenting strain of *S. cerevisiae* was used since *A. donax* has a high xylan content.[63] The respective highest overall ethanol yield and final ethanol concentration achieved using SHF were 0.27 g g^{-1} and 20.6 g L^{-1} compared with 0.24 g g^{-1} and 19.0 g L^{-1} with SSF.

Based on the already mentioned high xylan content, Scordia *et al.*[64] studied the production of bioethanol from the sugars contained in the giant reed (*Arundo donax* L.) hemicellulosic hydrolysate using *Scheffersomyces stipitis* CBS6054, a xylose fermenting yeast. The objective was to determine the optimum conditions to maximise sugar release and to minimise degradation products.

7.2.1.2.5 Alfalfa (*Medicago sativa* L.). Alfalfa is a perennial flowering plant in the Fabaceae family. It has considerable potential as a feedstock for production of fuels, feed and industrial materials. For competitive use of alfalfa as a biofuel feedstock, several studies have been carried out to enhance the conversion of polymers into fermentable sugars.

In the United States, alfalfa biomass was evaluated for biochemical conversion into ethanol using dilute acid and ammonia pretreatments.[65] The results showed bioethanol yields for alfalfa stems pretreated with dilute acid were significantly impacted by harvest maturity and lignin composition, whereas when pretreated with dilute ammonia, yield was solely affected by lignin composition. The use of a recombinant xylose-fermenting *Saccharomyces* strain to convert the ammonia pretreated alfalfa samples further increased ethanol yields to 232–278 L ton^{-1}. Another study described ethanol production from alfalfa fibre using SHF and SSF with and without liquid hot water pretreatment.[66] The research revealed that, for untreated alfalfa fibre, the use of *Candida shehatae* FPL-702 strain produced 5 and 6.4 g L^{-1} of ethanol with a yield of 0.25 and 0.16 g ethanol g^{-1} sugar by SHF and SSF, respectively. For the pretreated biomass and using SSF, the yeast strain produced 18.0 g L^{-1} ethanol (a yield of 0.45 g ethanol g^{-1} sugar). Using SHF, it produced 9.6 g L^{-1} ethanol, a yield of 0.47 g ethanol g^{-1} sugar.

7.2.1.2.6 Other Herbaceous Bioresources. Eastern gamagrass (*Trypsacum dactyloides*), a perennial grass, has been investigated as a new lignocellulosic crop for bioethanol production. With regard to its cellulosic fraction, results showed that eastern gamagrass has a comparable content (37%) to that of switchgrass *Panicum virgatum* (35.1%). With cellulose solvent-based lignocellulose fractionation (CSLF) pretreatment and subsequent enzymatic saccharification, 80.5–99.8% of cellulosic glucose was released from the gamagrass biomass compared with 73.5–87.1% for switchgrass. Furthermore, the fermentation of the gamagrass hydrolysate gave an ethanol yield of 0.496 g g^{-1} glucose.[67]

Bermuda grass (*Cynodon dactylon*) was also tested as a potential feedstock for fuel ethanol production because it has a relatively high cellulose and hemicellulose content.[68] Dilute sulfuric acid pretreatment of the herbaceous biomass followed by enzymatic hydrolysis of the cellulosic fraction was performed under various operating conditions. The analysis revealed that the highest conversion rate was 83.1% under the optimum conditions of 1.5% acid concentration and 90 min of exposure time. The effect of pretreatment was quite significant if the conversion rate of 83.1% for the pretreated biomass is compared with the ratio of the untreated grass (*i.e.* 26.4%).

In India, kans grass (*Saccharum sponteneum*) was proposed as a possible source for bioethanol production. The biomass was subjected to enzymatic hydrolysis to produce fermentable sugars and later fermented to bioethanol using *Saccharomyces cerevisiae*. Maximum sugar recovery was found to be 69.08 mg g^{-1} dry biomass with 20 FPU g^{-1} dry biomass of enzyme dosage (CMCase, cellulase and xylanase) under optimum conditions. The fermentation of the enzymatic hydrolysate gave an ethanol yield of 0.46 g g^{-1}.[69]

7.2.1.3 Bioethanol from Agricultural and Industrial Wastes

7.2.1.3.1 Sugarcane Bagasse. Sugarcane (*Saccharum* sp.) is a native plant to tropical regions of South Asia, belonging to the Poaceae family. The main product is sucrose, which accumulates in the stalk internodes. It contributes 60% of the raw sugar produced worldwide.[70] Thus in Brazil the sugarcane crop is used to produce sugar (38.7 million tons of sugar, a 2010 estimate) as well as bioethanol fuel (about 27 billion litres annually).[71] Although economically efficient, such double production potential was at the heart of the well-known controversy over the possible competition between biofuel and food (sugar) production.[72] To overcome this problem, several research teams have worked on the valorisation of sugarcane bagasse, a by-product of the sugar industry, to produce bioethanol.

With its cellulosic and hemicellulosic content of 60%[73] and an estimated worldwide production of 180 million tons,[74] sugarcane bagasse is an interesting feedstock for bioethanol production. In one study, sugarcane bagasse was pretreated by steam explosion at 205 and 215 °C and hydrolysed with cellulolytic enzymes. The hydrolysates were then subjected to two kinds of detoxification methods: chemically by overliming and enzymatically with phenoloxidase laccase. The analysis revealed that approximately 80% of the inhibiting phenolic compounds were specifically removed by the laccase treatment. The detoxified hydrolysates were fermented with the recombinant xylose-utilising *Saccharomyces cerevisiae*, with an ethanol yield of 0.18 g g^{-1} dry bagasse compared with 0.13 g g^{-1} dry bagasse in the case of the undetoxified substrate.[75]

Another pretreatment method combining NH_4OH–H_2O_2 and ionic liquid was developed for the recovery of cellulose from sugarcane bagasse.[76] Results showed that the regenerated substrate from the combined pretreatment significantly enhanced enzymatic digestibility with an efficiency of 91.4% after

12 h of hydrolysis, which was 64% higher than the efficiency observed for the regenerated bagasse after a single NH_4OH–H_2O_2 pretreatment. Using SSCF, an ethanol yield of 0.42 g g^{-1} was achieved with a corresponding fermentation efficiency of 94.5%.

7.2.1.3.2 Rice Straw. As a non-food agricultural residues resource, rice straw is a promising feedstock for bioethanol production. The biochemical composition of rice straw reveals a rich content with 32–47% of cellulose and 19–27% of hemicellulose.[77,78] The annual production of rice was estimated to be about 721 million tons in 2011 and each kilogram of grain harvested is associated with an average production of 1.325 kg of straw. Thus, an estimated production of 955 million tons per year of rice straw was deduced.[79]

Several studies have been conducted into the conversion of rice straw to bioethanol. A study pretreated the rice residue using ionic liquids to improve enzymatic hydrolysis and ethanol production. The pretreatments were carried out with two cellulose solvents: N-methyl morpholine N-oxide (NMMO) and 1-buthyl-3-methylimidazolium acetate (BMIM-OAc) at 120 °C for 1–5 h, with 5% of rice straw loading. After enzymatic hydrolysis, the conversion of the glucan content into glucose (*i.e.* hydrolysis yield) was complete for biomass treated with BMIM-OAc and 96% for samples treated with NMMO. The effect of the pretreatment phase was therefore significant as the conversion was only 27.7% for the untreated straw.[79]

In Japan, a research team investigated the feasibility of an ethanol production process based on cell recycling repeated batch SSF of alkali-pretreated rice straw using immobilised *Saccharomyces cerevisiae* cells.[80] The analysis revealed that, in batch SSF of 20% (w/w) rice straw, the ethanol yields based on the glucan content of the immobilised cells were slightly lower (76.9% of the theoretical yield) than that of free cells (85.2% of the theoretical yield). In repeated batch SSF of 20% (w/w) rice straw, stable ethanol production of approx. 38 g L^{-1} and an ethanol yield of 84.7% were obtained. The authors also confirmed that the immobilising carrier could be reused without disintegration or any negative effect on ethanol production ability.

Another study applied a combination of acid pretreatment with ultrasound and subsequent enzyme treatment. The highest conversion of lignocellulose in rice straw to sugar and, consequently, the highest ethanol production (0.42 g g^{-1}) was found after six days of fermentation using *S. cerevisae* yeast.[81]

7.2.1.3.3 Wheat Straw and Bran. Wheat is an annual cereal grain, originally from the Levant region of the Near East and Ethiopian Highlands. Currently, it is cultivated worldwide. This agricultural crop is grown for the grain portion of the plant used for food consumption. The residue or the rest of the plant, consisting of the stems, leaves, chaff and the underground root system, is called wheat straw. The biochemical composition of wheat straw made it a promising feedstock for bioethanol production. According to the National Renewable Energy Laboratory (Colorado), the chemical analysis of raw material showed the following dry weight composition: 30.2% cellulose and 22.3% hemicellulose, with (18.7% xylan, 2.8% arabinan and 0.8% galactan).[82]

In the United States, a study subjected wheat straw to dilute acid pretreatment at varied temperatures prior to enzymatic saccharification for the conversion of the cellulose and hemicellulose content to monomeric sugars. The maximum yield of monomeric sugars from wheat straw by dilute H_2SO_4 pretreatment and enzymatic saccharification (45 °C, pH 5.0, 72 h) using cellulase, β-glucosidase, xylanase and esterase was 565 ± 10 mg g^{-1}. The analysis showed that the detoxification of the acid and enzyme treated wheat straw hydrolysate by overliming reduced the fermentation time from 118 to 39 h for the case of SHF and from 136 to 112 h for SSF, and increased the ethanol yield from 13 to 17 g L^{-1}.[83]

In India, another study used microwave alkali pretreated wheat straw as a substrate for ethanol production using *Saccharomyces cerevisiae*. Optimisation of ethanol production from wheat straw enzymatic hydrolysate was studied using a sequential statistical optimisation process for several factors: pH, temperature, initial total reducing sugar concentration (TRS) and inoculum level. The results revealed that, at bioreactor level, the obtained ethanol production was 16.4 g L^{-1} with ethanol productivity of 0.45 g L^{-1} h^{-1} obtained at under the optimum conditions of pH 5.5, temperature 30 °C, inoculum level of 3.3% and TRS concentration of 6.5%.[84]

An Italian research team studied wheat bran for bioethanol production using newly isolated *Saccharomyces cerevisiae* strains. The evaluation of the efficiency of these stains was based on their fermentative ability in minimal media supplemented with high glucose and xylose concentrations. Among the strains having remarkable ethanol yields from glucose in a minimal broth with high sugars concentrations, *S. cerevisiae* MEL2 showed the highest glucose-to-ethanol conversion efficiency of 96% of the theoretical, compared with the 86% exhibited by the benchmark strain. The strain showed also promising fermentative activity on the whole unfiltered hydrolysates.[85]

7.2.1.3.4 Corn Stover. Corn stover consists of the leaves and stalks of maize (*Zea mays* L.). An important and relatively recent way to valorise corn stover is as a feedstock to produce bioethanol. Indeed, in the United States, the amount of corn stover that can be sustainably collected is estimated to be 80–100 million metric ton yr^{-1} of dry matter. The long-term demand for corn stover by non-fermentative applications is estimated to be about 20 million metric ton yr^{-1} of dry matter. Hence, 60–80 million metric ton yr^{-1} of the dry biomass should be available to fermentative routes. If 40% of the harvestable corn stover were made available for biofuel conversion, an annual ethanol production potential of 11 billion litres could be achieved.[86]

In China, Zhao and Xia[87] analysed the fermentability of three corn stover hydrolysates to optimise ethanol production. The results demonstrated that the use of enzymatic hydrolysate from alkaline-pretreated corn stover was more effective than the hydrolysates from acid pretreatment. Furthermore, after fermentation with free cells of *S. cerevisiae* ZU-10, 41.2 g L^{-1} ethanol was obtained within 72 h for the enzymatic hydrolysate from alkaline-pretreated corn stover containing 66.9 g L^{-1} glucose and 32.1 g L^{-1} xylose. Fermentation of the same hydrolysate was improved in productivity with immobilised cells of *S. cerevisiae* ZU-10 (in alginate beads), reaching 1.70 g L^{-1} h^{-1} compared with 0.572 g L^{-1} h^{-1} obtained by free cells. All glucose and 96.3% xylose were consumed within 24 h and 40.7 g L^{-1} ethanol was obtained with the ethanol yield on fermentable sugars of 0.411 g g^{-1}.

Another study compared the effect of SSF and SHF on ethanol production efficiency from steam-pretreated corn stover.[88] The enzymatic loading in the related experiments was 10 FPU g^{-1} of water-insoluble solids and the yeast concentration in SSF was 1 g L^{-1} (dry weight) of a spent sulfite liquor-adapted strain of *S. cerevisiae*. The results showed that SSF gave a 13% higher overall ethanol yield than SHF (72.4% *versus* 59.1% of the theoretical). In addition, the study revealed that in presence of inhibiting degradation products (in the case of whole slurry), SSF was shown to be a better process configuration than SHF.

7.2.1.3.5 Other Agricultural Wastes. Several other agricultural residues have been analysed for their potential to produce bioethanol. In the United States, barley hull was pretreated using aqueous ammonia in order to be converted into ethanol. The published results showed that, among the tested conditions, the best pretreatment conditions observed were 75 °C, 48 h, 15 wt.% aqueous ammonia and a solid to liquid ratio of 1 : 12, resulting in saccharification yields of 83% for glucan and 63% for xylan with 15 FPU g^{-1} glucan enzyme loading. The production process was based on SSCF using ammonia-treated barley hull and recombinant *Escherichia coli* (KO11). The process resulted in an ethanol concentration of 24.1 g L^{-1} at 15 FPU cellulose g^{-1} glucan loading, which corresponds to 89.4% of the maximum theoretical yield based on glucan and xylan.[89]

Another study used barley straw which contained 34.3% cellulose, 23.0% hemicellulose and 13.3% lignin. Several pretreatments (dilute acid, lime and alkaline peroxide) and enzymatic saccharification procedures were evaluated for the conversion of barley straw to monomeric sugars and ethanol.[90] The authors found that dilute acid and lime pretreatments followed by enzymatic saccharification generated 566 mg (88% yield) and 582 mg (91% yield) total sugars per g of barley straw, respectively. Furthermore, the yield of ethanol from the acid pretreated and enzymatically saccharified barley straw hydro-lysate was 11.4 g L^{-1} using a recombinant *E. coli* strain FBR5.

Mandarin (*Citrus unshiu*) peel waste was also investigated as a feedstock to produce ethanol. A Korean research team designed a pretreatment process using a fired burner and a horizontal cylinder rotating on an axis called 'bio-mass popping pretreatment'.[91] The mandarin peel was pretreated at 150 °C for 10 min without chemical treatment. Enzymatic hydrolysis of pretreated peel was performed (pH 4.8 at 45°C for 6 h) and the total saccharification rate was approximately 95.6%. Subsequent fermentation of the pretreated biomass hydrolysate (pH 5.0 at 30 °C for 12 h) in a laboratory bioreactor provided an ethanol yield of 90.6% compared with 78% at 36 h from raw mandarin peel.

The same 'popping' pretreatment procedure was also applied to another waste, coffee residue.[92] The enzymatic conversion rate to fermentable sugars was 85.6%. After SSF, the ethanol concentration (based on sugar content) was 15.3 g L^{-1}, providing therefore an ethanol yield of 87.2%.

7.2.1.3.6 Pulp and Paper Industry Wastes

7.2.1.3.6.1 Paper Sludge. Paper sludge is the residual pulp and ash derived from solid waste materials generated by the pulp and paper industry. It is considered among the largest solid waste streams with interesting characteristics, namely high carbohydrate content and the dispersion of its structure. Hence, this material could be used to produce new products almost without pretreatment.[93] Such properties are of great interest especially in the field of bioethanol production.

Many research works have investigated the conversion process of paper sludge to ethanol under industrially relevant conditions.[93,94] Thus, specific parameters have been tested to optimise the conversion process such as the search for new microorganisms for the conversion of carbohydrates or the evaluation of the technical and economic viability of converting paper sludge to ethanol on a full commercial scale.

In Japan, high-solid paper sludge was tested for its potential to produce renewable biofuel. The substrate was subjected to different pretreatment procedures: mechanical grinding by ball mill; chemical swelling by phosphoric acid; and sequential system with a ball mill and phosphoric acid. The combined method was more favourable with an enzyme saccharification rate and a reducing sugar productivity of 84.1% and 28.1 mg g^{-1} h^{-1}, respectively. SSF from the sequential pretreatment system provided a 81.5% ethanol conversion rate, the productivity being 1.27 g L^{-1} h^{-1} compared with untreated raw paper sludge, which gave a 54.3% ethanol conversion rate and 0.42 g L^{-1} h^{-1} productivity.[95]

Another study applied SSF for the conversion of paper sludge to alcohol using two yeast strains, *Saccharomyces cerevisiae* and *Kluyveromyces marxianus*. In addition, two types of SSF experiments were adopted, *viz.* isothermal SSF and SSF with temperature profiling. The results showed no significant observed difference between *S. cerevisiae* and *K. marxianus* when the SSF results were compared. The ethanol yields were in the range 0.31–0.34 g g^{-1} for both strains.[96]

7.2.1.3.6.2 Paper Wastes. Waste papers are an important part of the degradable fraction in municipal solid waste and have potential to be a promising feedstock for bioethanol production. Such potential comes from their abundance, the high carbohydrate content and the relatively easy accessibility to polysaccharides.[97] In addition, the utilisation of waste papers for bioethanol production may offer a useful and valuable route to managing these papers as an alternative or complement to recycling.

Used newspapers are among the most commonly known paper wastes. A number of studies have been performed to valorise such important waste. In India, the bioconversion of the carbohydrate component of newspaper (70% holocellulose) to sugars by enzymatic saccharification and its fermentation to ethanol was investigated using several enzymatic treatments. The results revealed that a cellulase enzyme system was found to de-ink the newspaper most efficiently. Compared with the batch mode, fed batch enzymatic saccharification of the newspaper increased the sugar concentration in hydrolysate from 14.64 g L^{-1} to 38.21 g L^{-1}. Furthermore, the batch and fed batch enzymatic hydrolysates when fermented with *Saccharomyces cerevisiae*, produced 5.64 g L^{-1} and 14.77 g L^{-1} ethanol, respectively.[98]

Waste office papers were also analysed as a substrate for bioethanol production. The wastes were treated with steam in a pressure vessel at 170–220 °C and then enzymatically saccharified and fermented to ethanol. The results showed that steam alone was not effective in raising yields, but acid-catalysed steam treatment resulted in yields of 460 L ton-, an increase of 29% over the untreated controls (91% of theoretical).[99]

Furthermore, waste money bills that are no longer legal tender are non-recyclable and are usually destroyed. Hence, a Korean research group used this cellulose-containing material for bioethanol fermentation. The objective was to reduce waste money bill management costs and make a profit from ethanol. The results from this study revealed that glucose production was enhanced by using dilute H_2SO_4 during pretreatment. Different incubation periods were tested for saccharification and subsequent bioethanol fermentation. The highest yield of glucose (41.90 g L^{-1}) was shown to increase by 27.20% and 25.90%, respectively, by increasing the reaction period by 30 min and by increasing the acid concentration by 0.5%. The amount of bioethanol obtained was 22.01 g L^{-1}; alcohol fermentation increased by 59.38%, 110.02% and 64.13%, respectively with 30 min of reaction period, 0.5% acid concentration and under anoxic condition with benzoic acid.[100]

7.2.2 Bioethanol from Algae and Seagrasses

7.2.2.1 Algae (Seaweeds)

The production of ethanol from algae is possible by converting the carbohydrate content to ethanol. Algae are among the best source for second-generation bioethanol due to the fact that they are rich in polysaccharides, in addition to their high area productivity, no competition with conventional crops for land and the capture of CO_2.[101]

The use of algal biomass as a source of bioethanol production has been as investigated in many studies. Kim *et al.*[102] studied several algae (*Ulva lactuca, Gelidium amansii, Laminaria japonica* and *Sargassum fulvellum*) for ethanol fermentation using ethanogenic *E. coli* KO11. The composition of algae was proven to be rich in protein (28–63%), carbohydrates (4–57%) and lipids (2–40%). The bacterial strain was able to utilise both mannitol (30%) and glucose (7%), and produced 0.4 g ethanol g^{-1} carbohydrate when cultured in *L. japonica* hydrolysate supplemented with Luria–Bertani medium and hydrolytic enzymes. In addition, the study proposed a procedure of acid hydrolysis followed by simultaneous enzyme treatment and inoculation with *E. coli* KO11 as a viable strategy to produce ethanol from marine algal biomass.

In Japan, three seaweeds, sea lettuce (*Ulva pertusa*), chigaiso (*Alaria crassifolia*) and agar weed (*Gelidium elegans*) were studied for bioethanol production.[103] The first finding was that little to no lignin was present in the seaweeds, making the polysaccharide content readily hydrolysable. Also, it was revealed that glucans were the only polysaccharides in sea lettuce and chigaiso that can be hydrolysed to fermentable sugars. An ethanol yield of more than 3% was obtained from these algae. Agar weed, on the other hand, was found to contain both galactan and glucan; hence the higher concentration of ethanol from agar weed (5.5%). The fermentation was performed using *S. cerevisiae* IAM 417.

In Indonesia, a research team investigated bioethanol production from the algae Spirogyra by means of fermentation using *Zymomonas mobilis* and *S. cerevisiae*. The carbohydrate content of Spirogyra could reach 64% dry weight.[104] After fermentation, the highest ethanol production of *Z. mobilis* was 9.70% ethanol (v/v) with addition of α-amylase enzyme at 0.09 g/50 ml for 96 h. The highest ethanol production of *S. cerevisiae* was 4.42% ethanol (v/v), also with addition of α-amylase enzyme at 0.06 g/50 ml for 96 h.[105]

Another study used the red algae *Gelidium corneum* to produce ethanol. Aqueous extracts obtained at 100–140 °C were subjected to saccharification, purification, fermentation and distillation to produce ethanol. An extraction process incorporating 5% sodium thiosulfate by dry weight of the algae provided optimal conditions for the production of a high ethanol yield, 10% (w/w) by dry weight of red algae.[106]

A brown macroalga, *Saccharina japonica,* was investigated for its potential to be a substrate for bioethanol production. First, the marine biomass was subjected to a low acid pretreatment using 0.06% (w/w) sulfuric acid at 170 °C for 15 min. A glucan content of 29% and an enzymatic digestibility of 84% were obtained for pretreated *S. japonica*. Subsequent SSF was conducted using *S. cerevisiae* DK 410362, along with cellulase and β-glucosidase. The related results showed that a bioethanol concentration of 6.65 g L^{-1} was obtained, which indicated an SSF yield of 67.41% based on the total available glucan of the pretreated macroalga.[107]

It is well-known that algae are used to produce alginate. What is less known is that the industrial-derived residues are causing serious environmental pollution due to their limited recycling capacity. In this context, a Chinese group tried to remediate this problem by converting the surplus by-product from the alginate extraction process, which contains a large amount of cellulosic materials, to valuable bioethanol. The residues from the brown algae *Laminaria japonica* were subjected to a dilute sulfuric acid pretreatment followed by enzymatic hydrolysis (cellulase and cellobiase). The results showed that the marine-derived resource is an interesting bioenergy resource, having a high content of cellulose (30.0 ± 0.1%) and little hemicellulose (2.2 ± 0.9%). The maximum yield of glucose reached 277.5 mg g^{-1} of biomass under the optimal condition of acid pretreatment (0.1% w/v, 121 °C, 1 h) followed by enzymatic hydrolysis (50 °C, pH 4.8, 48 h). After fermentation by *S. cerevisiae* at 30 °C for 36 h, the ethanol conversion rate of the concentrated hydrolysates reached 41.2%, corresponding to 80.8% of the theoretical yield.[108]

7.2.2.2 Seagrasses

Seagrasses, like algae, are marine biomasses but, botanically, they are completely distinct since the former are angiosperms (*viz.* flowering plants). Recently, some studies have been conducted to test this highly available biomass for bioethanol production. The studies on algae are much more abundant and elaborate than the ones on seagrasses. However, the biochemical composition of the seagrasses is quite interesting for bioethanol production and more studies should be performed in order to valorise such rich marine biomasses. For instance, the fibres of *Posidonia oceanica*, an endemic seagrass from the Mediterranean Sea, are composed of 38% cellulose and 21% hemicellulose.[109]

In India, a research team tried to produce bioethanol from fresh and semi-decayed leaves of *Cymodocea serrulata* seagrass using *Saccharomyces cerevisiae*. The marine biomass was subjected to stream explosion and oxalic acid hydrolysis. After the acid pretreatment, the content of total sugar was higher (0.071 μg g^{-1}) in the fresh seagrass leaves than in the semi-decayed ones (0.045 μg g^{-1}). In the case of steam explosion, the same behaviour was registered, *i.e.* the content of total sugar was maximum (0.014 μg g^{-1}) in fresh leaves and minimum (0.009 μg g^{-1}) in the semi-decayed leaves. After fermentation, it was found that the acid hydrolysis was better than steam explosion for ethanol production. Indeed, the acid-treated biomass revealed that the content of ethanol was maximum (0.047 ml g^{-1}) for the fresh leaves and minimum (0.033 mL g^{-1}) for the semi-decayed leaves. In the case of steam explosion, the content of ethanol was 0.001 mL g^{-1} in fresh leaves and 0.0004 ml g^{-1} in the semi-decayed ones.[110]

In Turkey, residues of the seagrass *Zostera marina* reach the coastlines and create a nuisance that needs to be dealt with. The disposal of these residues requires high costs. A research group proposed an interesting way to avoid this problem and valorise the marine biomass. Thus, a study was carried out to investigate the potential of *Z. marina* residues as a feedstock for bioethanol production in acid and enzymatic hydrolyses. The residues were firstly extracted using supercritical CO_2 where pressure was set to 250 bar at 80 °C. The raw biomass presented a composition of 27.4% cellulose, 16.9% hemicellulose and 22.4% lignin. Two kinds of pretreatment were tested using sulfuric acid and lime. The results indicated that supercritical CO_2 extraction also acted like a pretreatment method since it increased the availability of hemicelluloses and celluloses, thereby fulfilling two objectives at the same time.

Under optimised conditions, 58.24 g of reducing sugar per 100 g of dry biomass were reached by consecutive enzymatic and acid hydrolysis and 8.72%

bioethanol was produced by SSF which, according to the authors, shows the potential of *Z. marina* residues to be utilised for bioethanol production on an industrial scale.[111]

7.3 Biodiesel

7.3.1 General Aspects

Biodiesel is defined as a renewable fuel derived from natural oil sources such as vegetable oil, recycled waste cooking oil and animal fats. Rudolf Diesel is invented the first internal combustion engine run on vegetable oil.[112] Biomass is the most abundant renewable resource on Earth. Converting it into biodiesel could meet some of the human need for energy.

Biodiesel is a fatty acid methyl ester (FAME) produced by a chemical transesterification process. It is considered an excellent alternative fuel that can compete with conventional fossil diesel. However, it has a high cetane number (combustion quality) and a high viscosity making it difficult to use directly in a conventional diesel engine. Biodiesel is also better for the environment because it is composed of renewable materials and emits less greenhouse gas than petroleum diesel, thus reducing the health risks associated with air pollution.

The production of biodiesel has become a high priority in many countries around the world and especially in the United States, Brazil, the European Union and Asian countries. The attention paid to biodiesel production has to major objectives: establishing oil dependence and reducing CO_2 emissions. The feedstock varieties depend on the potential raw feedstocks in the country such as rapeseed and soybeans in the United States, and palm oil and jatropha in Asian countries.[113] In recent years, biodiesel from palm oil and jatropha has been identified as a promising non-edible oil source. Palm oil is the major oil produced in the world. Malaysia, one of the world's leading palm oil producer and exporters, is focusing on palm oil as a raw feedstock for biodiesel production. Figure 7.1 shows world palm oil production in 2009;[114] total production in that year was 45 million tonnes and the highest production was in

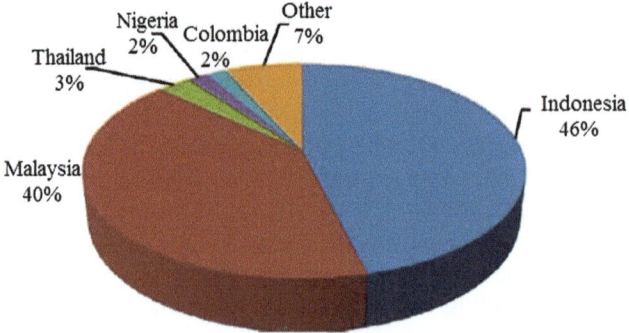

Figure 7.1   World palm oil production in 2009.[114]

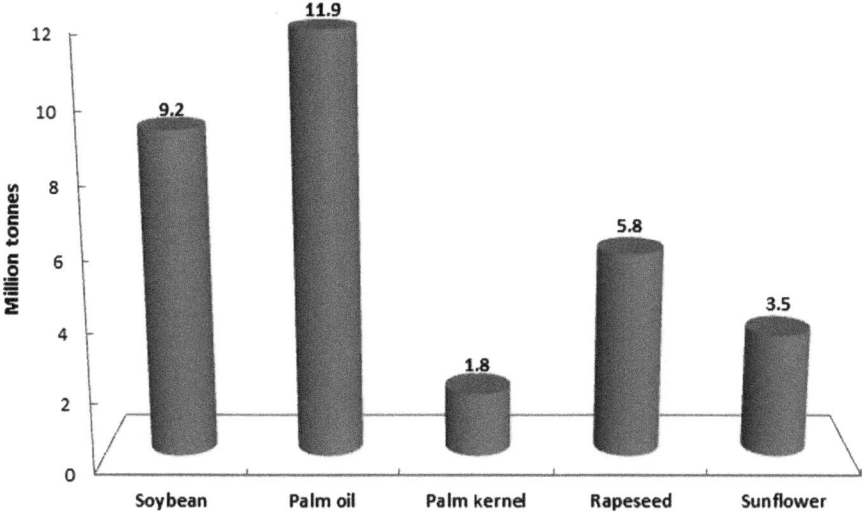

Figure 7.2 World production of major oil 2008/2009 estimates.[113]

South East Asia with 89% of total oil production. Figure 7.2 shows the world production of the major oilseeds 2008/2009 estimates;[113] palm oil is the highest amount followed by soybean oil.

7.3.2 Biodiesel Properties

The properties of biodiesel differ depending on the source of oil used. These properties are related to the chemical structure of oil—essentially the number of carbons and double bonds in the hydrocarbon chain. The substantial levels of highly unsaturated constituents in many oils also have important implications for biodiesel properties.

A wide variety of potential feedstocks for the production of biodiesel have been studied. These feedstocks can be classified into categories such as oilseed, animal fats, algal lipids and used cooking oil. Table 7.1 shows the fatty acid composition of a number of feedstocks. The composition of used cooking oil also called yellow grease was found to be the most variable because it is composed by various sources such as can oil, canola oil, sunflower oil and olive oil.

Vegetable and algal oils are composed of unsaturated and saturated fatty acids. These components determine the properties of some parameters of produced biodiesel such as cetane number, viscosity and cold flow properties (Table 7.2). High viscosity, low volatility and the reactivity of unsaturated hydrocarbon chains are generally disadvantageous for a good biodiesel production. The increase in saturation and chain length of the fatty acids increases the cetane number. Therefore, a low cetane number is generally associated with the saturated components.[131] The highest cetane number is obtained by oils rich in high saturated and monounsaturated fatty acids. Thus, based on the

Table 7.1 Composition profile of fatty acid fractions in various biodiesel feedstocks.

		Saturated fatty acid		Unsaturated fatty acid				Other acids	Ref.
		C16:0	C18:0	C16:1	C18:1	C18:2	C18:3		
Oil seeds	Rapeseed	3.5	0.9	0.1	54.1	22.3	–	19.1	112
	Sunflower	6.4	2.9	0.1	17.7	72.9	0.00	0.00	115
	Soybean	11.2	3.7	–	22.0	55.0	6.80	1.30	116
	Palm	47.9	4.23	0.04	37.0	9.07	0.26	1.50	117
	Jatropha	14.2	7.0	0.7	44.7	32.8	0.20	0.40	118
	Peanut	11.4	2.4	–	48.5	32.0	0.90	4.80	118
	Karanja	9.6	6.5	–	50.0	16.0	3.10	14.8	119
	Mahua	17.8	14	–	46.3	17.9	–	4.00	120
Algae	*C. humicola*	40.3	2.43	1.36	23.3	13.83	14.64	4.14	121
	C. sorokiniana	13.7	–	4.6	14.4	14.1	26.4	26.8	122
	C. vulgaris	66	5	–	7	11	12	0	123
	N. oculata	62	6	–	1	12	18	1	123
	C. calcitrans	26.3	2.6	27.5	4.5	0.8	0	38.3	124
	D. salina	19.4	1.5	1.7	5.3	6.2	38.7	27.2	124
Used cooking oil	Sunflower	6.8	3.7	0.4	22.8	65.2	0.1	1.00	115
	Soybean	11.5	4.0	–	24.5	53.0	7.0	0.00	125
	Palm	9.45	–	–	19.69	2.91	–	67.95	126
	Yellow grease	16.5	44.6	0.9	25.1	1.10	0.5	11.3	124

Table 7.2 Physico-chemical proprieties of biodiesel produced from different oil sources.

			Density / kg cm^{-3}	Kinematic viscosity at 311 K/mm^2 s^{-1}	Flash point/K	Cetane number	Ref.
Limits			–	1.9–6	>130	47	127
Diesel fuel			75–840	1.9–4.1	358	40–46	128
Biodiesel	Oil seeds	Rapeseed	880	4.30–5.83	–	49–50	129
		Sunflower	880	4.90	183	49	129
		Soybean	885	4.08	178	40	129
		Palm	860	4.42	164	62	129
		Jatropha	880	4.84	135	51.6–63	129
		Peanut	883	4.42	176	62	129
		Karanja	889	4.42	116–176	57.6	130
	Algal	Microalgae	864	5.20	124	–	118
	Cooking used oil	Yellow grease	882.5	4.55	–		112

results presented in Table 7.1, jatropha, karanja, rapeseed and palm should have the highest cetane numbers due to their high contents of saturated (C16:0) and monounsaturated (C18:1) fatty acids. The flash point can be attributed to the chain length of the oil. A low flash point depends on the presence of a chain length less than 12 carbons. A fatty acid with a high chain length such as C16 and C18 gives a high flash point.[118]

7.3.3 Biodiesel Sources of Production

It is important to search for new sources of fuel to reduce the world's dependency on petroleum diesel. The main potential sources for oil production are oilseeds, algae and animal fats. In this section, we focus on the biomass contribution as renewable and sustainable substitution to expensive petroleum. A distinction is made based on whether the extracted oil is edible or not. We have imposed this division on purpose to separate the potential oil sources based on a crucial criterion, *i.e.* human consumption. The idea is that a clear decision can be made in relation to a country's natural resources and the possibility of valorising marginal lands, on one hand, and the status of its dependency on regular diesel, on the other. Thus, for economic and environmental considerations, the transformation of non-edible oil into biodiesel should be the foremost choice to substitute petroleum-based diesel. However, given the major importance of such fuel substitution for the decades to come, the edible oils could be used if the country's natural resources are not very rich in oily feedstocks. The exploitation in this case should be carefully planned and based mainly on a gradual substitution of the regular diesel with the biodiesel produced from oil plants cultivated in marginal lands using high-yielding varieties and avoiding the use of chemicals in the cultivation cycle. Thus, the competition with the vital food crops remains minimal, and therefore, a reasonable balance could be managed between food and energy demands.

7.3.3.1 Edible Feedstocks

In past years the main concern in the world has been the recent increase in petroleum prices and its availability. Therefore many researchers have been interested in the use of vegetable oil fuels for diesel engines. This interest is attributed to species characterised by their high lipid content. Table 7.3 illustrates the oil content and the annual yield of various oilseeds and algae.

Edible feedstocks are an important source of oil selected for biodiesel production. These raw materials are characterised by their high potential oil content. The most well-known feedstocks are sunflower, soybean, colza, palm seed and coconut.

7.3.3.1.1 Sunflower. Sunflower, *Helianthus annuus*, is a plant native to the American continent. This plant is grown for its seeds rich with good quality oil. Sunflower is one of the main sources of edible oil in Europe. It is classified as the one of the principal oils used by the biodiesel industry.[132] Several researchers have investigated the potential and the quality of sunflower for oil production. The oil content of sunflower seeds is often more than 35% and the annual oil yield about 800 kg/ha.[133] Another interesting propriety of sunflower oil is that it contains mostly unsaturated fatty acid (18:2).[124]

Santorie *et al.*[133] have studied the processing costs of the vegetable oil and found that the costs represent about 6 to 10% for the sunflower oil. Yin *et al.*[134] have studied the different enhancing methods for biodiesel production from sunflower to decrease the consumption of time, energy, methanol and catalyst.

Table 7.3 Oil content and annual yield of cultivated oilseeds and algae.

Feedstock	Oil content[112,118,130]/ %	Annual oil yield per cultivated hectare[112, 113]/ g ha^{-1}
Sunflower	25–35	655–800
Rapeseed	38–46	1000
Palm	30–60	5000
Coconut	63–65	2260
Jatropha	30–40	1595
Soybean	15–20	520
Peanut	45–55	3000
Karanja	21–32	–
Mahua	35–42	–
Chlorella sp.	28–32	–
Cylindrotheca sp.	16–37	–
Botryococcus braunii	25–75	–

7.3.3.1.2 Soybeans. Soybeans, *Soja angustifolia* is an edible species of legume native to East Asia. It is generally grown for its beans, which are used for their oil content, 20% wt. (*cf.* Table 7.3). Therefore, it has been classified as a potential feedstock for biodiesel production. Soybean can grow in arid condition and even in mountain regions.[135] Like to sunflower, soybean oil contains more than 80% unsaturated fatty acids.[118]

Candeia *et al.*[136] investigated the effect of biodiesel concentration on the physicochemical and rheological properties of different blends of soybean biodiesel with diesel fuel. Yu *et al.*[137] tested the ultrasonic effect on the transesterification by Novozym 435 to produce biodiesel from soybean oil. The results showed that, under the optimum conditions, 96% yield of fatty acid methyl ester (FAME) could be achieved in 4 h. Furthermore, repeated use of Novozym 435 after five cycles showed no obvious loss in enzyme activity.

7.3.3.1.3 Rapeseed. Rapeseed, *Brassica napus*, is an edible seed grown for its oil production. It can be used for human consumption and as an animal feed, and in our case for biodiesel production. The important producers are the European Union, Canada, United States, Australia, China and India. Rapeseed is one of the most profitable crops available and has provided excellent sources of oil for biodiesel production. Rapeseed can yield about 36% to over 46% of oil and the major fatty acid content found belongs to the group of monounsaturated fats with 61.16% of oleic acid.[138] This species is considered one of the primary species used for the production of biofuel.[139]

7.3.3.1.4 Palm Seeds. Palm seeds are the most promising prospective feedstock compared with other oilseeds for biodiesel production. The advantageous of this tree are the culture's lower need in water, fertilisation and pesticides, along with the high production yield. Indonesia and Malaysia are the principal palm oil producers in the world.[113] Palm oil is a vegetable oil that could be used in cooking and in biodiesel production; the latter should be planned carefully to minimise competition with the former. In Japan, a research team studied the optimisation of the transesterification of palm oil and has evaluated its solvent extraction and fuel properties. The results showed that the optimised process gave a maximum ester yield of 97%.[117]

7.3.3.1.5 Coconut Palm. Coconut palm, *Cocos nucifera*, is a native tree to the coastal areas of South East Asia and possibly also to the Pacific coast of Central America. Coconut has an oil content of about 65% and a yield for about 2260 kg ha^{-1} (Table 7.3). The most abundant fatty acid content belongs to the saturated group: lauric acid (C12:0), myristic acid (C14:0) and palmitic acid (C16:0) with 49.70%, 19.30% and 11.10%, respectively.[140] Kumar *et al.*[141] studied the production of coconut oil ethyl ester using ultrasonic irradiation and found a biodiesel yield of 98%, as well as a reduction in the reaction time by 15–40 times compared with conventional batch processes.

Although all these edible feedstocks are classified as advantageous for their oil content and quality, their use for the production of biodiesel remains a concern given the alarming food shortages worldwide. To find a solution to this dilemma, many researchers have looked for found other alternatives, starting with the production of biodiesel from non-edible oils.

7.3.3.2 Non-edible Feedstocks

The contribution of non-edible oils from various bioresources was found to be promising for the biodiesel production. A wide variety of non-edible oil plants were and still available for investigation.

7.3.3.2.1 Jatropha. Jatropha, *Jatropha curcas*, is a plant belonging to the Euphorbiaceae family which is characterised by its seeds producing a significant amount of oil. The oil content in Jatropha seeds is about 40% and the yield production is about 1600 kg ha^{-1} (Table 7.3). Jatropha seeds are considered one of the most promising potential oil sources to produce biodiesel. This plant is widespread in arid, semi-arid and tropical regions. It is a resistant perennial tree that grows in marginal lands characterised by a rapid growth and an easy propagation.[142]

Many researchers have been interested in the Jatropha oil for biodiesel production. Qian et al.[143] studied the production of biodiesel from *Jatropha curcas* oil by two-phase solvent extraction. Kumar et al.[144] studied the transesterification of Jatropha oil with anhydrous methanol, ethanol and various mixtures of methanol and ethanol. In China, *Jatropha curcas* oil is selected as the best biodiesel feedstock for its non-edible oil content.[145]

7.3.3.2.2 Karanja. Karanja, *Pongamia pinnata*, belongs to the Leguminaceae family, native to part of India and Australia in humid and subtropical ecosystems.[130] This plant is characterised by rapid growth in various types of soil, along with a high tolerance to drought and salinity. The seeds are a good source of oil production and can produce up to 21–32% of oil (*cf.* Table 7.3).

Vivek and Gupta[119] studied the feasibility of karanja oil for the production of biodiesel and found that this non-edible raw material can be considered a good alternative to regular diesel fuel. In karanja oil, the oleic acid (C18:1) is the most abundant fatty acid with about 50%. The main disadvantage of karanja oil is the high free fatty acid (FFA) oil content. Naik et al.[127] adopted a dual process mechanism for the production of biodiesel from karanja oil containing FFA up to 20%. The results revealed a biodiesel yield from high karanja oil's FFA of about 97%.

7.3.3.2.3 Mahua. Mahua, *Madhuca indica*, is a tree native to India belonging to the Sapotaceae family. It is a fast-growing tree adapted to arid environments and cultivated in warm climates for its oilseeds. The seeds can produce an important quantity of oil (up to 42%) and can have an annual production of about 181 000 t in India.[120] Oleic acid (C18:1) is the major fatty acid present in mahua oil at 46.3%. Ghadge and Raheman[146] studied the production of biodiesel from mahua oil with a high FFA level. The process was conducted in two steps and gave a yield of 98% mahua-derived biodiesel. In another study, Jena *et al.*[120] tested a mixture of mahua and simarouba oils with a high FFA content to produce biodiesel and reached 90% of ester conversion.

7.3.3.2.4 Algae. Microalgae are unicellular or multicellular photosynthetic organisms. To produce biomass with high oil content, these organisms use light energy and carbon dioxide.[123] Recent research has focused on microalgae to produce biodiesel because of their high biomass productivity, photosynthetic efficiency and high oil content. Commonly, microalgae are non-edible feedstocks and thus can be cultivated without any competition with food crops, neither for land nor freshwater. Therefore, microalgae can be cultivated in non-arable lands using non-potable water and/or wastewater.[121] The use of algal oil for biodiesel production has been classified as third generation biodiesel. Algae can reach a high level of oil content such as 75% of dry basis for *Botryococcus braunii*.[118] The major constituents of most algal oils are unsaturated fatty acids such as palmitic acid (C16:0), along with significant amounts of highly unsaturated species (Table 7.1).

7.3.4 Production Process

Various technologies are being used to optimise the biodiesel production process. Basically, the main problems with vegetable oils are their high viscosity, low volatility, the reactivity of unsaturated hydrocarbon chains and the high contain of free fatty acids, which cause poor combustion in diesel engines. Thus, different methods and multiple chemical steps have been proposed such as dilution, micro-emulsification, pyrolysis, catalytic cracking and transesterification.[128]

This section describes the two major steps in biodiesel production: the extraction of vegetable oil from the seeds and its subsequent conversion into biodiesel (*i.e.* transesterified vegetable oil).

7.3.4.1 Oil Extraction

The processes extracting of vegetable oil from the seed depends on the physical and chemical characteristics of the raw material. The extraction process of vegetable oils is based on technologies that favor the separation of the fat from protein and designed to obtain high quality oil and produce high extraction yields. There are several techniques for extracting oil from oilseeds. Two common extraction technologies were essentially founded are solvent and pressure extractions.

7.3.4.1.1 Mechanical Extraction Process. Mechanical extraction is a pressure-based method widely used in the world.[147] It preserves the characteristics of the oilseed and the residual product. The use of a screw press is generally preferred because of its low cost production and energetic efficiency.[148] The seeds are fed into an extruder that compresses them, warming them in the process. A preliminary crushing action is often used before they reach the extruder for oilseeds with a high oil content. The mechanical screw press consists of a vertical feeder and a horizontal screw with increasing body diameter to exert pressure on the oilseeds as they advance along the length of the press. In a discontinuous press, the seeds are compressed under high pressure (4 ± 35 MPa) and temperatures close to 95 °C. In a continuous press, the pressures are around 40 MPa with temperatures up to 120–155 °C.[147] The oil is collected in a trough under the screw and the de-oiled cake is discharged at the end of the barrel. The press type depends on the raw material. Seeds containing >20% by weight of lipids are usually fully or partially treated by mechanical means.[133]

7.3.4.1.2 Chemical Extraction Process. Chemical extraction is a process based on placing oilseeds in contact with a solvent to allow the oil to dissolve in the solvent. This is the most efficient technique to recover oil from oilseeds. The efficiency of the process depends on the preparation of oilseed, temperature, operation mode and the equipment design. Because it is difficult to extract the oil inside the cell because the diffusion process is very slow, the raw material is pretreated to facilitate the diffusion process. The efficiency of the extraction is also influenced by the type of solvent.[133] The choice of solvent is based on the solubility of the oil in the selected solvent, cost and safety. Commonly, the solvents used for oil extraction include *n*-hexane, white spirit, trichloroethylene, carbon sulfide and some biosolvents. Currently, hexane is widely used as solvent for the extraction of vegetable oil.[149]

In general, there are three methods used for the chemical oil extraction process: immersion, percolation and mixed immersion–percolation.[150] The chemical extraction method ensures higher yields while the mechanical method can produce better oil quality. Thus, in practice, the two systems—mechanical and chemical procedures—are often combined to enhance the efficiency of oil extraction. In this context, Morshed *et al.*[151] studied different methods for oil extraction: mechanical press with and without solvent and cold percolation. The maximum oil content (49%) was obtained by the mechanical press with periodic addition of solvent. Evon *et al.*[152] investigated the feasibility of an aqueous process to extract sunflower seed oil using a co-rotating twin-screw extruder. The analysis showed that aqueous extraction of the oil was more efficient in the twin-screw extruder than the reference trial (batch reactor). The best oil extraction yield obtained was about 55% and the residual oil content of the cake meal was approximately 30%.

7.3.4.2 Transesterification

The transesterification procedure is the most commonly used method for converting vegetable oils into biodiesel and a good solution to the high viscosity problem. Generally, there are two ways to convert vegetable oils into biodiesel

Table 7.4 Comparison of different transesterification processes.[153]

Process	Biodiesel yield	Advantages	Disadvantages
Acidic catalyst	99% after 4 h of reaction	High yield production	–
Alkaline catalyst	99% after 2 h of reaction	High and rapid yield production Low temperature Low excess alcohol	Calcium foam formation
Enzymatic catalyst	95% after 105 h of reaction	Operated at room temperature	Slow reaction

by transesterification; the first method employs a catalyst and the second one is a non-catalyst option including supercritical processes and co-solvent systems.

Various catalysts are used to increase the yield and reaction rate, such as base catalysts like NaOH, KOH and NaMeO, acid catalysts that include H_2SO_4, H_3PO_4 and $CaCO_3$, and lipase enzymes.[112] Table 7.4 presents a comparison between acidic, alkaline and enzymatic catalyst transesterification.

Overall, biodiesel production follows a multi-step path. First, the catalyst is dissolved in the alcohol in a standard mixing machine. The oil is added with the mix and then placed in a closed reaction vessel. The reaction takes place when the temperature of the mix reaches 70 °C. After 1–8 h reaction time, the conversion of oils into esters should be complete. An excess of alcohol is usually used to ensure complete conversion. At the end of the process there are two outputs: glycerin and biodiesel. The separation of the two outputs is based on the difference in density of glycerol relative to biodiesel (glycerol is denser). Therefore separation is made using the gravimetric method by drawing off the glycerol from the bottom of the settling vessel. In some cases, it is possible to use a centrifuge to separate the two phases.[113]

The crude biodiesel needs to be purified before use. This purification is done in three steps: (i) neutralization; (ii) passage through an alcohol stripper (distillation); and (iii) washing. To neutralise the unreacted alkali catalyst and split any soaps, an acid solution is added to the crude biodiesel. The distillation process removes unreacted alcohol. The last step involves washing the remnants of the catalyst, soaps, salts, residual alcohol and free glycerol from the crude biodiesel.[153]

7.3.5 Environmental and Economic Considerations

Biodiesels derived from vegetable oil are becoming more attractive because of their substantial advantages compared with the regular fossil diesel. The obvious advantages of biodiesels are their availability and renewability, leading to a gradual reduction in the dependency on fossil petroleum. These benefits could be divided into environmental and economical ones.

7.3.5.1 Environmental Advantages

Biodiesel is considered ecofriendly because of its biodegradable nature. Also, it exhausts less emissions in terms of carbon monoxide, hydrocarbons, polycyclic aromatic hydrocarbons (PAHs) and nitrited polycyclic aromatic hydrocarbon compounds (nPAHS).[115]

The combustion of biodiesel alone can reduce 90% of unburned hydrocarbons and 75–90% of PAHs.[154] Candeia *et al.*[136] tested a blend of soybean biodiesel with diesel and found that, the higher the proportion of biodiesel in the blend, the lower the sulfur content and the emissions of CO_2 and sulfur oxides (SO_X). Therefore, the use of biodiesel as an alternative fuel should have important environmental benefits by minimising global air pollution and reducing emission levels of potential carcinogens.

7.3.5.2   Economic Advantages

It is important to study the cost of biodiesel production from feedstock oils and to compare it with the petroleum-based fuel in order to assess the viability of biodiesel production and its economic benefits. Several factors must be considered in the calculation of biodiesel production costs such as the feedstock, which is about 80% of the total production cost, labour, solvent and catalyst.[128]

The use of edible feedstock for biodiesel production has been proven not be economically competitive with diesel fuel because of the high cost involved in growing the corps (seeds, fertilisation, pesticides, harvesting, labour, *etc.*) and the conversion into biodiesel (solvents, catalysts, labour, *etc.*). The use of non-edible resources enhances the competitiveness of the derived biodiesel since it involves growing non-food crops in marginal land without agricultural treatments or even better using natural wastes or industrial by-products.

Hill *et al.*[155] calculated the inputs and outputs for biodiesel production from soybean oil. They found significant benefits if the soybean feedstock was produced with low agriculture input, was producible on land with low agricultural value, and required low-input energy to convert it to biodiesel. The biodiesel also has good proprieties compared with petroleum diesel fuel such as better lubricant proprieties, higher heating values, higher cetane number and higher flash point.[156]

7.4   Biomethane

7.4.1   General Aspects

Biogas is a carbon-neutral source of renewable energy with an interesting prospective in terms of both its energy efficiency and environmental impact. Vehicle fuel produced from biogas from manure, wastes and energy crops would fulfill the European Union's sustainability requirement from 2017 onwards to reduce greenhouse gas emissions by 60% compared with fossil fuels.[157] It has also been suggested that a major part of the EU's renewable energy target for 2020 will originate from bioenergy and at least 25% of bioenergy could be met with biogas produced from wet organic materials.[158]

Unlike bioethanol (high sugar content) or biodiesel (high oil content), no specific properties are required to qualify a resource as suitable for biogas production. That is why biogas can be produced from a broad range of feedstocks suitable for anaerobic digestion such as sludge from wastewater treatment plants,[159] municipal solid wastes,[160] cattle slurry and cheese whey,[161] dairy manure and food waste,[162] and many other resources. In this chapter, we focus on plantae and marine bioresources as potential feedstock for biogas production.

7.4.2 From Biogas to Biomethane

The main components of biogas are methane (CH_4) and carbon dioxide (CO_2), but typically biogas also contains trace compounds such as carbon monoxide, hydrogen sulfide and ammonia. At some production sites, biogas may also contain compounds such as siloxanes and aromatic, halogenated and other volatile organic compounds (VOCs).

In order to transform biogas into biomethane, two major steps are required: (i) a cleaning process to remove the trace components; and (ii) an upgrading process to adjust the calorific value. Upgrading is generally performed to meet the standards for use as vehicle fuel or for injection in the natural gas grid. After transformation, the final product, biomethane, typically containing 95–97% CH_4 and 1–3% CO_2, compared with a typical biogas containing 50–65% CH_4 and 30–45% CO_2.

Conventional biogas separative removal is based on physical and chemical absorption–adsorption. One of the absorption-based processes uses amines to remove CO_2. Thus, absorption and desorption columns are combined so that the amine solution can be continuously regenerated by heating the liquid-using steam. CO_2 is then stripped from the amine solution. The advantage of amine scrubbing is that high CH_4 purities (>95%) and low CH_4 losses (<0.1%) are achieved.[163] However, the application of amine scrubbing is energy-intensive as steam has to be supplied to regenerate the amine solution. One of the other methods is the use of water as a physical absorbent. Here the CO_2 is separated from the biogas by washing with water at high pressure. Generally, the biogas is introduced from the bottom of a tall vertical column, typically at a pressure of 1000–2000 kPa. Water is fed at the top of the column to achieve a gas–liquid counter flow.[164] The pressurised water scrubbing process operates using only water as a solvent, which is more secure than applying chemical solvents.

In addition to the liquid separation processes, adsorption can be applied to capture water vapour, CO_2 and H_2S on porous solid adsorbents such as zeolites and activated carbons. One of these methods is pressure swing adsorption (PSA), which uses a column filled with a molecular sieve—typically activated carbon, silica gel, alumina or zeolite—for the differential adsorption of the gases (*i.e.* CH_4 is allowed to pass through while CO_2 and H_2O are captured). The adsorption is generally performed under a relatively higher pressure (around 800 kPa) and desorption (regeneration) at lower pressures. PSA can be operated on the basis of equilibrium or kinetic selectivity, depending on the residence time in the column.[165] As well, for the water vapour removal, the adsorption was quite efficient using silica, alumina or other chemical components that can bind water molecules (*viz.* adsorption dryer). The gas is pressurised and passed through a column filled with the sorbent. Usually, two columns are used in parallel: one to adsorb water, while the other is being regenerated through decompression and heating.[166]

Other studies have presented membrane-based biogas separation (or upgrading) as an alternative to conventional upgrading technologies, which require significant amounts of energy and large equipment. A membrane-based separation is mainly designed to remove CO_2 from the CH_4 bulk, but trace components in the raw biogas such as hydrogen sulfide or water vapour could also be removed since they permeate through the membrane even faster than CO_2 if sufficient driving force for permeation is provided.[167]

For the case of H_2S removal, along with the previously mentioned methods, a biological alternative, biofiltration, has attracted interest among the scientific community in the past couple of decades.[168] For instance, a research team in Chile developed a biofiltration system using peat as solid support inoculated with *Thiobacillus thioparus* to treat a gaseous stream containing high concentrations of H_2S.[169] The utilisation of this biological method is increasing because it is cheaper than chemical cleaning and is also able to remove ammonia from the biogas. With the addition of air to the biological filters, the H_2S content can be decreased from 2000–3000 to 50–100 cm^3 m^{-3}.[166]

Of course each one of the aforementioned methods has its own advantages and drawbacks. Thus, the best way to enhance the overall efficiency of the biogas upgrading process is to combine some of them. For example, Esteves and Mota[170] studied the possibility of a new combined membrane–PSA system for gas separation. The PSA system's main advantage is its ability to achieve high purity, while membranes are clearly advantageous in terms of speed (when product purity requirements are less severe). This combined process contains a membrane module and a dual-bed PSA unit. The membrane performs most of the bulk separation to maximise the average driving force. Both permeated gas flow and residual gas flow are fed to the PSA system at different steps for higher CH_4 purity (gas) and enhanced CO_2 recovery.

7.4.3 Plantae and Marine Resources for Biomethane Production

7.4.3.1 Biomethane from Plantae

Several studies have investigated the biomethane production potential of different plants. The main advantages of these feedstocks are their availability, renewability and rich composition, and the fact that they do not compete with food crops over soil or water.

In Sweden, chipped and milled softwood spruce (*Picea abies*) was studied for its ability to produce methane. The woody biomass was pretreated with *N*-methylmorpholine-*N*-oxide (at 130 °C), prior to the six weeks' anaerobic digestion. The analysis showed that the untreated chips (10 mm) and milled (<1 mm) spruce resulted in 11 and 66 normalised mL (NmL) CH_4 g^{-1} raw wood, respectively. The best digestion results of pretreated chips and milled spruce were 125 and 245 NmL CH_4 g^{-1} raw material respectively.[171] In neighbouring Finland, reed canary grass (*Phalaris arundinacea* L.) was investigated for biomethane production at 35 °C. The biogas yield under optimised conditions was 8.26 mmol CH_4 per g (dry weight).[172] The same resource, reed canary grass, was also tested for biomethane production in Canada and where it was found that the average specific methane yield from reed canary grass was 0.187 normalised L (NL) CH_4 g^{-1} of volatile solids (VS). Such a yield corresponds to an average methane yield from reed canary grass seeded plots of 1.37 gigalitres (GL) ha^{-1}.[173] The same study analysed another herbaceous biomass, switchgrass (*Panicum virgatum* L.); the results revealed that the average specific methane yield from reed canary grass was 0.212 NL CH_4 g^{-1} VS, corresponding to a CH_4 yield of 0.91 GL ha^{-1}.

Shoots of willow (*Salix viminalis*) were also used as feedstock for biogas production. Indeed, Horn *et al.*[174] applied different steam explosion conditions to Salix chips and the effect of these pretreatments on the biogas production potential was evaluated. The results showed that the biomass pretreated at 220 and 230 °C initially showed low production of biogas; the authors related this to the impact of the inhibitors produced during the pretreatment phase. They then noticed that the microbial community was able to adapt (after 57 days) and showed a high final biogas production of 440 mL g^{-1} VS, corresponding to about 240 mL CH_4 g^{-1} VS under standard temperature and pressure (STP).

Parthenium hysterophorus L. is an aggressive and invasive weed of some crops. An interesting way to valorise such biomass is to use it as a feedstock for biomethane production. An Indian research team investigated the potential of this resource using NaOH as a pretreatment agent prior to the anaerobic fermentation. The results showed that a methane yield of 110 mL CH_4 g^{-1} VS and a volumetric methane productivity of 456 mL CH_4 L^{-1} day^{-1} were obtained for untreated *Parthenium* at 30 °C and after 10 days retention time. Using the NaOH-pretreated biomass, methane production jumped up to 883 mL CH_4 L^{-1} day^{-1} at 40 ° for 10 days.[175]

7.4.3.2 Biomethane from Agricultural and Industrial Wastes

7.4.3.2.1 Crops and Agricultural Residues. Several crops and their by-products have been thoroughly investigated for their aptitude to produce biomethane. In Poland, the production of methane from silage of four crop species, *Zea mays* L., *Sorghum saccharatum*, *Miscanthus* × *giganteus* and *Miscanthus sacchariflorus* was monitored. The results showed that, at comparable lignin concentrations in the feedstock, methane productivity for *M. sacchariflorus* (0.19 ± 0.08 L g^{-1} VS) was twice that of *Miscanthus* × *giganteus* (0.10 ± 0.03 L g^{-1} VS). The Polish researchers also proved that holocellulose conversion into biogas depends on the ratio of their concentration to the lignin content in the feedstock.[176]

Another energy crop with high biomass yields, *Cynara cardunculus* L., was studied for the production of biofuels. Its seeds are used for biodiesel production and, in a Portuguese study, the researchers tried to produce methane after anaerobic digestion of cynara stalks.[177] Minimum methane production was achieved for the untreated substrate yielding 0.3 L CH_4 g^{-1} VS. After mechanical, thermal and thermochemical pretreatments, a maximum methane yield of 0.5–0.6 L CH_4 g^{-1} VS was achieved depending on the selected pretreatment, with the thermochemical pretreatment using NaOH being the most efficient hydrolysis method.

Highly abundant corn straws were also subjected to several experiments to enhance the biomethane production potential of this biomass. One of the related research works used a biological pretreatment with new complex microbial agents (a mixture of yeasts and cellulolytic bacteria) to pretreat corn straw at ambient temperature (about 20 °C) in order to improve its biodegradability and anaerobic biogas production. These treatment conditions resulted in 33.07% more total biogas yield, 75.57% more methane yield and 34.6% shorter technical digestion time compared with the untreated sample.[178]

Oil palm empty fruit bunches, a lignocellulosic by-product of vegetable oil production industries mainly in Indonesia and Malaysia, were used as feedstock for biogas production. The analyses revealed that, after seven days of incubation, the yield of methane was 0.13 Nm^3 kg^{-1} VS for the untreated residue. Minor improvement was observed with phosphoric acid pretreatment with a yield of 0.15 Nm^3 kg^{-1} VS. The best result of methane production (0.24 Nm^3 kg^{-1} VS) was achieved using NaOH-pretreated material for 60 min. Hence, under optimal conditions, the initial methane yield improved by 85% compared with that of the untreated biomass (0.13 Nm^3 kg^{-1} VS).[179]

Another by-product of the oil extraction industry is sunflower oil cake. A Spanish team investigated the effect of hydrothermal pretreatment at different temperatures (from 25 to 200 °C) on the biomethane productivity of sunflower oil cake. After the pretreatment phase, digestion of the solid fractions in batch assays at 35 °C resulted in methane yields of 114 ± 9, 105 ± 7, 82 ± 7 and 53 ± 8 mL CH_4 g^{-1} of soluble chemical oxygen demand added (COD_{added}), respectively at 25, 100, 150 and 200 °C. For the liquid fraction, the corresponding methane yields were 276 ± 6, 310 ± 4, 220 ± 15 and 247 ± 10 mL CH_4 g^{-1} COD_{added}, respectively.[180]

In India, non-edible de-oiled seedcake obtained after the oil extraction of two species, nahua (*Madhuca indica*) and hingan (*Balanites aegyaptiaca*), were investigated for biogas production. Both seedcakes were found to have a mean biogas generation potential in the range 198–233 L kg^{-1} of by-products. The authors claimed that the biogas generated from these sources could make a significant contribution to cooking energy demand in rural areas.[181]

7.4.3.2.2 Industrial Wastes. Large quantities of industrial by-products have currently destined for landfill. However, based on their composition, a number of studies have been conducted to generate biogas from these wastes. In Italy, batch trials were carried out to asses the biogas productivity potential of grape and tomato transformation wastes. Their specific methane yields were 116 and 98 NL CH_4 kg^{-1} VS, for grape marcs and stalks, respectively.

As for the tomato skins and seeds, the methane production yield was 218 NL CH_4 kg^{-1} VS.[182]

Along with solid wastes, there are also liquid industrial wastes that could be used for biogas generation, especially in food processing industries. For instance, an Indian research team has studied the anaerobic digestion of wastewater from jam industries in a continuous reactor. The aim of this work was to produce biomethane as well as to stabilise the waste efficiency. The removal efficiency of total COD and soluble COD were 82% and 85%, respectively. The specific methane production was 0.28 m^3 kg^{-1} of COD removed per day.[183]

Another problematic, but interesting, wastes are paper residues. A number of studies have been conducted to monitor the biogas production potential of these lignocellulosic wastes. In Sweden, biogas production from paper tube waste was monitored in anaerobic batch digestion tests. Different pretreatment conditions were applied including steam explosion and non-explosive hydrothermal pretreatment, in combination with NaOH and/or H_2O_2, have been used to improve biogas production. Explosive pretreatment was found to be more effective than the non-explosive method, and gave the best results at 220 °C, and 10 min, with addition of both 2% NaOH and 2% H_2O_2. Digestion of the pretreated materials under these conditions produced a methane yield of 493 NmL g^{-1} VS, which was 107% more than the untreated materials.[184]

Another study tried to develop an alkali pretreatment process prior to the anaerobic digestion of pulp and paper sludge in order to improve methane productivity.[185] Experiments were carried out in mixed bioreactors; the optimal amount of NaOH for organics solubilisation in the pretreatment step was found to be 8 g NaOH per 100 g total solids (TS). The anaerobic digestion efficiency of the paper and pulp sludge, with and without pretreatment, was evaluated. The highest methane yield under optimal pretreatment condition was 0.32 m^3 CH_4 kg^{-1} VS, corresponding to an improvement of 183.5% compared with the untreated residue.

7.4.3.3 Biomethane from Marine Biomass

Based on their availability, high biomass production and their non-competitiveness towards food corps, several marine bioresources were investigated to asses their potential to produce biomethane. In Belgium, anaerobic digestibility of two microalgae, *Scenedesmus obliquus* and *Phaeodactylum tricornutum* was carried out under mesophilic and thermophilic conditions. The biomethane potential assays revealed that the highest methane yield (0.36 L CH_4 g^{-1} VS) of *P. tricornutum* biomass to be about 1.5 times higher than that of *S. obliquus* biomass (0.24 L CH_4 g^{-1} VS). The authors also proved that the hydrolysis of the algae cells was limiting the anaerobic processing of both microalgae.[186]

Two other marine algae *Macrocystis pyrifera* and *Durvillea Antarctica*, were investigated as feedstocks for biogas production. Anaerobic digestion of the algae and their blend 1 : 1 (w/w) was evaluated in a two-phase anaerobic digestion system. The results showed that 70% of the total biogas produced in the system was generated in the upflow anaerobic filter. Both algae species showed similar biogas productions of 180 (± 1.5) mL g^{-1} dry algae day^{-1}, corresponding to a biomethane concentration of around 65%.[187]

In Brazil, researchers assessed the biomass culture and biogas production from the microalga *Spirulina* LEB-18.[188] The marine biomass produced was used as substrate for biogas production in a 10 L semi-continuous anaerobic bioreactor at under 35 °C, inoculated with granular anaerobic sludge (VS 25 g L^{-1}). The obtained methane production from biomass was 0.79 L g^{-1}, corresponding to a biogas content of 77.67% (v/v).

Ulva lactuca (L.), a green macrolaga, has also been analysed for its potential to produce biomethane. The marine resource was cultivated in a land-based facility and showed a production potential of 45 TS ha^{-1} yr^{-1}. Biogas production from fresh and macerated *U. lactuca* yielded up to 271 mL CH$_4$ g^{-1} VS, which according to the authors, is in the range of the methane production from cattle manure and land-based energy crops like grass clover.[189]

As described in the sections on bioethanol and biodiesel production, several research works have focused on algae as a promising feedstock for biofuel production. But, like any other transformation-based process, there are always residues. One proposed solution, within an overall bioenergy vision, is to produce biogas from these algal biomass residues from bioethanol and biodiesel production processes.[190] For instance, a Korean team investigated the anaerobic digestibility of a algal bioethanol residue from saccharification and fermentation processes. A series of batch anaerobic digestion tests showed that the maximum methane yields of the saccharification residue and the fermentation residue were 239 L kg^{-1} VS and 283 L kg^{-1} VS, respectively. The authors also observed that 5-hydroxymethylfurfural (5-HMF), a saccharification by-product, could retard methanogenesis at over 3 g L^{-1}. However, the inhibition was prevented by increasing the cell biomass concentration.[191]

7.4.3.4   Biomethane from Food Wastes

Food wastes from households or markets are solid residues rich in carbohydrates, making them promising feedstocks for methane production. Several research studies have been conducted in this regard using diverse sources. In Denmark, a research group managed to produce biomethane from household solid waste. In the study, the production of methane was first performed in a one-stage process (*i.e.* only methane). Then, the process was developed into a combined two-stage one (*i.e.* hydrogen production as a first stage and methane as a second stage). The reported results revealed that the first hydrogen production stage generated 43 mL H$_2$ g^{-1} VS while methane production in the second stage was 500 mL CH$_4$ g^{-1} VS (7500 mL CH$_4$ day^{-1}). Thus, methane production yield was 21% higher than the one registered for the one-stage process (413 CH$_4$ g^{-1} VS or 6200 mL CH$_4$ day^{-1}).[192]

Another two-stage process of anaerobic digestion of potato wastes was developed in Canada to co-produce both hydrogen and methane.[193] A maximum gas production rate of 270 mL h^{-1} and an average of 119 mL h^{-1} were obtained from the hydrogen stage over 110 days. The maximum and average gas production rates observed from the methane reactor during the 74 days of semi-continuous flow operation were 187 and 141 mL h^{-1}, respectively, with an average methane concentration of 76%.

In Spain, a research team demonstrated that orange peel waste is a potentially valuable resource for methane generation. The substrate was initially treated by steam distillation to reduce the D-limonene content (70% removals were achieved with pretreatment). Then, the anaerobic digestion of orange peel waste was evaluated at laboratory and pilot scale under mesophilic and thermophilic conditions. The results showed that the highest methane yield of 0.27–0.29 L_{STP} CH_4 g^{-1} added COD was reached at pilot scale and thermophilic temperatures.[194]

Market wastes have also been analysed for their biogas production potential. In India, market wastes consisting of rotten vegetables, fruit skins, potatoes, onions, *etc.* were subjected to anaerobic digestion in a 25 L capacity laboratory-scale biogas plant. The data collected over a period of 75 days showed that maximum production of biogas was 35 L kg^{-1} day^{-1}, obtained at 20 days' hydraulic retention time.[195]

7.5 Perspectives

Despite many years of research and development, no real breakthroughs have been made as far as large-scale bioethanol production from lignocellulosic materials is concerned. Currently, bioethanol is still being produced from sugarcane and starch-containing substrates. Knowing the complex structure of lignocellulose, the major problem seems to be related to the crucial pretreatment step. Indeed, it is the most challenging and costly phase in the bioethanol production process from highly recalcitrant lignocellulosic materials.

The basic path forward will rely in general on consolidation of the diverse processing steps, both in the engineering and biological sense. For instance, microbial cells will be expected to conduct multiple conversion reactions with high efficiency and to remain robust to process conditions. These improvements require deeper understanding of cellular and metabolic processes, essentially in order to produce new generations of hydrolytic enzymes functioning near their theoretical limits.[196]

Recent research has focused on the optimisation of biodiesel production process to minimise the input and maximise the output. Yin *et al.*[134] studied several methods for biodiesel production from sunflower oil involving mechanical stirring and ultrasonic irradiation. The results showed that biodiesel production with ultrasound irradiation assistance is better than using only mechanical stirring.

Other researchers have paid attention to the use of enzymes as a biocatalyst for biodiesel production in order to substitute the conventional alkaline process. Gog *et al.*[197] used lipase for the conversion and purification of biodiesel and confirmed that the use of enzymes in biodiesel production ensures high productivity, several possibilities of reuse and a short reaction time. Thus, industrial improvements in the enzymatic production of biodiesel could be a viable option for the future. There should be a serious focus on the optimisation of biodiesel production from microalgae, starting with an efficient culture process in order to grow and harvest algae with high production yields and lipid content. Lam and Lee[198] investigated the potential of using organic fertiliser as an alternative nutrient source to cultivate *Chlorella vulgaris*. Numerous advantages in terms of environmental perspective and cost-effectiveness have been obtained using this approach.

However, it will be crucial to optimise the extraction technology to maximise the lipid extraction process. For instance, Halim *et al.*[199] examined the factors involved in lipid extraction from microalgae cells by studying the prospects for organic solvent and supercritical fluid extraction.

A critical analysis of the literature reveals there are strong possibilities to enhance the biogas production under field conditions. The use of certain inorganic and organic additives seems promising for enhancing biogas production. Among different types of biomass (plant and crop residues) used as additives, some have been found to enhance the gas production manifolds.[200]

Ultimately, the conversion of renewable bioresources into bioethanol, biodiesel and biomethane remains an attractive path for clean energy. Although developing the technology for cost-effective energy production is not for tomorrow, scientific research endeavours remain the solution. Each advance and each breakthrough will get us closer to the end of this long, challenging and vital path of providing humanity with an eco-friendly source of energy.

References

1. S. Kim and B. E. Dale, *Biomass Bioenergy*, 2004, **26**, 361.
2. M. Balat, *Energy Sources*, 2005, **27**, 569.
3. C. Kaya, C. Hamamci, A. Baysal, O. Akba, S. Erdogan and A. Saydut, *Renew. Energy*, 2009, **34**, 1257.
4. B. H. Hameed, L. F. Lai and L. H. Chin, *Fuel Process. Technol.*, 2009, **90**, 606.
5. C. A. Cardona, J. A. Quintero and I. C. Paz, *Bioresour. Technol.*, 2010, **101**, 4754.
6. M. Galbe and G. Zacchi in *Bioconversion of Forest and Agricultural Plant Residues*, ed. J. N. Saddler, CAB International, Wallingford, UK, 1993, pp. 291–319.
7. C. Wyman, *Ann. Rev. Energy Environ.*, 1999, **24**, 189.
8. L. R. Lynd, *Energy Environ.*, 1996, **21**, 403.
9. M. Kaltschmitt, G. A. Reingardt and T. Stelzer, *Biomass Bioenergy*, 1997, **12**(2), 121.
10. S. J. Duff and W. D. Murray, *Bioresour. Technol.*, 1996, **55**, 1.
11. C. N. Hamelinck, G. V. Hooijdonk and A. P. Faaij, *Biomass Bioenergy*, 2005, **28**, 384.

12. A. Milbrandt, *A Geographic Perspective on the Current Biomass Resource Availability in the United States*, Technical Report NREL/TP-560-39181, National Renewable Energy Laboratory, Golden, CO, 2005.
13. R. D. Perlack, L. L. Wright, A. F. Turnhollow, R. L. Graham, B. J. Stokes and D. C. Erbach, *Biomass as Feedstock for a Bioenergy and Bioproducts Industry: The Technical Feasibility of a Billion-ton Annual Supply*, Oak Ridge National Laboratory, Oak Ridge, TN, 2005.
14. M. E. Walsh, D. G. Ugarte, H. Shapouri and S. P. Slinsky, *Environ. Resour. Econom.*, 2003, **24**, 313.
15. L. Cagelli and F. Lefèvre, *For. Genet.*, 1995, **2**(3), 135.
16. R. S. Zalesny Jr., R. B. Hall, J. A. Zalesny, B. G. McMahon, W. E. Berguson and G. R. Stanosz, *Bioenergy Res.*, 2009, **2**, 106.
17. J. H. Kennedy. *Cottonwood: An American Wood*, US Department of Agriculture, Forest Service, Washington DC, 1985, pp. 1–8.
18. M. J. Negro, P. Manzanares, I. Ballesteros, J. M. Oliva, A. Cabanas and M. Ballesteros, *Appl. Biochem. Biotechnol.*, 2003, **105**, 87.
19. R. Gupta and Y. Y. Lee, *Biotechnol. Progress*, 2009, **25**, 357.
20. Z. J. Wang, J. Y. Zhu, R. S. Zalesny and K. F. Chen, *Fuel*, 2012, **95**, 606.
21. K. Wang, H. Y. Yang, F. Xu and R. C. Sun, *Bioresour. Technol.*, 2010, **102**, 4524.
22. M. Jeya, N. Nguyen, H. J. Moon, S. H. Kim and J. K. Lee, *Bioresour. Technol.*, 2010, **101**, 8742.
23. S. S. Jagtap, S. S. Dhiman, M. Jeya, Y. C. Kang, J. H. Choi and J. K. Lee., *Bioresour. Technol.*, 2012, **120**, 264.
24. R. M. Toivonenen and L. J. Tahvanainen, *Biomass Bioenergy*, 1998, **15**, 27.
25. A. L. Stephenson, P. Dupree, S. A. Scott and J. S. Dennis, *Bioresour. Technol.*, 2010, **101**, 9612.
26. P. Sassner, C. G. Martensson, M. Galbe and G. Zacchi, *Bioresour. Technol.*, 2008, **99**, 137.
27. S. J. Horn, M. M. Estevez, H. K. Nielsen, R. Linjordet and V. G. Eijsink, *Bioresour. Technol.*, 2011, **102**, 7932.
28. E. Sjostrom, *Wood Chemistry: Fundamentals and Applications*, Academic Press, San Diego, 2nd edn, 1993.
29. Y. Matsushita, K. Yamauchi, K. Takabe, T. Awano, A. Yoshinaga, M. Kato, T. Kobayashi, T. Asada, A. Furujyo and K. Fukushima, *Bioresour. Technol.*, 2010, **10**, 4936.
30. W. Wei, S. Wua and L. Liu, *Bioresour. Technol.*, 2012, **110**, 302.
31. S. McIntosh, T. Vancov, J. Palmer and M. Spain, *Bioresour. Technol.*, 2012, **110**, 264.
32. A. Romani, G. Garrote and J. C. Parajo, *Fuel*, 2012, **94**, 305.
33. A. von Schenck, N. Bergli and J. Uusitalo, *Appl. Energy*, 2013, **102**, 229.
34. W. J. Frederick, S. J. Lien, C. E. Courchene, N. A. DeMartini, A. J. Ragauskas and K. Iisa, *Bioresour. Technol.*, 2008, **99**, 5051.
35. T. Q. Lan, R. Gleisner, J. Y. Zhu, B. S. Dien and R. E. Hector, *Bioresour. Technol.*, 2013, **127**, 291.

36. Y. Baba, T. Tanabe, N. Shirai, T. Watanabe, Y. Honda and T. Watanabe, *Biomass Bioenergy*, 2011, **35**, 320.
37. Y. Yamashita, C. Sasaki and Y. Nakamura, *J. Biosci. Bioeng.*, 2010, **110**, 79.
38. S. Gonzalez-Garcia, C. M. Gasol, M. T. Moreira, X. Gabarrell, J. R. Pons and G. Feijoo, *Int. J. Life Cycle Assess.*, 2011, **16**, 465.
39. S. Gonzalez-Garcia, M. T. Moreira, G. Feijoo and R. Murphy, *Biomass Bioenergy*, 2012, **39**, 378.
40. M. Ozturk, A. Celik, A. Guvensen and E. Hamzaoglu, *For. Ecol. Manage.*, 2008, **256**, 510.
41. R. B. Santos, J. M. Lee, H. Jameel, H. M. Chang and L. A. Lucia, *Bioresour. Technol.*, 2012, **110**, 232.
42. W. G. Lee, J. S. Lee, C. S. Shin, S. C. Park, H. N. Chang and Y. K. Chang, *Appl. Biochem. Biotechnol.*, 1999, **77**, 547.
43. I. Kamei, Y. Hirota and S. Meguro, *Bioresour. Technol.*, 2012, **126**, 137.
44. D. A. Huggett, S. R. Leather and K. F. Walters, *Agric. For. Entomol.*, 1999, **1**, 143.
45. I. Lewandowski, J. C. Clifton-Brown, B. Andersson, G. Basch, D. G. Christian, U. Jorgensen, M. B. Jones, A. B. Riche, K. U. Schwarz, K. Tayebi and F. Teixeira, *Agron. J.*, 2003, **95**, 1274.
46. S. J. Kim, M. Y. Kim, S. J. Jeong, M. S. Jang and I. M. Chung, *Ind. Crops Prod.*, 2012, **38**, 46.
47. A. Sorensen, P. J. Teller, T. Hilstrom and B. K. Ahring, *Bioresour. Technol.*, 2008, **99**, 6602.
48. E. A. Heaton, T. Voigt and S. P. Long, *Biomass Bioenergy*, 2004, **27**, 21.
49. E. A. Heaton, F. G. Dohleman and S. P. Long, *Change Biology*, 2008, **14**, 1365.
50. K. P. Vogel, J. J. Brejda, D. T. Walters and D. R. Buxton, *Agron. J.*, 2002, **94**, 413.
51. J. Liu, M. L. Wang, B. Tonnis, M. Habteselassie, X. Liao and Q. Huang, *Bioresour. Technol.*, 2013, **135**, 39.
52. J. Xu, J. J. Cheng, R. R. Sharma-Shivappa and J. C. Burns, *Bioresour. Technol.*, 2010, **101**, 2900.
53. C. Somerville, *Science*, 2006, **312**, 1277.
54. W. R. Morrow, W. M. Griffin and H. S. Matthews, *Environ. Sci. Technol.*, 2006, **40**, 2877.
55. M. R. Schmer, K. P. Vogel, R. B. Mitchell and R. K. Perrin, *Proc. Natl. Acad. Sci. U. S. A.*, 2008, **105**, 464.
56. T. Aysu, *Biomass Bioenergy*, 2012, **41**, 139.
57. M. Finell, M. Arshadi and R. Gref, *Biomass Bioenergy*, 2011, **35**, 1097.
58. A. Kallioinen, J. Uusitalo, K. Pahkala, M. Kontturi, L. Viikari, N. von Weymarn and M. Siika-aho, *Bioresour. Technol.*, 2012, **123**, 669.
59. M. F. Digman, K. J. Shinners, M. D. Casler, B. S. Dien, R. D. Hatfield, H. G. Jung, R. E. Muck and P. J. Weimer, *Bioresour. Technol.*, 2010, **101**, 5305.
60. B. S. Dien, M. D. Casler, R. E. Hector, L. B. Iten, N. N. Nichols, J. A. Mertens and M. A. Cotta, *Intl. J. Low-Carbon Technol.*, **7**, 338.

61. R. E. Perdue, *Econ. Bot.*, 1958, **12**, 368.
62. A. A. Shatalov and H. Pereira, *Ind. Crops Prod*, 2002, **15**, 77.
63. M. Aska, K. Olofsson, T. Di Felice, L. Ruohonen, M. Penttila, G. Liden and L. Olssona, *Process Biochem.*, 2012, **47**, 1452.
64. D. Scordia, S. L. Cosentino, J. W. Lee and T. W. Jeffries, *Biomass Bioenergy*, 2011, **35**, 3018.
65. B. S. Dien, D. J. Miller, R. E. Hector, R. A. Dixon, F. Chen, M. McCaslin, P. Reisen, G. Sarath and M. A. Cotta, *Bioresour. Technol.*, 2011, **102**, 6479.
66. H. K. Sreenath, R. G. Koegel, A. B. Moldes, T. W. Jeffries and R. J. Straub, *Process Biochem.*, 2001, **36**, 1199.
67. X. Gea, V. S. Green, N. Zhang, G. Sivakumara and J. Xu, *Process Biochem.*, 2012, **47**, 335.
68. Y. Sun and J. J. Cheng, *Bioresour. Technol.*, 2005, **96**, 1599.
69. R. Katari and S. Ghosh, *Bioresour. Technol.*, 2011, **102**, 9970.
70. L. Grivet and P. Arruda, *Curr. Opin. Plant Biol.*, 2001, **5**, 122.
71. H. V. Amorim, M. L. Lopes, J. V. Oliveira, M. S. Buckeridge and G. H. Goldman, *Appl. Microbiol. Biotechnol.*, 2011, **91**, 1267.
72. M. Gauder, S. Graeff-Honninger and W. Claupein, *Appl. Energy*, 2011, **88**, 672.
73. A. Pandey, C. R. Soccol, P. Nigam and V. T. Soccol, *Bioresour. Technol.*, 2000, **74**, 69.
74. S. Kim and B. E. Dale, *Biomass Bioenergy*, 2004, **26**, 361.
75. C. Martin, M. Galbe, C. F. Wahlbom, B. Hahn-Hägerdal and L. J. Jönsson, *Enzyme Microb. Technol.*, 2002, **31**, 274.
76. Z. Zhu, M. Zhu and Z. Wu, *Bioresour. Technol.*, 2012, **119**, 199.
77. J. Lee, *J. Biotechnol.*, 1997, **56**, 1.
78. K. Karimi, S. Kheradmandinia and M. J. Taherzadeh, *Biomass Bioenergy*, 2006, **30**, 247.
79. N. Poornejad, K. Karimi and T. Behzad, *Ind. Crops Prod.*, 2013, **41**, 408.
80. I. Watanabe, N. Miyata, A. Ando, R. Shiroma, K. Tokuyasu and T. Nakamura, *Bioresour. Technol.*, 2012, **123**, 695.
81. N. Yoswathana, P. Phuriphipat, P. Treyawutthiwat and M. N. Eshtiaghi, *Energy Res. J.*, 2010, **1**, 26.
82. I. Ballesteros, M. J. Negro, J. M. Oliva, A. Cabanas, P. Manzanares and M. Ballesteros, *Appl. Biochem. Biotechnol.*, 2006, **129**, 496.
83. B. C. Saha, L. B. Iten, M. A. Cotta and Y. V. Wu, *Process Biochem.*, 2005, **40**, 3693.
84. A. Singh and N. R. Bishnoi, *Bioresour. Technol.*, 2005, **96**, 843.
85. L. Favaro, M. Basaglia, W. H. van Zyl and S. Casella, *Appl. Energy*, 2013, **102**, 170.
86. L. K. Kadam and J. D. McMillan, *Bioresour. Technol.*, 2003, **88**, 17.
87. J. Zhao and L. Xia, *Fuel Process. Technol.*, 2010, **91**, 1807.
88. K. Ohgren, R. Bura, G. Lesnicki, J. Saddler and G. Zacchi, *Process Biochem.*, 2007, **42**, 834.
89. T. H. Kim, F. Taylor and K. B. Hicks, *Bioresour. Technol.*, 2008, **99**, 5694.

90. B. C. Saha and M. A. Cotta, *New Biotechnol.*, 2010, **27**, 10.
91. I. S. Choi, J. H. Kim, S. G. Wi, K. H. Kim and H. J. Bae, *Appl. Energy*, 2013, **102**, 204.
92. I. S. Choi, S. G. Wi, S. B. Kim and H. J. Bae, *Bioresour. Technol.*, 2012, **125**, 132.
93. L. Kang, W. Wang and Y. Y. Lee, *Appl. Biochem. Biotechnol.*, 2010, **161**, 53.
94. Z. Fan and L. R. Lynd, *Bioprocess Biosyst. Eng.*, 2007, **30**, 27.
95. Y. Yamashita, C. Sasaki and Y. Nakamura, *Carbohydr. Polym.*, 2010, **79**, 250.
96. Z. Kadar, Z. Szengyel and K. Réczey, *Ind. Crops Prod.*, 2004, **20**, 103.
97. L. Wang, M. Sharifzadeh, R. Templer and R. J. Murphy, *Appl. Energy*, 2012.
98. R. C. Kuhad, G. Mehta, R. Gupta and K. K. Sharma, *Biomass Bioenergy*, 2010, **34**, 1189.
99. E. Capek-Ménard, P. Jollez, E. Chornet, M. Wayman and K. Doan, *Biotechnol. Lett*, 1992, **14**, 985.
100. M. M. Islam Sheikh, C. H. Kim, J. Y. Lee, S. H. Kim, G. C. Kim, J. Y. Lee, S. W. Shim and J. W. Kim, *Food Bioprod. Process.*, 2013, **91**, 60.
101. D. Sahoo, G. Elangbam and S. S. Devi, *Phykos*, 2012, **42**, 32.
102. N. J. Kim, H. Li, K. Jung, H. N. Chang and P. C. Lee, *Bioresour. Technol.*, 2011, **102**, 7466.
103. M. Yanagisawa, K. Nakamura, O. Ariga and K. Nakasaki, *Process Biochem.*, 2011, **46**, 2111.
104. E. W. Becker, *J. Biotechnol. Adv.*, 2006, **25**, 207.
105. S. Sulfahri, S. Mushlihah, E. Sunarto, M. Y. Irvansyah, R. S. Utami and S. Mangkoedihardjo, *J. Basic Appl. Sci. Res.*, 2011, **1**, 589.
106. M. H. Yoon, Y. W. Leeb, C. H. Leeb and Y. B. Seo, *Bioresour. Technol.*, 2012, **126**, 198.
107. J. Y. Lee, P. Li, J. Lee, H. J. Ryu and K. K. Oh, *Bioresour. Technol.*, 2013, **127**, 119.
108. L. Ge, P. Wang and H. Mou, *Renew. Energy*, 2011, **36**, 84.
109. M. C. Ncibi, V. Jeanne-Rose, B. Mahjoub, C. Jean-Marius, J. Lambert, J. J. Ehrhardt, Y. Bercion, M. Seffen and S. Gaspard, *J. Hazard. Mater.*, 2009, **165**, 240.
110. S. Ravikumar, R. Gokulakrishnan, M. Kanagavel and N. Thajuddin, *Ind. J. Sci. Technol.*, 2011, **4**, 1087.
111. M. Pilavtepe, S. Sargin, M. S. Celiktas and O. Yesil-Celiktas, *J. Supercrit. Fluids*, 2012, **68**, 117.
112. A. Karmakar, S. Karmakar and S. Mukherjee, *Bioresour. Technol.*, 2010, **101**, 7201.
113. S. Mekhilef, S. Siga and R. Saidur, *Renew. Sustain. Energy Rev.*, 15, 1937, **2011**.
114. H. C. Ong, T. M. Mahlia, H. H. Masjuki and R. S. Norhasyima, *Renew. Sustain. Energy Rev.*, 2011, **15**, 3501.
115. A. Demirbas, *Energy Convers. Manage.*, 2003, **44**, 2093.

116. W. Cao, H. Han and J. Zhang, *Fuel*, 2005, **84**, 347.
117. E. Crabbe, C. Nolasco-Hipolito, G. Kobayashi, K. Sonomoto and A. Ishizaki, *Process. Biochem.*, 2001, **37**, 65.
118. M. Balat, *Energy Convers. Manage.*, 2011, **52**, 1479.
119. Vivek and A. K. Gupta, *J. Sci. Ind. Res.*, 2004, **63**, 39.
120. P. C. Jena, H. Rahem, G. V. P. Kumar and R. Machavaram, *Biomass Bioenergy*, 2010, **34**, 1108.
121. S. Chaichalerm, P. Pokethitiyook, W. Yuan, M. Meetam, K. Sritong, W. Pugkaew, K. Kungvansaichol, M. Kruatrachue and P. Damrongphol, *Appl. Energy*, 2012, **89**, 296.
122. Y. Zheng, Z. Chi, B. Lucker and S. Chen, *Bioresour. Technol.*, 2012, **103**, 484.
123. A. Converti, A. A. Casazza, E. Y. Ortiz, P. Perego and M. Del Borghi, *Chem. Eng. Prog.*, 2009, **48**, 1146.
124. S. Hoekman and A. Broch, *Renew. Sustain. Energy Rev.*, 2012, **16**, 143.
125. K. G. Georgogianni, A. P. Katsoulidis, P. J. Pomonis and M. G. Kontominas, *Fuel Process. Technol.*, 2009, **90**, 671.
126. W. N. Omar and N. A. Amin, *Biomass Bioenergy*, 2011, **35**, 1329.
127. M. Naik, L. C. Meher, S. N. Naik and L. M. Das, *Biomass Bioenergy*, 2008, **32**, 354.
128. A. Demirbas., *Energy Convers. Manage.*, 2009, **50**, 923.
129. D. Y. Leung, X. Wu and M. K. Leung, *Appl. Energy*, 2010, **87**, 1083.
130. L. C. Meher, S. N. Naik and L. M. Das, *J. Sci. Ind. Res.*, 2004, **63**, 913.
131. J. van Gerpen, B. Shanks, R. Pruszko, D. Clements and G. Knothe, *Biodiesel Production Technology, August 2002–January 2004*, NREL/SR-510-36244, National Renewable Energy Laboratory, Golden, Colorado, 2004.
132. A. L. Stephenson, J. S. Dennis and S. A. Scott, *Process Saf. Environ. Protect.*, 2008, **86**, 427.
133. G. Santori, G. Di Nicola, M. Moglie and F. Polonara, *Appl. Energy*, 2012, **92**, 109.
134. X. Yin, H. Ma, Q. You, Z. Wang and J. Chang, *Appl. Energy*, 2012, **91**, 320.
135. F. Qiu, Y. Li, D. Yang, X. Li and P. Sun, *Appl. Energy*, 2011, **88**, 2050.
136. R. A. Candeia, M. C. DSilva, J. R. Carvalho, M. G. Brasilino, T. C. Bicudo, I. M. Santos and A. G. Souza, *Fuel*, 2009, **88**, 738.
137. D. Yu, L. Tian, H. Wu, S. Wang, Y. Wang, D. Ma and X. Fang, *Process. Biochem.*, 2010, **45**, 519.
138. N. Zapata, M. Vargas, J. F. Reyes and G. Belmar, *Ind. Crops Prod.*, 2012, **38**, 1.
139. T. Thamsiriroj and J. D. Murphy, *Energy Fuels*, 2010, **24**, 1720.
140. A. Bouaid, M. Martinez and J. Aracil, *Bioresour. Technol.*, 2010, **101**, 4006.
141. D. Kumar, G. Kumar, Poonam and C. P. Singh, *Ultrason. Sonochem.*, 2010, **17**, 555.

142. A. B. Chhetri, M. S. Tango, S. M. Budge, K. C. Watts and M. R. Islam, *Int. J. Mol. Sci.*, 2008, **9**, 169.

143. J. Qian, H. Shi and Z. Yun, *Bioresour. Technol.*, 2010, **101**, 7025.

144. D. Kumar, G. Kumar, R. Johari and P. Kumar, *Ultrason. Sonochem.*, 2012, **19**, 816.

145. C. Y. Yang, Z. Fang, B. Li and Y. F. Long, *Renew Sustain. Energ. Rev.*, 2012, **16**, 2178.

146. S. V. Ghadge and H. Raheman, *Biomass Bioenerg.*, 2005, **28**, 601.

147. J. Singh and P. C. Bargale, *J. Food Eng.*, 2000, **43**, 75.

148. C. Ofori-Boateng, L. K. Teong and L. Jitkang, *Energy Convers. Manage.*, 2012, **55**, 164.

149. T. Nguyen, L. Do and D. A. Sabatini, *Fuel*, 2010, **89**, 2285.

150. L. C. Meher, V. S. S. Dharmagadda and S. N. Naik, *Bioresour. Technol.*, 2006, **97**, 1392.

151. M. Morshed, K. Ferdous, M. R. Khan, M. S. I. Mazumder, M. A. Islam and M. T. Uddin, *Fuel*, 2011, **90**, 2981.

152. P. Evon, V. Vandenbossche, P. Y. Pontalier and L. Rigal, *Ind. Crops Prod.*, 2007, **2**, 351.

153. S. A. Basha, K. R. Gopal and S. Jebaraj, *Renew. Sustain. Energ. Rev.*, 2009, **13**, 1628.

154. N. N. Yusuf, S. K. Kamarudin and Z. Yaakub, *Energy Convers. Manage.*, 2011, **52**, 2741.

155. J. Hill, E. Nelson, D. Tilman, S. Polasky and D. Tiffany, *Proc. Natl. Acad. Sci. U. S. A.*, 2006, **103**, 11206.

156. N. Boz, M. Kara, O. Sunal, E. Alptekin and N. Degirmenbasi, *Turk. J. Chem.*, 2009, **33**, 1.

157. European Union, Directive 2009/28/EC of the European Parliament and of the Council of 23 April 2009 on the promotion of the use of energy from renewable sources and amending and subsequently repealing Directives 2001/77/EC and 2003/30/EC, Off. J. Eur. Union, 2009, L140, 16.

158. J. B. Holm-Nielsen, T. Al Seadi and P. Oleskowicz-Popiel, *Bioresour. Technol.*, 2009, **100**, 5478.

159. E. Athanasoulia, P. Melidis and A. Aivasidis, *Renew. Energy*, 2012, **47**, 147.

160. M. Melikoglu, *Renew. Sustain. Energy Rev.*, 2013, **19**, 52.

161. E. Comino, V. A. Riggio and M. Rosso, *Bioresour. Technol.*, 2012, **114**, 46.

162. H. M. El-Mashad and R. Zhang, *Bioresour. Technol.*, 2010, **101**, 4021.

163. M. Scholz, T. Melin and M. Wessling, *Renew. Sustain. Energy Rev.*, 2013, **17**, 199.

164. M. Persson, *Evaluation of Upgrading Techniques for Biogas*, Report SGC 142, Swedish Gas Centre, Malmo, 2003.

165. V. G. Gomes and M. M. Hassan, *Sep. Purif. Technol.*, 2001, **24**, 189.

166. E. Ryckebosch, M. Drouillon and H. Vervaeren, *Biomass Bioenergy*, 2011, **35**, 1633.

167. J. Hao, P. Rice and S. Stern, *J. Membrane Sci.*, 2002, **209**, 177.

168. G. Leson and A. M., *J. Air Waste Manage. Assoc.*, 1991, **41**, 1045.
169. P. Oyarzun, F. Arancibia, C. Canales and G. E. Aroca, *Process Biochem.*, 2003, **39**, 165.
170. I. A. Esteves and J. P. Mota, *Desalination*, 2002, **148**, 275.
171. A. Teghammar, K. Karimi, I. S. Horvath and M. J. Taherzadeh, *Biomass Bioenergy*, 2012, **36**, 116.
172. A. M. Lakaniemi, P. E. Koskinen, L. M. Nevatalo, A. H. Kaksonen and J. A. Puhakka, *Biomass Bioenergy*, 2011, **35**, 773.
173. D. Masse, Y. Gilbert, P. Savoie, G. Belanger, G. Parent and D. Babineau, *Bioresour. Technol.*, 2011, **102**, 10286.
174. S. J. Horn, M. M. Estevez, H. K. Nielsen, R. Linjordet and V. G. Eijsink, *Bioresour. Technol.*, 2011, **102**, 7932.
175. V. N. Gunaseelan, *Biomass Bioenergy*, 1994, **6**, 391.
176. E. Klimiuk, T. Pokoj, W. Budzynski and B. Dubis, *Bioresour. Technol.*, 2010, **101**, 9527.
177. I. Oliveira, J. Gominho, S. Diberardino and Elizabeth Duarte, *Ind. Crops Prod*, 2012, **40**, 318.
178. W. Zhong, Z. Zhang, Y. Luo, S. Sun, W. Qiao and M. Xiao, *Bioresour. Technol.*, 2011, **102**, 11177.
179. D. C. Nieves, K. Karimi and I. S. Horvath, *Ind. Crops Prod.*, 2011, **34**, 1097.
180. V. Fernandez-Cegri, M. A. De la Rubia, F. Raposo and R. Borja, *Bioresour. Technol.*, 2012, **123**, 424.
181. N. V. Deshpande, N. W. Kale and S. J. Deshmukh, *Energy Sustain. Develop.*, 2012, **16**, 363.
182. E. Dinuccio, P. Balsari, F. Gioelli and S. Menardo, *Bioresour. Technol.*, 2010, **101**, 3780.
183. S. Mohan and N. Sunny, *Bioresour. Technol.*, 2008, **99**, 210.
184. A. Teghammar, J. Yngvesson, M. Lundin, M. J. Taherzadeh and I. S. Horvath, *Bioresour. Technol.*, 2010, **101**, 1206.
185. L. Yunqin, W. Dehan, W. Shaoquan and W. Chunmin, *J. Hazard. Mater.*, 2009, **170**, 366.
186. C. Zamalloa, N. Boon and W. Verstraete, *Appl. Energy*, 2012, **92**, 733.
187. A. V. Fernandez, G. Vargas, N. Alarcon and A. Velasco, *Biomass Bioenergy*, 2008, **32**, 338.
188. J. A. Costa, F. B. Santana, M. R. Andrade, M. B. Lima and D. T. Franck, *J. Biotechnol.*, 2008, **136**, 402.
189. A. Bruhn, J. Dahl, H. B. Nielsen, L. Nikolaisen, M. B. Rasmussen, S. Markager, B. Olesen, C. Arias and P. D. Jensen, *Bioresour. Technol.*, 2011, **102**, 2595.
190. F. De Paoli, A. Bauer, C. Leonhartsberger, B. Amon and T. Amon, *Bioresour. Technol.*, 2011, **102**, 6621.
191. J. H. Park, J. J. Yoon, H. D. Park, D. J. Lim and S. H. Kim, *Bioresour. Technol.*, 2012, **113**, 78.
192. D. Liu, D. Liu, R. J. Zeng and I. Angelidaki, *Water Res.*, 2006, **40**, 2230.

193. H. Zhu, A. Stadnyk, M. Beland and P. Seto, *Bioresour. Technol.*, 2008, **99**, 5078.
194. M. A. Martin, J. A. Siles, A. F. Chica and A. Martin, *Bioresour. Technol.*, 2010, **101**, 8993.
195. D. R. Ranade, T. Y. Yeole and S. H. Godbole, *Biomass*, 1987, **13**, 147.
196. M. C. Ncibi, *Recent Patents Chem. Eng.*, 2010, **3**, 165.
197. A. Gog, M. Roman, M. Tos, C. Paizs and F. D. Irimie, *Renew. Energy*, 2012, **39**, 10.
198. M. K. Lam and K. T. Lee, *Appl. Energy*, 2012, **94**, 303.
199. R. Halim, M. K. Danquah and P. A. Webley, *Biotechnol. Adv.*, 2012, **30**, 709.
200. Y. Santosh, T. R. Sreekrishnan, S. Kohli and V. Rana, *Bioresour. Technol.*, 2004, **95**, 1.

CHAPTER 8

Hydrogen Production from Biomass Derivatives over Heterogeneous Photocatalysts

KATSUYA SHIMURA[a] AND HISAO YOSHIDA*[b]

[a] Biomass Refinery Research Centre, National Institute of Advanced Industrial Science and Technology, AIST Chuugoku Centre, Hiroshima 739-0046, Japan; [b] Department of Interdisciplinary Environment, Graduate School of Human and Environmental Studies, Kyoto University, Kyoto 606-8501, Japan
*Email: yoshida.hisao.2a@kyoto-u.ac.jp

8.1 Introduction

The development of environmentally benign, renewable and sustainable energy production is one of the most important subjects for the near future. Hydrogen is a storable energy carrier with the high energy content and non-polluting nature. Furthermore, it can be effectively converted into electricity by a fuel cell or into motive power by a hydrogen-fuelled engine without any emission other than water. But although hydrogen is an attractive alternative energy source, about 96% of the hydrogen supplied currently is derived from fossil fuels such as natural gas (49%), crude oil (29%) and coal (18%) by means of thermal chemical processes and gasification at high temperature.[1] Hydrogen produced from fossil fuels cannot be regarded as really an environmentally benign fuel because it takes a very long time to regenerate fossil fuels and the consumption of the fossil fuel increases the concentration of carbon dioxide in the atmosphere contributing to

RSC Green Chemistry No. 25
Biomass for Sustainable Applications: Pollution Remediation and Energy
Edited by Sarra Gaspard and Mohamed Chaker Ncibi
© The Royal Society of Chemistry 2014
Published by the Royal Society of Chemistry, www.rsc.org

global warming. For the realisation of a sustainable society, hydrogen needs to be produced from renewable resources and natural energy.

In recent years, biomass (*e.g.* plants, starch and oil) and its derivatives (*e.g.* ethanol, glycerol, sugars and methane) have attracted much attention as possible hydrogen sources. They are also regarded as a renewable resource. If the biomass and its derivatives are consumed for hydrogen production with carbon dioxide formation, the produced carbon dioxide can be converted again into biomass through plant photosynthesis. This means that the carbon dioxide produced from the biomass should not, in principle, contribute to global warming (*i.e.* it is carbon-neutral) when the consumption of the biomass does not exceed the natural capacity for conversion of carbon dioxide to biomass.

Currently two major approaches, *i.e.* thermal gasification and biological hydrogen production by fermentation, are extensively studied as methods to convert biomass into hydrogen. Although these are promising hydrogen production methods, there are major problems to be solved for practical appreciation. For example, the former method generally requires high reaction temperatures at 1073–1273 K (*i.e.* consuming considerable amounts of energy) and the reaction rate of the latter method is quite low (*i.e.* low productivity).

Photocatalytic hydrogen production from water and biomass derivatives is another possible hydrogen production method from biomass.[2,3] This system is very attractive since hydrogen can be produced at room temperature using sunlight and a photocatalyst. Research on photocatalytic hydrogen production from biomass began in the early 1980s. Since then various attempts have been made to achieve efficient hydrogen evolution. In this chapter, we introduce the background, fundamentals, history and recent developments in these photocatalytic systems for hydrogen production.

8.2 Hydrogen Production Methods from Biomass other than Photocatalysis

Before describing photocatalytic hydrogen production from biomass, two major approaches, *i.e.* thermal gasification of biomass and fermentative hydrogen production, are introduced briefly for comparison. In thermal gasification, biomass is partially oxidised and reformed at high temperature to produce a gas containing hydrogen, carbon monoxide, carbon dioxide, methane and so on. Although a high reaction temperature and special reaction setup are required for gasification, this method has the advantage of being able to be applied to any kind of biomass. In hydrogen fermentation, biomass is converted into hydrogen by bacteria under anaerobic conditions. The fermentation proceeds at ambient temperature and thus the reaction rate is usually low.

8.2.1 Thermal Gasification of Biomass

The thermal gasification of biomass is largely divided into two categories, *i.e.* high temperature gasification and supercritical gasification. In high temperature gasification, biomass can be thermally decomposed in the presence of an oxidant such as oxygen and/or water according to eqn (8.1):

$$C_xH_yO_z + H_2O/O_2 \rightarrow H_2, CO, CO_2, CH_4, C_{2+}, \text{ tar and } C(s) \qquad (8.1)$$

The products obtained in the synthesis gas are hydrogen, carbon monoxide and carbon dioxide, together with unwanted products such as hydrocarbons including methane, tar and carbon. The gasification of biomass is usually carried out at high temperature (*e.g.* 1273 K) because large amounts of condensable organic compounds (tar) and methane are produced by gasification at low temperature. It has been reported that temperatures >1523 K were required to obtain synthesis gas with a composition free from tar and methane in the equilibrium without using any catalysts.[4] The formation of tar sometimes becomes a major problem because it can causes blockages in the process equipment and the deactivation of the catalyst.

The use of a suitable catalyst can suppress the formation of tar, promote the reforming of methane and lower the reaction temperature. Catalysts such as dolomite, olivine and supported metals have been examined for this reaction system. Dolomite[5] (a mixture of magnesium and calcium carbonate containing iron) and olivine[6] (a mixture of magnesium and iron silicate) are natural minerals and effective for the removal of tar. The iron content of these minerals can play an important role in their catalytic activity. Supported Ni catalysts have been frequently applied because of their comparatively low price and high catalytic activity.[7] Ni catalysts serve not only to remove tar and methane but also to adjust the synthesis gas composition *via* a water–gas shift reaction. Precious metal catalysts such as $Rh/CeO_2/SiO_2$ were also reported to show a high activity.[8] Details of thermal gasification of biomass are reviewed in ref. 9 and 10.

In supercritical water gasification, biomass is decomposed to small molecules due to their high solubility and reactivity to the supercritical water. Biomass containing a lot of water is generally not suitable for high temperature gasification and is better suited for supercritical gasification from the viewpoint of energy efficiency. Supercritical water gasification can be roughly divided into two types: one is carried out at higher temperatures between 823 and 973 K, and the other is at temperatures <823 K. Under the former condition, decomposition of biomass proceeds without the presence of catalyst; it was reported that a solution containing 11 wt% glucose can be completely gasified at 973 K using a flow reactor.[11] Under the latter condition, however, transition metal catalysts such as Ni (*e.g.* Ni/Al_2O_3) and Ru (*e.g.* Ru/TiO_2) are used for the reaction.[12,13] Biomass such as cellulose and lignin could be completely gasified in such a low temperature, although the catalyst was deactivated during the reaction.[13] The current situation regarding supercritical water gasification of biomass is reviewed in ref. 14 and 15.

In the supercritical water, the reactions shown in eqns (8.2) to (8.4) proceed to yield mainly hydrogen, carbon dioxide and methane, with cellulose being used as an example of biomass:

$$C_6H_{10}O_5 + H_2O \rightarrow 6CO + 6H_2 \quad \Delta H_{298K}^0 = 310 \text{ kJ mol}^{-1} \qquad (8.2)$$

$$CO + H_2O \rightarrow CO_2 + H_2 \quad \Delta H^0_{298K} = -41\,kJ\,mol^{-1} \qquad (8.3)$$

$$CO + 3\,H_2 \rightarrow CH_4 + H_2O \quad \Delta H^0_{298K} = -206\,kJ\,mol^{-1} \qquad (8.4)$$

From the thermodynamic equilibrium, hydrogen was dominantly obtained at high temperature, while methane was the main product at low temperature.[16]

8.2.2 Fermentation of Biomass

Some routes of biomass fermentation including hydrogen fermentation and methane fermentation are shown in Figure 8.1.

In the hydrogen fermentation, large organic compounds such as carbohydrates, proteins and fatty oils are hydrolysed to form smaller molecules such as monosaccharides, amino acids and higher fatty acids. The produced monosaccharides and amino acids are then decomposed to low molecular weight organic acids, hydrogen and carbon dioxide, as representatively shown in eqn (8.5):

$$C_6H_{12}O_6 + 2\,H_2O \rightarrow 2\,CH_3COOH + 2\,CO_2 + 4\,H_2 \qquad (8.5)$$

In the hydrogen fermentation of glucose (eqn (8.5)), only four moles of hydrogen at a maximum can be produced per one mole of glucose due to the formation of organic acid as a by-product. To enhance the energy efficiency of the hydrogen fermentation, the produced organic acids are further converted

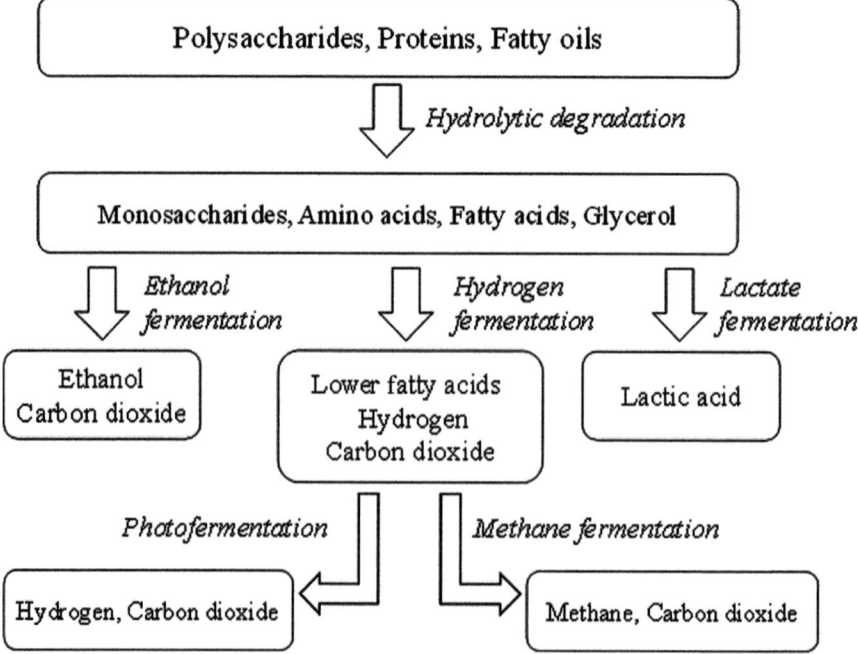

Figure 8.1 Some routes of biomass fermentation.

into methane by methane fermentation or hydrogen by photofermentation as representatively shown in eqns (8.6) and (8.7):

$$CH_3COOH \rightarrow CH_4 + CO_2 \tag{8.6}$$

$$2\,CH_3COOH + 4\,H_2O + \text{Light energy} \rightarrow 8\,H_2 + 4\,CO_2 \tag{8.7}$$

In methane fermentation, *ca.* 70% of the methane is produced from acetic acid and the rest is produced by the reduction of carbon dioxide by hydrogen (eqn (8.8)):

$$CO_2 + 4\,H_2 \rightarrow CH_4 + 2H_2O \tag{8.8}$$

Details of hydrogen production from biomass by fermentation have been reviewed by several researchers.[1,17,18] Large reaction equipment is generally required for biomass fermentation due to the low reaction rate. The treatment of the fermentation residues is also a serious problem.

8.2.3 Steam Reforming of Methane

As described above, methane is usually formed as a by-product when hydrogen is produced from biomass by the thermal gasification or fermentation of biomass. Methane is further converted into hydrogen by steam reforming and the water–gas shift reaction to obtain the maximum amount of hydrogen from the biomass.

The steam reforming of methane (SRM) (eqn (8.9)) using a heterogeneous catalyst such as Ni or Ru catalysts has been industrially employed for hydrogen production from methane.[19,20] To obtain the maximum yield of hydrogen from methane, a further reaction of the produced carbon monoxide with water is carried out using both Fe–Cr and Cu–Zn catalysts as shown in eqn (8.10), *i.e.* the water–gas shift reaction.

$$CH_4 + H_2O \rightarrow 3\,H_2 + CO \quad \Delta G^0_{298K} = 142\,kJ\,mol^{-1} \tag{8.9}$$

$$CO + H_2O \rightarrow H_2 + CO_2 \quad \Delta G^0_{298K} = -29\,kJ\,mol^{-1} \tag{8.10}$$

The positive and large Gibbs free energy change for SRM requires a high temperature typically >1073 K, even in the presence of these catalysts. Steam reforming is also applied to the hydrogen production from other hydrocarbons or alcohols.

8.3 Photocatalytic Hydrogen Production from Water and Biomass Derivatives

To help realise a sustainable society, hydrogen must be produced from renewable resources and natural energy. Photocatalytic water splitting ($H_2O \rightarrow H_2 + \frac{1}{2}O_2$) is an ideal method, since hydrogen can be produced from water by using solar energy. Following the discovery of the Honda–Fujishima effect,[21] water splitting with photoenergy and photocatalysts has been widely studied all over the world.[22–24] Various semiconductor materials (*e.g.* metal oxide, oxynitride and sulfide), co-catalysts (*e.g.* NiO_x, RuO_2 and Rh–Cr) and their modifications (*e.g.* metal ion doping, and combination with another semiconductor or photosensitiser) have been examined. Nowadays, many photocatalytic systems driven by visible light have been developed to utilise the main part of the solar spectrum.[25–27] However, the criteria for their practical application have not yet been achieved.

Photocatalytic hydrogen production from water and biomass derivatives is also an attractive hydrogen production method. In this system, biomass derivatives and water can be converted to hydrogen and carbon dioxide by photoenergy and a photocatalyst (Figure 8.2). The carbon dioxide formed can be transformed into carbohydrates by photosynthesis in plants, and the grown biomass can be degraded into small organic compounds (derivatives) such as ethanol and methane by fermentation. The hydrogen formed is used to produce electricity using a fuel cell or to supply motive power using a hydrogen-fuelled engine with the formation of water only.

In the early 1980s, Kawai and Sakata reported that hydrogen could be efficiently produced from water and many kinds of organic compounds, *i.e.* active carbon,[28] methanol,[29] ethanol,[30] saccharides,[31] amino acids,[32] polymers[32] and raw biomass[32] (also fossil fuels[33]) over TiO_2 photocatalysts loaded with Pt and/ or RuO_2 co-catalysts. Sato and White examined hydrogen production from water vapour and organic compounds such as carbon monoxide,[34] ethylene[35] and carbon[36] over Pt/TiO_2 photocatalysts. Harada *et al.*[37] reported that hydrogen could be produced from water and lactic acid over Pt/TiO_2 and Pt/ CdS photocatalysts. Following these early reports, hydrogen production from water and methanol has been frequently applied as the test reaction to confirm a photocatalyst's potential for hydrogen production.[22–24]

In the present century, photocatalytic hydrogen production from water and biomass derivatives has become an active research field. Kondarides and colleagues examined photocatalytic hydrogen production from aqueous solutions containing not only organic compounds such as azo dyes,[38] alcohols and organic acids,[39] but also biomass (glycerol and carbohydrates)[40] over Pt/TiO_2 photocatalysts. They confirmed the stoichiometry of the products, and clarified that hydrogen and carbon dioxide were produced from these compounds and

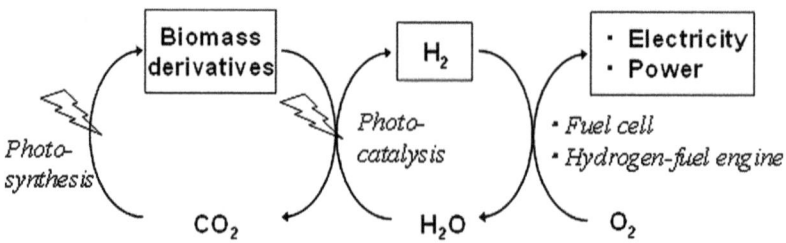

Figure 8.2 Carbon neutral cycle with photocatalysis.

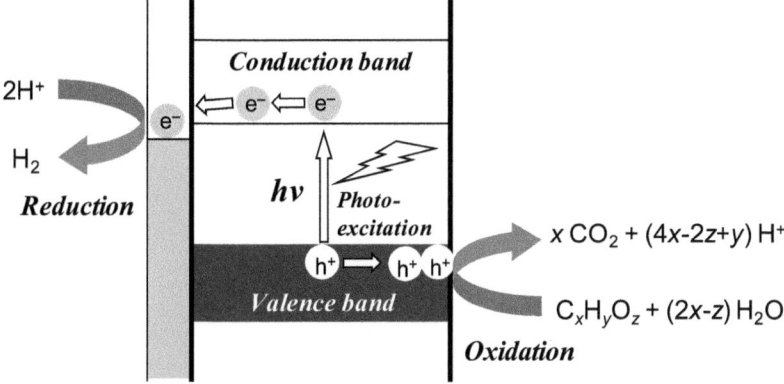

Figure 8.3 Photocatalytic steam reforming of biomass.

biomass at room temperature and atmospheric pressure as shown in Figure 8.3 and eqn (8.11):

$$C_xH_yO_z + (2x - z)H_2O \rightarrow \left(2x - z + \frac{y}{2}\right)H_2 + xCO_2 \qquad (8.11)$$

Various attempts have been carried out to enhance the efficiency of these photocatalytic systems. In this section, we review photocatalytic hydrogen production from water and biomass derivatives, especially ethanol, glycerol, saccharides, organic acid and methane.

8.3.1 Thermodynamic and Electrochemical Perspectives

Before going to the main part of the review, we first present an overview of the thermodynamic and electrochemical basics of the reactions in photocatalytic hydrogen production from water and biomass derivatives. Table 8.1 shows the changes in Gibbs free energy at 298 K for the reactions of water vapour with some small organic compounds and for water splitting.[41] A reaction can proceed preferably when the change of Gibbs free energy is negative. However, a reaction with a positive Gibbs free energy change is thermodynamically unfavourable. For example, the changes in Gibbs free energy for the steam reforming reactions of methane, carbon, ethane and ethanol (Table 8.1, entries 1, 2, 9 and 10) are largely positive. This means that these reaction systems require an energy supply and that part of the energy is stored as chemical potential in the products. This type of reaction can be called an 'uphill' reaction. In the steam reforming reactions of methanol, carbon monoxide, formaldehyde, glycerol and glucose (Table 8.1, entries 4, 6, 8, 17 and 18), on the other hand, the changes in Gibbs free energy are negative. This means that these systems must lose the chemical potential energy. This type of reaction can be called a 'downhill' reaction. We note that the changes in Gibbs free energy for hydrogen production from water and organic compounds are obviously smaller than that for water

Table 8.1 Change in Gibbs free energy at 298 K for the reactions of hydrogen production.[a]

Entry	Chemical equation	$\Delta G_{298K}^0 /kJ\ mol^{-1}$
1	$CH_4 + 2\ H_2O(g) \rightarrow CO_2 + 4\ H_2$	114
2	$C + 2\ H_2O(g) \rightarrow CO_2 + 2\ H_2$	63
3	$CH_3OH(g) \rightarrow HCHO(g) + H_2$	59
4	$CH_3OH(g) + H_2O(g) \rightarrow CO_2 + 3\ H_2$	−4
5	$HCHO(g) + H_2O(g) \rightarrow HCOOH(g) + H_2$	−20
6	$CO(g) + H_2O(g) \rightarrow CO_2 + H_2$	−29
7	$HCOOH(g) \rightarrow CO_2 + H_2$	−43
8	$HCHO(g) + H_2O(g) \rightarrow CO_2 + 2\ H_2$	−63
9	$C_2H_6 + 4\ H_2O(g) \rightarrow 2\ CO_2 + 7\ H_2$	158
10	$C_2H_5OH(g) + 3\ H_2O(g) \rightarrow 2\ CO_2 + 6\ H_2$	65
11	$C_2H_5OH(g) \rightarrow CH_3CHO(g) + H_2$	36
12	$CH_3CHO(g) + 3\ H_2O(g) \rightarrow 2\ CO_2 + 5\ H_2$	30
13	$CH_3CHO(g) + H_2O(g) \rightarrow CH_3COOH(g) + H_2$	−13
14	$CH_3COOH(g) \rightarrow CO_2 + CH_4$	−71
15	$CH_3CH(OH)COOH(s) + 3\ H_2O(g) \rightarrow 3\ CO_2 + 6\ H_2$	25
16	$CH_3CH(OH)CH_3(g) \rightarrow CH_3COCH_3(g) + H_2$	21
17	$C_3H_8O_3(l) + 3\ H_2O(g) \rightarrow 3\ CO_2 + 7\ H_2$	−20
18	$C_6H_{12}O_6(s) + 6\ H_2O(g) \rightarrow 6\ CO_2 + 12\ H_2$	−85
19	$H_2O(g) \rightarrow H_2 + 1/2\ O_2$	229
20	$H_2O(l) \rightarrow H_2 + 1/2\ O_2$	237

[a]The changes in Gibbs free energy were calculated from the reported values of standard Gibbs energy of formation.[41]

Table 8.2 Electrochemical reactions with corresponding oxidative potential.[a]

Entry	Reaction	$E_{ox}^0 (pH = 0) /V\ vs.\ NHE$
1	$CH_3CHO + H_2O + 2\ h^+ \rightarrow CH_3COOH + 2\ H^+$	−0.12
2	$CO + H_2O + 2\ h^+ \rightarrow CO_2 + 2\ H^+$	−0.12
3	$HCHO + H_2O + 4\ h^+ \rightarrow CO_2 + 4\ H^+$	−0.07
4	$C_6H_{12}O_6 + 6\ H_2O + 24\ h^+ \rightarrow 6\ CO_2 + 24\ H^+$	−0.02
5	$H_2 + 2\ h^+ \rightarrow 2\ H^+$	0.00
6	$C_3H_8O_3 + 3\ H_2O + 14\ h^+ \rightarrow 3\ CO_2 + 14H^+$	0.00
7	$C_2H_6O_2 + 2\ H_2O + 10\ h^+ \rightarrow 2\ CO_2 + 10H^+$	0.01
8	$CH_3OH + H_2O + 6\ h^+ \rightarrow CO_2 + 6\ H^+$	0.03
9	$C_2H_5OH + 3\ H_2O + 12\ h^+ \rightarrow 2\ CO_2 + 12\ H^+$	0.08
10	$C_3H_7(OH) + 5\ H_2O + 18\ h^+ \rightarrow 3\ CO_2 + 18\ H^+$	0.10
11	$C_3H_6O + 5\ H_2O + 16\ h^+ \rightarrow 3\ CO_2 + 16\ H^+$	0.10
12	$CH_3CH(OH)CH_3 + 5\ H_2O + 18\ h^+ \rightarrow 3\ CO_2 + 18\ H^+$	0.11
13	$2\ CH_3COOH + 2\ h^+ \rightarrow C_2H_6 + 2\ CO_2 + 2\ H^+$	0.12
14	$CH_4 + 2\ H_2O + 8\ h^+ \rightarrow CO_2 + 8\ H^+$	0.17
15	$C_2H_5OH + 2\ h^+ \rightarrow CH_3CHO + 2\ H^+$	0.19
16	$H_2O + 2\ h^+ \rightarrow 1/2\ O_2 + 2\ H^+$	1.23

[a]The oxidative potentials are taken from ref. 42–44. NHE = normal hydrogen electrode.

splitting (Table 8.1, entries 19 and 20). Thus, it is expected that these reactions can produce hydrogen more efficiently than water splitting.

The oxidative potentials for the reaction of various organic compounds are between −0.12 and 0.19 V (Table 8.2).[42–44] These values are more negative than that of water (1.23 V). This means that these organic compounds can be more easily oxidised than water by oxidant or the photoformed holes in the valence band of typical metal oxide semiconductors such as TiO_2 to release proton.

8.3.2 Hydrogen Production from Water and Alcohols/Acids/ Sugars over TiO_2 Photocatalyst

TiO_2 has been most frequently applied catalyst for photocatalytic hydrogen production from water and biomass derivatives. So far, various examinations have been carried out such as analysis of the reaction mechanism, the effect of the reaction condition, development of active co-catalysts and morphology controls. Below we give various examples of hydrogen production from water and alcohols/acids/sugars over TiO_2 photocatalyst.

8.3.2.1 Reaction Mechanism

Photocatalytic hydrogen production from water and ethanol over TiO_2 was first reported by Kawai and Sakata.[30] Hydrogen, acetaldehyde, acetic acid, methane and carbon dioxide were produced. At the early stage of irradiation, a considerable amount of acetaldehyde was produced but carbon dioxide was not obtained. These results indicated that the reactions shown in eqns (8.12) and (8.13) took place successively, followed by further oxidation as shown in eqn (8.6):

$$C_2H_5OH \rightarrow CH_3CHO + H_2 \qquad (8.12)$$

$$CH_3CHO + H_2O \rightarrow CH_3COOH + H_2 \qquad (8.13)$$

Assuming that the reactions shown in eqns (8.12), (8.13) and (8.6) went to completion, the molar ratio of methane to hydrogen would be expected to be 0.5. However, the ratio was actually *ca.* 0.1 even after a long reaction time (>800 h), suggesting that the complete oxidation of acetaldehyde shown in eqn (8.14) might proceed simultaneously.

$$CH_3CHO + 3H_2O \rightarrow 5H_2 + 2CO_2 \qquad (8.14)$$

Some mechanisms have been proposed for the dehydrogenation of ethanol (eqn (8.12)). The first mechanism is that ethanol is directly oxidised by the photoexcited holes to acetaldehyde and two protons (H^+), and the protons are reduced by the photoexcited electrons according to eqns (8.15) and (8.16):[45]

$$CH_3CH_2OH + 2h^+ \rightarrow CH_3CHO + 2H^+ \qquad (8.15)$$

$$2H^+ + 2e^- \rightarrow H_2 \qquad (8.16)$$

The second mechanism is that ethanol, dissociatively, is adsorbed on the surface of TiO_2. The adsorbed ethoxy species ($\bullet OCH_2CH_3$) on TiO_2 is directly oxidised by the photoformed holes, while the proton is reduced on the metal co-catalyst by the photoformed electrons according to eqns (8.17) to (8.19):[46,47]

$$CH_3CH_2O-H + Ti(s)-O(s) \rightarrow CH_3CH_2O-Ti(s) + H-O(s) \qquad (8.17)$$

$$CH_3CH_2O^{\bullet} + h^+ \rightarrow CH_3CHO + H^+ \qquad (8.18)$$

$$H^{\bullet} + H^+ + e^- \rightarrow H_2 \qquad (8.19)$$

where Ti(s) and O(s) show the surface Ti and O atoms on TiO_2.

The third mechanism is that the active species originating from water promote the activation of alcohols, for example, as shown in eqns (8.16), (8.20) and (8.21):[46,48]

$$H_2O + h^+ \rightarrow H^+ + {}^{\bullet}OH \qquad (8.20)$$

$$CH_3CH_2OH + {}^{\bullet}OH + h^+ \rightarrow CH_3CHO + H_2O + H^+ \qquad (8.21)$$

When the reaction was carried out in an aqueous alcohol solution containing HCl and H_2SO_4 or over a photocatalyst with many acid sites, acetal was obtained as the major product in the solution.[49,50] The acetal would be produced through an acid-catalysed reaction between ethanol and photo-catalytically produced acetaldehyde (eqn (8.22)).

$$2\,CH_3CH_2OH + CH_3CHO \rightarrow CH_3CH(OCH_2CH_3)_2 + H_2O \qquad (8.22)$$

Recently, it was reported that the ethanol coupling reaction (eqn (8.23)) simultaneously proceeded when the Pt/TiO_2 photocatalyst was photoirradiated in an aqueous ethanol.[51]

$$2\,CH_3CH_2OH \rightarrow CH_3CH(OH)CH(OH)CH_3 + H_2 \qquad (8.23)$$

The crystal phase and pore structures of TiO_2 photocatalysts largely influenced the product selectivity; 2,3-butanediol was produced at very high selectivity (96.6%) when rutile TiO_2 without small pores was used as photocatalyst.[51]

The mechanism for photocatalytic hydrogen production from water and glycerol has not been examined in detail. However, Daskalaki and Kondar-ides[52] confirmed that the reaction proceeded with the formation of intermediates such as methanol and acetic acid, and eventually resulted in the complete conversion of glycerol to hydrogen and carbon dioxide (eqn (8.24)).

$$C_3H_8O_3 + 3\,H_2O \rightarrow 7\,H_2 + 3\,CO_2 \qquad (8.24)$$

Kawai and Sakata[31] also examined photocatalytic hydrogen production from liquid water and carbohydrate $(C_6H_{12}O_6)_n$ such as saccharose ($n = 2$), starch ($n \approx 100$) and cellulose ($n \approx 1000–5000$) over $RuO_2/TiO_2/Pt$ photocatalyst. The obtained gas products were mainly hydrogen and carbon dioxide, with a small amount of methanol and ethanol (<0.2% of the total amount). Carbon monoxide, methane and oxygen were not obtained. The atomic ratio of produced hydrogen to carbon dioxide was $3:1$ to $4:1$ at the initial stage, becoming $2:1$ in the stoichiometric ratio. The excess formation of hydrogen at the beginning of the reaction could be due to the dehydrogenation of carbohydrate with the production of $>$C$=$O, $-$CH$=$O or $-$COOH groups.[53] The terminal hydroxymethyl group of sugars would be sequentially oxidised into carbon dioxide (eqn (8.25)), and finally all carbons would be converted to carbon dioxides:

$$R-CH_2OH \rightarrow R-CHO \rightarrow R-COOH \rightarrow R-H + CO_2 \qquad (8.25)$$

where R is $C_5H_9O_5$.

For photocatalytic hydrogen production from water and organic acid (*e.g.* acetic acid and propionic acid), mechanisms have been proposed which begin with the reaction of water and organic acid to form hydrogen, carbon dioxide and alcohol, followed by steam reforming of the produced alcohol, as representatively shown in eqns (8.26) and (8.27):[54,55]

$$CH_3COOH + H_2O \rightarrow H_2 + CO_2 + CH_3OH \qquad (8.26)$$

$$CH_3OH + H_2O \rightarrow 3H_2 + CO_2 \qquad (8.27)$$

During the reaction of water and organic acid, the photo-Kolbe reaction often proceeds simultaneously to produce alkane and carbon dioxide (eqns (8.6) and (8.28)):

$$2CH_3COOH \rightarrow C_2H_6 + 2CO_2 + H_2 \qquad (8.28)$$

8.3.2.2 Reactivity of Substrates

Photocatalytic hydrogen production rates from water and ethanol/glycerol/ sugars have been compared with those from water and other organic chemicals such as alcohol, aldehyde and aliphatic acid.

When the hydrogen production rates from various neat alcohols were compared over Pt/TiO$_2$, the hydrogen production rate decreased in the following order: methanol $\approx$ ethanol $>$ 1-propanol $\approx$ 2-propanol $>$ 1-butanol.[47] A clear correlation was obtained between the hydrogen production rate and the polarity of the alcohol. Similar results were reported for hydrogen production from aqueous C1–C4 alcohols over Pt/TiO$_2$ photocatalysts.[56–59] Thus, the dissociative adsorption of alcohol over Pt/TiO$_2$ shown in eqn (8.17) appears to mainly govern the reaction rate. Another possibility is that, with increasing carbon chain length and branches, the steric hindrance in the molecule reduces the hydrogen production rate.

The photocatalytic hydrogen production rate from an aqueous solution of methanol was compared with those from the oxidation products, *i.e.* formaldehyde and formic acid, over a SrTiO$_3$ photocatalyst.[60] The hydrogen production rate decreased in the following order: formic acid (33.3 μmol h^{-1}) $>$ formaldehyde (8.1 μmol h^{-1}) $>$ methanol (4.8 μmol h^{-1}). This means that the rates for the consecutive reactions of the methanol oxidation are fast. In other words, the complete oxidation of methanol by water should proceed relatively easily. However, the oxidation products from ethanol showed an opposite reactivity to those from methanol. The hydrogen production rates from aqueous acetaldehyde and acetic acid over a Pt/TiO$_2$ photocatalyst were a third and an eighth as large, respectively, as that from aqueous ethanol.[57] This suggests that the complete oxidation of ethanol to hydrogen and carbon dioxide would be rather difficult and that the complete oxidation of 2-propanol into carbon dioxide would be more difficult. Actually it was reported that almost equivalent moles of hydrogen and acetone were photocatalytically produced from 2-propanol over Pt/TiO$_2$[46,61] and CdS[62] even in the presence of water. These results show that the development of a highly active photocatalyst would be required for the application of alcohols other than methanol.

The hydrogen production rates from water and C$_3$-polyols (glycerol, propyleneglycol and isopropanol) were compared over a Pt/TiO$_2$ photocatalyst.[63] The reaction rate increased with increasing the number of OH groups included in the substrate: glycerol > propyleneglycol > isopropanol. This is because the hydrogen atoms bonding to hydroxylated carbon atoms would be more easily extracted to produce molecular hydrogen.

When the photocatalytic hydrogen production rates from aqueous disaccharides (lactose, cellobiose and maltose) solution were compared with these from aqueous solution of polysaccharides (starch and cellulose) over a Pt/TiO$_2$ photocatalyst, the rates for the disaccharides were much higher than those for the polysaccharides.[40] The reactivity of carbohydrates decreased with increasing the molecular weight, since carbohydrates with a large molecular weight were insoluble in water. It was also reported that the rate of hydrogen evolution from α-D-glucose was higher than that from β-D-glucose although β-D-glucose was dominant (*ca.* 67%) in equilibrium in an aqueous glucose solution at room temperature.[64]

8.3.2.3 *Effect of Reaction Conditions*

The reaction conditions such as the concentration of the substrate, pH of the solution and the reaction temperature are important factors determining the hydrogen production rate.

It is obvious that the concentration of biomass derivatives largely influences the reaction rate. With increasing concentration of substrates, the hydrogen production rate first increased and then decreased in many cases.[46,54,56,65] The increase can be understood by considering the chemical kinetics. However, the addition of excess substrates would disturb the adsorption of water and thus decrease the photocatalytic activity.

The solution pH also influences the hydrogen production rate. Table 8.3 shows the optimum pH for the hydrogen production from water and various organic compounds. An acid condition is suitable for the reaction of organic acid, while a neutral condition is preferable for alcohols and glycerol. The hydrogen production rate from sugars tends to increase in a basic solution. The solution pH can influence various factors such as the surface electric charge of the semiconductor, the chemical state of the substrate, the agglomeration of the photocatalyst particles, the redox potential of water and organic compounds,

Table 8.3 Optimum pH for photocatalytic hydrogen production from liquid water and various organic compounds over Pt/TiO$_2$.

Organic compound	Optimum pH	Ref.
Acetic acid	1	54
Propionic acid	3	55
Methanol	6	56
Ethanol	6	39
Glycerol	8–10	52
Glucose	11	66

and the band bending of the semiconductor, which may influence the hydrogen production rate.[48,56,66–68]

Thermal acceleration of the photocatalytic reaction rate has been reported by many researchers.[34,36,46.47,52,57,65,69–72] Thermal activation energies of these photocatalytic systems varied with the types of reactions, semiconductors and co-catalysts, or the concentration of organic compounds. However they were typically less than 30 kJ mol^{-1}, which was lower than the activation energy for the conventional catalytic reaction. Thus, thermal energy would influence only the mild activation steps such as the desorption of products[52,65] and the migration of the photoexcited carriers.[69,70]

The influence of the electrolyte NaCl was also examined for hydrogen production from aqueous ethanol over Pt/TiO$_2$, since producing hydrogen from natural seawater would be desirable from the viewpoint of practical application.[73] When the concentration of NaCl was lower than 0.1 mol L^{-1}, the hydrogen production rate increased with an increase in NaCl concentration. This is because Na$^+$ adsorbed on TiO$_2$ promoted the adsorption of ethanol. However, further increase in NaCl concentration reduced the reaction rate due to the negative effect of Cl$^-$ ions.

8.3.2.4 Effect of Co-catalyst

Loading metal co-catalyst is the most popular way to increase the activity of the semiconductor photocatalyst. When the hydrogen production rates from aqueous ethanol were compared between four kinds of metals (Pt, Rh, Pd and Ni) as a co-catalyst loaded by a photodeposition method on TiO$_2$ photocatalyst, the activity decreased in the following order: Pt/TiO$_2$ > Rh/TiO$_2$ > Pd/TiO$_2$ > Ni/TiO$_2$ > TiO$_2$.[30] Pt/TiO$_2$ showed the highest quantum yield (38% at 380 nm), which was more than 40 times higher than that for TiO$_2$ alone (0.9%). The Pt co-catalyst was effective in increasing the activity even when the metal was loaded by impregnation[47] and sonochemical methods.[74] The high photocatalytic activity of the Pt-loaded sample originated from the enhancement of the electron–hole separation at the interface between the semiconductor and metal, and/or the catalytic activity of the Pt co-catalyst for ethanol oxidation and hydrogen evolution.

Rates for dehydrogenation of 2-propanol over the metal (Pt, Rh, Pd, Au, Ni, Cu, Ag, Ru and Ir) loaded TiO_2 exponentially increased with increasing work function of the metal co-catalyst.[75] Thus, it is suggested that the metal nanoparticles provide an electric field gradient in the band structure of the semiconductor through the metal–semiconductor junction, causing an efficient electron–hole separation.

Non-noble metals have also been examined as co-catalysts of TiO_2, since these metals are less expensive than noble metals. The hydrogen production rate from aqueous ethanol over a CoO_x/TiO_2 photocatalyst prepared by an impregnation method was much higher than that over the pure TiO_2 (0.13 µmol h^{-1}), although it decreased by a factor of 50% (from 16.1 to 8.0 µmol h^{-1}) during the long reaction time (240 h) due to leaching of the co-catalyst.[76] Some TiO_2 photocatalyst samples with Cu, Ni or Ag co-catalyst prepared by an *in situ* photodeposition method were also examined for the reaction. Although no detectable hydrogen was produced over the bare TiO_2, it was produced with these metal-loaded TiO_2 samples, especially the Cu-loaded sample.[65] The high activity of Cu/TiO_2 is likely to have originated from the adjustable junction or strong interaction between the Cu co-catalyst and the TiO_2 surface, which would promote electron migration from the conduction band of TiO_2 to the metal nanoparticles. Not only these metal nanoparticles but also metal oxide species can also function as the co-catalyst. Loading of RuO_2 increased the photocatalytic activity of TiO_2 for hydrogen production from aqueous ethanol, although the activity of the RuO_2/TiO_2 photocatalyst was lower than those of Pt/TiO_2 and Pd/TiO_2.[77] It was suggested that the RuO_2 would also work as a hydrogen-evolution catalyst.

CuO_x is an excellent co-catalyst for hydrogen production from water and biomass derivatives.[78–82] A TiO_2 photocatalyst loaded with 1.3 wt% of CuO by an impregnation and calcination method (CuO/TiO_2) showed 129 times higher activity than that of the bare TiO_2 for hydrogen production from aqueous glycerol.[79] For a CuO_x/TiO_2 photocatalyst, it was found that a higher concentration of Cu(I) led to an increase in photocatalytic activity but that of Cu(II) produced a detrimental effect.[82]

8.3.2.5   *Particle Size, Morphology and Polymorphs of TiO_2*

The particle size, morphology and polymorphs of a TiO_2 photocatalyst are also important factors determining its activity.

The hydrogen production rate from water and ethanol over a Pt/TiO_2 photocatalyst was found to be largely influenced by the particle sizes of Pt and TiO_2.[83] Loading very small amount of Pt (0.01 monolayer on the TiO_2 surface) and use of TiO_2 with a particle size <0.4 µm increased the activity remarkably.

As for the polymorphs, an anatase TiO_2 showed high photocatalytic activity for hydrogen production from 2-propanol, while a rutile TiO_2 showed negligible activity.[61] The low activity of the rutile is attributed to the band structure. The flat band potential of rutile exists at almost the same level to the

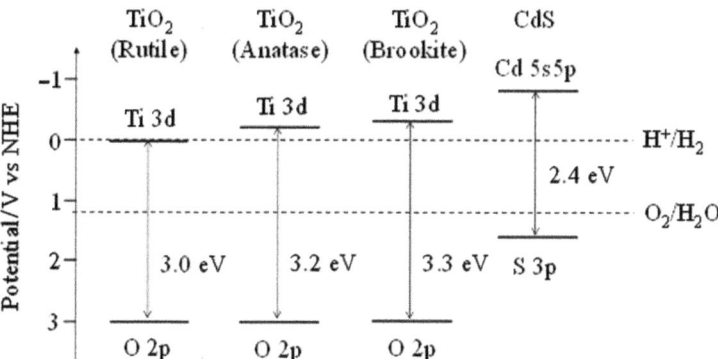

Figure 8.4 Band structure of TiO$_2$ polymorphs and CdS at pH $= 0$.[84-87]

reduction potential of protons, while that of anatase is shifted cathodically by *ca*. 0.2 eV[84] (*cf*. Figure 8.4). Thus, the energy of photoexcited electrons in the conduction band of anatase would be higher than that for rutile. The activity of amorphous TiO$_2$ was also negligible, since it would contain many recombination sites for the photoexcited electrons and holes.[88]

The activity of brookite TiO$_2$ was equal to or higher than those of a commercial TiO$_2$ (P-25, the famous commercial TiO$_2$ sample consisting both anatase and rutile, Degussa) and anatase TiO$_2$.[85,89,90] The flat band potential of the brookite TiO$_2$ was more cathodically shifted by *ca*. 0.1 V than that of the anatase TiO$_2$ (Figure 8.4). Therefore, this is one possible reason for the high activity of the brookite sample. However, some rutile TiO$_2$ photocatalysts showing high photocatalytic activity for hydrogen production have been recently reported. Acid treatment of titanate nanotubes with 0.1 mol dm^{-3} H$_2$SO$_4$ at 298 K formed rutile TiO$_2$ nanoparticles with a good crystallinity, a high specific surface area (*ca*. 250 m^2 g^{-1}) and a small particle size (*ca*. 3 nm).[91] The activity of the prepared rutile TiO$_2$ for hydrogen production from aqueous ethanol was comparable to that of P-25. A rutile-rich TiO$_2$ sample containing a small amount of anatase, which was prepared by an impregnation of P-25 with Na$_2$SO$_4$ followed by calcination at 973 K, showed six-fold higher activity than P-25 for hydrogen production from methanol.[92] High activity is attributed to the improvement of the surface-phase junctions between anatase and rutile, together with the enhancement of the crystallinity of TiO$_2$. TiO$_2$ prepared by a similar thermal treatment showed 3–5 times higher activity than P-25 for the hydrogen production from aqueous glycerol and that from aqueous glucose.[93]

TiO$_2$(B), another polymorph of TiO$_2$, has also been examined for the photocatalytic dehydration of ethanol.[94-96] Some TiO$_2$(B) samples were prepared by a hydrothermal treatment of anatase TiO$_2$ powder in 10 M NaOH, followed by washing with acid and calcination at various temperatures. The hydrothermal treatment at 387 K provided TiO$_2$(B) nanotubes, while that at 403 K produced TiO$_2$(B) nanofibres. With increasing calcination temperature, TiO$_2$(B) transformed into anatase nanoparticles. The TiO$_2$(B) nanotubes calcined at 673 K, consisting of a mixture of $TiO_2(B)$ nanotubes and anatase nanoparticles, showed the highest activity, which was 20% higher than that over the P-25 TiO_2.

Pt/TiO_2 nanosheets with exposed (001) facets were prepared by a hydrothermal route in a $Ti(OC_4H_9)_4$–HF–H_2O mixed solution, followed by photodeposition of Pt nanoparticles.[97] The sample loaded with 2 wt% Pt showed the highest activity (334 μmol h^{-1}), which was much higher than a Pt/P-25 TiO_2 sample (223 μmol h^{-1}). The high photocatalytic activity is likely to have originated from the effect of surface fluorination and exposed (001) facets.

The photocatalytic hydrogen production from aqueous alcohols over mesoporous TiO_2 has been examined by various researchers.[56,57,65,98] For example, mesoporous TiO_2 synthesised by a surfactant-assisted templating sol-gel method from laurylamine hydrochloride, tetraisopropyl orthotitanate and acetyl acetone showed high activity for hydrogen production from aqueous methanol (331 μmol h^{-1}), which was much higher than those over the commercial TiO_2 samples, P-25 (31 μmol h^{-1}) and ST-01 (84 μmol h^{-1}).[98]

8.3.2.6   *Visible Light Driven TiO_2 Photocatalyst*

TiO_2 is an excellent photocatalyst due to its high activity, low cost and excellent resistance to chemical and photochemical corrosion during the photocatalytic reaction. However, it cannot utilise sunlight efficiently, since it can only adsorb ultraviolet (UV) light due to the relatively large band gap (Figure 8.4). Recently, various attempts such as doping, photosensitisation and the utilisation of localised surface plasmon resonance on metal nanoparticles have been carried out to develop TiO_2 photocatalysts driven in visible light (Figure 8.5).

Doping of non-metals (N, S, C, *etc.*)[99–101] or transition metals with partially filled d orbitals (Cr, Ni, Rh *etc.*)[102–104] can lower the bandgap of the semiconductor, since the doped ions can form impurity levels in the forbidden band of the semiconductor (Figure 8.5(a)). TiO_2 photocatalysts doped with sulfur,[105] nitrogen,[106–108] boron,[108] iron[109] or nickel[109] were examined for hydrogen production from water and biomass derivatives. For example, TiO_2 thin film doped with sulfur by atmosphere-controlled pulsed laser deposition was examined for hydrogen production from pure ethanol under the irradiation of both UV and visible (vis) light.[105] UV-vis and X-ray photon spectroscopy (XPS) studies demonstrated that the sulfur atoms substituted for the oxygen atoms in the TiO_2 lattice to induce visible light sensitivity. Although sulphur doping reduced the crystallinity of TiO_2, the photocatalytic activity of the S-doped TiO_2 was more than twice higher than that of the bare TiO_2. Furthermore, nickel deposition by the chemical vapour reductive deposition method drastically increased the activity several times as much as the Ni-free thin film.

Photosensitisation with dyes or quantum dot is another possible method to make a TiO_2 photocatalyst active in visible light. A Pt/TiO_2 photocatalyst modified with a heteropoly blue (HPB) showed activity for hydrogen production from aqueous glycerol under visible light.[110] Upon visible light irradiation, the excited electrons in the HPB migrated to the TiO_2, enabling

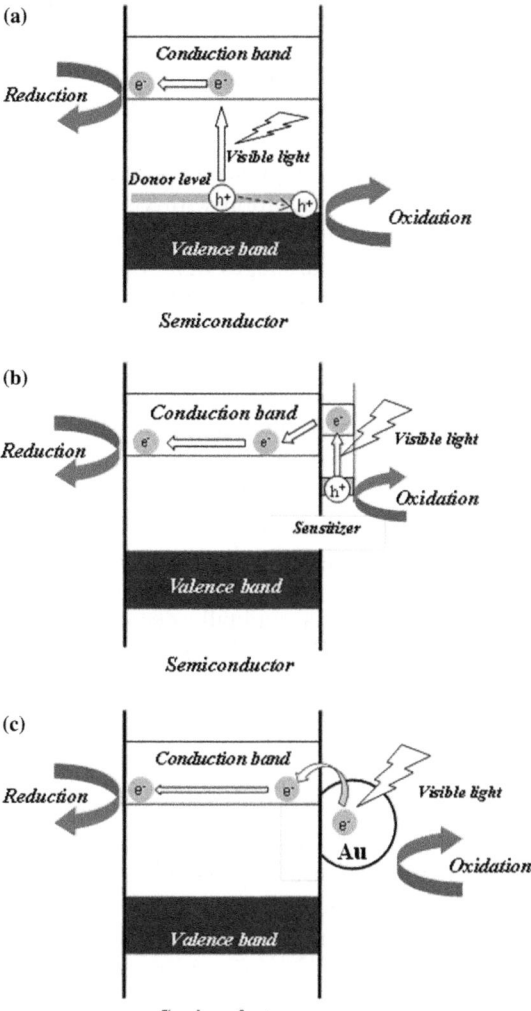

Figure 8.5 Visible light responsive photocatalysts with: (a) metal or non-metal ion doping; (b) photosensitisation with dye or quantum dot; and (c) localised surface plasmon resonance of metal nanoparticles.

hydrogen production (Figure 8.5(b)). CdS-sensitised Pt/TiO$_2$ nanosheets with exposed (001) facets showed a high quantum yield (13.9% at 420 nm) for hydrogen production from aqueous lactic acid under visible light.[111] Furthermore, the CdS/Pt/TiO$_2$ nanosheets showed stable activities for many recycling tests.

Recently, Au/TiO$_2$ plasmonic photocatalysts have been reported to be active for hydrogen production from aqueous alcohol solutions under visible light irradiation.[112–114] Electrons are excited in the Au metal nanoparticles by the localised surface plasmon resonance with the visible light absorption and partly transfer to the conduction band of attached semiconductors such as TiO_2. Then, the separated electrons and holes contribute to the reductive and oxidative reactions, respectively (Figure 8.5(c)). An Au/TiO_2 photocatalyst prepared by a multi-step photodeposition method had a strong adsorption in the visible light region and showed a high and constant activity for hydrogen production from water and various organic compounds (*e.g.* methanol, ethanol, 2-propanol and glycerol) under irradiation with visible light.[113] Factors influencing the activity of the Au/TiO_2 photocatalysts have also been examined.[114] Au nanoparticles with short rod-like shapes were found to be more effective for the reaction than those with a spherical shape because of the higher efficiency for electron transfer from the Au nanoparticles to the conduction band of TiO_2. Aggregates of the Au nanoparticles were not, however, appropriate for the reaction which derived from the low efficiency of the electron transfer. As for TiO_2, large particles of anatase were preferable.

Anpo and colleagues[59] developed an oxygen-deficient TiO_2 film on a Ti metal substrate that functioned upon visible light irradiation (Vis-TiO_2). The Vis-TiO_2 was prepared by a radio frequency magnetron sputtering deposition method and examined for hydrogen production from aqueous alcohols in an H-type glass container. The Vis-TiO_2 has a unique anisotropic structure in that the concentration of O^{2-} ions gradually decreases from the top surface (O/Ti = 2.00) to the inside bulk (O/Ti = 1.93). The separate evolution of pure hydrogen from the Ti metal side deposited with Pt co-catalyst could proceed under visible light irradiation.

8.3.2.7 *Modification of TiO₂ Photocatalyst*

Doping with metal ions such as alkali, alkaline earth and some lanthanide ions hardly influences the band structure of the semiconductor. However, doping with these ions can sometimes increase the activity, since it promotes the migration of photoexcited carriers or has a positive effect on the particle size and morphology of the photocatalyst.[115–117] For example, a Pt/TiO_2 sample doped with 0.5 mol% of Gd showed 1.4 times higher activity for hydrogen production from aqueous methanol than the undoped Pt/TiO_2.[58] Pt/TiO_2 loaded with 0.5 wt% graphene showed 1.7 times higher activity than unmodified Pt/TiO_2 for hydrogen production from aqueous methanol.[118] Due to its two-dimensional p-conjugation structure, graphene served as an acceptor of the photogenerated electrons from the conduction band of TiO_2, resulting in the effective separation of photogenerated electrons and holes.

8.3.3 Hydrogen Production from Water and Alcohols/Acids/ Sugars over Photocatalysts other than TiO₂

Photocatalysts such as sulfide, oxide, oxynitride and nitride, as well as their combination, have been also studied for hydrogen production from water and biomass derivatives.

8.3.3.1 Sulfide Photocalyst

Sulfide is an attractive material for photocatalytic hydrogen production since many sulfide photocatalysts can utilise visible light, and the potentials of the conduction band and valence band edges satisfy the reductive potential of proton to hydrogen and the oxidative potential of water and organic compounds, respectively. Sulfide photocatalysts have been frequently applied for hydrogen production from biomass derivatives as well as from water including S^{2-} and SO_3^{2-} ions[22-24] although photocorrosion has occasionally been a problem.

CdS has been most extensively studied among sulfide photocatalysts. CdS has a small bandgap (*ca.* 2.4 eV) and high chemical potential of its conduction band (-0.9 V *vs.* NHE) as shown in Figure 8.4.[87] Thus, it is one of the promising materials for hydrogen production under visible light irradiation. A Pt/CdS photocatalyst prepared by a photodeposition method in a H_2PtCl_6 solution was examined for the hydrogen production from aqueous solutions of various alcohols.[48] The photocatalytic activity was largely enhanced by thermal pretreatment at high temperature (*ca.* 773 K). Although the deposited Pt species was initially in the PtS state that is less active than the metallic state, the thermal treatment reduced it to form active Pt metal particles. Thermal pretreatment was also effective for the RhO_x/CdS photocatalyst.[87]

A MoS_2/CdS photocatalyst showed higher activity for hydrogen production from water and lactic acid than a Pt/CdS photocatalyst[119,120] due to not only the excellent catalytic activity of MoS_2 for hydrogen evolution, but also the improved junction between MoS_2 and CdS. It was also reported that loading NiS on CdS largely increased the hydrogen production rate from aqueous lactic acid.[121] A NiS/CdS photocatalyst prepared by a hydrothermal method showed a high quantum efficiency of 51.3% under 420 nm light.

The combination of a CdS photocatalyst with other materials (*e.g.* zeolite and mesoporous materials,[122] $KNbO_3$[68,123] and graphene oxide[124]) was effective in increasing photocatalytic activity and suppressing photocorrosion. For example, a composite of CdS and graphene oxide loaded with Pt co-catalyst showed 10 times higher activity for hydrogen production from aqueous ethanol than bare CdS under the irradiation of both UV and visible light.[124] Furthermore, this composite photocatalyst did not deactivate at least for 16 h of reaction time. Effective electron transfer from CdS to graphene oxide and Pt would suppress the recombination of photoexicited electron–hole pairs, leading to high hydrogen evolution efficiency and strong photostability.

As examples of other sulfide photocatalysts, solid-solutions of CdS and other materials ($Cd_xZn_{1-x}S$[67] and CdS_xSe_{1-x}[125]), Zn-based sulfides ($ZnIn_2S_4$[126] and ZnO/ZnS core/shell nanorods[72]) were examined for hydrogen production from water and biomass derivatives.

The chemistry of hydrogen production from water and biomass derivatives over CdS and related photocatalysts is different to that over TiO_2 photocatalysts. The difference in the surface property and the band structure between TiO_2 and CdS (Figure 8.4) may cause the difference in activities for hydrogen production from water and various organic compounds. For example, the hydrogen production rate from aqueous alcohol solutions over a Ni/NiO/ KNbO$_3$/CdS photocatalyst decreased in the following order: 2-propanol > ethanol > methanol.[68] For a Pt/CdS photocatalyst, the order of the photo-catalytic activity was: ethanol > methanol ≈ *n*-propanol ≈ *n*-butanol > *iso*-butanol > *tert*-butanol.[48] For an MoS$_2$/CdS photocatalyst, it was: lactic acid ≫ glycerol > ethylene glycol > ethanol > methanol.[120] Different activities among these three photocatalysts may originate from the kinds of co-catalyst or the structure of CdS. It is noted that, unlike the TiO$_2$, the hydrogen pro-duction rate from aqueous methanol was quite small over some CdS-based photocatalysts.

8.3.3.2 Oxide Photocatalysts

The alkaline earth titanates have been well researched for hydrogen production from aqueous alcohols and sugars.[60,62,71,127,128] The hydrogen production rate from aqueous 2-propanol over alkaline earth titanates prepared by a solid-state reaction method decreased in the following order: SrTiO$_3$ (30 μmol h^{-1}) > anatase-TiO$_2$ (15 μmol h^{-1}) > CaTiO$_3$ (13 μmol h^{-1}) > BaTiO$_3$ (6 μmol h^{-1}).[60] Pt/CaTiO$_3$[127] and Pt/BaTi$_4$O$_9$[128] samples prepared by a polymerised complex method showed much higher activity for hydrogen production from aqueous ethanol than those prepared by a solid-state reaction method. Furthermore, the activity of the CaTiO$_3$ photocatalyst was drastically improved by Zr-doping.

Ta-based oxide photocatalysts (*e.g.* alkali tantalates[129,130] and Ta$_2$O$_5$ nanotube[131]) have also been examined. For hydrogen production from aque-ous glucose solution over La-doped alkali tantalates, the activity decreased in the following order: NaTaO$_3$ ≫ KTaO$_3$ > LiTaO$_3$.[129] NiO-loading increased the hydrogen production rate over NaTaO$_3$, which was more effective than Pt-loading. Ni-loaded La-doped NaTaO$_3$ was effective for hydrogen pro-duction from aqueous methanol.[130] Among three kinds of Ni co-catalysts (Ni metal, NiO and Ni/NiO with a core-shell structure), Ni metal was the most effective co-catalyst for hydrogen production from aqueous methanol, al-though Ni/NiO with a core-shell structure was suitable for water splitting.

As V-based photocatalysts, Pt-loaded Bi$_x$Y$_{1-x}$VO$_4$[132] and nanorods of vanadium dioxide (VO$_2$) with body-centred-cubic structure[133] have been re-ported. Although BiVO$_4$ did not satisfy the reduction potential of protons to hydrogen,[134] formation of the solid solution with YVO$_4$ changed the band structure and enabled hydrogen production from aqueous glucose under visible light.[132] The photocatalyst with a Bi/Y ratio of 1 : 1 showed the highest activity. Unlike TiO$_2$ and sulfide photocatalysts, an acid solution (pH = 3) was suitable for effective hydrogen production over Pt/ Bi$_x$Y$_{1-x}$VO$_4$.

Sn-based photocatalysts (*e.g.* SnO$_2$[135] and Pt-loaded ZnSn(OH)$_6$ nano-cube[136]) were recently examined. SnO$_2$, which was prepared by co-precipitation and calcination at 473 K, showed an activity for hydrogen production from ethanol aqueous solution under visible light.[135] The prepared SnO$_2$ samples contained Sn^{2+} impurities, which formed the donor level above the valence band of SnO$_2$ and enabled the adsorption of visible light. However, the catalysts gradually deactivated because Sn^{2+} was oxidised by photoformed holes during the reaction.

Other base metal catalysts, Fe_2O_3,[137] $LaFeO_3$[138] and F-doped Co_3O_4[139] were reported for the hydrogen production from water and biomass derivatives. Although the conduction band edge of Fe_2O_3 does not satisfy the potential for the hydrogen production, Mcfarland and colleagues developed Ti-doped Fe_2O_3 thin films modified by CoF_3 for hydrogen production from aqueous glucose.[137] The modification of Fe_2O_3 surface with CoF_3 negatively shifted the flat-band potential of Fe_2O_3, which enabled he hydrogen production without an extra bias. The incident photon to current efficiency (IPCE) was 3.7% at 400 nm.

8.3.3.3 Others

Some oxynitrides (TaON, $Y_2Ta_2O_5N_2$ and $LaTaON_2$) were examined for the hydrogen production from aqueous ethanol under visible light irradiation.[140–142] In these cases, Ru-loaded samples showed much higher activity than the Pt-loaded samples; this is attributed to the good contact with Ru and oxynitride, and a remarkable improvement was achieved when both Pt and Ru were present.[141,142] A composite of InN and TiN was also active for this reaction under irradiation of both UV and visible light.[143]

Yoneyama and colleagues modified Si powder (n-type semiconductor, bandgap is 1.1 eV) with polypyrrole and Ag by simultaneous photodeposition, followed by further photodeposition of Pt on the Ag particles,[43] and found that the Si powder coated with polypyrrole and platinised Ag was more stable in an aqueous ethanol solution and more active for hydrogen production than the bare Si sample, where the quantum efficiency was 2.1% at 550 nm.

Highly dispersed photocatalysts[144] were also active for hydrogen production from aqueous alcohol.[145–147] In this type of photocatalyst, the active species such as Ti and W are supported by insulating materials such as silica and one metal cation is coordinated by some oxygen anions on the surface of the support. Photoexcitation takes place at the isolated metal sites, as representatively shown in eqn (8.29):

$$Ti^{4+} - O^{2-} + photon \rightarrow Ti^{3+} - O^- \tag{8.29}$$

Mesoporous W-MCM-48 with a three-dimensional pore system showed detectable hydrogen evolution from aqueous methanol under UV light irradiation, although bulk WO_3 was not active for this reaction.[145] Ti-MCM-48 with the cubic phase exhibited higher activity than Ti-MCM-41 with the hexagonal phases for hydrogen production from methanol aqueous solution.[147] It was confirmed that the tetrahedral-coordinated Ti species possessed much higher photocatalytic efficiency than the octahedral ones.

Photocatalytic evolution of hydrogen from an aqueous ethanol solution of polytungstate was studied in the presence of colloidal metal catalysts.[148] UV-vis and electron spin resonance (EPR) spectra of an irradiated polytungstate solution containing ethanol indicated the subsequent formation of one- and two-electron reduced decatungstates, *i.e.* $[W_{10}O_{32}]^{5-}$ and $[W_{10}O_{32}]^{6-}$, respectively. The addition of colloidal platinum increased the hydrogen production rate. The proposed reaction mechanism involved the reduction of polytungstate by two electrons and its subsequent re-oxidation yielding hydrogen in the presence of colloidal platinum.

8.3.4 Photocatalytic Steam Reforming of Methane

The technology to convert methane into hydrogen is very important because methane is often obtained as the ultimate product from biomass. However, except for combustion, methane conversion usually requires much energy (high temperature) since it is the most stable chemical compound among the hydrocarbons. It has been reported that methane can be photocatalytically activated around room temperature or under mild conditions with or without an oxidant molecule such as oxygen, water or carbon dioxide.[149] The photocatalytic reaction of water and methane over some metal oxides has also been reported by a number of research groups.[150–152] Although hydrogen was produced as the minor by-product, they needed high pressure (1.0 or 10.1 MPa),[150] an electron transfer reagent (methyl viologen),[150] strong UV laser beam[151] or deep UV light (165 or 180 nm).[152]

Recently, we found that the photocatalytic steam reforming of methane (PSRM) shown in eqn (8.30) could proceed over Pt-loaded semiconductor photocatalysts around room temperature at atmospheric pressure:[153]

$$CH_4 + 2\,H_2O(g) \rightarrow 4\,H_2 + CO_2 \qquad (8.30)$$

This is an uphill reaction (Table 8.1, entry 1) so part of the photoenergy can be stored in the chemical potential of hydrogen through this photocatalytic reaction.

When Pt/TiO_2 was photoirradiated in a flow of water vapour and methane around room temperature, hydrogen and carbon dioxide were the main products, and only trace amounts of ethane and carbon monoxide were observed.[154] After an induction period, the molar ratio of hydrogen to carbon dioxide became close to 4. Thus, the main reaction is suggested to be that shown in eqn (8.30). The reaction intermediate, possibly described as $[CH_2O]_{n,ad}$, was formed on the catalyst surface and reacted with water to produce hydrogen and carbon dioxide. The formation of the surface intermediates depended on the activity of the photocatalyst, the methane concentration in the reaction gas, and the incident light intensity. The moderate accumulation of the surface intermediates considerably enhanced the entire reaction rate.

The hydrogen production rate over $Pt/NaTaO_3$:La was two times higher than that over Pt/TiO_2 and the reaction selectively lasted for a long time without deactivation.[155] The activity of $NaTaO_3$ was much influenced by the crystallite size, doping metal ions and the co-catalyst. The highest activity was obtained over the $NaTaO_3$ doped with La^{3+}, where a moderate amount of La^{3+} substituted for Na^+ without distorting the crystal structure of $NaTaO_3$, and would increase the density and the mobility of carriers to improve the activity. A moderate amount of Pt nanoparticles loaded on $NaTaO_3$: La most enhanced the activity among the examined metal particles. A heat treatment after the addition of Pt co-catalyst, which probably improved the metal–semiconductor junction, achieved high and stable photocatalytic activity. The apparent quantum yield over $Pt/NaTaO_3$: La was 30% at 240–270 nm.

On $Pt/CaTiO_3$ photocatalysts, it is noted that even in the presence of methane both PSRM and water decomposition proceeded.[156] With an increase in the loading amount of Pt, the hydrogen production rate in PSRM increased. Furthermore, the hydrogen production rate through water decomposition (with oxygen production) in the flow of water vapour and methane was higher than that in the flow of water. It is suggested that the activated methane species or the reaction intermediates accelerate water decomposition or suppress its reverse reaction. When $CaTiO_3$ samples were prepared by three preparation methods, *i.e.* co-precipitation, homogeneous precipitation and solid-state reaction methods,[157] the obtained samples had different particle sizes, shapes, crystal defects and impurity phases. The highest activity for PSRM was obtained over the sample prepared by the solid-state reaction method from anatase TiO_2. This success originated from the large size of crystallites and few crystal defects, which are important factors for PSRM.

Ga_2O_3 photocatalyst was also active for PSRM.[158] The activity was largely influenced by the crystal structure and surface area of the Ga_2O_3 samples as well as by the co-catalyst. A Pt-loaded β-Ga_2O_3 photocatalyst with a high specific surface area of 10–20 m^2 g^{-1} was effective for PSRM without deactivation for a long time. The activity of the Pt/β-Ga_2O_3 sample was further enhanced by the addition of metal ions either into the bulk or on the surface. The improvement of bulk processes such as carrier migration mainly enhanced the rate of the water activation, while the improvement of surface properties promoted methane activation. Thus, simultaneous modification in the bulk with Mg ions and on the surface with In_2O_3 realised further improvement.

The hydrogen production rate over a Ga_2O_3 photocatalyst increased with an increase in the reaction temperature before reaching *ca.* 343 K after which it became constant.[159] A pseudo Arrhenius plot for the production rate in the lower temperature range showed a straight line and the thermal activation energy was typically less than *ca.* 10 kJ mol^{-1}, which varied with the irradiation light intensity, the crystallite size of Ga_2O_3 and the loading of metal co-catalyst. This study revealed that the lower thermal activation energy was achieved by a higher intensity of the irradiation light, the larger crystallite size of Ga_2O_3 particles and the larger particle size of the co-catalyst. Higher thermal activation energy was achieved with the surface defects and/or co-catalyst alloying. It was suggested that the thermal energy promoted the migration of photoexcited carriers both in the bulk of Ga_2O_3 and at the metal–semiconductor junction from the conduction band of Ga_2O_3 to the metal co-catalyst.

Ga_2O_3 photocatalysts doped with Zn were prepared by a homogeneous precipitation method.[160] In low loading samples such as 0.5 mol%, Zn^{2+} ions substituted for Ga^{3+} ions in the Ga_2O_3 photocatalyst, without changing the crystallite size, the specific surface area, the bandgap or the photoabsorption of the Ga_2O_3 photocatalyst. The substitutionally doped Zn^{2+} ions enhanced the activity in PSRM although they increased the thermally activation energy in photocatalysis. These results suggest that the substitutionally doped Zn^{2+} ions reduced the number of original trap sites in the Ga_2O_3 photocatalyst and provided smooth migration of the photoexcited electrons and holes. In the highly doped samples prepared by the present method, the Zn ions existed as the $ZnGa_2O_4$ species and reduced the photocatalytic activity. Several experiments revealed that the surface $ZnGa_2O_4$ species on Ga_2O_3 increased the activity, although the $ZnGa_2O_4$ species in the bulk reduced the activity.

A Rh-loaded $K_2Ti_6O_{13}$ sample prepared by an oxidative photodeposition method showed two times higher activity than the Pt-loaded one and promoted PSRM selectively without deactivation for many hours.[161,162] X-ray absorption fine structure spectroscopy (XAFS) and transmission electron microscopy (TEM) measurements showed that small Rh metal particles and large Rh oxide ones co-existed on the highly active Rh-loaded photocatalysts. The photocatalytic activity tests for hydrogen evolution and oxygen evolution from each aqueous solution of sacrificial reagent (methanol and silver nitrate, respectively) revealed that Rh metal and oxide co-catalysts cooperatively promoted the reductive and oxidative reactions, respectively. As a result, the Rh co-catalyst increased the entire photocatalytic activity as a bifunctional co-catalyst. The mixture of rhodium metal and oxide co-catalysts also enhanced the activity of a $Na_2Ti_6O_{13}$ photocatalyst.[162]

8.4   Conclusions and Outlook

Photocatalytic hydrogen production from water and biomass derivatives such as ethanol, glycerol, sugars and methane is an attractive reaction which can convert biomass into hydrogen at room temperature using photocatalyst and photoenergy (solar light). Since research on photocatalytic hydrogen production from biomass derivatives began in the early 1980s, various approaches have been taken in order to improve the efficiency of these photocatalytic systems, such as development of various photocatalysts and co-catalysts, modification of photocatalysts, morphology control of photocatalysts, optimisation of reaction conditions, and clarification of the reaction mechanisms. As a result, many efficient photocatalytic systems driven in UV light and several promising systems driven in visible light have been developed as summarised in Table 8.4. Further efforts will be made in order to realise the practical application of these photocatalytic systems, such as the development of more active photocatalysts driven in the visible light, the expansion of these photocatalytic systems to raw biomass (including biomass with much contamination) and improvement of the photoreactor.

Compared with other hydrogen production methods using biomass such as thermal gasification and hydrogen fermentation, photocatalytic hydrogen production from biomass derivatives has two significant merits. First, the reaction proceeds at room temperature under the irradiation of sunlight. This means that the conversion of biomass into hydrogen consumes less energy

Table 8.4 Quantum yield for photocatalytic hydrogen production from water and organic compounds.

Entry	Organic compound	Photocatalyst	Wavelength / nm	Quantum yield / %	Ref.
1	Ethanol	Pt/TiO$_2$ (rutile)	380	38	30
2		Pt/CaTi$_{0.93}$Zr$_{0.07}$O$_3$	365	13	127
3		Pt/BaTi$_4$O$_9$	365	12	128
4		VO$_2$	UV light	39	133
5		Pt(Ag)/Si/ polypyrrole	550	2.1	43
6		Pt/N-dope TiO$_2$	365	3.6	107
			312	12.3	
7	Ethanol	Ru/TaON	$420 < \lambda < 500$	2.1	140
	Methanol			0.8	
8	Ethanol	CuO$_x$/TiO$_2$ (anatase)	365	27	81
	Glycerol			29	
9	Ethanol	Pt/TiO$_2$ (P-25)	365	50	40
	Glycerol			70	
	Glucose			63	
10	Methanol	Pt/SrTiO$_3$	UV light	1.9	71
11		LaFeO$_3$	$\lambda \geq 420$	8.1	138
12	Isopropanol	Ni/NiO/KNbO$_3$/ CdS	$\lambda \geq 400$	8.8	123
13	Lactic acid	Pt/TiO$_2$ (rutile)	360	71	37
		Pt/CdS	440	38	
14		MoS$_2$/CdS	420	7.3	120
15		NiS/CdS	$\lambda \geq 420$	51	121
16		CdS-sensitised Pt/ TiO$_2$	420	13.9	111
17	Glycerol	CuO/TiO$_2$	365	13.4	79
18		ZnO/ZnS	UV light	22	72
19	Propionic acid	Pt/TiO$_2$	365	1.7	55
20	Saccharose	RuO$_2$/TiO$_2$/Pt	380	1.5	31
21	Methane	Pt/NaTaO$_3$:La	240–270	30	155
22		Pt/Ga$_2$O$_3$	254	7.9	158

especially compared with the thermal chemical process. A special reaction setup is also not required, since the reaction proceeds under mild and ambient conditions. Secondly, any kind of biomass can be converted completely into hydrogen and carbon dioxide.[28–40] This is a big advantage compared with the biological process.

Therefore, photocatalytic hydrogen production from biomass is very advantageous as one of the methods for producing environmentally benign hydrogen. We expect this photocatalytic system will become one of the important hydrogen production methods in the near future.

References

1. S. M. Raj, S. Talluri and L. P. Christopher, *Bioenergy Res.*, 2012, **5**, 515.
2. K. Shimura and H. Yoshida, *Energy Environ. Sci.*, 2011, **4**, 2467.

3. M. Cargnello, A. Gasparotto, V. Gombac, T. Montini, D. Barreca and P. Fornasiero, *Eur. J. Inorg. Chem.*, 2011, 4309.
4. G. van Rossum, S. R. Kersten and W. P. van Swaaij, *Ind. Eng. Chem. Res.*, 2007, **46**, 3959.
5. D. Sutton, B. Kelleher and J. R. Ross, *Fuel Process. Tech.*, 2001, **73**, 155.
6. L. Devi, M. Craje, P. Thune, K. J. Ptasinski and F. J. Janssen, *Appl. Catal. A*, 2005, **294**, 68.
7. M. Inaba, K. Murata, M. Saito and I. Takahara, *Energy Fuels*, 2006, **20**, 432.
8. M. Asadullah, T. Miyazawa, S. Ito, K. Kunimori and K. Tomishige, *Appl. Catal. A*, 2003, **246**, 103.
9. M. M. Yung, W. S. Jablonski and K. A. Magrini-Bair, *Energy Fuels*, 2009, **23**, 1874.
10. D. A. Bulushev and J. R. Ross, *Catal. Today*, 2011, **171**, 1.
11. I. G. Lee, M. S. Kim and S. K. Ihm, *Ind. Eng. Chem. Res.*, 2002, **41**, 1182.
12. D. C. Elliott, T. R. Hart and G. G. Neuenschwander, *Ind. Eng. Chem. Res.*, 2006, **45**, 3776.
13. D. C. Elliott, L. J. Sealock, Jr. and E. G. Baker, *Ind. Eng. Chem. Res.*, 1993, **32**, 1542.
14. Y. Guo, S. Z. Wang, D. H. Xu, Y. M. Gong, H. H. Ma and X. Y. Tang, *Renew. Sustain. Energy Rev.*, 2010, **14**, 334.
15. P. Azadi and R. Farnood, *Int. J. Hydrogen Energy*, 2011, **36**, 9529.
16. A. Yamaguchi, N. Hiyoshi, O. Sato, K. K. Bando, M. Osada and M. Shirai, *Catal. Today*, 2009, **146**, 192.
17. T. Keskin, M. Abo-Hashesh and P. C. Hallenbeck, *Bioresour. Technol.*, 2011, **102**, 8557.
18. A. J. Guwy, R. M. Dinsdale, J. R. Kim, J. Massanet-Nicolau and G. Premier, *Bioresour. Technol.*, 2011, **102**, 8534.
19. M. A. Pena, J. P. Gomez and J. L. Fierro, *Appl. Catal. A*, 1996, **144**, 7.
20. J. N. Armor, *Appl. Catal. A*, 1999, **176**, 159.
21. A. Fujishima and K. Honda, *Nature*, 1972, **238**, 37.
22. A. Kudo and Y. Miseki, *Chem. Soc. Rev.*, 2009, **38**, 253.
23. X. Chen, S. Shen, L. Guo and S. S. Mao, *Chem. Rev.*, 2010, **110**, 6503.
24. K. Maeda, *J. Photochem. Photobiol. C*, 2011, **12**, 237.
25. K. Maeda, K. Teramura, D. Lu, T. Takata, N. Saito, Y. Inoue and K. Domen, *Nature*, 2006, **440**, 295.
26. Y. Sasaki, H. Nemoto, K. Saito and A. Kudo, *J. Phys. Chem. C*, 2009, **113**, 17536.
27. K. Maeda, M. Higashi, D. Lu, R. Abe and K. Domen, *J. Am. Chem. Soc.*, 2010, **132**, 5858.
28. K. Kawai and T. Sakata, *Nature*, 1979, **282**, 283.
29. T. Kawai and T. Sakata, *J. Chem. Soc. Chem. Commun.*, 1980, **15**, 694.
30. T. Sakata and T. Kawai, *Chem. Phys Lett.*, 1981, **80**, 341.
31. T. Kawai and T. Sakata, *Nature*, 1980, **286**, 474.
32. T. Kawai and T. Sakata, *Chem. Lett.*, 1981, 81.
33. K. Hashimoto, T. Kawai and T. Sakata, *J. Phys. Chem.*, 1984, **88**, 4083.

34. S. Sato and J. M. White, *J. Am. Chem. Soc.*, 1980, **102**, 7206.
35. S. Sato and J. M. White, *Chem. Phys. Lett.*, 1980, **70**, 131.
36. S. Sato and J. M. White, *J. Phys. Chem.*, 1981, **85**, 336.
37. H. Harada, T. Sakata and T. Ueda, *J. Am. Chem. Soc.*, 1985, **107**, 1773.
38. A. Patsoura, D. I. Kondarides and X. E. Verykios, *Appl. Catal. B*, 2006, **64**, 171.
39. A. Patsoura, D. I. Kondarides and X. E. Verykios, *Catal. Today*, 2007, **124**, 94.
40. D. I. Kondarides, V. M. Daskalaki, A. Patsoura and X. E. Verykios, *Catal. Lett.*, 2008, **122**, 26.
41. Chemical Society of Japan, *Kagakubinran Kisohen*, Maruzen, Tokyo, 3rd edn, 1984.
42. V. P. Indrakanti, J. D. Kubicki and H. H. Schobert, *Energy Environ. Sci.*, 2009, **2**, 745.
43. Y. Taniguchi, H. Yoneyama and H. Tamura, *Chem. Lett.*, 1983, 269.
44. D. V. Esposito, R. V. Forest, Y. Chang, N. Gaillard, B. E. McCandless, S. Hou, K. H. Lee, R. W. Birkmire and J. G. Chen, *Energy Environ. Sci.*, 2012, **5**, 9091.
45. G. R. Bamwenda, S. Tsubota, T. Nakamura and M. Haruta, *J. Photochem. Photobiol. A*, 1995, **89**, 177.
46. I. Ait-Ichou, M. Formenti, B. Pommier and S. J. Teichener, *J. Catal.*, 1985, **91**, 293.
47. Y. Z. Yang, C. H. Chang and H. Idriss, *Appl. Catal. B*, 2006, **67**, 217.
48. Z. Jin, Q. Lin, X. Zheng, C. Xi, C. Wang, H. Zhang, L. Feng, H. Wang, Z. Chen and Z. Jiang, *J. Photochem. Photobiol. A*, 1993, **71**, 85.
49. B. Ohtani, M. Kakimoto, S. Nishimoto and T. Kagiya, *J. Photochem. Photobiol. A*, 1993, **70**, 265.
50. C. H. Lin, C. H. Lee, J. H. Chao, C. Y. Kuo, Y. C. Cheng, W. N. Huang, H. W. Chang, Y. M. Huang and M. K. Shih, *Catal. Lett.*, 2004, **98**, 61.
51. H. Lu, J. Zhao, L. Li, L. Gong, J. Zheng, L. Zhang, Z. Wang, J. Zhang and Z. Zhu, *Energy Environ. Sci.*, 2011, **4**, 3384.
52. V. M. Daskalaki and D. I. Kondarides, *Catal. Today*, 2009, **144**, 75.
53. M. R. St. John, A. J. Furgala and A. F. Sammells, *J. Phys. Chem.*, 1983, **87**, 801.
54. X. Zheng, L. Wei, Z. Zhang, Q. Jiang, Y. Wei, B. Xie and M. Wei, *Int. J. Hydrogen Energy*, 2009, **34**, 9033.
55. L. Wei, X. Zheng, Z. Zhang, Y. Wei, B. Xie, M. Wei and X. Sun, *Int. J. Energy Res.*, 2012, **36**, 75.
56. T. Sreethawong, T. Puangpetch, S. Chavadej and S. Yoshikawa, *J. Powder Sources*, 2007, **165**, 861.
57. W. Sun, S. Zhang, Z. Liu, C. Wang and Z. Mao, *Int. J. Hydrogen Energy*, 2008, **33**, 1112.
58. M. Zalas and M. Laniecki, *Solar Energy Mater. Solar Cells*, 2005, **89**, 287.
59. S. Fukumoto, M. Kitano, M. Takeuchi, M. Matsuoka and M. Anpo, *Catal. Lett.*, 2009, **127**, 39.

60. B. Zielinska, E. Borowiak-Palen and R. J. Kalenczuk, *Int. J. Hydrogen Energy*, 2008, **33**, 1797.
61. S. Nishimoto, B. Ohtani, H. Kajiwara and T. Kagiya, *J. Chem. Soc. Faraday Trans.*, 1985, **81**, 61.
62. K. Domen, S. Naito, T. Onishi and K. Tamaru, *Chem. Lett.*, 1982, 555.
63. X. Fu, X. Wang, D. Y. Leung, Q. Gu, S. Chen and H. Huang, *Appl. Catal. B*, 2011, **106**, 681.
64. M. Zhou, Y. Li, S. Peng, G. Lu and S. Li, *Catal. Commun.*, 2012, **18**, 21.
65. A. V. Korzhak, N. I. Ermokhina, A. L. Stroyuk, V. K. Bukhtiyarov, A. E. Raevskaya, V. I. Litvin, S. Y. Kuchmiy, V. G. Ilyin and P. A. Manorik, *J. Photochem. Photobiol. A*, 2008, **198**, 126.
66. X. Fu, J. Long, X. Wang, D. Y. C. Leung, Z. Ding, L. Wu, Z. Zhang, Z. Li and X. Fu, *Int. J. Hydrogen Energy*, 2008, **33**, 6484.
67. S. Peng, Y. Peng, Y. Li, G. Lu and S. Li, *Res. Chem. Intermed.*, 2009, **35**, 739.
68. J. Choi, S. Y. Ryu, W. Balcerski, T. K. Lee and M. R. Hoffmann, *J. Mater. Chem.*, 2008, **18**, 2371.
69. F. H. Hussein and R. Rudham, *J. Chem. Soc. Faraday Trans.*, 1984, **80**, 2817.
70. F. H. Hussein and R. Rudham, *J. Chem. Soc. Faraday Trans.*, 1987, **83**, 1631.
71. T. Puangpetch, T. Sreethawong, S. Yoshikawa and S. Chavadej, *J. Mol. Catal. A*, 2009, **312**, 97.
72. H. X. Sang, X. T. Wang, C. C. Fan and F. Wang, *Int. J. Hydrogen Energy*, 2012, **37**, 1348.
73. Y. Li, F. He, S. Peng, D. Gao, G. Lu and S. Li, *J. Mol. Catal. A*, 2011, **341**, 71.
74. Y. Mizukoshi, Y. Makise, T. Shuto, J. Hu, A. Tominaga, S. Shironita and S. Tanabe, *Ultrason. Sonochem.*, 2007, **14**, 387.
75. Y. Nosaka, K. Norimatsu and H. Miyama, *Chem. Phys. Lett.*, 1984, **106**, 128.
76. Y. Wu, G. Lu and S. Li, *J. Photochem. Photobiol. A*, 2006, **181**, 263.
77. T. Sakata, K. Hashimoto and T. Kawai, *J. Phys. Chem.*, 1984, **88**, 5214.
78. N. L. Wu and M. S. Lee, *Int. J. Hydrogen Energy*, 2004, **29**, 1601.
79. J. Yu, Y. Hai and M. Jaroniec, *J. Colloid Interface Sci.*, 2011, **357**, 223.
80. K. Lalitha, G. Sadanandam, V. D. Kumari, M. Subrahmanyam, B. Sreedhar and N. Y. Hebalkar, *J. Phys. Chem. C*, 2010, **114**, 22181.
81. V. Gombac, L. Sordelli, T. Montini, J. J. Delgado, A. Adamski, G. Adami, M. Cargnello, S. Bernal and P. Fornasiero, *J. Phys. Chem. A*, 2010, **114**, 3916.
82. Y. Wu, G. Lu and S. Li, *Catal. Lett.*, 2009, **133**, 97.
83. T. Sakata, T. Kawai and K. Hashimoto, *Chem. Phys. Lett.*, 1982, **88**, 50.
84. L. Kavan, M. Gratzel, S. E. Gilbert, C. Klemenz and H. J. Scheel, *J. Am. Chem. Soc.*, 1996, **118**, 6716.
85. T. A. Kandiel, A. Feldhoff, L. Robben, R. Dillert and D. W. Bahnemann, *Chem. Mater.*, 2010, **22**, 2050.
86. A. D. Paola, M. Bellardita, R. Ceccato, L. Palmisano and F. Parrino, *J. Phys. Chem. C*, 2009, **113**, 15166.

87. G. Lu and S. Li, *J. Photochem. Photobiol. A*, 1996, **97**, 65.
88. B. Ohtani, Y. Ogawa and S. Nishimoto, *J. Phys. Chem. B*, 1997, **101**, 3746.
89. H. Kominami, Y. Ishii, M. Kohno, S. Konishi, Y. Kera and B. Ohtani, *Catal. Lett.*, 2003, **91**, 41.
90. G. L. Chiarello, A. D. Paola, L. Palmisano and E. Selli, *Photochem. Photobiol. Sci.*, 2011, **10**, 355.
91. D. V. Bavykin, A. N. Kulak, V. V. Shvalagin, N. S. Andryushina and O. L. Stroyuk, *J. Photochem. Photobiol. A*, 2011, **218**, 231.
92. Y. Ma, Q. Xu, X. Zong, D. Wang, G. Wu, X. Wang and C. Li, *Energy Environ. Sci.*, 2012, **5**, 6345.
93. Q. Xu, Y. Ma, J. Zhang, X. Wang, Z. Feng and C. Li, *J. Catal.*, 2011, **278**, 329.
94. H. L. Kuo, C. Y. Kuo, C. H. Liu, J. H. Chao and C. H. Lin, *Catal. Lett.*, 2007, **113**, 7.
95. C. H. Lin, J. H. Chao, C. H. Liu, J. C. Chang and F. C. Wang, *Langmuir*, 2008, **24**, 9907.
96. F. C. Wang, C. H. Liu, C. W. Liu, J. H. Chao and C. H. Lin, *J. Phys. Chem. C*, 2009, **113**, 13832.
97. J. Yu, L. Qi and M. Jaroniec, *J. Phys. Chem. C*, 2010, **114**, 13118.
98. T. Sreethwong, Y. Suzuki and S. Yoshikawa, *J. Solid State Chem.*, 2005, **178**, 329.
99. R. Asahi, T. Morikawa, T. Ohwaki, K. Aoki and Y. Taga, *Science*, 2001, **293**, 269.
100. T. Ohno, T. Mitsui and M. Matsumura, *Chem. Lett.*, 2003, **32**, 364.
101. H. Irie, Y. Watanabe and K. Hashimoto, *Chem. Lett.*, 2003, **32**, 772.
102. D. W. Hwang, H. G. Kim, J. S. Lee, J. Kim, W. Li and S. H. Oh, *J. Phys. Chem. B*, 2005, **109**, 2093.
103. A. Kudo and M. Sekizawa, *Chem. Commun.*, 2000, 1371.
104. R. Konta, T. Ishii, H. Kato and A. Kudo, *J. Phys. Chem. B*, 2004, **108**, 8992.
105. M. Yoshinaga, K. Yamamoto, N. Sato, K. Aoki, T. Morikawa and A. Muramatsu, *Appl. Catal. B*, 2009, **87**, 239.
106. G. Halasi, I. Ugrai and F. Solymosi, *J. Catal.*, 2011, **281**, 309.
107. M. Wu, J. Hiltunen, A. Sapi, A. Avila, W. Larsson, H. Liao, M. Huuhtanen, G. Toth, A. Shchukarev, N. Laufer, A. Kukovecz, Z. Konya, J. Mikkola, R. Keiski, W. Su, Y. Chen, H. Jantunen, P. M. Ajayan, R. Vajtai and K. Kordas, *ACS Nano*, 2011, **5**, 5025.
108. N. Luo, Z. Jiang, H. Shi, F. Cao, T. Xiao and P. P. Edwards, *Int. J. Hydrogen Energy*, 2009, **34**, 125.
109. T. Sun, J. Fan, E. Liu, L. Liu, Y. Wang, H. Dai, Y. Yang, W. Hou, X. Hu and Z. Jiang, *Powder Technol.*, 2012, **228**, 210.
110. N. Fu and G. Lu, *Catal. Lett.*, 2009, **127**, 319.
111. L. Qi, J. Yu and M. Jaroniec, *Phys. Chem. Chem. Phys.*, 2011, **13**, 8915.
112. C. G. Silva, R. Juarez, T. Marino, R. Molinari and H. Garcia, *J. Am. Chem. Soc.*, 2011, **133**, 595.

113. A. Tanaka, S. Sakaguchi, K. Hashimoto and H. Kominami, *Catal. Sci. Technol.*, 2012, **2**, 907.
114. H. Yuzawa, T. Yoshida and H. Yoshida, *Appl. Catal. B*, 2012, **115–116**, 294.
115. T. Ishihara, H. Nishiguchi, K. Fukamachi and Y. Takita, *J. Phys. Chem. B*, 1999, **103**, 1.
116. H. Kato, K. Asakura and A. Kudo, *J. Am. Chem. Soc.*, 2003, **125**, 3082.
117. T. Takata and K. Domen, *J. Phys. Chem. C*, 2009, **113**, 19386.
118. P. Cheng, Z. Yang, H. Wang, W. Cheng, M. Chen, W. Shangguan and G. Ding, *Int. J. Hydrogen Energy*, 2012, **37**, 2224.
119. X. Zong, H. Yan, G. Wu, G. Ma, F. Wen, L. Wang and C. Li, *J. Am. Chem. Soc.*, 2008, **130**, 7176.
120. X. Zong, G. Wu, H. Yan, G. Ma, J. Shi, F. Wen, L. Wang and C. Li, *J. Phys. Chem. C*, 2010, **114**, 1963.
121. W. Zhang, Y. Wang, Z. Wang, Z. Zhong and R. Xu, *Chem. Commun.*, 2010, **46**, 7631.
122. S. Y. Ryu, W. Balcerski, T. K. Lee and M. R. Hoffmann, *J. Phys. Chem. C*, 2007, **111**, 18195.
123. S. Y. Ryu, J. Choi, W. Balcerski, T. K. Lee and M. R. Hoffmann, *Ind. Eng. Chem. Res.*, 2007, **46**, 7476.
124. P. Gao, J. Liu, S. Lee, T. Zhang and D. D. Sun, *J. Mater. Chem.*, 2012, **22**, 2292.
125. S. Kambe, M. Fujii, T. Kawai, S. Kawai and F. Nakahara, *Chem. Phys. Lett.*, 1984, **109**, 105.
126. Y. Li, J. Wang, S. Peng, G. Lu and S. Li, *Int. J. Hydrogen Energy*, 2010, **35**, 7116.
127. W. Sun, S. Zhang, C. Wang, Z. Liu and Z. Mao, *Catal. Lett.*, 2007, **119**, 148.
128. W. Sun, S. Zhang, C. Wang, Z. Liu and Z. Mao, *Catal. Lett.*, 2008, **123**, 282.
129. X. Fu, X. Wang, D. Y. Leung, W. Xue, Z. Ding, H. Huang and X. Fu, *Catal. Commun.*, 2010, **12**, 184.
130. H. Husin, W. Su, H. Chen, C. Pan, S. Chang, J. Rick, W. Chuang, H. Sheu and B. Hwang, *Green Chem.*, 2011, **13**, 1745.
131. R. V. Goncalves, P. Migowski, H. Wender, D. Eberhardt, D. E. Weibel, F. C. Sonaglio, M. J. Zapata, J. Dupont, A. F. Feil and S. R. Teixeira, *J. Phys. Chem. C*, 2012, **116**, 14022.
132. D. Jing, M. Liu, J. Shi, W. Tang and L. Guo, *Catal. Commun.*, 2010, **12**, 264.
133. Y. Wang, Z. Zhang, Y. Zhu, Z. Li, R. Vajtai, L. Ci and P. M. Ajayan, *ACS Nano*, 2008, **2**, 1492.
134. A. Kudo, K. Omori and H. Kato, *J. Am. Chem. Soc.*, 1999, **121**, 11459.
135. J. Long, W. Xue, X. Xie, Q. Gu, Y. Zhou, Y. Chi, W. Chen, Z. Ding and X. Wang, *Catal. Commun.*, 2011, **16**, 215.
136. X. Fu, D. Y. Leung, X. Wang, W. Xue and X. Fu, *Int. J. Hydrogen Energy*, 2011, **36**, 1524.

137. Y. S. Hu, A. Kleiman-Shwarsctein, G. D. Stucky and E. W. Mcfarland, *Chem. Commun.*, 2009, 2652.
138. K. M. Parida, K. H. Reddy, S. Martha, D. P. Das and N. Biswal, *Int. J. Hydrogen Energy*, 2010, **35**, 12161.
139. A. Gasparotto, D. Barreca, D. Bekermann, A. Devi, R. A. Fischer, P. Fornasiero, V. Gombac, O. I. Lebedev, C. Maccato, T. Montini, G. V. Tendeloo and E. Tondello, *J. Am. Chem. Soc.*, 2011, **133**, 19362.
140. M. Hara, J. Nunoshige, T. Takata, J. N. Kondo and K. Domen, *Chem. Commun.*, 2003, 3000.
141. M. Liu, W. You, Z. Lei, G. Zhou, J. Yang, G. Wu, G. Ma, G. Luan, T. Takata, M. Hara, K. Domen and C. Li, *Chem. Commun.*, 2004, 2192.
142. M. Liu, W. You, Z. Lei, T. Takata, K. Domen and C. Li, *Chin. J. Catal.*, 2006, **27**, 556.
143. Y. Kuo and K. J. Klabunde, *Appl. Catal. B*, 2011, **104**, 245.
144. H. Yoshida, *Curr. Opin. Solid State Mater. Sci.*, 2003, **7**, 435.
145. D. Zhao, A. Rodriguez, N. M. Dimitrijevic, T. Rajh and R. T. Koodali, *J. Phys. Chem. C*, 2010, **114**, 15728.
146. G. Liu, X. Wang, X. Wang, H. Han and C. Li, *J. Catal.*, 2012, **293**, 61.
147. R. Peng, D. Zhao, N. M. Dimitrijevic, T. Rajh and R. T. Koodali, *J. Phys. Chem. C*, 2012, **116**, 1605.
148. M. I. Rustamov, N. Z. Muradov, A. D. Guseinova and Y. V. Bazhutin, *Int. J. Hydrogen Energy*, 1988, **13**, 533.
149. L. Yuliati and H. Yoshida, *Chem. Soc. Rev.*, 2008, **37**, 1592.
150. C. E. Taylor, *Catal. Today*, 2003, **84**, 9.
151. M. A. Gondal, A. Hameed, Z. H. Yamani and A. Arfaj, *Chem. Phys. Lett.*, 2004, **392**, 372.
152. F. Sastre, V. Fornes, A. Corma and H. García, *J. Am. Chem. Soc.*, 2011, **133**, 17257.
153. H. Yoshida, S. Kato, K. Hirao, J. Nishimoto and T. Hattori, *Chem. Lett.*, 2007, **36**, 430.
154. H. Yoshida, K. Hirao, J. Nishimoto, K. Shimura, S. Kato, H. Itoh and T. Hattori, *J. Phys. Chem. C*, 2008, **112**, 5542.
155. K. Shimura, S. Kato, T. Yoshida, H. Itoh, T. Hattori and H. Yoshida, *J. Phys. Chem. C*, 2010, **114**, 3493.
156. K. Shimura and H. Yoshida, *Energy Environ. Sci.*, 2010, **3**, 615.
157. K. Shimura, H. Miyanaga and H. Yoshida, *Stud. Surf. Sci. Catal.*, 2010, **175**, 85.
158. K. Shimura, T. Yoshida and H. Yoshida, *J. Phys. Chem. C*, 2010, **114**, 11466.
159. K. Shimura, K. Maeda and H. Yoshida, *J. Phys. Chem. C*, 2011, **115**, 9041.
160. K. Shimura and H. Yoshida, *Phys. Chem. Chem. Phys.*, 2012, **14**, 2678.
161. K. Shimura, H. Kawai, T. Yoshida and H. Yoshida, *Chem. Commun.*, 2011, **47**, 8958.
162. K. Shimura, H. Kawai, T. Yoshida and H. Yoshida, *ACS Catal.*, 2012, **2**, 2126.

CHAPTER 9

Nanoporous Carbons for High Energy Density Supercapacitors

PIERRE-LOUIS TABERNA[a] AND SARRA GASPARD*[b]

[a] Université Paul Sabatier, CIRIMAT UMR CNRS 5085, 118 route de Narbonne, 31062 Toulouse, France; [b] Laboratory COVACHIM-M2E EA 3592, Université des Antilles et de la Guyane, BP 250, 97157 Pointe-à-Pitre Cedex, France
*Email: sarra.gaspard@univ-ag.fr

9.1   Introduction

One of the promising solutions to the greenhouse effect is to develop sustainable energy. Significant progress has therefore been made in developing improved electrochemical storage systems. Carbon materials play a significant role in electrochemical energy storage. In particular, carbon electrodes are used in lithium-ion batteries and also, more widely, in supercapacitor electrodes (needless to say that most of the batteries use at least carbon as an conductive additive).

Supercapacitors, and more especially, electrochemical double layer capacitors (EDLCs), have attracted increasing attention over the past 10 years. So far, most of the commercial products found on the market are carbon-based devices. This growing interest is linked to the fact that this system can deliver higher energy than conventional dielectric capacitors and more power than batteries; another important feature is the high cyclability (more than several million of cycles). In an overwhelming context of energy saving, energy harvesting issues can be perfectly addressed by EDLCs. Thus, due to their power ability and cyclability, they are recommended for automotives,

RSC Green Chemistry No. 25
Biomass for Sustainable Applications: Pollution Remediation and Energy
Edited by Sarra Gaspard and Mohamed Chaker Ncibi
© The Royal Society of Chemistry 2014
Published by the Royal Society of Chemistry, www.rsc.org

tramways, buses,[1] wind turbines and electricity load levelling; that is to say in any fields where an energy boost and energy recovery is needed. For instance, energy brake recovery is one of the applications of EDLCs. Nevertheless, to fulfil industrial demand, EDLCs lack energy density; most of the applications need high power for several tens of seconds. This is why, these last few years, most of the development has focused on developing new materials and concepts. Recently, important breakthroughs were achieved using nanoporous carbons (highly microporous, *i.e.* having a pore diameter below 2 nm and even less). For this purpose different precursors have been extensively studied such as seaweeds, lignocellulosic material and other carbide-derived carbons.

9.2 Basics Concerning EDLCs

Two main kinds of supercapacitors are studied depending of the active material of their electrodes: pseudo-capacitors, using oxide based materials;[2–6] and EDLCs, using activated carbons. Since almost all the commercial products are symmetric carbon-based supercapacitors (EDLCs), only this kind of system is addressed in this chapter.

An EDLC is made of two carbon-based electrodes [usually composed of a current collector—aluminium or stainless steel depending on the electrolyte— and an activated carbon film—an activated carbon mixed with an organic binder such as polytetrafluoroethylene (PTFE), polyvinylidene fluoride (PVDF) or carboxymethylcellulose (CMC), and a conductive additive such as carbon black][7] soaked in an electrolyte (aqueous or organic). As it is depicted in Figure 9.1, the energy storage of such system is purely capacitive. This means that under a

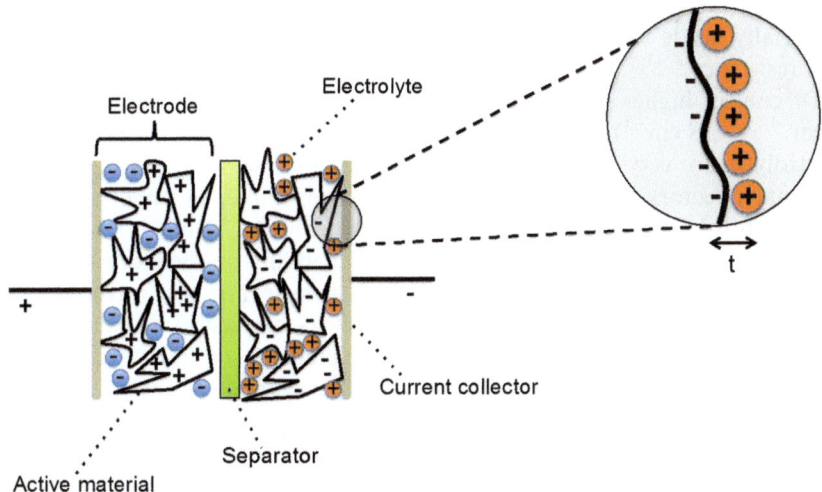

Figure 9.1 Schematic view of a supercapacitor. Two electrodes made from an activated carbon paste are linked, usually with an organic polymer laminated onto a current collector, are insulated from each other by a separator. The assembly is soaked in an electrolyte.

certain voltage polarisation there is a charge accumulation at each electrode (see zoom-in picture of the interface in Figure 9.1); charges come from the ions of the electrolyte, which migrate close to the electrode surface. A first-order description of this interface, which is known as the electrochemical double layer, was described by Helmholtz in the 19th century. Roughly, it can be said that cations approach the negative electrode and anions the positive one.

From this description, the capacitance, C, of an EDLC is the combination of two series capacitances C_+ and C_- respectively for the positive electrode and the negative electrode. Equation (9.1) gives the capacitance an EDLC cell:

$$C = \frac{C_+C_-}{C_+ + C_-} \tag{9.1}$$

Two other important characteristics concerning supercapacitors are the working voltage and the series resistance because they impact directly on the energy density and power density of such a system. Equations (9.2) and (9.3) define what are called the maximum specific energy and the maximum specific power, respectively:

$$E_{max} = \frac{\frac{1}{2}CV_0^2}{weight} \tag{9.2}$$

$$P_{max} = \frac{V_0^2}{4r \cdot weight} \tag{9.3}$$

where V_0 is the maximum working voltage and r is the series resistance. Since higher voltages are required to obtain high performance, most of today's commercial devices use an organic electrolyte. Many parameters impact the series resistance, but the most important are the electrolyte resistance (especially inside the separator) and the current collector-active material contact resistance.[8–11]

Of course, higher conductivities are obtained in aqueous electrolytes (few S cm^{-1} *vs.* mS cm^{-1}), leading to lower resistance. However, highly conductive electrolytes are very corrosive, thus prescribing the use of expensive and heavy current collectors such as stainless steel, for example; in an organic-based electrolyte, aluminium is usable as the current collector. Table 9.1 summarises the main characteristics of supercapacitors depending on the electrolyte used.

Since NEC's first supercapacitor for power electronics was unveiled in the early 1970s, the performance of supercapacitors has increased considerably. Table 9.2 provides an overview of the characteristics of today's commercial products. From this table, it is seen that the average specific energy and power are about 5 Wh kg^{-1} and 10 kW kg^{-1}, respectively.

For this reason, supercapacitors, in terms of performance, are set between batteries and dielectric capacitors: they can sustain high power pulses from a few microseconds (ms) up to a few seconds. To date, 95% of commercial products are built using activated carbon in an organic electrolyte. Since all the energy is stored at the electrode–electrolyte interface, manufacturers have developed high surface area activated carbons (mostly from the physical activation of coconut

Table 9.1 Comparison of properties of supercapacitors in aqueous and organic electrolyte.[12]

Property	Organic electrolyte	Aqueous electrolyte
Operating voltage	2.3–2.7 V	0.6–0.7 V
Conductivity	≈ 0.02–0.06 mS cm^{-1}	≈ 1 S cm^{-1}
Maximum capacitance	100–150 F g^{-1}	250–300 F g^{-1}
Technical, economic characteristics and security	Construction in inert atmosphere, unfriendly for environment, cheap current collectors (aluminium, *etc.*)	Easy construction, environment-friendly, expensive current collectors (stainless steel, *etc.*)

shell) (from 1000 to 2000 m^2 g^{-1}) in order to increase the amount of electrostatic charge stored as much as possible. Electrode capacitance ranging from 80 to 100 F g^{-1} is currently achieved in commercial supercapacitors.

9.3 Toward Higher Energy Density

Although today's supercapacitors meet the power requirement, energy density is not yet enough for many transportation applications which require a utilisation time of up to more than several tens of seconds. So far, supercapacitors fall short in supplying energy for longer than a few seconds.

Most of the research teams in the world are striving to increase the energy density by tuning the active material used in the electrodes. Three main ways are possible:

- **Carbon functionalisation.** Adding an electrochemical group to the surface of the group such as a heteroatom[12–16] allows a higher capacitance to be obtain; the main drawback of such approach is the dramatic decrease in the cyclability[17] which is not suitable for transportation applications.
- **Hybridation.** Here a faradaic electrode (battery-like electrode) is combined with a EDLC electrode; this combination can be either in parallel or in series.[18–20] Well-balanced higher voltages are reachable with such an approach but again electrochemical reactions occurring at the faradaic electrode lead to a significant decrease in span life.
- **Nanostructuration of the active material/electrolyte interface.**[21] In this case the target is to maintain the way the charges are stored at the electrodes and increase the efficiency of this charge storage. This approach is very promising since no electrochemical reactions are involved, retaining the high cycle ability and power ability even at temperature as low as $-40\,°C$. Thus, the challenge is to obtain advanced carbon materials leading to higher energy density.

9.4 Active Material

As shown in eqn (9.2), one way of increasing the specific energy is to increase the working voltage by selecting a suitable electrolyte. Ionic liquids are

Table 9.2 Comparison of performances of commercial supercapacitors.

Manufacturer	Operating voltage /V	Capacitance /F	Series resistance/mΩ	Maximum, specific energy/ Wh kg^{-1}	Maximum power density/ kW kg^{-1}	Weight/ kg	Volume/ L
Batscap	2.7	2680	0.2	4.2	18.2	0.5	0.57
Maxwell	2.7	2800	0.48	4.5	8.8	0.48	0.32
Panasonic	2.5	1200	1	2.3	4.6	0.34	0.25
NessCap	2.7	5000	0.24	4.3	8.5	0.89	0.71
Nippon Chemicon	2.5	2400	0.8	4	3.8	0.52	
LS Cable	2.8	3200	0.25	3.7	12.4	0.63	0.47

well-known for such purposes[22] but at the detriment of the specific power; room temperature conductivities are five times lower than ones exhibited by a standard organic electrolyte (mixture of acetonitrile and tetraethylamonium tetrafluoroborate salt). So, to increase the specific capacitance without compromising the power ability so much, one research line is to increase the capacitance, C. The capacitance is described using eqn (9.4):

$$C = \varepsilon\varepsilon_0 \frac{S}{e} \qquad (9.4)$$

where S is the surface area of the electrode/electrolyte interface, ε_0 is the vacuum permittivity, ε is relative permittivity of the solvent and e is the average charge separation distance.

In traditional capacitors (dielectric), the capacitance is limited by the thickness (e) of the dielectric material sandwiched between the two electrodes. The thinnest go down to few hundreds of nanometres.[23] At the opposite end of the range, the thickness is of the order of 1–2 nm (even less in aqueous electrolytes) in the case of an organic electrolyte in an EDLC. Since in such systems, the dielectric is replaced by an electrolyte, the charge distance separation is assumed to be the electrochemical double layer (EDL), where average thickness takes into account the ion size and the surrounding solvent molecules (solvated ions). Activated carbons are ideal materials because they exhibit a large surface area (1000 m^2 g^{-1} to 3000 m^2 g^{-1}) enabling high capacitance. Usually, typical capacitance values range between 100 and 200 F g^{-1}.[17] A wide range of high surface area carbon fulfils the electrochemical grade: activated carbon (AC),[24,25] multi- and single-walled carbon nanotubes (MWCNT and SWCNT),[26–28] carbon nanofibres (CNF),[29,30] graphene[31–33] and carbon-derived carbon (CDC).[21]

9.4.1 Activated Carbons

This kind of carbon[24,25] is obtained in 2–3 steps, which are a pyrolysis (carbon formation) and (carbon formation), an activation (oxidation leading to build up of a pore network) and a thermal post-treatment for the creation or removal of particular surface groups. Precursors used can be wood (saw dusts, bamboo, etc.),[34,35] plant matter (rice straw, sunflower seeds, seaweed, apple pulp, etc.),[36–39] petroleum sources, phenolic resins,[40] fruit stones, fruit shell,[41,42] etc. Among these precursors the most widely used by the manufacturers is coconut shell for its low cost and high specific capacitance (50 F cm^{-3} and 120 F g^{-1}).

The activation process is also applied to enhance already high surface area carbons such as graphene or carbon nanofibres.[32,43] The precursor is also important in that it will determine the amount of impurities, which can degrade electrochemical performance because of irreversible faradaic reactions. Usually activated carbons exhibit a surface area ranging from 1000 to 2700 m^2 g^{-1}. The maximum capacitance obtained for such material is 150 F g^{-1} and 300 F g^{-1} for an organic or aqueous based electrolyte, respectively.

The porous network is responsible for the performance of the EDLC. It is believed that, most of the time, a microporous network and a mesoporous

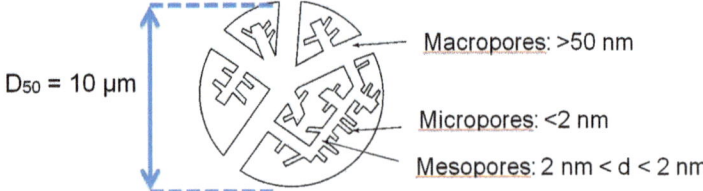

Figure 9.2   Schematic view of an activated carbon grain and its porous network.

network are necessary to obtain improved performance—micropores for the capacitance and mesopores to increase the kinetics of the electrodes (high current rate). Figure 9.2 shows a sketch of the porous network within an activated carbon grain:

9.4.2   Activated Carbon Fibres

Activated carbon fibres[29,30,40,43,44] are obtained from the carbonisation and activation of textile fibre that can be woven or not. Such carbons do not exhibit the hierarchical structuration of the porosity as observed for activated carbons. Micropores are accessible through interfibre macroporosity, meaning that a higher current rate capability can be reached thanks to quicker ion adsorption/desorption. Nevertheless, further activation of such carbon fibre can lead to mesoporosity and functional groups. This kind of carbon ranges in its specific surface area from 600 up to 2500 m^2 g^{-1} leading to a maximal capacitance of 100–120 F g^{-1} in an organic electrolyte. After exfoliation, huge capacitances are obtainable (500 F g^{-1}) in H$_2$SO$_4$ based electrolytes. Although such carbons can lead to high capacitances they suffer from a poor specific density, *i.e.* <0.4 g cm^{-3}.

9.4.3   Templated Porous Carbons

This synthesis route[45–48] has been extensively studied since it enables a way to tune the porosity. The principle of this method is to use for instance, a zeolite, mesoporous silicas (such as SBA-15) and oxide particles (MgO for instance) as a template. This template is impregnated with a furfuryl alcohol, a sucrose solution or another kind of carbon source. After carbonisation, the template is removed. Zeolite-based carbon leads to microporous carbon exhibiting a high capacitance (up to 150–170 F g^{-1} in an organic electrolyte); however, silica templates have been extensively studied since well-structured mesoporous carbons are obtained. This kind of carbon allows mesoporosity to be obtained with an average pore size ranging from 2 to 10 nm, which is believed to be perfectly adapted to solvated ion size and able to sustain high charge/discharge rates.

9.4.4   Carbon Aerogels and Xerogels

These types of carbons[49–51] are obtains from the pyrolysis of a resorcinol–formaldehyde aerogel/xerogel. They are composed of nanoparticles (a few nanometres in size) covalently bonded to form a mesoporous network. This aerogels/xerolgels can be also activated to create microporosity leading to capacitance ranging from 100 to 170 F g^{-1} (170 to 530 m^2 g^{-1}).

9.4.5 Carbon Nanotubes

Carbon nanotubes[26,27,52] exhibit a full accessible surface area, since only the outer surface of the nanotubes is involved for charge accumulation. The main advantages of such materials are their high conductivity and the fast ionic transport throughout the inter-nanotube space; the nanotubes are often bundled.[53] To improve nanotube-based supercapacitors, one strategy is to unbundled the nanotubes which can be realised by direct growth from a surface (usually a current collector) by chemical vapour deposition (CVD).[27] Although nanotube-based supercapacitors exhibit relatively good power ability and have a surface area as high as 2000 m^2 g^{-1}, they suffer from a poor specific density, leading to poor volumetric capacitance—even though gravimetric capacitance obtained can be as high as 180 F g^{-1} in aqueous electrolyte.[54] So far, no inner surface utilisation has been efficiently obtained (inner volume of carbon nanotubes).

9.4.6 Onion-like Carbons

Onion-like carbons[55,56] are formed by vacuum annealing 5 nm nano diamond powder at 1200–2000°C. As carbon nanotubes, only the outer surface is fully accessible to the electrolyte ions. The surface area of such compounds (up to 500 m^2 g^{-1}) is in the mesoporosity range since it developed within the inter-space between carbon particles. They exhibit a high power capability but a poor specific capacitance (30–40 F g^{-1}).

9.4.7 Carbon Obtained from Biomass

Carbonaceous materials extracted from biomass such as cassava peel waste,[57] coffee bean waste,[58] coffee shells,[59] cherry stone,[42] bamboo,[60] fir woods,[61] banana fibres[41] and sugar cane bagasse[62] have been used as supercapacitor carbon electrodes.[63] This kind of material is often used as active material after carbonisation in nitrogen or argon followed by an activation step. Usually, they exhibit considerable surface functionality giving them an extra capacitance. Table 9.3 summarises the principal results reported in the literature.[41,42,58–62] These results highlight the interesting and valuable materials that can be obtained from biomass and especially from organic waste. Most of the results reported here were obtained in aqueous electrolyte giving extra capacitance, often related to a pseudo-capacitive behaviour, but which is expected to degrade the cyclability.

9.4.7.1 Activated Carbons from Wood Residues and Derivatives

Wood by-products from the wood forestry, woodwork and joinery industries offer inexpensive and renewable source of precursors for the production of

Table 9.3   Comparison of physical and electrochemical properties of different carbonised biomass precursors.

Material	Capacitance $/F\,g^{-1}$	BET surface area $/m^2\,g^{-2}$	Electrolyte
Cassava peel	153	1352	H_2SO_4
Coffee beans waste	368	1029	H_2SO_4
Coffee shells	150	842	KOH
Cherry stones	232	1292	H_2SO_4
Bamboo	68	1251	H_2SO_4, KOH
Firewood	197	2821	H_2SO_4
Banana fibre	74	1079	Na_2SO_4
Sugar cane bagasse	300	1452	H_2SO_4
Rubber wood sawdust	138	912	H_2SO_4
Waste celtuce	421	3404	KOH
Argan (*Argania spinosa*)	259–355	2100	H_2SO_4

carbon materials. Wu et al.[64] prepared activated carbons by the steam activation of firewoods at 900 °C with a holding tine ranging from 1 to 7 h. The specific surface area ranged from 528 to 1131 $m^2\,g^{-1}$ and the total porous volume from 0.354 to 0.868 $cm^3\,g^{-1}$ (with an average pore diameter, mean pore diameter, Dp, between 2.68 and 3.04 nm). The sample treated for 1 h had the highest percentage of micropores (66.7%), while the sample steam treated for 7 h exhibited the lowest one (47.6%). The highest specific capacitance values were found for the carbon activated for 7 h, the value being estimated from cyclic voltammetric curves of 114 F g^{-1} in $NaNO_3$, and 142 F g^{-1} in HNO_3 and H_2SO_4. The authors concluded that the observed increase in double-layer capacitance arose mainly from the development of mesopores. Later, they prepared activated carbon by the activation of firewood using KOH (in a 1 : 1 weight ratio) as activating agent, and pyrolysis at 780 °C for 1 h. The sample prepared had a specific surface area of 1064 $m^2\,g^{-1}$, a pore value of 0.608 $cm^3\,g^{-1}$, a Dp value of 2.2 nm and micropore percentage of 90.8%. The sample showed a remarkable capacitive performance in H_2SO_4 due to excellent reversibility and high specific capacitance (180 F g^{-1} measured at 10 mV s^{-1}), which was larger than the typical value of 100 F g^{-1} generally found for activated carbons with specific surface areas around 1000 $m^2\,g^{-1}$.

Bamboo-based activated carbons were prepared using different precursor to KOH ratios giving samples OBK1 (1 : 1), OBK2 (1 : 2), OBK3 (1 : 3) and OBK4 (1 : 4).[60] Nitrogen adsorption experiments showed that the development of micropores was nearly complete at two-fold amount of KOH addition, and further addition of KOH resulted in the formation of larger pores. The Brunauer–Emmett–Teller (BET) surface area ranged from 1010 to 1400 $m^2\,g^{-1}$, the pore volume from 0.123 to 0.708 $cm^3\,g^{-1}$, and the micropore percentage from 67.6 to 88.7%. Using an aqueous electrolyte (30 wt. % H_2SO_4), the bamboo-based activated carbons could be classified into two groups on the basis of the rate capability: OBK1 and OBK2 (group 1); and OBK3 and OBK4 (group 2). The group 2 with a relatively higher mesopore fraction had a higher capacitance at the high discharge current density region. Using the non-aqueous electrolyte tetraethylammonium tetrafluoroborate (Et_4NBF_4) 1 M with a larger ion radii (0.74 nm for Et_4N^+) in an aprotic solvent, 1 M propylene carbonate, the OBK3 sample with the highest mesopore fraction provided the maximum specific capacitance. The authors concluded that the difference in the results between the aqueous and organic systems implied that the ion-transfer diffusion with organic solvents was hindered by narrow pores.

Fibres collected from banana stem were used as activated carbon precursors and impregnated with KOH or $ZnCl_2$ for five days at a precursor to chemical agent ratio of 1 : 1 by weight.[65] Pyrolysis was then carried out at 800 °C for 1 h under flowing nitrogen. The best surface area obtained for the carbons was 686 $m^2 g^{-1}$ and 1097 $m^2 g^{-1}$ after treatments with KOH and $ZnCl_2$, respectively. The pore diameter was 2.5 and 2.3 nm, for the samples prepared with KOH and $ZnCl_2$, respectively. Galvanostatic charge–discharge studies were performed in the 1 M Na_2SO_4 electrolyte. The specific reversible capacitance for $ZnCl_2$ treated sample was found to be 74 F g^{-1} in the first cycle and 65 F g^{-1} after 500 cycles with a coulombic efficiency of 88%. For the 10% KOH treated sample, it was 66 and 46 F g^{-1} for the first and 500th cycle, respectively, with an efficiency of 70%.

9.4.7.2 Activated Carbons from Nuts, Fruit Shells and Peels

There are a large number of studies in the literature describing the preparation of activated carbons from agricultural wastes and their application as electrodes for supercapacitors. Pistachio shell[66] were used for the preparation of activation by steam and KOH activation giving P-H_2O-AC and P-W-AC samples, with a BET surface area of 1009 and 1096 $m^2 g^{-1}$ respectively, a pore volume of 0.667 and 0.608 $cm^3 g^{-1}$, respectively, and a micropore percentage of 66.7 and 90.8%, respectively. The specific capacitance values of all activated carbons prepared measured in $NaNO_3$, HNO_3 and H_2SO_4 were *ca.* 180–85 F g^{-1} (at 10 mVs^{-1} in various electrolytes) comparable with those of samples with similar BET surface areas.

Activated carbons were prepared by pyrolysis of coffee shells: porogen ratio 1 : 20 by weight.[59] The carbon sample had a BET surface area of 842 $m^2 g^{-1}$. Carbons derived by pyrolysis of coffee shells treated with $ZnCl_2$ had a total pore volume of 0.46 $cm^3 g^{-1}$ and a micropore percentage of 36%; the mean pore diameter was 2.2 nm. Activated carbon was used as an electrode material with KOH 6M as the electrolyte, allowing a specific capacitance of about 150 F g^{-1} to be determined.

Hulicova-Jurcakova and colleagues studied the role of phosphorous in the electrochemical behaviour of phosphorus-rich activated carbon using data obtained from two co-polymer activated carbons compared with data obtained from a fruit stone activated carbon. The latter sample was prepared by pyrolysis of the phosphoric acid treated fruit stones at 800 °C and presented a surface area of 1055 $m^2 g^{-1}$, a total pore volume of 0.55 and a micropore percentage of 70%. The specific capacitances were determined in 1 M H_2SO_4 as electrolyte. This sample had a very high electrochemical stability that was explained by the absence of oxygen containing groups such as quinone at the AC surface. The authors also showed that the content of phosphorus (% P) and the surface area of pores (0.65–0.83 nm) were the most significant parameters influencing the capacitance value.

Olivares-Marín *et al.*[42] showed that cherry stones wastes can be recycled as activated carbons for electrode material in supercapacitors. Activated carbons were prepared by impregnation of cherry stones with an aqueous solution of different activating agents such as H_3PO_4, $ZnCl_2$ or KOH, and heated in N_2 at temperatures between 400 and 900 °C. KOH activation of this precursor at 800–900 °C led to large specific surface areas of the activated carbons between 1100 and 1300 m^2 g^{-1} with average pore sizes around 0.9–1.3 nm. These pore sizes allowed electrolyte ions to access the porous network and, at a low current density, a capacitance as high as 230 F g^{-1} in 2 M H_2SO_4 aqueous electrolyte and 120 F g^{-1} in the aprotic medium 1 M $(C_2H_5)_4NBF_4$/acetonitrile to be achieved. Furthermore, high performance was also achieved at high current densities, showing that this cherry stones activated carbon competed well with commercial carbons used in supercapacitors.

Sunflower seed shells were impregnated by Li *et al.*[36] with KOH solution and then pyrolysed at temperatures ranging from 600 to 800 °C, with a holding time of 1 h, for preparation of activated carbon samples. The authors showed that, for the same activation temperature, the use of higher concentrations of KOH solution resulted in a larger BET surface. In addition the total pore volume increased when the KOH to carbonised shell weight ratio increased from 1 to 4; the mesopore volume followed the same trend. The micropore volume reached a maximum at a KOH to shell ratio of 3. For higher amounts of KOH, the micropore volume was lower due to micropore collapse. The sample prepared at an activation temperature of 700 °C and impregnation ratio of 15 wt% exhibited the best performance, with a specific capacitance of 244 F g^{-1} at a current density of 0.25 A g^{-1}, and 171 F g^{-1} at a current density of 10 A g^{-1}, in 30 wt% KOH electrolyte. This was attributed to the presence of mesopores favouring ionic diffusion.

9.4.7.3 Activated Carbons from Agro-industrial By-products

As described in Chapter 2, wastes from agro-industries are a rich and low-cost source for the production of activated carbon.

Activated carbons were prepared using bagasse as precursor with a $ZnCl_2$ to sugar cane bagasse weight ratios of 0, 1, 2 and 3.5.[62] The mixture was stirred at room temperature for 4 h. Carbonisation was carried out at 900 °C for 1 h. The sample prepared with a $ZnCl_2$ to bagasse ratio of 3.5 had a BET surface area of 1788 m^2 g^{-1}, a pore volume of 1.74 cm^3 g^{-1} and a micropore percentage of 11%; it showed the most stable electrochemical performance at fast charge–discharge rates with a specific capacitance of 230 F g^{-1} in a 1M H_2SO_4 electrolyte solution.[62]

A study dedicated to the preparation of activated carbons from cotton stalks found that an activated carbon prepared with a cotton stalk to phosphoric acid mass ratio of 1 : 4 at an activation temperature of 800 °C for 2 h had interesting properties.[67] This sample had a BET surface area of 1481 cm^2 g^{-1} and a micropore volume of 0.0377 cm^3 g^{-1}. Electrochemical measurements carried out in 1M Et$_4$NBF$_4$ electrolyte measured a capacitance of 114 F g^{-1}.

Carbons with different porous structures were prepared from rice husks under different activation conditions.[68] Rice husks were carbonised at 400 °C under a nitrogen flow for 4 h. The material obtained was heated in the presence of KOH or NaOH at 400 °C for 0.2–1.0 h, and finally at temperatures between 650 and 850 °C for 0.5–2 h to activate the mixture. Due to the difference between the KOH and NaOH activation mechanism, the porous carbons prepared by NaOH activation had a larger pore diameter than the carbons prepared by KOH activation. Indeed, because of the higher boiling point (883 °C) of sodium than potassium (758 °C), the sodium could not enter the interior of the carbon structure, resulting in a strong outer activation. Thus, larger pore sizes were obtained with NaOH activation. Moreover, a longer precalcination process time led to a higher water loss and weaker steam gasification. This lowered ablation of the exterior of the carbon structure and resulted in lower mesoporosity. Samples prepared with the longest precalcination time had a small mesoporosity. The porous carbon prepared by KOH activation had a well-developed microporosity and lower pore size. The BET surface areas were between 1886 and 2721 m^2 g^{-1}, the pore volume between 1.88 and 0.98 cm^3 g^{-1} and the micropore percentage between 18 and 32% for samples prepared by NaOH activation. BET surface areas between 1392 and 1930, pore volumes between 0.70 and 0.97 cm^3 g^{-1} and micropore percentages between 50 and 98% were obtained for samples prepared by KOH activation. The porous carbons prepared with NaOH activation gave larger double layer capacitance reaching 200 F g^{-1}, and the specific capacitance of the carbons was not linearly correlated to the surface area.

Wheat straw and potato starch based activated carbon spheres (PACS) were prepared by stabilisation of potato starch at 210 °C for 12 h in an air atmosphere, and the samples then pyrolysed for 1 h under N$_2$. The material obtained was mixed with KOH and then carbonised under N$_2$ at 800 °C.[69] The carbonisation temperature and the ratio of KOH to PACS varied. High BET surface areas ranging from 1419 to 2342 m^2 g^{-1} were obtained, with total pore volumes ranging from 0.74 to 1.24 cm^3 g^{-1} and micropore percentages from 87 to 96%. A remarkably high capacitance was obtained for the sample with the highest surface area using 6 M KOH as electrolyte.

Argan (*Argania spinosa*) seeds from an endemic species of the southwest region of Morocco used for oil production were used for activated carbon preparation using KOH.[70] The seeds were first carbonised at 300 °C and then mixed with KOH. The sample obtained was first heated at 60 °C for 12 h and then at 110 °C. The pyrolysis was finally performed at 800 °C for 2 h. The activated carbons prepared were further superficially modified by treatment with ammonium peroxydisulfate (sample AKO) and melamine (sample AKN) to introduce surface oxygen and nitrogen functionalities, respectively. The AKN sample had a surface area of 2062 m^2 g^{-1} and a pore volume of

0.95 cm^3 g^{-1}. The highest capacitance obtained was 355 F g^{-1} at 125 mA g^{-1} with a 93% retention capacitance at 1 A g^{-1}, attributable to the large surface area, well-developed micro–mesopore texture, and nitrogen content of the activated carbon. The oxygen-rich activated carbon showed the lowest capacitance due to surface carboxyl groups hindering electrolyte diffusion into the pores.

Camellia oleifera shells (COS) have also been used to prepare activated carbons. Activation was realised by first mixing COS in 2 M NaOH solution at 100 °C for 12 h.[36] The sample obtained was then activated using ZnCl$_2$ at an impregnation ratio of ZnCl$_2$ to COS of 2, 3, 4, respectively, and then carbonised at 500, 600 and 700 °C. The activated carbon obtained at the activation temperature of 600 °C and impregnation ratio of 3 had a high surface area of 1935 m^2 g^{-1}, a total pore volume of 1.02 cm^3 g^{-1}, 31% of micropores, and a mean pore diameter of 2.1 nm. Electrochemical investigations showed maximum specific capacitance values of 374 and 266 F g^{-1} in 1 M H$_2$SO$_4$ and 6 M KOH electrolytes at 0.2 A g^{-1}, respectively.

9.4.7.4   Activated Carbon Fungal Biomass

In a review, Wang and Kaskel[71] described the use of natural fungi genus, *Agaricus*, for the preparation of activated carbons. Specific surface areas of up to 2264 m^2 g^{-1} were obtained by KOH activation. Fungi-based char show a bulky morphology (Figure 9.3(a)), whereas AC-4, prepared with a KOH to char mass ratio of 4 had large cavities with pore sizes of 1 to 10 mm (Figure 9.3(b)), implying a significant morphological change during the activation. The activated carbons demonstrate type I isotherms (Figure 9.3(c)), typical of microporous materials, and exhibited narrow pore size distribution (PSD) mainly centred at 0.8 nm. Increasing the mass ratio of KOH to char led to broadening of the micropores and the formation of larger micropores (1.3–1.9 nm) was observed for activated carbons prepared with a KOH to char mass ratio of 1 : 4 (Figure 9.3). The resulting fungal-based carbons show an excellent specific capacitance of 240 F g^{-1} in an organic electrolyte.

9.4.7.5   Activated Carbons from Seaweeds

Carbon materials can be prepared by simple pyrolysis of seaweeds with different physicochemical characteristics. Alginic acid, a fibrous substance located in the cellular wall of seaweeds, offers interesting properties such as a high oxygen content after pyrolysis. Thus, the oxygen trapped in these carbon electrodes can allow them to adsorb ionic electrolytes more easily. Seaweeds constitute a natural source of carbon that is very accessible and cheap. The chemical composition of seaweeds depends on the species and varies sometimes with the season, they generally contain carbonate, proteins, lipids, fibre and ash. Particular carbons can be achieved from seaweeds, depending on their chemical composition and the pyrolysis conditions; it is possible to obtain high specific energy material by varying either the functional group content (for aqueous electrolyte) or the micropore to mesopore ratio.

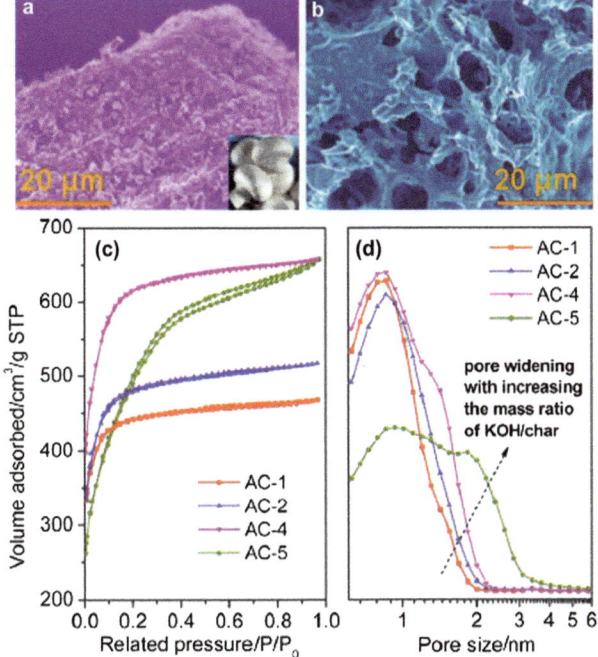

Figure 9.3 SEM images of (a) the fungal-based char and (b) AC-4, and N_2 adsorption isotherms (c) and PSD (d) of the activated carbons prepared using different KOH to char ratios (1, 2, 4 and 5). The inset in (a) is the photograph of Agaricus.

Raymundo *et al.*[38,72,73] studied the capacitive properties of different carbons from seaweeds which do not exceed $1300 \text{ m}^2 \text{ g}^{-1}$. They prepared carbons from different seaweeds by a unique carbonisation step between 600 °C and 900 °C under nitrogen. The samples obtained exhibited surface areas between 15 and $1300 \text{ m}^2 \text{ g}^{-1}$. The high-density materials obtained contained a high amount of oxygenated surface functionalities and were excellent as supercapacitor electrodes in an aqueous medium. Their volumetric ranged from 79 to 264 F g^{-1} with tetraethyl ammonium tetrafluoroborate (TEABF$_4$) in acetonitrile as electrolyte.

Table 9.4 summarises the principal physical properties depending on the decomposition temperature. From this table it can be seen that carbons with a moderate specific surface area can be produced by pyrolysis of seaweeds without any further activation step, exhibiting quite large nanoporosity. Figure 9.4 shows the cyclic voltammetry in 1 M H$_2$SO$_4$, of respectively, carbon obtained from carbonisation at 600 °C (SW600), at 900 °C (SW900) and a commercial carbon (Maxsorb).

As observed from Figure 9.4, the seaweed carbons led to a high capacitance compared with standard carbon (Maxsorb). Gravimetric capacitance as high as 260 F g^{-1} were obtained for SW600, whereas only 175 F g^{-1} and 119 F g^{-1} were

Table 9.4 Comparison of physical properties of seaweed-based carbon depending on the applied thermal treatment.

Sample	$S_{BET}(N_2)/$ $m^2\,g^{-1}$	$S_{DR}(CO_2)/$ $m^2\,g^{-1}$	Oxygen content/ at%	Nitrogen content/ at%
SW600	746	1001	9.5	2.3
SW900	1090	1448	7.1	1.6

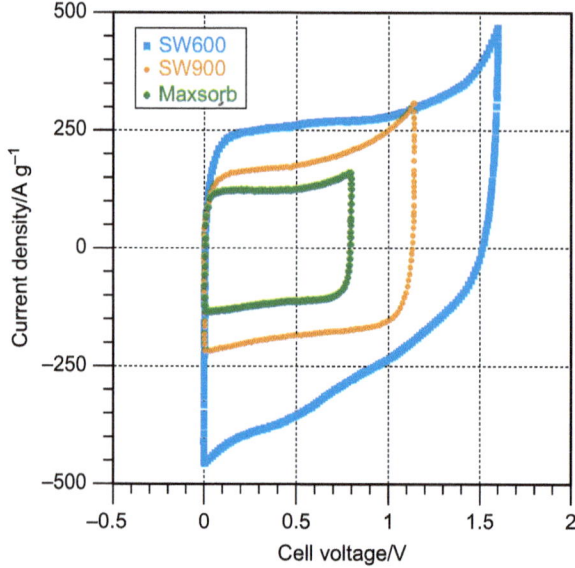

Figure 9.4 Cyclic voltammograms of SW600-, SW900- and Maxsorb-based electrodes in 1 mol L^{-1} H$_2$SO$_4$. Two-electrode cell; scan rate: 2 mV s^{-1}; electrodes from active material (90 wt%) and PVDF (10wt%).

obtained for SW900 and Maxsorb, respectively. The nanotexture nature of this type of material not only allows high capacitance to be obtained but also, because of the particular functional group, a higher voltage window. For example, a SW600 based supercapacitor can reach a cell voltage of 1.4 V, which 40% higher than conventional carbon in such aqueous electrolyte. The volumetric Ragone plot presented in Figure 9.5 demonstrates the advantages of SW600 compared with typical activated carbons. As seen here, the optimised surface functionalities, together with the improved cell voltage, lead to higher energy density and power density. In aqueous electrolyte (H$_2$SO$_4$) using such material, one order of magnitude more power density and energy density than commercial product is achieved.

Pintor et al.[74] used a brown Caribbean seaweed *Turbinaria turbinata*, as activated carbon precursor. The samples were prepared by pyrolysis of the seaweed at temperatures ranging from 600 to 900 °C. The sample prepared at 800 °C showed the highest BET surface area of 812 m^2 g^{-1}, a pore volume of

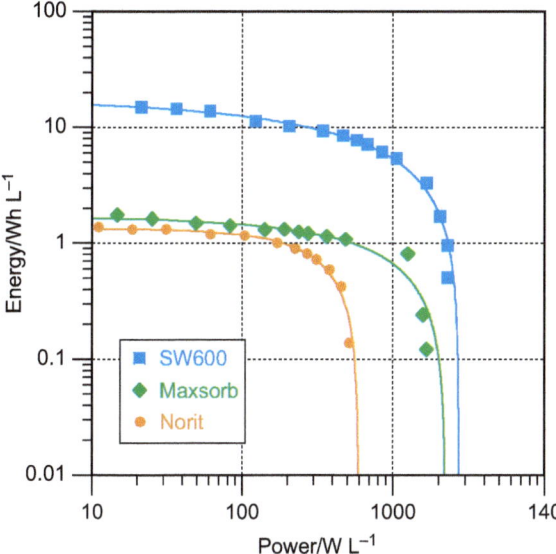

Figure 9.5 Volummetric Ragone plots that compare the performance of supercapacitors with SW600, Maxsorb and Norit based electrodes (90wt% active material and 10 wt% PVDF) in 1 mol L^{-1} H$_2$SO$_4$. Data obtained from galvanostatic charge–discharge cycling with current densities varying from 50 mA g^{-1} to 50 A g^{-1}.

0.27 cm^3 g^{-1}, a micropore percentage of 38%, a specific capacitance of 74.5 F g^{-1}, an EPR of 0.5 Ω.cm^2 and an ionic resistivity of 0.76 Ω.cm^2 in 1.5 M TEABF4 in acetonitrile. Nevertheless, even though such materials give quite high performances in aqueous media, they do not exhibit superior performance in organic electrolyte. In addition, due to their surface functionalities, it is still difficult to obtain cyclability as high as millions of cycles; ongoing work is endeavouring to improve this downside.

9.4.8 Carbide-derived Carbons

Carbons derived from metal carbides[21,75,76] are heat treated at different temperatures Equation (9.5) gives the general chemical reaction:

$$MC + nCL_2 = MCl_{2n} + C \qquad (9.5)$$

Depending on the annealing temperature, surface areas from 1000 to 2000 m^2 g^{-1} are obtained; fine tuning of the average pore diameter is also achievable. Such carbons allow capacitance from 100 F g^{-1} up to 150 F g^{-1} to be achieved in an organic electrolyte. Nevertheless, the main issue concerning these carbons is the utilisation of chlorine, which makes low-cost carbon fabrication difficult to scale up.

9.5   Issue of Surface Area

To increase the capacitance of carbons as much as possible, it has been believed for over 10 years that the BET surface area of carbon was mandatory, as was having only mesopores. The former because the higher the surface, the more ions are expected to be stored. The latter because solvated ions, which are involved in electrochemical double layer formation at the surface of the carbon, are in the nanometre range; the Stokes radius of ions in organic solvent is usually between 1 and 2 nm.[77] Figure 9.6 represents the variation of the capacitance according to ref. 21.

As observed from Figure 9.6, a trend line shows a correlation between mesoporosity and specific capacitance. That why many researchers have attempted to develop mesoporous carbon by templating methods to achieve a narrow pore size distribution [45,46] Hence, the only way to increase further the charge density, and thus the capacitance, is to increase the surface area. Figure 9.7 shows the variation of the gravimetric capacitance with the BET surface area.[78]

For a low surface area, the specific capacitance increases when the surface area does (Figure 9.7). Nevertheless, it is also noticeable that, beyond a certain specific surface area, the capacitance does not increase anymore. The activation, which leads to an increase in porous volume, leads first to the development of micropores; increasing the activation time makes the micropores collapse into mesopores. Additionally, as it is observed in Figure 9.8, the increase in specific capacitance is directly linked to the microporous volume determined from CO_2 adsorption.[79] The achievement of highly microporous

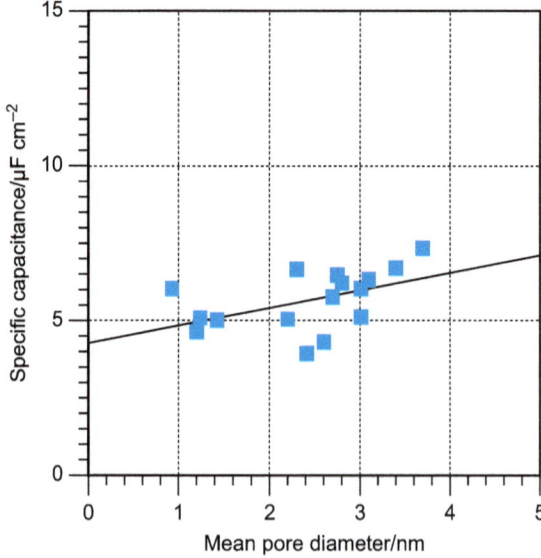

Figure 9.6   Specific capacitance *vs.* pore diameter taken from ref. 22. The black line represents the trend line for a series of mesoporous activated carbons.

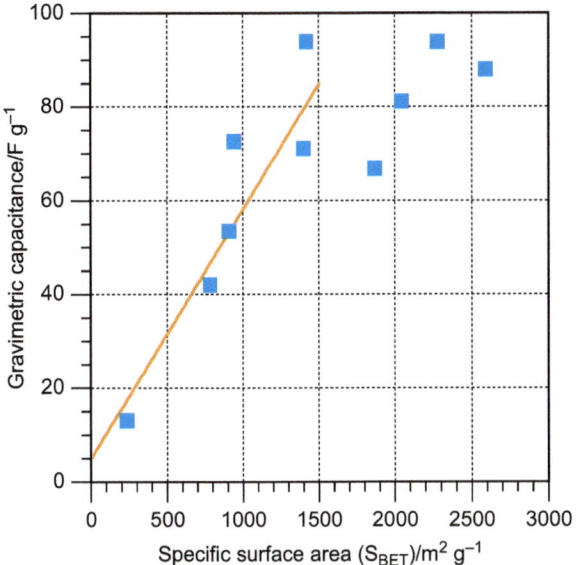

Figure 9.7 Gravimetric capacitance vs. specific surface area.[63]

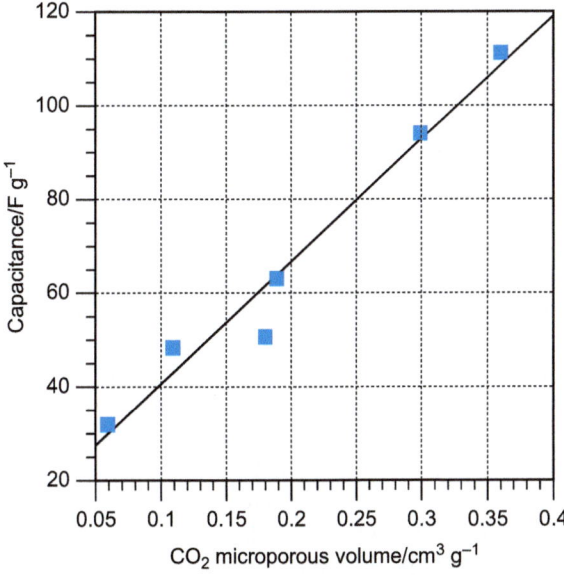

Figure 9.8 Gravimetric capacitance in organic media of nanostructured carbons synthesized with different carbon precursors and template *vs.* their micropore volume determined by CO_2 adsorption.[49]

carbon appears to be essential to obtain a higher specific capacitance. However, the pore size distribution (PSD) of conventional activated carbons is quite wide; achieving a narrower PSD would lead to higher specific capacitances.

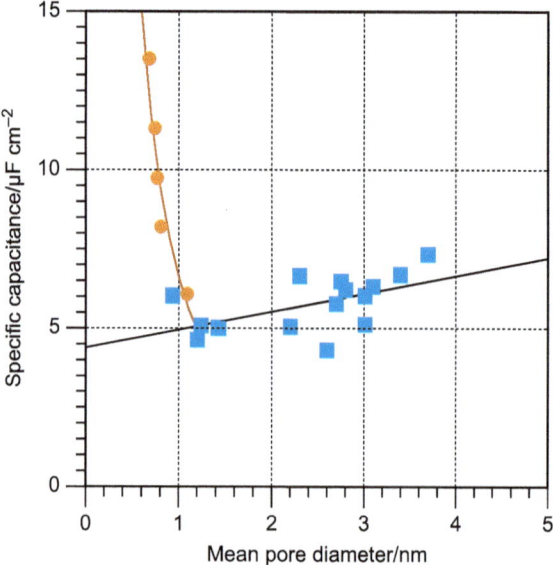

Figure 9.9   Specific capacitance *vs.* mean pore diameter.[22] Red circles (●) represent the capacitance extracted from CDC.

In this context, carbide-derived carbons (CDCs) could be suitable for such a purpose.[74,73]

9.6   Influence of Pore Diameter

Figure 9.9 presents the evolution of the specific capacitance ($\mu F\ cm^{-2}$) with average pore size. As observed, below 1 nm, a dramatic increase in the specific capacitance occurs although it corresponds to an average pore size well below that of solvated ions. A partial or complete desolvation is suspected here. This observation raises questions about how the electrochemical double layer builds up in such small pores. In addition, Figure 9.10 shows that, at this scale (pore diameter <1 nm), the specific capacitance is proportional to the reciprocal diameter. This means that the capacitance, at this scale, is proportional to the charge separation distance rather than to the specific surface area (see eqn (9.4)). So far, materials for supercapacitors have been tweaked to make them fit solvated ion size. These results, for the first time, seem to indicate that active material structure could shape the electrochemical double layer. A solvation shell distortion of the ions may force them enter the pores rather than, as thought so far, a repelling of the ions, leading to poor specific capacitance.

To address this point, different electrochemical tests were performed using room temperature ionic liquids (RTILs). These compounds are salts that melt at room temperature or below,[22,80] enabling them to be used as a solvent-free electrolyte. Largeot *et al.*[81] characterised CDCs in ethylmethyl imidazolium bis(trifluorosulfonate imide) ionic liquid (EMI-TFSI) (Figure 9.11).

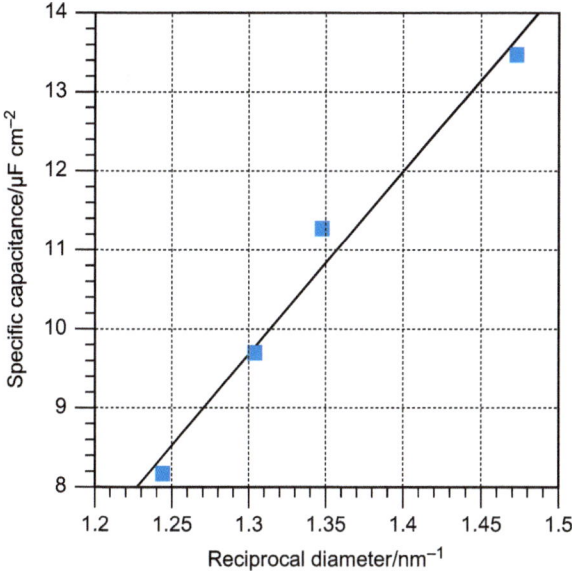

Figure 9.10 Specific capacitance *vs.* the reciprocal diameter. The trend line (black line in the plot) shows a linear relationship between the capacitance and reciprocal diameter.

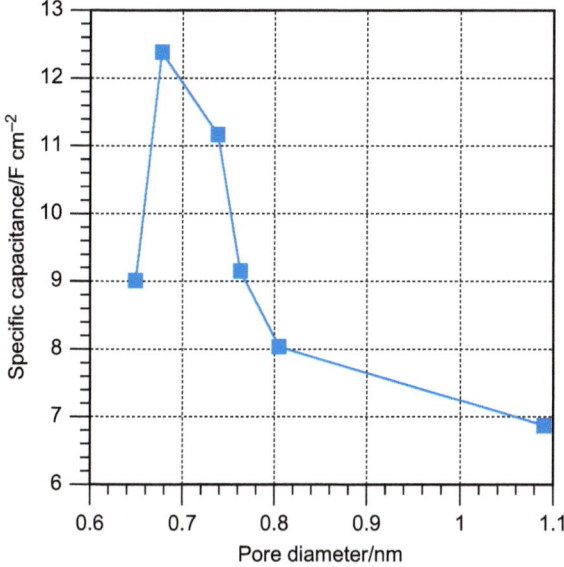

Figure 9.11 Specific capacitance *vs.* pore diameter. A maximum is obtained when pore size is close to ion size, here EMI-TFSI.[68]

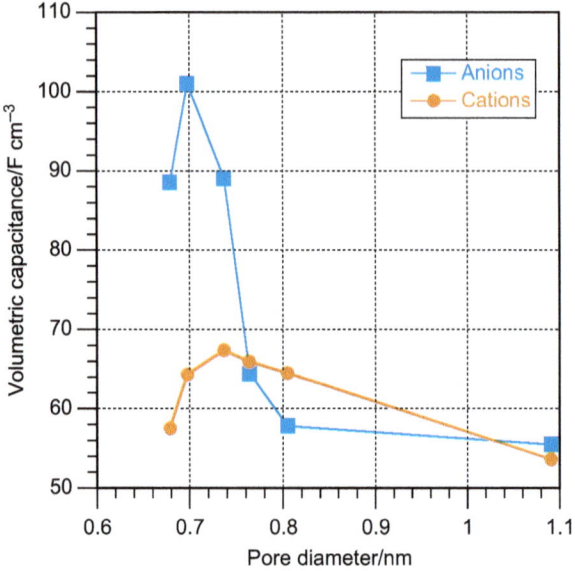

Figure 9.12   Volumetric capacitance *vs.* pore diameter in TEA-BF$_4$ in acetonitrile. The experiment was done using a three electrode electrochemical cell. Such set-up allows measuring separately the capacitance of each electrode.[69]

The maximum capacitance, when ion size is close to the average pore size of the activated carbon, was obtained. This means that, to obtain improved energy density, the ion size and pore size need to fit each other as closely as possible.

Moreover, Figure 9.12, representing the specific capacitance variation with the average pore size, for tests carried out in a standard electrolyte (TEA-BF4 in acetonitrile) shows that, when cations and anions have different sizes, each electrode has to be chosen with the right pore size.[82]

To obtain optimised supercapacitors, it must therefore be stressed that asymmetric supercapacitors are preferred. In this way, it is possible to realise supercapacitors exhibiting twice the specific energy of a standard device, assembled with traditional electrodes made from coconut shell microporous activated carbon in a symmetric manner.

9.7   Application: Bulky CDC Layers and Micro Supercapacitors

To go even further, the carbon layer can be made by converting a sputtered metal carbide layer into carbon by chlorination. In fact, a layer can be deposited on a silicon substrate enabling the realisation of micro devices.[83] The advantage is the formation of a bulky carbon film without any porosity. Two key parameters need to be controlled to obtain the targeted layer. For instance, a CDC-based carbon layer was realised by converting a titanium carbide film on a glassy

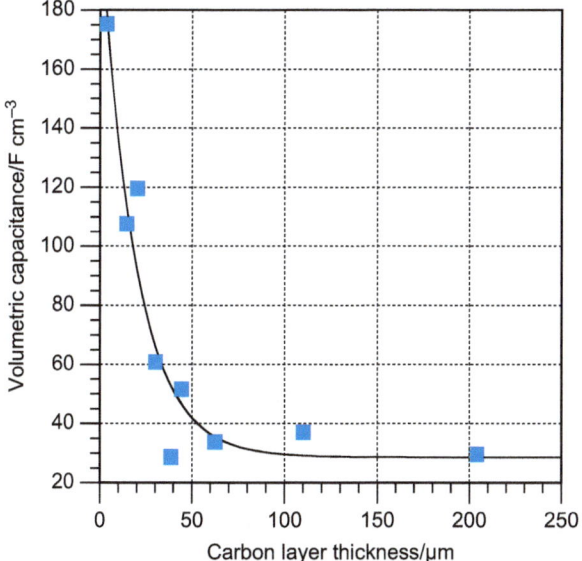

Figure 9.13 Volumetric capacitance variation with carbon film thickness. Capacitance obtained in 1.5 M NEt_4BF_4 in acetonitrile. CDC (Carbide Derived Carbon) film was realised by 500°C chlorination temperature leading to an average pore diameter of 0.7 nm.

carbon. The goal was to determine the intrinsic electrochemical properties of such films. The variation of the volumetric capacitance with carbon layer thickness is plotted in Figure 9.13. The capacitance is more or less constant from 200 μm down to few microns; it is around 50 F cm^{-3}, which is in the range of traditional activated carbons (coconut shell based activated carbons).

Nevertheless, when the thickness is less than few microns a sharp increase in the volumetric capacitance is observed leading to values as high as 180 F cm^{-3}, *i.e.* up to more than three times that of traditional activated carbons. Although many key issues remain to be resolved before a complete device is obtained, these results are encouraging and will make possible the realisation of micro-supercapacitors for microsystem applications, since high energy density are also required to get improved runtime.

9.8 Charge Storage Mechanism Study

It has been seen that an improved supercapacitor could be obtained by making the carbon electrode more compatible, that is to say, the bare ion size as close as possible to the carbon pore size. Also, since carbon structure seems to have a huge impact on the electrochemical double layer formation, different studies have begun with the aim of getting a better understanding of the physico-chemical mechanisms occurring at the interface. For this purpose, different *in situ* techniques have been carried out to tackle this issue such as nuclear magnetic resonance (NMR), molecular dynamic simulation (MDS) and electrochemical quartz crystal microbalance (EQCM).

9.8.1   Electrochemical Quartz Crystal Microbalance (EQCM)

Aurbach and colleagues studied the ionic fluxes in microporous carbons as a function of electrode surface charge.[84] They also tried to assess the solvation numbers of electrolytic ions confined in carbon nanopores. For their studies, they used a carbon film sprayed onto a gold electrode supported on an oscillating quartz. The purpose of this work was to measure the mass variation while the electrode is adsorbing ions. Figure 9.14 presents the adsorbed mole quantity variation with injected charge into the electrode; the black dashed line represents the theoretical line calculated from the Faraday law.

For instance, different variations were observed when comparing the adsorption of different alkaline and alkaline earth cations in a nanoporous carbon (YP17 from Kuraray, Japan) An increasing slope was observed from $Ba^{2+} > Ca^{2+} > Mg^{2+}$, which was ascribed to a larger solvation numbers. Table 9.5 lists the solvation numbers of ions confined in carbon nanopores for different aqueous electrolytes

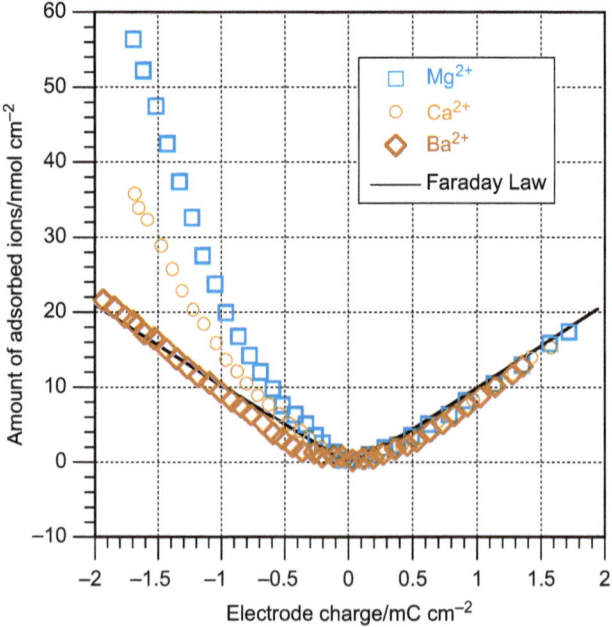

Figure 9.14   Adsorbed mole quantity variation with injected charge into the electrode (YP17 activated carbon sprayed onto a gold coated EQCM). Amount of adsorbed ions are measured for different aqueous electrolyte: $BaCl_2$, $MgCl_2$, $CaCl_2$. The black line is a reference baseline calculated from Faraday law assuming there is no solvation.[71]

Table 9.5 Solvation numbers (SN) of ions confined in nanoporous carbon, obtained from EQCM.

Type of ion	Solvation number in pores	Solvation number in bulk electrolyte
Li^+	2.6	22
Na^+	2.2	13
K^+	1.3	7
Cs^+	0.5	6
Mg^{2+}	5.8	36
Ca^{2+}	3.7	29
Ba^{2+}	2.8	
F^-	1.4	6.7
Cl^-	0.6	6.4
Br^-	0.05	5.9
I^-	0	5.8

From these results, it can be seen that in any case there at least a partial desolvation of the ions while entering the nanopore, since solvation numbers are each time less than the ones from bulk solution hydration number. To go further in this study, Aurbach and colleagues tried to determine the ionic fluxes in microporous carbons.[85] They carried out measurements with different concentration of CsCl to determine ion fluxes. They used an activated carbon spray coated quartz to analyse the gravimetric signal under polarisation. Figure 9.15 shows cyclic voltammetries and relative molar amount of ions *vs.* electrode charge plots. A deviation is observed compared with the theoretical curve (plotted using Faraday law). Near p_{zmc} [zero mass change, defined as the potential, separating the cationic and anionic fluxes into carbon micropores, which is quite close to the potential of zero charge (PZC)], the slope is lower than expected due to mixed cation-anion fluxes.[83] This effect is emphasised when the bulk solution concentration increased because of stronger ion pairing.

For larger charge densities, Aurbach and colleagues observed a slope equal to the theoretical one, indicating that the electrode became perm-selective. An extra feature was also observed for more diluted bulk concentration and high charge density, *i.e.* the slope exceeds the theoretical value. The latter was assumed to be linked to micropore deformation leading to excessive adsorption of ions, leading in turn to a concomitant flux of solvent molecules.

9.8.2 Nuclear Magnetic Resonance (NMR)

^{11}B NMR spectroscopy has been used to investigate the sorption of BF_4^- in highly porous carbon.[86] This NMR study was carried out *in situ* to observe ion binding to the carbon surface in real-time. First, to identify the chemical environment of each BF_4^- on the surface of uncharged carbon electrodes, successive volumes of electrolyte were added to the electrode (Figure 9.16). For 2 μL added electrolyte, a broad peak appears at −4.2 parts per million (ppm). When the electrolyte amount is increased, a second peak is observed at 1.7 ppm (increasing from 5 μL to 10 μL); a third one (at 0.6 ppm) appears when the electrolyte amount reaches 15 μL. The low chemical shifted peak, the medium

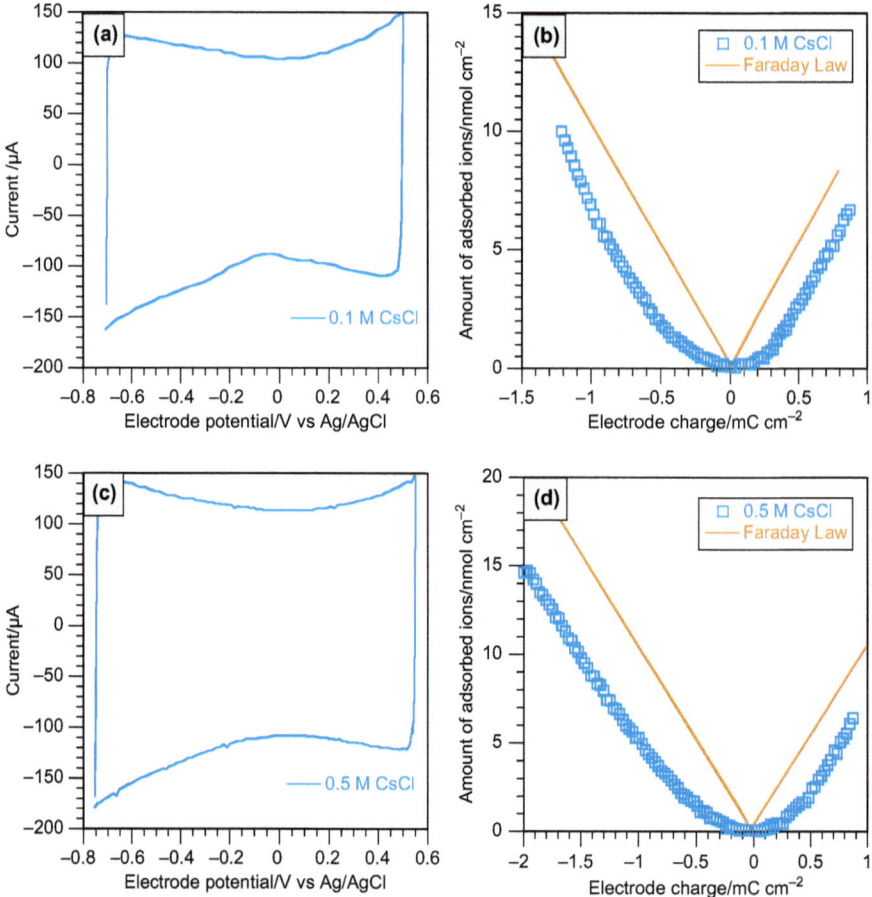

Figure 9.15   CVs of porous carbon electrodes measured in aqueous solutions of CsCl of different concentrations as indicated (a and c) at a scan rate of 50 mV s^{-1}, and the related molar amounts of ions *vs.* electrode charge plots (b and d, respectively). The red line is the evolution calculating from Faraday law.

one and the high chemical shifted one have been ascribed to strongly bound BF_4^- ions on the carbon, free electrolyte and weakly bound ions, respectively. The latter are in dynamic exchange with the strongly bound ions, and are located in the outer sorption layer (mesopores, for instance), as confirmed by two-dimensional (2D) NMR (data not presented). It was also noticed that the spin-lattice relaxation times (T1) fell significantly for molecules adsorbed on a carbon surface.

In situ NMR spectroscopy was then performed to study peak evolution upon electrode polarisation (from -2 to 2 V). Figure 9.17 shows the evolution of the peaks with electrode polarisation.

From 0 to $+2$ V (charging, BF_4^- insertion), the high frequency peak gets narrower and moves toward higher frequencies. Whereas from 0 to -2 V, the

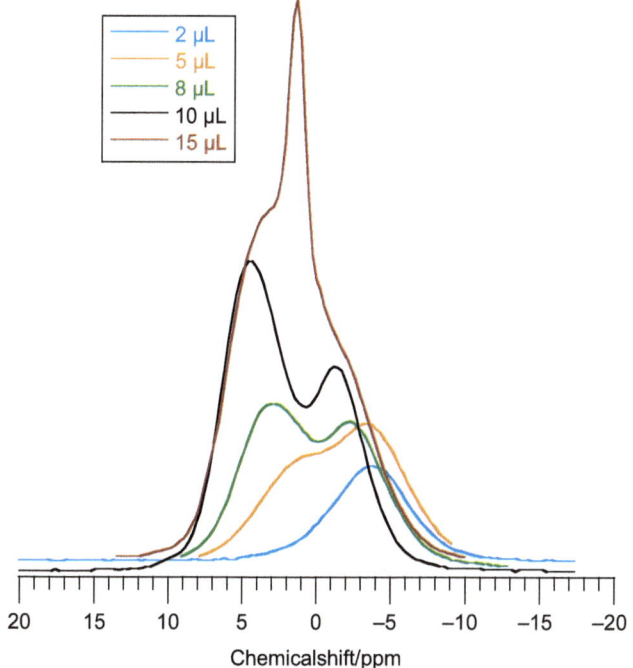

Figure 9.16 ^{11}B NMR spectra of YP17 carbon films with different amounts of 1.5 M TEA-BF$_4$ in acetonitrile, in plastic bags.

high frequency peak also get narrower but shifts toward lower frequencies until it disappears completely. This observation is emphasised by Figure 9.18, which shows the evolution of the integrated area with polarisation increase of the electrode, which is actually the evolution of adsorbed ion amount with the electrode polarisation.

As seen here, the total amount increases with the polarisation (from −2 to 2 V). This, together with the observations described above, means that the total of strongly bound ions increased with polarisation. Also, 2D NMR (data not presented here) revealed that a dynamic exchange exists between weakly bound ions (mesopores) and strongly bound ions. This NMR study showed that, upon charging, both the concentration of the anions changes, as does the nature and strength of the anion binding to the surface. These results could also give some clues about the high current drawn from nanoporous carbons, although the ionic conductivity should be low.

To go further into details in the description of ion adsorption into nanoporous carbons, molecular dynamic simulation was performed by Merlet *et al.*[87]

9.8.3 Molecular Dynamic Simulations

The huge capacitance increase in nanoporous carbon has attracted particular interest in the field of energy storage especially to uncover the underlying

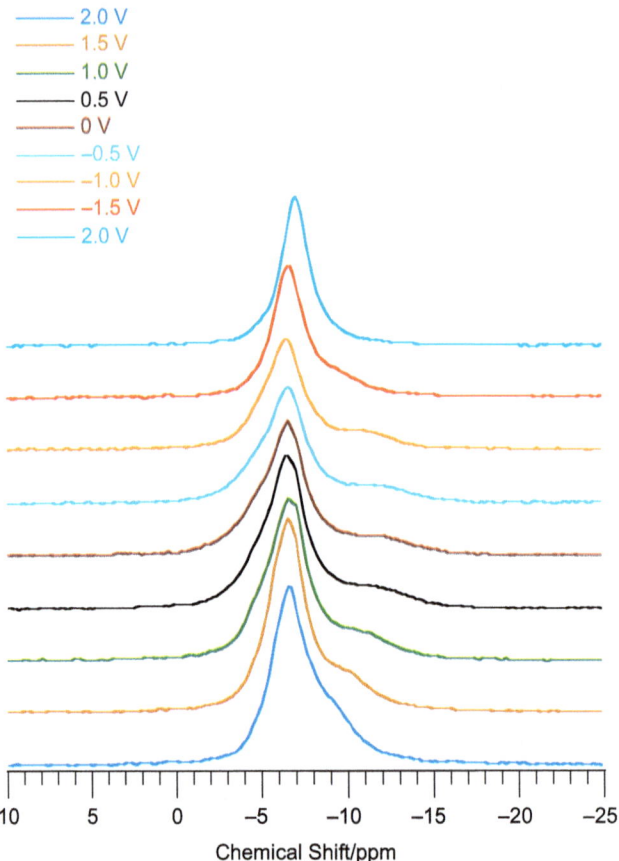

Figure 9.17 *In situ* ^{11}B NMR spectra of YP17 super capacitor held at different voltages (from −2 to 2 V).

molecular phenomena at the origin of the anomalous capacitance increase.[21] From a traditional point of view, it is well-established that the ions adopt a multi-layered structure at the surface of a planar electrode. For a planar electrode, it is now well-known that an over screening effect occurs, leading to a build-up of successive ionic layers compensating the electrode charge. However, the description of electrode/electrolyte interface is not that clear.

Several models are proposed: lining up of the ions in cylindrical pore[88] and superionic state model using a slit pore model.[89] However, these models have so far fallen short in forecasting the high capacitance experimentally observed in CDC-based nanoporous carbons. Most of the models have led to capacitance ranging from 0.5 to 3 μF cm^{-2}, that is to say, 6–40 F g^{-1} for a full electrochemical cell.

Merlet *et al.*[87] proposed a MDS to describe a CDC electrode in a ionic liquid (BMI-PF6) using a different approaches based on two key features—a realistic atomistic structure of CDC depicted elsewhere by Palmer *et al.*[90] and the

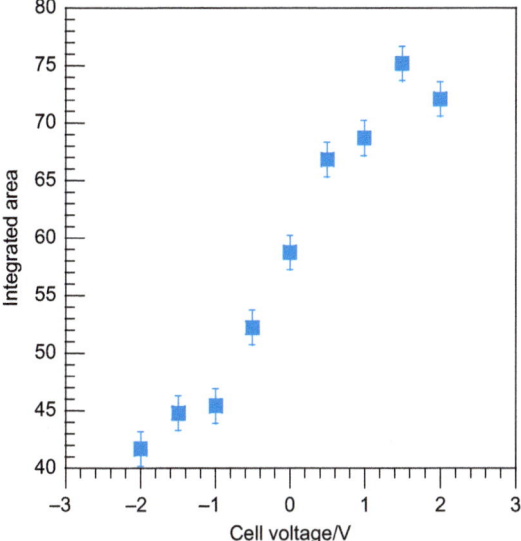

Figure 9.18 Integrated area *vs.* cell voltage. The integration represents the total amount of charge (BF_4^-) adsorbed in the electrode. An increase is observed while the cell voltage rises.

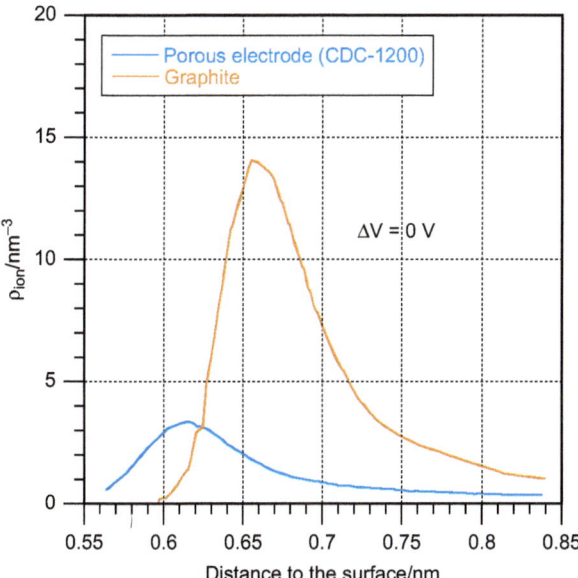

Figure 9.19 Density profiles normal to the electrode surface for graphite and CDC material. Ionic density profiles for the two types of ions and a polarization of 0 V; the distance is given with respect to the surface accessible to an argon atom probe, with the origin set to the position of the carbon atoms.[74]

polarisation of the electrode atoms by the ionic charge. Using a coarse grain model of the ionic liquid, the MDS gives capacitance values of up to 125 F g^{-1}, which is in the range of what was measured experimentally. In this study, ions interact highly with the structure of the carbon and undergo the same coulomb ordering effects that occur in the bulk. Under polarisation, the average co-ordination number decreases inside the electrode (from seven in the bulk to four inside the electrode), and an exchange of ions between the electrodes and the bulk electrolyte is observed, leading to an anion amount increase at the positive electrode, and the opposite for the negative one. By comparing a planar graphite electrode and a CDC-based electrode (nanoporous carbon), Figure 9.19 shows that both cations and anions are closer to the electrode surface in porous electrode (−0.07 nm).

Since the capacitance is directly linked to the reciprocal distance between two charged plates (see eqn (9.4)), it is now clearer that a shorter carbon ion dis-tance, observed in the case of nanoporous carbons, could be behind the origin of the anomalous capacitance in porous CDCs. To go further, Figure 9.20 depicts the total surface charge accumulated across the liquid-side of the interface normalised by the surface charge electrode.

Figure 9.20 compares a graphite electrode with a porous electrode. An over screening behaviour, whatever the polarity of the electrode, is observed for the

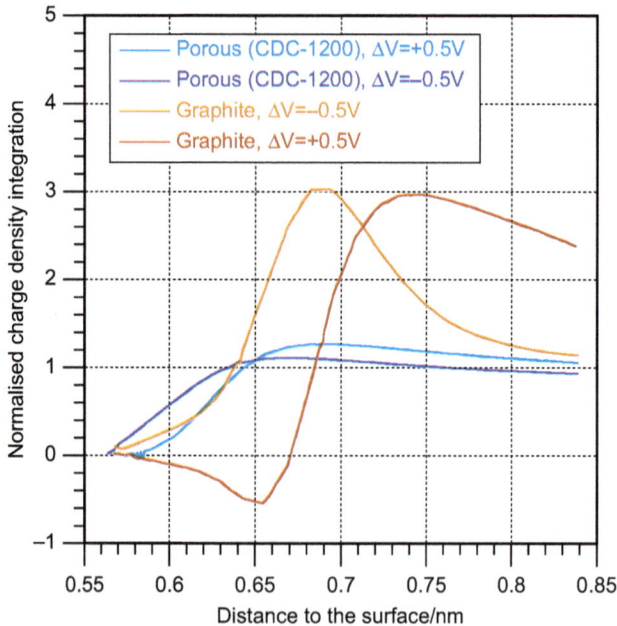

Figure 9.20   Integral of the charge density of the ionic layer normalised by the electrode surface charge (for 0.5 or −0.5V). The function reaches a value of 1 when the two quantities are equal. No over screening effect is observed for the CDC-1200 (carbide defined in ref. 77), contrarily to the graphite electrode.

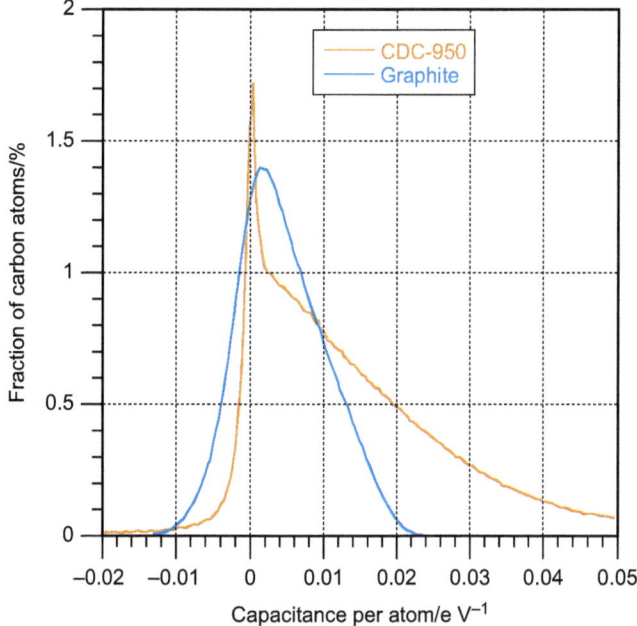

Figure 9.21 Distribution in capacitance per carbon atom for graphite and nanoporous electrode (CDC-950, see ref. 77). A broader distribution for porous carbon is observed leading to a higher average capacitance due to a closer charge distance separation between electrode and ions.

graphite electrode due to ionic correlations. This behaviour leads to a less efficient charge compensation, together with a larger carbon-ion distance. Whereas for the porous electrode, no over screening effect is observed leading to only one ion layer being closer to the carbon surface. As a consequence, Figure 9.21 shows that a larger average charge storage per carbon is observed. Indeed, a broader and more asymmetric distribution is observed for porous carbons compared with graphite.

9.9 Conclusions

Ongoing studies are seeking to refine even further the description of the electrode–electrolyte interface, especially the influence of the solvent. From these findings, it has become clearer that the structure of the carbon, together with the choice of electrolyte, are the keys to obtaining optimised high energy density supercapacitors.

Today's research projects are devoted to the optimisation and study of the electrode–electrolyte interface to obtain both better performance and a better mechanism understanding. For instance, Lin and colleagues recently devised a carbon–carbon supercapacitor exhibiting a 130 F g^{-1}, with an extended temperature range spanning from $-50\,^\circ$C up to $100\,^\circ$C.[91,92]

References

1. J. R. Miller and P. Simon, *Science*, 2008, **321**, 651–652.
2. T. Brousse, P.-L. Taberna, O. Crosnier, R. Dugas, P. Guillemet, Y. Scudeller, Y. Zhou, F. Favier, D. Bélanger and P. Simon, *J. Power Sources*, 2007, **173**, 633–641.
3. E. Machefaux, T. Brousse, D. Bélanger and D. Guyomard, *J. Power Sources*, 2007, **165**, 651–655.
4. F. Moser, L. Athouël, O. Crosnier, F. Favier, D. Bélanger and T. Brousse, *Electrochem. Commun.*, 2009, **11**, 1259–1261.
5. J. Santos-Peña, O. Crosnier and T. Brousse, *Electrochim. Acta*, 2010, **55**, 7511–7515.
6. Y. K. Zhou, M. Toupin, D. Bélanger, T. Brousse and F. Favier, *J. Phys. Chem. Solids*, 2006, **67**, 1351–1354.
7. L. Bonnefoi, P. Simon, J. F. Fauvarque, C. Sarrazin and A. Dugast, *J. Power Sources*, 1999, **79**, 37–42.
8. C. Portet, P. L. Taberna, P. Simon and C. Laberty-Robert, *Electrochim. Acta*, 2004, **49**, 905–912.
9. Z. Jim P., *J. Power Sources*, 2004, **137**, 158–162.
10. J. P. Zheng and Z. N. Jiang, *J. Power Sources*, 2006, **156**, 748–754.
11. H. Gualous, D. Bouquain, A. Berthon and J. M. Kauffmann, *J. Power Sources*, 2003, **123**, 86–93.
12. E. Frackowiak, G. Lota, J. Machnikowski, C. Vix-Guterl and F. Béguin, *Electrochim. Acta*, 2006, **51**, 2209–2214.
13. W. Kim, M. Y. Kang, J. B. Joo, N. D. Kim, I. K. Song, P. Kim, J. R. Yoon and J. Yi, *J. Power Sources*, 2010, **195**, 2125–2129.
14. G. Lota, B. Grzyb, H. Machnikowska, J. Machnikowski and E. Frackowiak, *Chem. Phys. Lett.*, 2005, **404**, 53–58.
15. J. Machnikowski, B. Grzyb, J. V. Weber, E. Frackowiak, J. N. Rouzaud and F. Béguin, *Electrochim. Acta*, 2004, **49**, 423–432.
16. X. Yang, D. Wu, X. Chen and R. Fu, *J. Phys. Chem. C*, 2010, **114**, 8581–8586.
17. A. Pandolfo and A. Hollenkamp, *J. Power Sources*, 2006, **157**, 11–27.
18. S. R. Sivakkumar and A. G. Pandolfo, *Electrochim. Acta*, 2012, **65**, 280–287.
19. D. Cericola, P. Novák, A. Wokaun and R. Kötz, *Electrochim. Acta*, 2011, **56**, 8403–8411.
20. A. D. Pasquier, I. Plitz, J. Gural, F. Badway and G. G. Amatucci, *J. Power Sources*, 2004, **136**, 160–170.
21. J. Chmiola, G. Yushin, Y. Gogotsi, C. Portet, P. Simon and P. L. Taberna, *Science*, 2006, **313**, 1760–1763.
22. A. Balducci, R. Dugas, P. L. Taberna, P. Simon, D. Plée, M. Mastragostino and S. Passerini, *J. Power Sources*, 2007, **165**, 922–927.
23. J.-B. Lim, Y. H. Jeong, S. Nahm, J.-H. Paik, H.-J. Sun and H.-J. Lee, *J. Eur. Ceram. Soc.*, 2007, **27**, 2871–2874.
24. Q. Deyang, *J. Power Sources*, 2002, **109**, 403–411.
25. J. Gamby, P. L. Taberna, P. Simon, J. F. Fauvarque and M. Chesneau, *J. Power Sources*, 2001, **101**, 109–116.

26. C. Du and N. Pan, *J. Power Sources*, 2006, **160**, 1487–1494.
27. S. Wei, W. P. Kang, J. L. Davidson and J. H. Huang, *Diamond Relat. Mater.*, 2008, **17**, 906–911.
28. X. Lu, H. Dou, B. Gao, C. Yuan, S. Yang, L. Hao, L. Shen and X. Zhang, *Electrochim. Acta*, 2011, **56**, 5115–5121.
29. Y. Korai, S. Ishida, F. Watanabe, S.-H. Yoon, Y.-G. Wang, I. Mochida, I. Kato, T. Nakamura, Y. Sakai and M. Komatsu, *Carbon*, 1997, **35**, 1733–1737.
30. M. H. Al-Saleh and U. Sundararaj, *Carbon*, 2009, **47**, 2–22.
31. D. Sun, X. Yan, J. Lang and Q. Xue, *J. Power Sources*, 2013, **222**, 52–58.
32. Y. Zhu, S. Murali, M. D. Stoller, K. J. Ganesh, W. Cai, P. J. Ferreira, A. Pirkle, R. M. Wallace, K. A. Cychosz, M. Thommes, D. Su, E. A. Stach and R. S. Ruoff, *Science*, 2011, **332**, 1537–1541.
33. A. Yu, I. Roes, A. Davies and Z. Chen, *Appl. Phys. Lett.*, 2010, **96**, 253105.
34. C. Kim, J.-W. Lee, J.-H. Kim and K.-S. Yang, *Korean J. Chem. Eng.*, 2006, **23**, 592–594.
35. E. Taer, M. Deraman, I. A. Talib, A. A. Umar, M. Oyama and R. M. Yunus, *Curr. Appl. Phys.*, 2010, **10**, 1071–1075.
36. X. Li, W. Xing, S. Zhuo, J. Zhou, F. Li, S.-Z. Qiao and G.-Q. Lu, *Bioresour. Technol.*, 2011, **102**, 1118–1123.
37. A. H. Basta, V. Fierro, H. El-Saied and A. Celzard, *Bioresour. Technol.*, 2009, **100**, 3941–3947.
38. M. P. Bichat, E. Raymundo-Piñero and F. Béguin, *Carbon*, 2010, **48**, 4351–4361.
39. G. Dobele, T. Dizhbite, M. V. Gil, A. Volperts and T. A. Centeno, *Biomass Bioenergy*, 2012, **46**, 145–154.
40. C. Ma, Y. Song, J. Shi, D. Zhang, M. Zhong, Q. Guo and L. Liu, *Mater. Lett.*, 2012, **76**, 211–214.
41. Y. Lv, L. Gan, M. Liu, W. Xiong, Z. Xu, D. Zhu and D. S. Wright, *J. Power Sources*, 2012, **209**, 152–157.
42. M. Olivares-Marín, J. A. Fernández, M. J. Lázaro, C. Fernández-González, A. Macías-García, V. Gómez-Serrano, F. Stoeckli and T. A. Centeno, *Mater. Chem. Phys.*, 2009, **114**, 323–327.
43. V. Barranco, M. A. Lillo-Rodenas, A. Linares-Solano, A. Oya, F. Pico, J. Ibañez, F. Agullo-Rueda, J. M. Amarilla and J. M. Rojo, *J. Phys. Chem. C*, 2010, **114**, 10302–10307.
44. B. C. Kim, J. S. Kwon, J. M. Ko, J. H. Park, C. O. Too and G. G. Wallace, *Synth. Met.*, 2010, **160**, 94–98.
45. A. Fuertes, G. Lota, T. Centeno and E. Frackowiak, *Electrochim. Acta*, 2005, **50**, 2799–2805.
46. A. B. Fuertes, *J. Mater. Chem.*, 2003, **13**, 3085.
47. B. Liu, H. Shioyama, H. Jiang, X. Zhang and Q. Xu, *Carbon*, 2010, **48**, 456–463.
48. C. Vix-Guterl, S. Saadallah, K. Jurewicz, E. Frackowiak, M. Reda, J. Parmentier, J. Patarin and F. Beguin, *Mater. Sci. Eng. B*, 2004, **108**, 148–155.

49. J. Li, X. Wang, Q. Huang, S. Gamboa and P. J. Sebastian, *J. Power Sources*, 2006, **158**, 784–788.
50. J. M. Skowroński and M. Osińska, *Curr. Appl. Phys.*, 2012, **12**, 911–918.
51. M. Lazzari, F. Soavi and M. Mastragostino, *J. Power Sources*, 2008, **178**, 490–496.
52. R. H. Baughman, A. A. Zakhidov and W. A. de Heer, *Science*, 2002, **297**, 787–792.
53. P.-L. Taberna, G. Chevallier, P. Simon, D. Plée and T. Aubert, *Mater. Res. Bull.*, 2006, **41**, 478–484.
54. E. Frackowiak, S. Delpeux, K. Jurewicz, K. Szostak, D. Cazorla-Amoros and F. Béguin, *Chem. Phys. Lett.*, 2002, **361**, 35–41.
55. C. Portet, G. Yushin and Y. Gogotsi, *Carbon*, 2007, **45**, 2511–2518.
56. D. Pech, M. Brunet, H. Durou, P. Huang, V. Mochalin, Y. Gogotsi, P.-L. Taberna and P. Simon, *Nat. Nanotechnol.*, 2010, **5**, 651–654.
57. A. E. Ismanto, S. Wang, F. E. Soetaredjo and S. Ismadji, *Bioresour. Technol.*, 2010, **101**, 3534–3540.
58. T. E. Rufford, D. Hulicova-Jurcakova, Z. Zhu and G. Q. Lu, *Electrochem. Commun.*, 2008, **10**, 1594–1597.
59. M. R. Jisha, Y. J. Hwang, J. S. Shin, K. S. Nahm, T. Prem Kumar, K. Karthikeyan, N. Dhanikaivelu, D. Kalpana, N. G. Renganathan and A. M. Stephan, *Mater. Chem. Phys.*, 2009, **115**, 33–39.
60. Y.-J. Kim, B.-J. Lee, H. Suezaki, T. Chino, Y. Abe, T. Yanagiura, K. C. Park and M. Endo, *Carbon*, 2006, **44**, 1592–1595.
61. F.-C. Wu, R.-L. Tseng, C.-C. Hu and C.-C. Wang, *J. Power Sources*, 2004, **138**, 351–359.
62. T. E. Rufford, D. Hulicova-Jurcakova, K. Khosla, Z. Zhu and G. Q. Lu, *J. Power Sources*, 2010, **195**, 912–918.
63. E. Taer, M. Deraman, I. A. Talib, A. Awitdrus, S. A. Hashmi and A. A. Umar, *Int. J. Electrochem. Sci.*, 2011, **6**, 3301–3315.
64. F.-C. Wu, R.-L. Tseng, C.-C. Hu and C.-C. Wang, *J. Power Sources*, 2006, **159**, 1532–1542.
65. V. Subramanian, C. Luo, A. M. Stephan, K. S. Nahm, S. Thomas, B. Wei, *J. Phys. Chem. C.*, 2007, **111**, 7527–7531.
66. F. C. Wu, R.-L. Tseng, C.-C. Hu, C.-C. Wang, *J. Power Sources*, 2005, 302–309.
67. M. Chen, X. Kang, T. Wumaier, J. Dou, B. Gao, Y. Han, G. Xu, Z. Liu and L. Zhang, *J Solid State Electrochem*, 1–8.
68. Y. Guo, J. Qi, Y. Jiang, S. Yang, Z. Wang and H. Xu, *Mater. Chem. Phys.*, 2003, **80**, 704–709.
69. S. Zhao, C.-Y. Wang, M.-M. Chen, J. Wang and Z.-Q. Shi, *J. Phys. Chem. Solids*, 2009, **70**, 1256–1260.
70. A. Elmouwahidi, Z. Zapata-Benabithe, F. Carrasco-Marin and C. Moreno-Castilla, *Bioresour. Technol.*, 2012, **111**, 185–190.
71. J. Wang and S. Kaskel, *J. Mater. Chem.*, 2012, **22**, 23710–23725.
72. E. Raymundo-Pinero, F. Leroux and F. Beguin, *Adv. Mater.*, 2006, **18**, 1877.

73. E. Raymundo-Pinero, M. Cadek and F. Beguin, *Adv. Funct. Mater.*, 2009, **19**, 1032–1039.
74. M.-J. Pintor, C. Jean-Marius, V. Jeanne-Rose, P.-L. Taberna, P. Simon, J. Gamby, R. Gadiou and S. Gaspard, CR Chimie, 2013, **16**, 73–79.
75. M. Arulepp, J. Leis, M. Lätt, F. Miller, K. Rumma, E. Lust and A. F. Burke, *J. Power Sources*, 2006, **162**, 1460–1466.
76. J. Chmiola, G. Yushin, R. Dash and Y. Gogotsi, *J. Power Sources*, 2006, **158**, 765–772.
77. C.-M. Yang, Y.-J. Kim, M. Endo, H. Kanoh, M. Yudasaka, S. Iijima and K. Kaneko, *J. Am. Chem. Soc.*, 2007, **129**, 20–21.
78. O. Barbieri, M. Hahn, A. Herzog and R. Kötz, *Carbon*, 2005, **43**, 1303–1310.
79. C. Vix-Guterl, E. Frackowiak, K. Jurewicz, M. Friebe, J. Parmentier and F. Béguin, *Carbon*, 2005, **43**, 1293–1302.
80. T. Abdallah, D. Lemordant and B. Claude-Montigny, *J. Power Sources*, 2012, **201**, 353–359.
81. C. Largeot, C. Portet, J. Chmiola, P.-L. Taberna, Y. Gogotsi and P. Simon, *J. Am. Chem. Soc.*, 2008, **130**, 2730–2731.
82. J. Chmiola, C. Largeot, P.-L. Taberna, P. Simon and Y. Gogotsi, *Angew. Chem., Int. Ed.*, 2008, **47**, 3392–3395.
83. J. Chmiola, C. Largeot, P.-L. Taberna, P. Simon and Y. Gogotsi, *Science*, 2010, **328**, 480–483.
84. M. D. Levi, S. Sigalov, G. Salitra, R. Elazari and D. Aurbach, *J. Phys. Chem. Lett.*, 2011, **2**, 120–124.
85. S. Sigalov, M. D. Levi, G. Salitra, D. Aurbach and J. Maier, *Electrochem. Commun.*, 2010, **12**, 1718–1721.
86. H. Wang, T. K.-J. Koester, N. M. Trease, J. Segalini, P.-L. Taberna, P. Simon, Y. Gogotsi and C. P. Grey, *J. Am. Chem. Soc.*, 2011, **133**, 19270–19273.
87. C. Merlet, B. Rotenberg, P. A. Madden, P.-L. Taberna, P. Simon, Y. Gogotsi and M. Salanne, *Nature Materials*, 2012, **11**, 306–310.
88. J. Huang, B. G. Sumpter and V. Meunier, *Angew. Chem., Int. Ed.*, 2008, **47**, 520–524.
89. S. Kondrat and A. Kornyshev, *J. Phys.-Condes. Matter*, 2011, 23.
90. J. C. Palmer, A. Llobet, S.-H. Yeon, J. E. Fischer, Y. Shi, Y. Gogotsi and K. E. Gubbins, *Carbon*, 2010, **48**, 1116–1123.
91. W.-Y. Tsai, R. Lin, S. Murali, L. Li Zhang, J. K. McDonough, R. S. Ruoff, P.-L. Taberna, Y. Gogotsi and P. Simon, Nano Energy.
92. R. Lin, P.-L. Taberna, S. Fantini, V. Presser, C. R. Perez, F. Malbosc, N. L. Rupesinghe, K. B. K. Teo, Y. Gogotsi and P. Simon, *J. Phys. Chem. Lett.*, 2011, **2**, 2396–2401.

Subject Index